Emerging Innovations in Microwave and Antenna Engineering

Jamal Zbitou
University of Hassan 1st, Morocco

Ahmed Errkik
University of Hassan 1st, Morocco

A volume in the Advances in Computer and Electrical Engineering (ACEE) Book Series

IGI Global
DISSEMINATOR OF KNOWLEDGE

Published in the United States of America by
IGI Global
Engineering Science Reference (an imprint of IGI Global)
701 E. Chocolate Avenue
Hershey PA, USA 17033
Tel: 717-533-8845
Fax: 717-533-8661
E-mail: cust@igi-global.com
Web site: http://www.igi-global.com

Library of Congress Cataloging-in-Publication Data

Names: Zbitou, Jamal, 1976- editor. | Errkik, Ahmed, 1960- editor.
Title: Emerging innovations in microwave and antenna engineering / Jamal Zbitou and Ahmed Errkik, editors.
Description: Hershey,PA : Engineering Science Reference, [2019] | Includes bibliographical references.
Identifiers: LCCN 2018029249| ISBN 9781522575399 (h/c) | ISBN 9781522575405 (eISBN)
Subjects: LCSH: Microwave devices--Design and construction. | Microwave devices--Technological innovation. | Microwave antennas--Design and construction. | Microwave antennas--Technological innovation.
Classification: LCC TK7876 .E57 2019 | DDC 621.3841/35--dc23 LC record available at https://lccn.loc.gov/2018029249

This book is published in the IGI Global book series Advances in Computer and Electrical Engineering (ACEE) (ISSN: 2327-039X; eISSN: 2327-0403)

British Cataloguing in Publication Data
A Cataloguing in Publication record for this book is available from the British Library.

Advances in Computer and Electrical Engineering (ACEE) Book Series

Srikanta Patnaik
SOA University, India

ISSN:2327-039X
EISSN:2327-0403

Mission

The fields of computer engineering and electrical engineering encompass a broad range of interdisciplinary topics allowing for expansive research developments across multiple fields. Research in these areas continues to develop and become increasingly important as computer and electrical systems have become an integral part of everyday life.

The **Advances in Computer and Electrical Engineering (ACEE) Book Series** aims to publish research on diverse topics pertaining to computer engineering and electrical engineering. **ACEE** encourages scholarly discourse on the latest applications, tools, and methodologies being implemented in the field for the design and development of computer and electrical systems.

Coverage

- Sensor Technologies
- Microprocessor Design
- Digital Electronics
- Circuit Analysis
- Analog Electronics
- Chip Design
- Applied Electromagnetics
- Optical Electronics
- Computer science
- VLSI Fabrication

IGI Global is currently accepting manuscripts for publication within this series. To submit a proposal for a volume in this series, please contact our Acquisition Editors at Acquisitions@igi-global.com or visit: http://www.igi-global.com/publish/.

The Advances in Computer and Electrical Engineering (ACEE) Book Series (ISSN 2327-039X) is published by IGI Global, 701 E. Chocolate Avenue, Hershey, PA 17033-1240, USA, www.igi-global.com. This series is composed of titles available for purchase individually; each title is edited to be contextually exclusive from any other title within the series. For pricing and ordering information please visit http://www.igi-global.com/book-series/advances-computer-electrical-engineering/73675. Postmaster: Send all address changes to above address.

Titles in this Series

For a list of additional titles in this series, please visit: www.igi-global.com/book-series

Algorithms, Methods, and Applications in Mobile Computing and Communications
Agustinus Borgy Waluyo (Monash University, Australia)
Engineering Science Reference • copyright 2019 • 297pp • H/C (ISBN: 9781522556930) • US $205.00 (our price)

Exploring Critical Approaches of Evolutionary Computation
Muhammad Sarfraz (Kuwait University, Kuwait)
Engineering Science Reference • copyright 2019 • 390pp • H/C (ISBN: 9781522558323) • US $215.00 (our price)

Applications of Security, Mobile, Analytic, and Cloud (SMAC) Technologies for Effective Information Processing and Management
P. Karthikeyan (Thiagarajar College of Engineering, India) and M. Thangavel (Thiagarajar College of Engineering, India)
Engineering Science Reference • copyright 2018 • 300pp • H/C (ISBN: 9781522540441) • US $215.00 (our price)

Smart Grid Analytics for Sustainability and Urbanization
Zbigniew H. Gontar (SGH Warsaw School of Economics, Poland)
Engineering Science Reference • copyright 2018 • 306pp • H/C (ISBN: 9781522539964) • US $215.00 (our price)

Vehicular Cloud Computing for Traffic Management and Systems
Jyoti Grover (Manipal University Jaipur, India) P. Vinod (SCMS School of Engineering and Technology, India) and Chhagan Lal (Manipal University Jaipur, India)
Engineering Science Reference • copyright 2018 • 281pp • H/C (ISBN: 9781522539810) • US $235.00 (our price)

Multidisciplinary Approaches to Service-Oriented Engineering
Mehdi Khosrow-Pour, D.B.A. (Information Resources Management Association, USA)
Engineering Science Reference • copyright 2018 • 389pp • H/C (ISBN: 9781522559511) • US $245.00 (our price)

Cyber-Physical Systems for Next-Generation Networks
Joel J. P. C. Rodrigues (National Institute of Telecommunications (Inatel), Brazil & Instituto de Telecomunicações, Portugal & University of Fortaleza (UNIFOR), Brazil) and Amjad Gawanmeh (Khalifa University, UAE)
Engineering Science Reference • copyright 2018 • 293pp • H/C (ISBN: 9781522555100) • US $210.00 (our price)

Augmented Reality for Enhanced Learning Environments
Gerardo Reyes Ruiz (Universidad Autónoma del Estado de México, Mexico) and Marisol Hernández Hernández (Universidad Autónoma del Estado de México, Mexico)
Information Science Reference • copyright 2018 • 322pp • H/C (ISBN: 9781522552437) • US $195.00 (our price)

IGI Global
DISSEMINATOR OF KNOWLEDGE

701 East Chocolate Avenue, Hershey, PA 17033, USA
Tel: 717-533-8845 x100 • Fax: 717-533-8661
E-Mail: cust@igi-global.com • www.igi-global.com

This book is dedicated to our wives and to our children:
Rayane, Nadine
Yahya, Meriem, and Roukia

Table of Contents

Section 3
Microwave and RF Active Circuits

Detailed Table of Contents

In this chapter, a detailed study of the one-dimension electromagnetic band gap (1D-EBG) structures and their application in a directional antenna design are presented. To improve the ability and analyze and understand the behavior of 1D-EBG, three techniques of analysis are developed. The results show that the periodicity of the dielectric permittivity makes it possible to stop the waves propagation in certain frequency bands. A comparison between the different methods shows an excellent agreement. An evolution of the transmission coefficient of a structure consisting of six layers with a cavity of thickness equal a wavelength in the middle of the structure, shows that there is a peak of transmission which is formed at the center frequency of the band gap and reflects a resonance phenomenon. This phenomenon of frequency filtering is exploited for the design of a directive EBG antenna by introducing an excitation to the cavity center.

Chapter 6
Miniaturization and Reduction of Mutual Coupling Between Antennas Arrays Using DGS and Planar EBG Structures 192
Ahmed Ghaloua, University of Hassan 1st, Morocco
Jamal Zbitou, University of Hassan 1st, Morocco
Larbi El Abdellaoui, University of Hassan 1st, Morocco
Mohamed Latrach, ESEO, France

As the size of the antenna often has a significant influence on overall dimensions of the wireless system, its reduction in size becomes a significant challenge. The objective of this chapter is to present new contributions made for reducing the size of the antenna array while maintaining excellent performance. An overview of the antenna array is introduced. Then, two designed and fabricated antenna arrays with compact size and good performances are exposed. The first microstrip patch antenna array is miniaturized using a novel shape of defected ground structure (DGS) etched in the ground plane of each radiating element of the antenna array. While the second one is two antenna arrays which are separated by two magnetic walls of a planar compact electromagnetic band gap (EBG) structure, with the aim to miniature and to reduce the mutual coupling between them, keeping both the antenna arrays separation smaller than 0.6λ5.8GHz. A full-wave electromagnetic analysis had achieved to evaluate the electrical performances of the proposed structures by using HFSS and CST-MWS.

Chapter 7
Design of New Reconfigurable and Miniature Microstrip Planar Antennas 223
Ridouane Er-Rebyiy, University of Hassan 1st, Morocco
Jamal Zbitou, University of Hassan 1st, Morocco
Abdelali Tajmouati, University of Hassan 1st, Morocco
Mohamed Latrach, ESEO, France

Recently, wireless communication systems have developed rapidly and have become more mobile and small. This necessarily requires the adequacy of its design. The materials used should be as much as possible small and at a lower cost. So low cost with reduced volume and low weight are some of the major challenges that must be faced by the designer of modern telecommunications equipment. The microstrip planar antenna is an attraction and the most crucial component used in mobile systems, and its miniaturization is one of the research challenges. This chapter focuses on the concept of miniature reconfigurable antennas by presenting and discussing the state of art with contributions in designing reconfigurable miniature printed antennas.

Preface

Nowadays, radio communications have brought fundamental changes in our life. Thus, mobile telephony has brought us the possibility to communicate easily and to have access to information networks. Such an evolution has been made possible through advances in electronics and techniques. The book of *Emerging Innovations in Microwave and Antenna Engineering* represents the efforts of many researchers working in Microwave field. The objective of this book is to share experiences and research studies in the areas of microwave engineering and antennas design.

This book of research provides theoretical studies on propagation and microwave design of passive and active devices. It covers important studies introducing modern practical antenna and microwave circuits design and theory. The state of art of each research field is given in each chapter which can be an ideal book for university antenna courses, training courses, engineers, PhD students, and researchers.

The themes covered in this book are divided into three sections:

The first section is "Antennas, Electromagnetic Theory, Design, and Applications." It contains chapters around the design of antennas more particularly the design of planar antennas miniature patches, frequency reconfigurable and slots antennas as well as the use of EBG in the design of antennas. Given the nature of patch antennas that have a narrow bandwidth, there are also studies that illustrate and present the design of new fractal antennas Ultra-Wide Band in microstrip and coplanar technology. In the same section, we find a study on the design of miniature antenna arrays with the reduction of the mutual coupling between the radiating elements. Then, we will find a second section, "Antenna Design for RFID Applications," reserved to the RFID domain and more particularly to the design of the antennas for the RFID system, whether for the reader or the TAG. And at the end of this chapter we will find a study around the design of a new structure of microwave amplifier with a detailed theoretical study based on transistors used in industry.

The book is organized into 11 chapters. A brief description of each of the chapters follows:

Chapter 1 is around the manufacturing of RF system particularly the millimeter antenna design which presents a challenge study due to the complex structures and also to the cost of such circuits. Therefore, this chapter comes with a study on the use of Additive Manufacturing (AM) which presents a 3D Printing technology, as mentioned by Johan et al. (2016), Fezai et al. (2015), Li et al; (2013) and Kumar, Louzir and Naour (2017). As known the standard technologies which are used to achieve RF components are limited in producing microwave circuits presenting complex shapes. As an example of future use of AM technology, NASA has mentioned that in the upcoming future the deep-space capsule Orion will carry hundreds of parts made up of AM technology. The AM technology will permit to achieve and to fabricate such complex shapes in one single time-phase which has an impact on mass, cost and RF performance. As an example of AM technology, we will find in this chapter an example of designing steerable front-end (FE) array antenna architecture dedicated to mmWave for the 5G network and the adaptive cruise control (ACC) applications.

Chapter 2 is discussing the practical design aspects relative to the different applications of CSRR structures (Pendry, Holden, Robbins & Stewart, 1999). Among the applications of metamaterial which attracted scientists and researchers are centered around microstrip antennas (Balanis, 2005) and other microwave passive and active circuits. The complementary split ring resonator (CSRR) is one of the most commonly used metamaterial structures. This structure has a negative value of its effective permittivity near its resonant frequency. In the first part of this chapter, a theoretical study of the CSRR structure is presented along with the analytical and experimental techniques for the measurement. The analytical and the experimental techniques are compared by discussing the obtained results. The second part of this chapter is presenting a study on the CSRR structure associated to a microstrip antenna in order to enhance the performances in terms of its gain and bandwidth. Another application of CSRR is developed by using CSRR structure in the design of dual band antennas. Therefore, the design of antennas with CSRR structures as integrated filtering elements is presented. Finally, the design considerations for CSRR based dual band antennas are covered with examples.

Chapter 3 presents a theoretical study on slot antennas by presenting the background, analytical formulations, and calibration routines. As an examples of slot antennas discussed in this chapter we find:

- Annular slot antenna (You, 2015);
- Open-ended rectangular slot antennas (Da Silva & McPhun, 1978);
- Waveguide-fed slot antennas (Gilbert, 2007); and
- Leaky-wave coaxial waveguide (Wang, Li, Zhang, & Chen, 2011).

Actually, the design of most antennas is based on the use of electromagnetic solvers. However, analytical formulations are useful for the pre-design of the antenna; also numerical simulation requires long period of computation time compared to the analytical calculation. Analytical calculations permit to predict the dimension and shape of the slot or slotted array on the antenna. After the prediction of antenna dimensions, we will need just minor adjustment to optimize its performance in numerical simulation. Another study which is so important is the antenna calibration process that is crucial for the antenna measurements. Some one-port calibration methods are described in detail.

Chapter 4 deals with the miniaturization of printed antennas particularly planar antennas array. Several studies of printed structures have been collected in a special issue of Pozar (1992) and Bahl (1980) and in Hall (1981). The use of printed antennas has become almost universal in all mobile communication systems. The first part of this chapter presents the theoretical study on patch antenna by introducing the different models of the patch and the feeding techniques for printed antennas. In the same part, we find a study on antennas array discussing the different topologies of antenna arrays and techniques for feeding a microstrip antenna array. The second part is about the miniaturization techniques for printed antennas by using:

- Modification of geometry (Meandering, Fractal, PIFA and Slot antennas);
- Materials (dielectric and magnetic materials); and
- Metamaterials which are materials artificially designed have simultaneously negative permittivity and permeability values.

The last part of this chapter discusses the introduction of new design examples of miniature microstrip antennas array for wireless communications applications. This section presents three new studies:

- The first one is on the design of a miniature microstrip patch antenna array with defected ground structure.
- The second study is on a new design of a wideband miniature antenna array.
- The last one is on the design of an antenna array miniaturization utilizing DGS and DMS techniques.

The different structures were validated into simulation by using tow electromagnetic solvers which permit to validate and to verify the different performances of antennas arrays as radiation pattern, matching input impedance and small dimensions.

Chapter 5 is about a detailed study of the one-dimension electromagnetic band gap (1D-EBG) structures and their application in a directional antenna design. EBG structures have attracted many researches in recent years, they are constructed as regular arrays of materials with different refractive indices, these structures can be calcified as follows:

- One-dimensional (1D) (Joannopoulos, Johnson, Winn, & Meade, 2011; Moghaddam, Entezar, & Adl, 2013);
- Two-dimensional (2D) (Baccarelli, Paulotto, & Di Nallo, 2007); and
- Three-dimensional (3D) (Mizeikis, Seet, Juodkazis, & Misawa, 2004) structures.

This chapter is centered on the 1D-EBG, the aim is to analyze and to understand the behavior of this structure. The obtained results demonstrate that the periodicity of the dielectric permittivity permit to reject the waves propagation in certain frequency bands. The chapter is composed from 4 parts:

- In the first part, a study of the propagation of electromagnetic waves in periodic structures permitting to solve Maxwell's equations leads to the formulation of the dispersion equation to determine the bandgaps frequencies (Collin, 1960; Cheney, & Kincaid, 2012);
- The second part will present the formulation of Eigen value problems for the radiation field in the EBG structures;
- The third par will discuss the S parameters results which will permit to determine the electromagnetic band gaps, and:
- The last part will discuss an application of the EBG structures for directive antenna applications.

Chapter 6 will deal with the miniaturization of antennas array (Aris, Ali, Rahman, & Pasya, 2016) and the reduction of mutual coupling between antennas array. The aim of this chapter is to present new contributions made for reducing the size of the antenna array by saving good performances in terms of radiation and matching input impedance.

In this chapter two studies are conducted to design and to fabricate antenna arrays with compact size and good performances. The methodology followed to achieve such antenna arrays is explained and detailed taking into account the introduction of theoretical study. The first antenna array structure is miniaturized by using a Defected Ground Structure (DGS) technique. Before achieving such antenna array, a detail study on the design of the basic microstrip patch antenna which is the elementary antenna used in the final proposed antenna array. This antenna element is based on circular shape microstrip antenna; this design is introduced from this state of the art to the achievement of the proposed antenna by using an electromagnetic solver. After this step, the first antenna array design is described and discussed, this array is fed with a microstrip feed. The final antenna array is validated at 5.8GHz and by using DGS technique the proposed antenna array is shifted to function at 2.45 GHz. The use of the defection in the ground plane of each radiating element of the antenna array, permits to reduce the antenna array size up to 82.46% as compared with the antenna array without DGS resonating at 2.45GHz. The second study is about the design two antennas array with reduction of mutual coupling (Ghaloua, Zbitou, Latrach, Abdellaoui, El Errkik, & Mandry, 2017). The two antennas arrays are separated by two magnetic walls of a planar (EBG) structure, the goal is to miniature and to reduce the mutual coupling between them, with the antennas arrays separation smaller than 0.6λ5.8GHz. A full-wave electromagnetic Analysis had achieved to evaluate the electrical performances of the proposed structures by using two electromagnetic solvers. Before the validation of the EBG technique used, a detail study on the achievement of antenna array fed by coaxial probe is presented, to validate an antenna array around 5.8 GHz. And by introducing EBG structures, the reduction mutual coupling is enhanced. The design achieves a high isolation of about 49dB between the two antennas arrays separated by an edge-to-edge distance of $0.29\lambda_{5.8GHz}$ at 5.8GHz.

Chapter 7 is centered around two studies. The first one is the miniaturization of antennas (Huff, 2006; Mosallaei, 2002; Bernhard, 2003) which can be done by reducing the size and keeping the fidelity of performance characteristic. In literature the different existing techniques are based on adding inductance, altering the antenna geometry, using additional components, or modifying the material characteristics (Roy, 2014). The second study is about reconfigurable antennas (PIN, FET, MEMS, etc.), which can be classified into different categories:

- The frequency reconfigurable antenna;
- Radiation pattern reconfigurable antenna; and
- Polarization reconfigurable antenna.

The chapter is divided into two parts. The first one will be about the most common used techniques of antenna miniaturization and a comprehensive study of the frequency agility techniques. The second part will focus on some contributions and current results of new miniature antennas by using Defected Ground Structure (DGS) technique, and other contributions of reconfigurable antennas based on PIN diodes. The first contribution is around the miniaturization of microstrip patches antennas, the technique used is the use of DGS structure which is incorporated, in the ground plane in the aim to shift the resonance frequency from 5.8 GHz to 2.5 GHz. The achieved miniaturization is up to 83%.

The second contribution is the design and realization of the reconfigurable antennas, where we find two studies:

- The first one presents a simple U-shaped geometry of a planar miniature reconfigurable antenna. The proposed antenna is based on PIN diodes in order to make the U-shaped antenna frequency reconfigurable.
- The second one discusses the design of a new structure miniaturized by using DGS method and in order to overcome the problem of the reduced bandwidth, the insertion of PIN Diodes has been used.

Chapter 8 describes a comparison study between the techniques of coplanar waveguide (CPW) and microstrip line applied to antenna in the Ultra-wide band by analyzing the different parameters achieved into simulation and fabrication. The fractal geometry has been used to achieve such antennas. It's defined as rough or fragmented shape that can be subdivided to copies of the original form with reduced sizes, therefore the geometry in different scales looks in identic view. Mandelbrot was introduced the first mathematical shape called 'Fractal Curve' in order to define a family of geometrical which remains without description in the standard Euclidean geometry. (Mandelbrot, 1975; Manimegalai, Raju, & Abhaikumar, 2009). The fractal geometry has been chosen to design the radiating patch of both types of antennas by including two electromagnetic solvers based on two different numerical methods. The parameters S11, Current density, gain and the radiation pattern have been achieved into simulation and measurement in the frequency range 3.1 – 10.6GHz released by the Federal Communications Commission (FCC) as a commercial UWB. The photolithographic technique, the network analyzer and the anechoic chamber have been involved to perform the fabrication and the measurement of the validated microstrip and CPW antennas. The two antennas present good matching input impedance at 50Ω. The fractal geometry has been used to design the radiating patch of both antennas in order to benefit from the self-similarity behavior. For the microstrip line has been chosen with a tapered part and a partial ground plane, while the CPW antenna has been achieved by using a transmission line with two partial ground planes in the same side of the substrate.

Chapter 9 aims the applications of Radio Frequency identifications (RFID) which is an automatic wireless technology for data gathering that has been deployed in a various ranges of applications including logistics, transport, animal tagging and asset tracking offering the capabilities of the identification in different weather conditions, the identification of many items simultaneously, identifying items within a vicinity between a few centimeters to several meters and enabling the off line-of-sight communication, hence RFID is becoming more attractive (Finkenzeller, 2010). The key of the RFId system is the antenna which is considered as an essential element, critical and very important in the performance of the whole RFID system due to many reasons (Rao & Nikitin, 2005):

- Antenna permits to recover and transfer power to the chip;
- The detection range and accuracy are directly dependent on the performance of tag antenna; and
- Reducing tag fabrication cost by using low cost substrates compatible with those manufacturing process are employed.

In this chapter, new microstrip antennas for RFID system are presented:

- The first section deals with the design of two dual-band antennas for handheld RFID readers. The Theory and principle characteristics of microstrip antenna and the fundamentals of multiband behavior are presented. The two proposed antennas are designed to operate in the ISM "Industrial Scientific and Medical" band at 2.45 and 5.8 GHz bands.
- The second section is dedicated to antennas for RFID Tag, the first part of this section deals with the introduction of the matching techniques for Tag antenna design. Then, the presentation of two examples of Tag antenna, the first one is a flexible antenna mounted on paper substrate and the second one is designed for the identification of metallic object.

Chapter 10 concerns the design of miniature antennas for RFID application, in the ISM "Industrial, Scientific, and Medical" band at 915 MHz and 2.45 GHz, by using two techniques, the first technique is the use of slots inserted into the microstrip antenna, and the second technique is the use of the fractal structure, in the end both techniques are used together in one structure to get the benefit of each technique at the same time. The first section of this chapter presents the theoretical part concerning the design of the patch antenna (Fang, 2010; Guha & Yahia, 2011), by introducing the different formulations of dimensions and using Transmission-Line Model. This part discusses the feeding methods which can be used for feeding planar antennas. The second section is dedicated to the state of art of RFID Antennas. In the last section, we have the different contributions presented and discussed:

- The first contribution is on the design of a miniature microstrip antenna with symmetric slots at 2.45GHz for RFID reader. The design of this antenna is based on a rectangular microstrip structure.
- The second design is a miniature microstrip antenna with symmetric slots at 915MHz for RFID reader.
- The third one is a miniature fractal microstrip antenna at 2.45 GHz for RFID reader, this structure is designed by using the fractal koch curve structure.
- The last contribution is a Miniature Fractal Microstrip Antenna with L Slots at 2.45 GHz for RFID Tag. This antenna is developed by using both techniques: fractal and slot techniques.

Chapter 11 is about the active microwave circuits which is the microwave power amplifier design. The aim of this chapter is to introduce and to explain the theoretical part and the concept of a microwave power amplifier (PA) (Pozar, 2009; Steer, 2009). PA is one of the blocks, that has a large effect on the overall performance of communication system especially in transmitter system and their design is decided by the parameters of the transistor selected. The chapter is divided into three parts:

1. Part 1 provides background theory relevant to the design of microwave power amplifier, like power amplifier classification, power gain, load line, output power and stability are introduced. A special emphasis on differentiating PAs from conventional small-signal amplifiers will be necessary in order to comprehend the specific requirements for this type of circuit.
2. Part 2 describes the matching techniques for PAs. Matching networks are designed in order to make the device behave in an optimal way according to a specific goal.
3. Part 3 utilizes the tools developed in parts 1 and 2 to analyze and to design the proposed microwave power amplifier with a microstrip technology intended to wireless applications. It presents a detailed process for designing a new structure of microwave single-stage power amplifier, suitable for LTE applications at 2.6GHz. This design is performed by employing a GaAs FET transistor.

Jamal Zbitou
University of Hassan 1st, Morocco

Ahmed Errkik
University of Hassan 1st, Morocco

REFERENCES

Aris, M. A., Ali, M. T., Rahman, N. H. A., & Pasya, I. (2016). Development of antenna array using Defected Ground Structure. In *2016 10th European Conference on Antennas and Propagation (EuCAP)* (pp. 1–5). Academic Press. 10.1109/EuCAP.2016.7481577

Baccarelli, P., Paulotto, S., & Di Nallo, C. (2007). Full-wave analysis of bound and leaky modes propagating along 2D periodic printed structures with arbitrary metallisation in the unit cell. *IET Microwaves, Antennas & Propagation, 1*(1), 217–225. doi:10.1049/iet-map:20050321

Bahl, I. J., & Bhartia, P. (1980). *Microstrip antennas*. Dedham, MA: Artech House.

Balanis, C. A. (2005). Antenna Theory: Analysis and Design. Hoboken, NJ: John Wiley & Sons Inc.

Bernhard. (2003). Reconfigurable antennas and apertures: State-of-the-art and future outlook. *Proceedings of SPIE Conference on Smart Electronics, MEMS, BioMEMS, and Nanotechnology, 5055*, pp. 1–9.

Da Silva, E. F., & McPhun, M. K. (1978). Calibration techniques for one-port measurement. *Microwave Journal, 21*(6), 97–100.

Fang, D. G. (2010). *Antenna Theory and Microstrip Antennas*. Taylor and Francis Group, LLC.

Finkenzeller, K. (2010). *RFID Handbook: Fundamentals and Applications in Contactless Smart Cards and Identification*. New York: John Wiley and Sons. doi:10.1002/9780470665121

Ghaloua, A., Zbitou, J., Latrach, M., El Abdellaoui, L., Errkik, A., & Mandry, R. (2017). Mutual coupling reduction and miniaturization arrays antennas using new structure of EBG. In *2017 International Conference on Wireless Technologies, Embedded and Intelligent Systems (WITS)* (pp. 1–5). Academic Press. 10.1109/WITS.2017.7934593

Gilbert, R. A. (2007). Waveguide Slot Antenna Arrays. In J. L. Volakis (Ed.), *Antenna Engineering Handbook* (pp. 9-1–9-36). McGraw-Hill.

Guha, D., & Yahia, M.M. (2011). *Microstrip and Printed Antennas: New Trends, Techniques and Application*. John Wiley & Sons, Ltd.

Hall, P. S., Wood, C., & James, J. R. (1981). *Microstrip antenna theory and design*. London: IEE Press.

Huff, G. H., & Bernhard, J. T. (2006). Integration of packaged RF-MEMS switches with radiation pattern reconfigurable square spiral microstrip antennas. *IEEE Transactions on Antennas and Propagation, 54*(2), 464–469. doi:10.1109/TAP.2005.863409

Joannopoulos, J. D., Johnson, S. G., Winn, J. N., & Meade, R. D. (2011). *Photonic crystals: molding the flow of light*. Princeton University Press.

Johan, S., William, F., Aurélien, P., Olivier, T., Nicolas, D., Bila, S., ... Gramond, R. P. (2016). Plastic and Metal Additive Manufacturing Technologies for Hyper Frequency Passive Components up to Ka-band. *46th European Microwave Conference*, London, UK.

Kumar, R., Louzir, A., & Naour, J. Y. L. (2017). Flexible RFID-Helix Antenna for Sensing the Home Electrical Device Activity. *European Conference on Antenna & Propagation (EuCAP2017)*, Paris, France.

Li, Y., Iskander, M. F., Zhang, Z., & Feng, Z. (2013). A new low-cost leaky wave coplanar waveguide continuous transverse stub antenna array using metamaterial-based phase shifters for beam steering. *IEEE Transactions on Antennas and Propagation*, *61*(7), 3511–3518. doi:10.1109/TAP.2013.2257649

Mandelbrot, B.B. (1975). *Les objets fractals: forme hasard et dimension.* Nouvelle bibliothèque scientifiques.

Manimegalai, B., Raju, S., & Abhaikumar, V. (2009). A multi fractal cantor antenna for multiband wireless applications. *IEEE Antennas and Wireless Propagation Letters*, *8*, 359–362. doi:10.1109/LAWP.2008.2000828

Mizeikis, V., Seet, K. K., Juodkazis, S., & Misawa, H. (2004). Three-dimensional woodpile photonic crystal templates for the infrared spectral range. *Optics Letters*, *29*(17), 2061–2063. doi:10.1364/OL.29.002061 PMID:15455780

Pendry, J. B., Holden, A. J., Robbins, D. J., & Stewart, W. J. (1999). Magnetism from conductors and enhanced nonlinear phenomena. *IEEE Transactions on Microwave Theory and Techniques*, *47*(11), 2075–2084. doi:10.1109/22.798002

Pozar, D. M. (1992). Microstrip antennas. *Proceedings of the IEEE*, *80*(1), 79–81. doi:10.1109/5.119568

Pozar, D. M. (2009). *Microwave engineering*. John Wiley & Sons.

Rao, S., Nikitin, P. V., & Lam, S. F. (2005). Antenna design for UHF RFID Tags: A review and a practical application. *IEEE Transactions on Antennas and Propagation*, *53*(12), 3870–3876. doi:10.1109/TAP.2005.859919

Steer, M. (2009). Microwave and RF design: A systems approach. *Sciences et Techniques (Paris).*

Wang, J. H., Li, Y. J, Zhang, Z., Chen, M. (2011). Radiation mechanism and polarization properties of leaky coaxial cables. *Radio Science, 46*(RS2009), 1-12.

You, K. Y. (2015). *RF Coaxial Slot Radiators: Modeling, Measurements, and Applications*. Artech House.

Acknowledgment

Editors would like to thank all those who have contributed to the success of this Handbook, especially authors who contributed with high quality chapters as well as the reviewers who helped us by their reviewing of the different chapters submitted.

A special thanks goes to all the staff of IGI global and particularly to Amanda Fanton for the patience and guidelines given along all stages of the book production.

The editors hope that this book will be useful for scientists, researchers, and PhD students.

Jamal Zbitou
University of Hassan 1st, Morocco

Ahmed Errkik
University of Hassan 1st, Morocco

Section 1

Antennas, Electromagnetic Theory, Design, and Applications

Chapter 1
Additive Manufacturing of Steerable Antenna Systems for 5G and Adaptive Cruise Control Applications

Rupesh Kumar
XLIM, France

Cyrille Menudier
XLIM, France

ABSTRACT

This chapter gives an insight view of the evolving additive manufacturing technology and its application for designing steerable front-end (FE) array antenna architecture dedicated to mmWave for the 5G network and the adaptive cruise control (ACC) applications. The manufacturing of RF system is a challenging work because it comes with complex designs, a long realization time, and high cost as well as weight. Technically, all these challenges could be addressed up with the use of additive manufacturing technology. It allows an increased flexibility in the manufacturing of complex designs such as steerable beamforming and other advanced RF products. Traditional machining technologies used to manufacture RF products are limited in their ability to produce complex shape. The AM technology allows to realize the entire designs in one single time-phase which positively impacts mass, cost, realization time, assembly quality and RF performance.

INTRODUCTION

In general, 5G new radio (NR) network based on the mmWave spectrum are bound to offer a much-enhanced wireless network connectivity and performance compared to the already existing 4G networks. The prospects of IoT (Internet of Thing), 5G (Peterson & Schnaufer, 2018), and radar applications (Kumar, Cousin & Huyart, 2014) such as the development of Autonomous Cruise Control (ACC discussed by Lacher, 2009 and Ramasubramanian & Ginsburg, 2009 in their works) system are pushing the real-world

DOI: 10.4018/978-1-5225-7539-9.ch001

interface close to the digital world by exploring some completely new possibilities at millimeter wave spectrum due to the availability of large contiguous bandwidth. The Front-End (FE) antenna design and its production are going to be a key factor in the success of mmWave services. Moreover, the digitally steerable beamform front-end design is going to replicate the same theme as of a DAC/ADC (that is available in the electronics designs for digital-to-analog and vice-versa), in between the real world and the dedicated machines such as human interface with the 5G network in wireless communication and with the autonomous vehicle in ACC systems. Nevertheless, such developments are going to bring a better experience of the human-machine establishment by enabling an enhanced wireless connectivity for the deployment of various applications through the secure and the reliable data available for the end user, irrespective of their situation, position, and time in space.

However, the manufacturing of RF system, especially a steerable front-end antenna design at millimeter spectrum, is a challenging research-work because it comes with some complex designs, along realization time, and high cost as well as weight. Technically, all these challenges could be addressed up with the use of Additive Manufacturing (AM) which is also known as 3D Printing technology, as discussed by the authors Johan et al. (2016), Fezai et al. (2015), Li et al; (2013) and Kumar, Louzir and Naour (2017). It allows an increased flexibility in the manufacturing of complex product designs such as steerable beamforming and other qualified advanced RF products (waveguides, filters, antenna feed chains, feed for parabolic reflector), array antennas and integrated components. Traditional machining technologies used to manufacture RF products are limited in their ability to produce products with complex shapes. To circumvent this limitation, complex products are often assembled out of many simpler subcomponents. The AM technology will not have such constraints, allowing it to realize the entire designs in one single time-phase which positively impacts mass, cost, realization time, assembly quality and RF performance.

The steerable beam with large spectrum bandwidth provides a dedicated wireless connectivity which is an important enabling factor for the high data capacity of the Non-Standalone 5G NR specifications realized by the 3GPP in Lisbon, December 2017. The one-dimensional beam steering is achieved by feeding a number of linearly equidistance antenna elements (linear array) with constant amplitude as well as constant relative phase differences. Similarly, the two-dimensional steering is achieved by introducing the constant relative phase shift in relative two-dimensional linearly placed antenna elements. Such one-or-two-dimensional array helps in achieving a highly-directive FE antenna array designs. The directivity directly influenced by the number of antenna elements in the array. The improved directivity enhances the coverage range of the signal and simultaneously it also avoids the unintentional interference through better beamwidth resolution. Moreover, it also helps in compensating the higher path loss associated with the millimeter spectrum. To compensate for these losses, the beamforming concept has been introduced to achieve a high antenna gain for a very high effective isotropic radiated power (EIRP). This approach helps in achieving longer radial coverage area for 5G network. Especially, in ACC the emitted power is much higher compared to the 5G and hence it causes more dissipation of heat in FE design. Therefore, the understanding of the heat dissipation requires some level of attention and to use a suitable semiconductor technology like Silicon Germanium (SiGe), Gallium Nitride (GaN) from the implementation perspectives. An overview of different cutting-edge semiconductor technologies is presented in this chapter related to ACC and 5G applications.

Also, the highly-directive beams allow spectral reuse through the spatial diversity. The well-known different architecture for the beam steering concept is explained here. The same concept of steering is also used in the ACC applications. Therefore, an example of the array antenna designs and its feeding network for the AM is presented in this chapter. The dedicated frequency band (76 GHz – 77 GHz, Sickinger, Weissbrodt & Vossiek, 2017) for ACC is used for sensing other cars and objects far ahead. However, the advanced driver assistance systems (ADAS) demand much larger bandwidths, up to 81 GHz, to enable the beam steering for 360-degree radar vision around the vehicle for a fine resolution. The design and fabrication of such a broadband FE array antenna design at millimeter spectrum is always a challenge for microwave researcher and engineers. In fact, because of the availability of the continuous frequency band at millimeter spectrum, the demand is surging up. However, the component cost and weight, as well as the designing time, are some real challenges which still exist. In this regard, the additive manufacturing of microwave designs is a potential solution which is presented in this chapter. A comparative study of different AM processes is explained in this chapter.

ADDITIVE MANUFACTURING

The additive manufacturing or the 3D printing is becoming an acceptable technology for rapid prototyping and manufacturing of standard and qualified complex RF systems. Although, it has been used for making prototypes across a range of different industries for many years, but its prospect of mass production is something that making it more acceptable. The most pertinent advantage of this technology is to directly transform the computer-assisted design into the ready to use prototype without further allowing any traditional methods such as molds, machining. The focus is to avoid the conventional formative and subtractive fabrication. It helps in prototyping very quickly and cheaply than the traditional assembly lines that require more investment.

The AM is destined to have a profound impact on the production line across all industries, from small devices to very complex systems for space applications, from sophisticated defense equipment to portable mobile for civilian used, and so on. Logically, the AM can be assumed to be at the forefront of the fourth industrial revolution which is justified by some exceptional growth prediction by the leading market research agencies. According to Statista, the global market survey portal, the global market for AM and services is expected to grow to almost 50 billion U.S. dollars by 2025 from 5.9 billion in 2015. Wintergreen Researcher, involved in analysis and forecasting of business opportunities, has predicted the global market for metallic AM for aerospace is to reach 21 billion dollars in 2024 from 1.9 billion dollars in 2016. The same trend is also published by the Gartner, an American consulting firm, the AM market is poised to reach 11billion U.S. by 2022 with a compounded annual growth rate of 27% from 2016 to 2022 (Alexander, 2017). Such trends will only lead to more investment in the AM.

There are many different factors driving the AM market growth, however, the most dominating factor is the advantage of fast and cheap fabrication. This gives more flexibility in the fabrication of complex qualified RF designs and shapes. Such freedom of complex design enables in achieving the better RF performance which is not possible with the traditional technologies such as machining which has limited ability to design and to produce a complex design in shapes. In the conventional technology, a complex shape is often produced from the assembling of many subcomponents that are fabricated as a separate design. Such time-consuming process is easily avoided with the introduction of AM and the ability to produce a complex shape as a one single element shape positively impact mass, cost, fabrication-time, assembly quantity, and RF performance. The reduced weight of the system compared to the conventionally produced one has leveraged a whole set of new opportunity in aerospace industries. With increasing acceptance of the AM technology in aerospace, it will be a game changer for the design of complex and integrated antenna structures particularly in Ku- and Ka-band. Recently NASA has mentioned that in the upcoming future the deep-space capsule Orion will carry hundreds of parts made up of AM technology. Industries like Stratasys, SWISSto12, Optisys are some early investors who are investing heavily in the fabrication of qualified RF system and antenna components from C- to W-band (4 to 110 GHz).

In a simple way, the Additive Manufacturing refers to a process by which a physical 3D system-or-device can be built with the primary aide of computer design. It builds the system layer by layer without further requirement of molds or machining. The commonly used materials for building the prototype are plastics, resins, metals, multicolor sandstone, ceramics, and wax.

In RF microwave field, the conductivity of the built system's surface must be further modified for supporting the wave signals in accordance with the signal wave's propagation law. Therefore, the electric characteristics may be incorporated on the surface of the built system by putting a metallic layer all over the surface. The thickness of this layer should be applied while considering the skin-depth phenomena at the frequency of interest. Here it is interesting to know its influent on the AM because it has a significant effect on millimeter wave propagation. A thin layer of silver paint can make a significant improvement of the RF system as most of the currents flow just within a few micron-depth at the surface. This happens because the alternating current does not flow uniformly with respect to the cross-section of a conductive element such as a wire. The current has the highest density near the conductor's surface and decreases exponentially with the depth from the surface. The computation of the skin-depth depends upon the frequency of the alternating current signal. We can drive a practical formula for the skin-depth (Tsuchiya & Onodera, 2015; Manke, Kardjilov, Strobl, Hilger, & Banhart, 2008) as:

$$\delta = \sqrt{\frac{\rho}{\pi f \mu_0 \mu_r}}$$

where δ is skin-depth in meters

μ_r is the relative permeability of the conductor

ρ is the resistivity of the medium in Ω-m

f is the frequency in Hz

At mmWave, the skin-depth becomes small for conductors. This depth must be in good agreement with the applied thickness in the AM process. The skin-depth or the thickness of the conductive surface needs to be less than of few micrometers. For better understanding, the skin-depths of some common elements at different frequencies, 30 GHz, 78.5 GHz, and 100 GHz, are tabulated in Table 1.

Skin depth decreases with the higher frequency as tabulated above. The thinner surface helps in reducing the weight of the fabricated RF design compared to the conventional approach such as machining technologies. Some commonly known different types of AM technologies are described in this chapter.

Some well-known AM technologies are fused deposition modelling (FDM) (Cadman, Zhang & Vardaxoglou, 2016), selective laser sintering (SLS) (Oscar, Peverini, Lumia, Caliganano, Addamo, Lorusso, … Virone, 2017), stereolithography (SLA) (Liu, Gong, and Chappell, 2004), and polyjet technologies

Table 1. Skin depth at three different frequencies

Conductor	Skin Depth (μm) @ 30 GHz	Skin Depth (μm) @ 78.50 GHz	Skin Depth (μm) @ 100 GHz
Copper	0.3764	0.2327	0.2062
Aluminum	0.4734	0.2927	0.2593
Gold	0.4349	0.2688	0.2382
Silver	0.3660	0.2262	0.2005
Nickel	0.0310	0.01918	0.0169

(He, Chahal & Papapolymerou, 2017). These technologies are briefly explained in this chapter. Some more technologies are also known but are not yet so popularized, therefore, the present chapter has not presented those, such as digital light processing, direct metal laser sintering (DMLS), selective deposition lamination (SDL), binder jetting, electron beam melting (EBM), laminated object manufacturing (LOM), selective laser melting (SLM), CLIP and fusion jet. The different AM processes are defined according to the ASTM International Committee F42 nomenclature for distinguishing technologies and processes. Here, the same guidelines have been followed for the readers and following technologies are mentioned below:

- **Binder Jetting**: Uses a liquid bonding agent which is deposited to fuse the powder materials.
- **Direct Energy Deposition:** Utilizes the focused thermal energy such as a laser to fuse materials by melting as they are being deposited. This process mainly uses the metallic material for the fabrication.
- **Material Extrusion**: Performs the disposition of the material through a nozzle. The FDM comes in this process.
- **Material Jetting**: Uses droplets of material such as a photopolymer, thermoplastic material for selective deposition. Inkjet process is a good example of this process.
- **Powder Bed Fusion:** Has thermal energy, provided by a laser or an electron beam, for fusing the selective regions of a powder bed. SLS, LS, and EBM are some examples of this process. This method utilizes for metals and polymers.
- **Sheet Lamination**: Involves the bonding of the sheets of material to form the shape.
- **Vat Photopolymerization**: Is an AM process in which liquid photopolymer in a vat is cured by light-activated polymerization. Lithography based AM processes (DLP, SLA) are a good example of this process.

Some commonly used additive manufacturing examples are:

- **Fused Deposition Modelling:** This is the most widely used process for AM. FDM printer uses a thermoplastic filament which is heated to its melting point and then deposited, layer by layer, to create a three-dimensional object. This is invented by Scott Crump, the founder of Stratasys Ltd. The initial design is developed by the CAD which then used by the 3D printer for prototyping. During printing, the materials (modeling and scaffolding) is converted into plastic threads or filaments which are deposited through an extrusion nozzle to build a layer by layer three-dimensional shape. The nozzle uses to melt the plastic thread through heating. The commonly used printing and support materials for FDM are butadiene styrene (ABS), polycarbonate (PC), polyetherimide (PEI), and water-soluble wax, brittle thermoplastics, like polyphenylsulfone (PPSF). The printer has two basic movements, horizontal and vertical. Cube, Replicator, and Mojo are some examples of FDM printers.

- **Selective Laser Sintering:** In this process, tiny particles of plastic, ceramic or glass are fused together to form a three-dimensional object. The heating is done by a high-power laser. The commonly used material is nylon. The focus is a good example of SLS printer.
- **Stereolithography:** This involves a stereolithography apparatus (SLA) which converts liquid plastic into a three-dimensional object. The process is patented by Charles Hull, co-founder of 3D Systems, in 1986. SLA has four main parts: a tank that carries the liquid plastic (photopolymer), a perforated platform that is lowered into the tank, an ultraviolet laser and a computer controlling platform. SLA creates a highly accurate shape.
- **Inkjet/Polyjet Technology:** The combined benefits of lithography and material jetting are used to develop an inkjet lithographic 3D-printer. It consists of several hundred nozzles for depositing small droplets of photopolymer. After deposition of each layer, the ultraviolet lamp cures the fresh layer and the process is repeated for the layer by layer deposition.

Since the AM uses the layer by layer approach, the built surface is not uniform and has sharp protrusion and recesses at multiple scales along with open pores transition into closed ones. The main challenges associated with the AM process are the surface texture of the printed objects, surface integrity, and degradation of the dimensions. Especially the surface texture which represents the geometrical irregularities present at a surface. It is measured as a Ra value which represents across the layer. At millimeter waves, the surface irregularities become comparable to the wavelengths which result in random scattering of the propagating mmWave signals. Such random scattering results in the increased level of cross-polarization. It also increases the surface resistance and it affects the radiating parameter, gain, in case of an antenna. The surface texture can be measured by a Stylus profilometer. In layer by layer approach, the corrosion changes the surface texture of the objects which leads to the performance degradation of the RF devices. This requires an advanced surface treatment and surface plating. For example, SWISSto12 are using metallic materials such as titanium combined with advanced surface treatment. Also, the bonding of the metal layer with the 3D printed objects should be strong enough to withstand any mechanical stresses. Also, the heat dissipation is a major concern for handling high-power RF components, mainly feeding and radiating structures of an antenna. Therefore, the printed objects must have good operational temperature range. Other challenges, like the viscosity of the molten thread, a heating process in a 3D printer and so, also need special attention to produce qualified RF systems. The AM gives freedom to design complex system structures, this is explained with the design of front-end array antenna designs for 5G and ACC applications. First, the architecture of the front-end array is explained and then as an example, the front-end design is presented for ACC application over 76-81 GHz.

STEERABLE FRONT-END ARRAY ANTENNA ARCHITECTURE

One of the main goals of 5G is to deliver a much higher data rate (>10 Gbps) compared to the existing 4G networks. Such high data rate will enable may use cases and services which will bring the virtual world closer to the real world. However, the sub 6 GHz carrier band is already congested and there is no possibility to use more bandwidth for increasing the data capacity as it directly contributes. To achieve the peak data capacity, the 5G network is advocating the use of available wide band at the mmWave spectrum, particularly between 24 and 30 GHz.

Although this will assist in increasing the data capacity, the use of mmWave spectrum introduces a high free space path loss, and consequently, this reduces signal wireless range. This can be solved with a very high effective isotropic radiated power (EIRP). In the U.S., the Federal Communications Commission (FCC) has defined very high EIRP limits for the 5G new radio network which is tabulated in Table 2.

The challenge is to design the front end (FE) transceiver systems that can achieve the power limit with the low cost, size, weight and the expected power budget by the mmWave carriers. Also, at mmWave, the wireless communication is mainly established through the line-of-sight (LOS) propagation which is different from the sub 6 GHz wireless transmission where the diffraction and reflection play a significant role. Therefore, to establish the LOS propagation between two devices, the steerable beamforming becomes one of the main features in the 5G NR network. Therefore, the proper FE architecture and the RF semiconductor are key to unlock the unlocked mmWave benefits. There are many factors like antenna gain, noise figure which contributes towards high EIRP.

Given the specifications required of an array antenna, the first step is to determine the size of the antenna-aperture because it is one system parameter that defines the antenna gain. However, in ACC application the resolution plays an important role, therefore, the array beamwidth may become a determining factor. In addition, there are other factors which also influence the aperture size, however, in this chapter, the more emphasis is given about the realization of a complex array antenna design from the Additive Manufacturing perspective.

Now, the steerable antenna design can be broadly classified into two types, Passive Array and Active Array, respectively. This classification is solely based on where the amplifier is in use with respect to the antenna element. Typical architectures of the array are shown in Figure 1.

The analog architecture, Figure 1 (a), is considered as a fundamental design for a steerable array antenna. Each antenna element has tunable phase through which a constant phase difference across the antenna element is simultaneously introduced at a frequency which results in the formation of the wavefront at an angle θ (theta) to the antenna's axis. The θ is steered by varying the phases at each antenna element. The scheme of beam steering is depicted in Figure 2.

The effective EIRP is dominated by the average power of each antenna element. Therefore, if the average power of each element is enhanced as shown by the active array then it will have more EIRP compared to the analog.

The one-dimensional array is only capable of steering the beam in a single dimension. However, the two-dimensional steering is possible with the two-dimensional arrays and so this increases the complexity of the array architecture. Further, if the two-dimensional array is sub-divided in the sub-square panel then it would form multiple steerable beams. Depending upon the application, such soft modification is

Table 2. FCC power limits for 28 and 39 GHz bands

Base Station	75 dBm/100 MHz
Mobile Station	43 dBm

Figure 1. Front End Array Architecture: a) passive array, b) active array

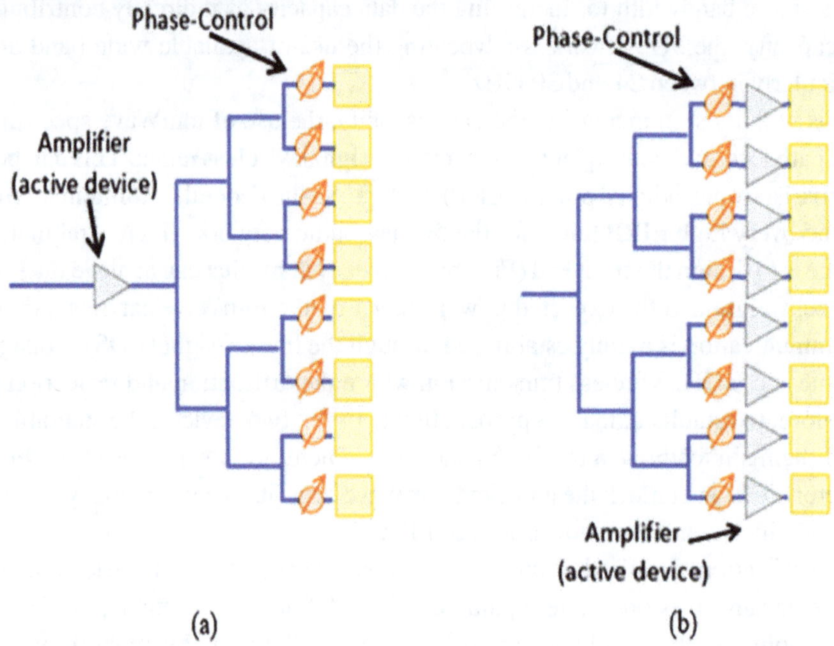

Figure 2. Beam Steering Scheme

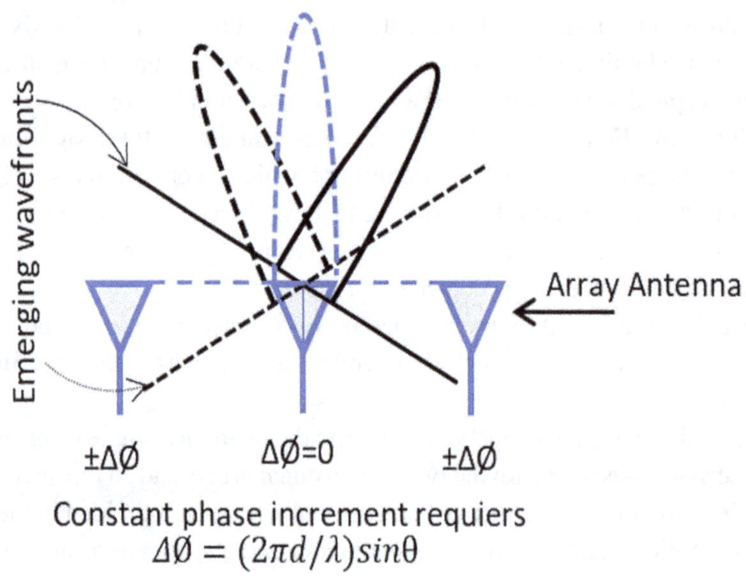

easily introduced with the embedded control ICs (responsible for varying phase and amplitude control) so that desired beam can be formed. Generally, the active beamforming architecture is categorized into three sub-categories: analog, digital and hybrid. The comparative advantages of the active beamforming are tabulated in Table 3.

Table 3. Advantages of active beamforming

Analog	Digital	Hybrid
Simple hardware implementation	Flexible with a number of beams and nulls	Flexible with a number of beams and nulls
Hardware can fit within the lattice (silicon array)	The high number of beams	The high number of beams
Beam benefits from full array gain	The beam can be changed dynamically	The beam can be changed dynamically
Lowest system DC power	Each beam benefits from the full array gain	Hardware can fit within the lattice (silicon array)

Based on the manufacturing process, the array is divided into two group: column-feed array and all-silicon array.

The column-feed array is shown in Figure 3.

In this architecture, the control ICs are mounted outside the array and drive a single column with a single gain and phase control. A typical corporate feeding structure is deployed between control ICs and antenna radiating element. High feed loss occurs in this architecture due to feed routing which results in power dissipation and high junction temperature. The power dissipation is directly influenced by the number of active elements in the array. The high dissipation also lowers down the overall efficiency of the front-end array antenna system. Typically, the designing of the efficient array design with minimum power dissipation is considered as a challenging work for researchers and engineers. Therefore, it is important to use a reliable semiconductor technology that must have high-temperature operability. The reliability of Silicon-Germanium (SiGe) technology becomes poor with the temperature excess of 150°C, however, the Gallium-Nitride (GaN) has better reliability and can support good performance up to 225°C. The advantage of 75°C gives an upper hand to GaN over SiGe for the passively-cooled array

Figure 3. Column-fed array design

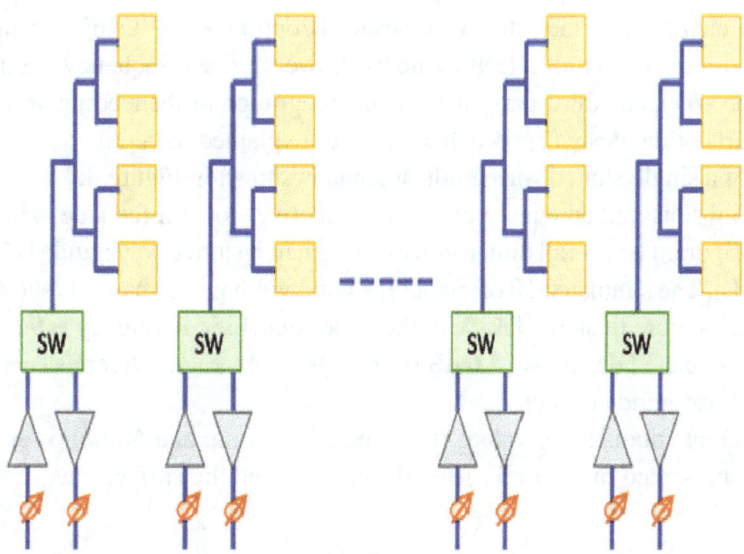

designs such as front-end array antenna design. All silicon arrays support the control ICs within the array lattice and hence it has low feed loss.

As an example of additive manufacturing concept of the front-end array antenna design, the antenna designing and its characterization for autonomous cruise control application over 76-81 GHz frequency band is presented in next section of this chapter.

FRONT-END ANTENNA DESIGN OF ACC AS ADDITIVE MANUFACTURING

The radar has been traditionally associated with the military but with the development in manufacturing as well as semiconductor technologies, it has added many new applications for the civilian safety purposes. One such application is Advanced Driver Assistance Systems (ADAS) for autonomous cruise control to enhance the road safety as well as for boosting the logistic supply through road transport system. The ADAS mainly consists of three types, short-range radar (79 to 81 GHz), medium-range radar (77 to 81 GHz), and long-range radar (76 to 77 GHz), respectively. Together, it occupies the mmWave spectrum from 76 to 81 GHz. The front-end array antenna design has active beamforming architecture for beam steering.

In this chapter, a slotted waveguide array antenna (SWAA) is presented for easy and cheap production under the AM process. The WR10 (2.54mm X 1.27mm, fundamental cutoff frequency 59 GHz) standard has been used for the designing of the 16-element array structure. The SWAA structure is designed as the initial step involves in the AM, that is, a computer-aided structure. Here, CST Microwave Studio has been used for the optimization of the SWAA in W band. Same frontend design would be used for three different radars (short-rage radar, medium-range radar and long-range radar) over 76 to 81 GHz while connecting with the dedicated RF bandpass filter over three different frequency bands.

The optimized CST 3D design of the waveguide antenna is fabricated using any one of the previously discuss AM process in the form of a plastic prototype. The silver metallic paint can be used for the mmWave propagation. The considered metallic thickness of waveguide is equal to 12 μm which is in accordance with the skin-depth as already explained in the starting of this chapter. The design approach of SWAA is subdivided into three parts: first part is the designing of an elementary slotted waveguide antenna as a single element and then the array design is optimized by using several single elements, second-part deals with the design of a feeding network over desired frequency band as per the matching bandwidth of the array, and third-part involves the integration of the feeding network with the array design. All three parts of the design approach are further explained in detail.

The optimized of a single slotted waveguide antenna is shown in Figure 4.

It has 20 radiating slots and an equal number of inductive slots on the side wall. The input port is shown in the above figure. The overall dimension of a single unit slotted waveguide is 2.54mm X 1.27mm X 42.6 mm (WxHxL). The simulated 10 dB matching bandwidth performance is shown in Figure 5. The matching bandwidth is more than 20 GHz with the upper matching frequency is 90 GHz.

The gain of the simulated design is 17.6 dB over 75-90 GHz. Figure 6 represents optimized gain of the designed slotted waveguide design.

Analysis of different conductivity values (between 0.125 S/μm and 50 S/μm) of the silver metallic paint is done and is presented in Figure 7. The thickness of metallic surface was kept unchanged.

Figure 4. Single Slotted Waveguide Array Design over W band

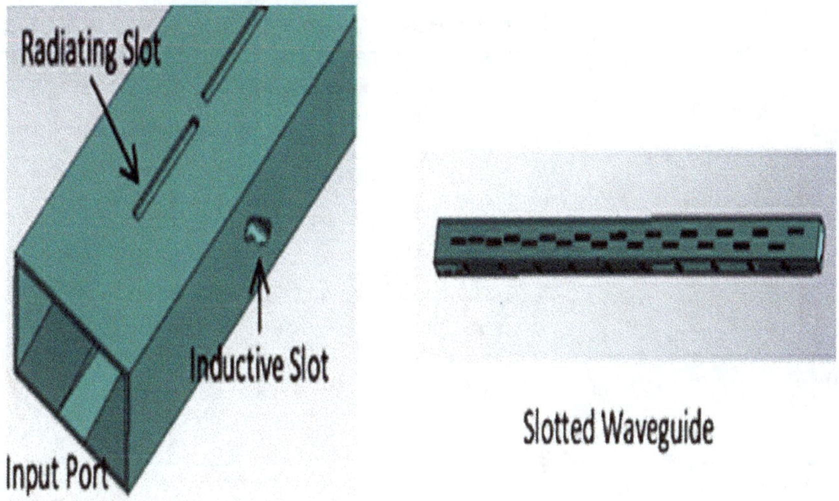

Figure 5. Matching Bandwidth Performance of Slotted Waveguide Antenna

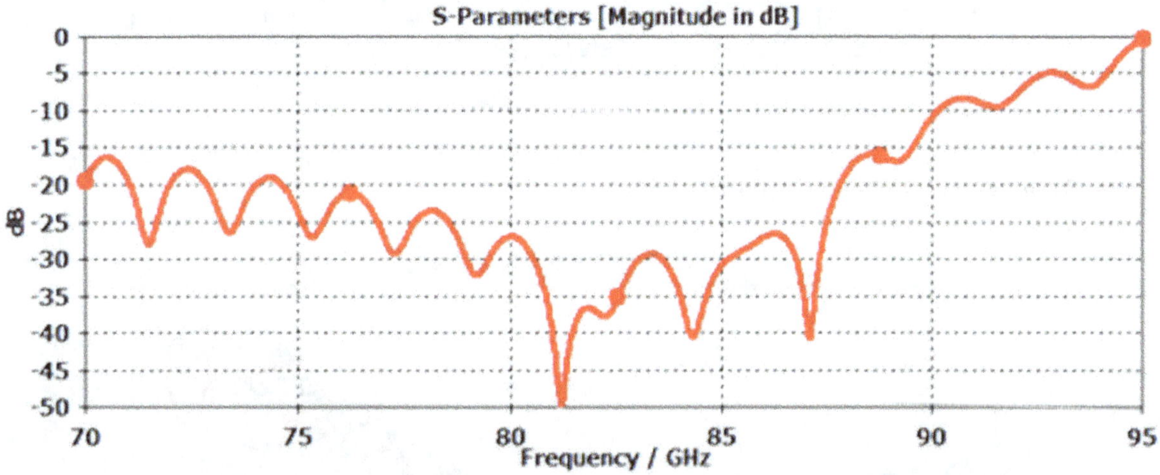

The matching bandwidth has more than 20 GHz with the higher cutoff frequency at 90 GHz. The simulated design is validated for the wide range of conductive values which brings more freedom in the fabrication process.

After achieving the satisfactory performance in terms of matching bandwidth and the gain over the W band, the array design is investigated. The optimized single waveguide is replicated to form an array of 16 elements. The SWAA design is presented in Figure 8. The designed slotted waveguide array antenna is shown in Figure 8.

The simulated array has 31 dB directivity in the W band. The 10 dB matching bandwidth has good resonance with the single element design. The array is connected to the front-end architecture through a feeding network.

Figure 6. Simulated Gain of the Single Slotted Waveguide Antenna

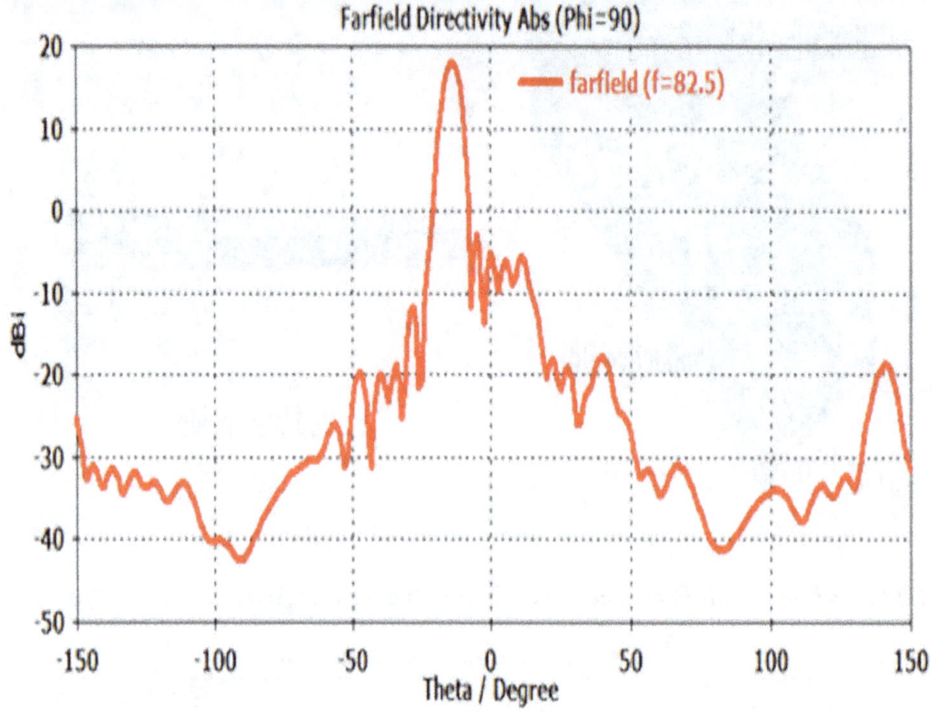

Figure 7. Analysis of Different Conductive Values

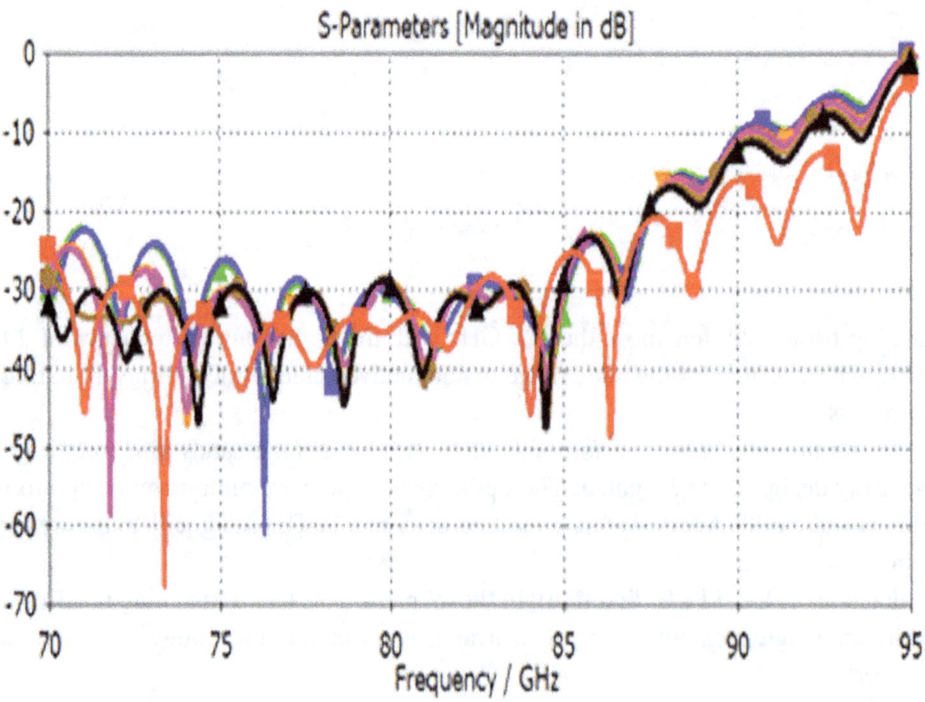

Figure 8. Slotted Waveguide Array Antenna

In a broad sense, an antenna feed can be defined as an impedance transformer between the FE and the radiating part of an antenna. In this way, it helps in optimum transfer of the radio waves from the electrical parameters such as current in the transmit chain and conversely does in the receive chain. The antenna feed usually incorporates all the components such as switches, diplexer, for connecting the array elements of an antenna to the Front-end design. In case of beam steering array antenna design, its complexity increases many times compared to a single element antenna design. Such complex designs would be addressed by the Additive Manufacturing process. As an example, the 1:16 corporate waveguide feed network over W band is explained in the next segment.

For achieving the beam steering phenomena over a desired frequency band of interest through an array design, a feed network must fulfill two important conditions, that is, to feed equal amplitudes as well as constant relative phase difference into relative elements of an array. This can be designed based on the concept of 1:2 splitter which divides the amplitude and the phase of a signal over specified frequency bands at the input port into two equal proportions at two output ports.

A simple W-band, 1:2 waveguide splitter design is presented in Figure 9.

Based on the above design, a corporate feed network is designed for the 1:16 (input: output) channels. It is depicted in Figure 10. The front-, side- and free-views are shown here. The 1:2 splitter is connected at 4 successive stages to form a 1:16 splitter.

The simulated parameters are shown in Figure 11-13. The return loss is lower than standard -10 dB benchmark for the impedance matching over almost entire W band.

Figure 9. Splitter design at W band

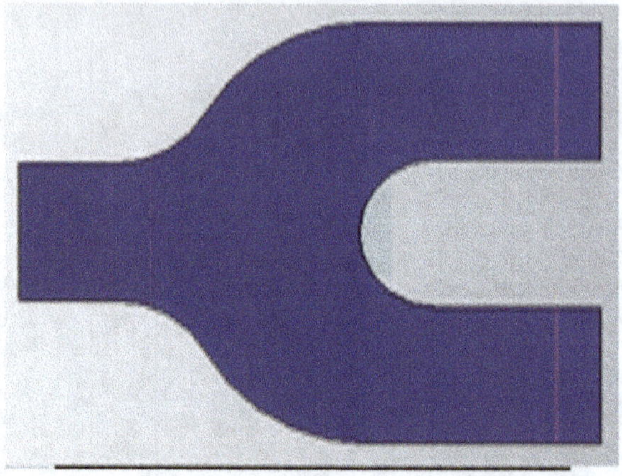

Figure 10. A 1:16 Feeding Network over W Band

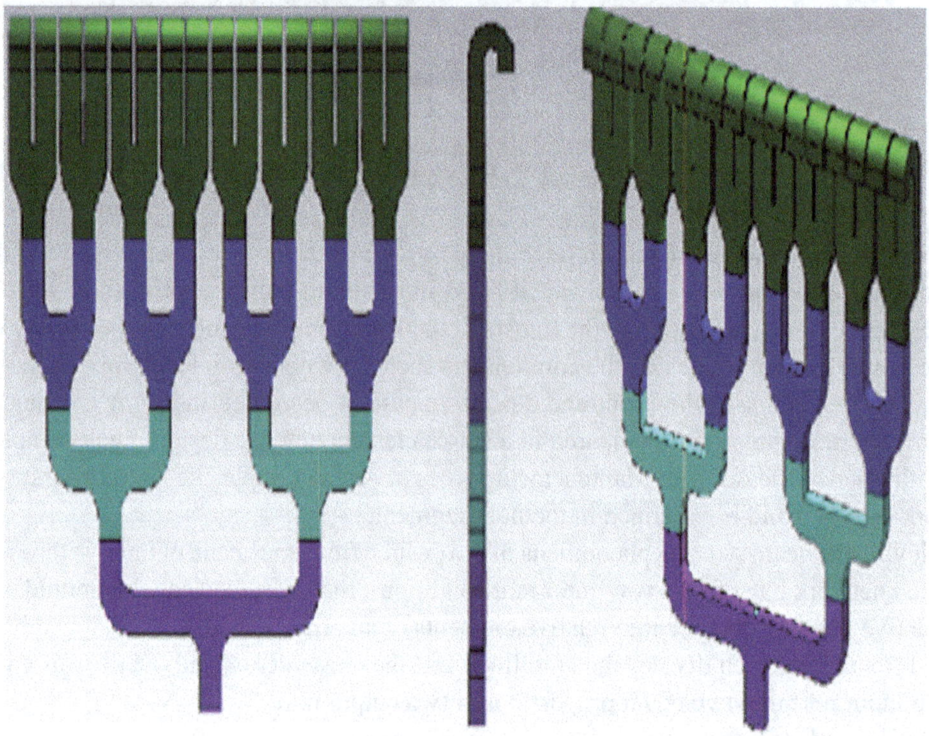

At all output ports, amplitudes and phases are divided into equal proportions as marked in the Figure 12-13. The variation of output amplitudes is within 0.4 dB (between -12.25 dB and -12.65 dB) over 75 to 90 GHz.

The output phase at each output port is equal in magnitude.

The integrated design of the slotted waveguide array antenna with its feeding network is presented in Figure 14. The overall dimension of integrated design is 49mm x 47.5mm x 3.6mm (LxWxH). The

Figure 11. Matching Bandwidth of Feeding Network

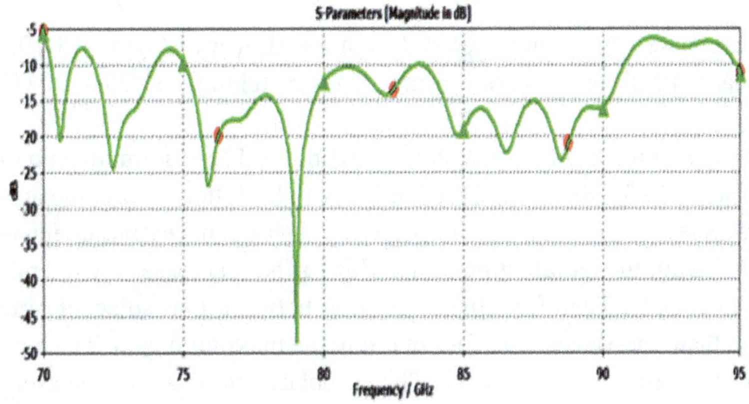

Figure 12. Variation in Amplitude at the Output ports of Feeding Network

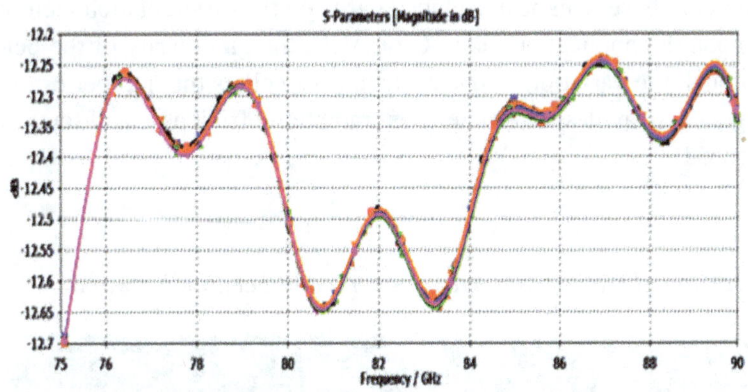

Figure 13. Phases at the Output ports of Feeding Network

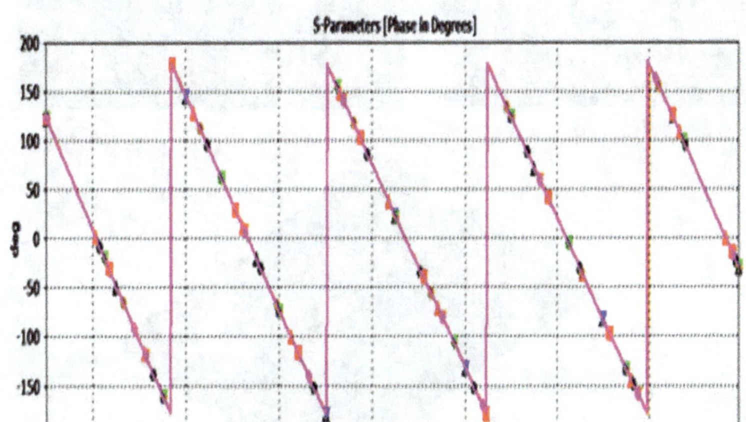

front, back, and side views are shown in the upper row whereas the free view of the design is shown in the lower layer.

The simulated parameters of the integrated design are shown in Figure 15-19. The return loss is lower than the standard -10 dB benchmark for impedance matching over 75 to 90 GHz frequency band in the W band.

The simulated array antenna gain is presented in Figure 16. The maximum gain of the array antenna is 31.3 dB which is around 13.5 dB more than that of a single slotted waveguide array antenna.

More than one-wavelength of central spacing between any two successive antenna elements increases the side lobe level (SLL) in the radiation patterns. This further decreases the overall antenna gain performance over the frequency. Therefore, the separation between two adjacent branches of a feeding network must be less than one-wavelength for obtaining the maximum gain. The cross-polarization and the co-polarization are presented in Figure 17. The co-polarization gain of the array is 31 dB whereas the cross-polarization is less than -33 dB.

The beam steering phenomena are shown in Figure 18 for different values of the relative phase difference of 60 and -60 degrees.

Appropriate values of phases can be introduced based on the different architecture explained in Section 2. Depending upon the application, like 5G or ACC, the granularity of the beam steering can be easily controlled through the variation of phase change as well as the number of elements in an array. The presented integrated array design efficiencies, radiation efficiency, and total radiation efficiency are presented in Figure 19.

Figure 14. Integrated Slotted Waveguide Array Antenna and Feeding Network

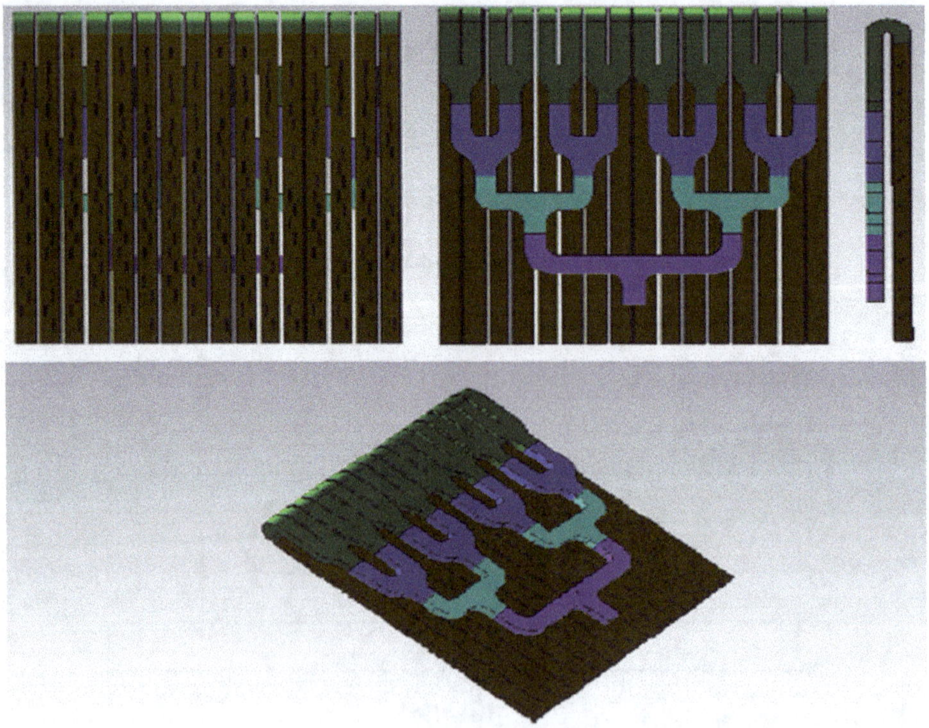

Figure 15. Matching Bandwidth of Integrated Array Antenna Design

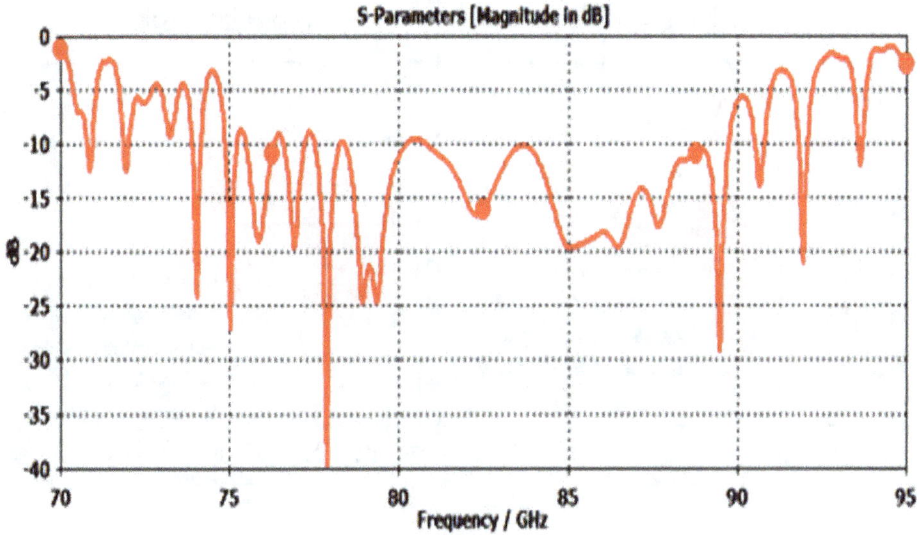

The radiation efficiency is between -0.8 and -0.4 dB over the frequency band, 75-90 GHz. The total radiation efficiency is between -1.2 and -0.54 dB over the frequency band, 75-89 GHz.

The designing of steerable beamforming antenna array incorporates other electrical components such as a switch, high-power amplifier, which need further deep analysis with respect to the semiconductor technologies (GaN, SiGe, etc.) from the overall product fabrication. Some important factors in this regard are presented in the next section of this chapter.

Figure 16. Array Gain over Frequency

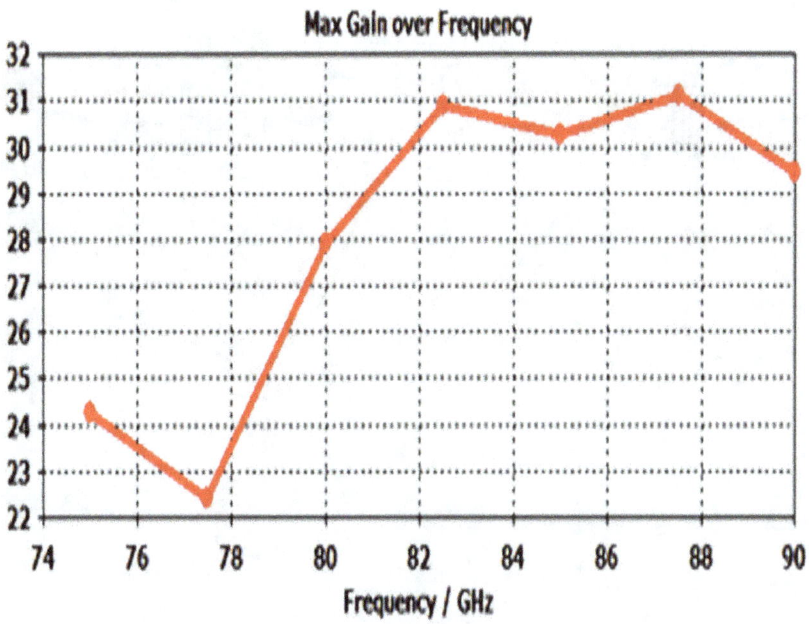

Figure 17. Cross- and Co-polarization gains of the array

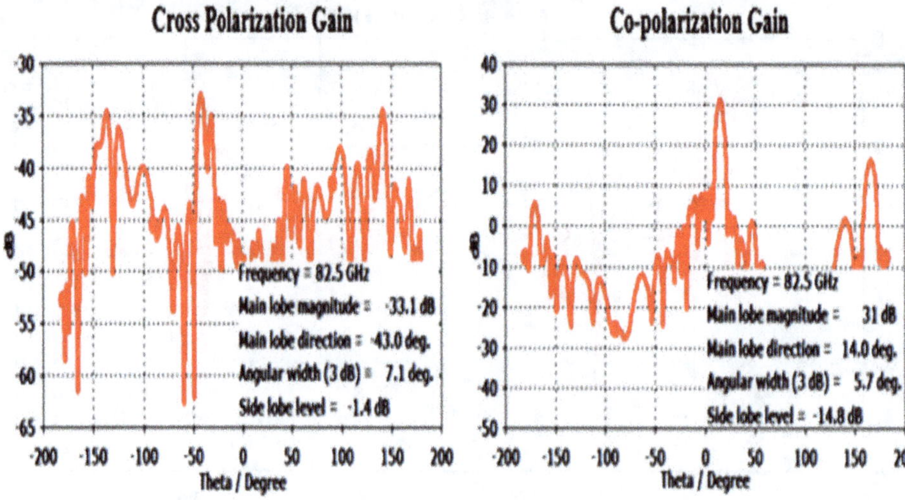

Figure 18. Beam Steering

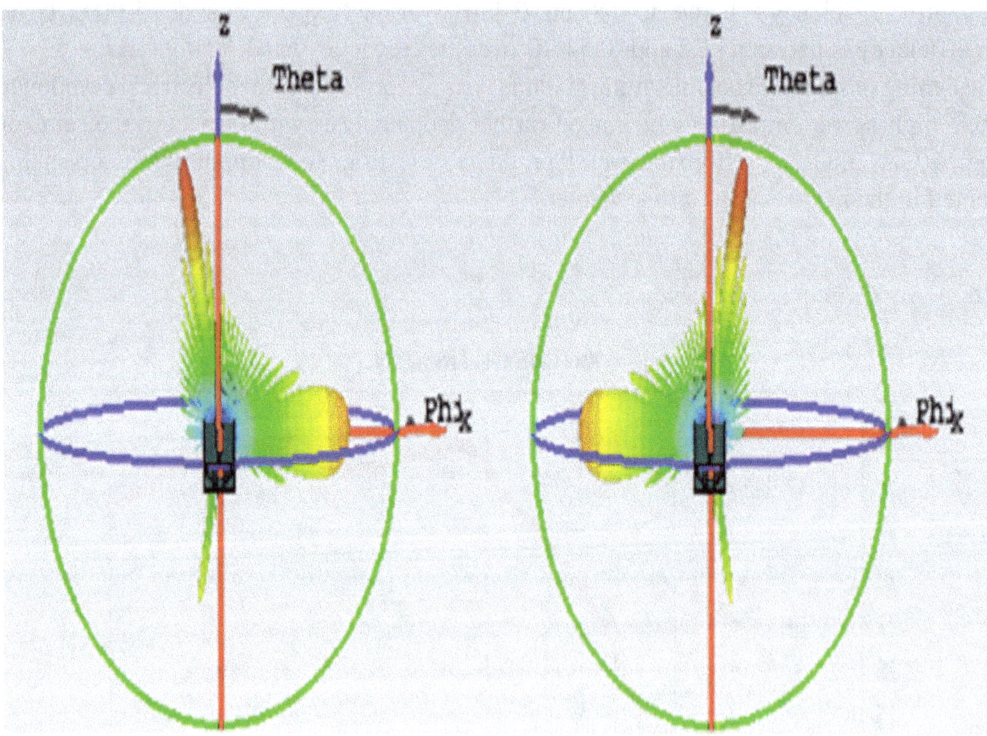

Figure 19. Array Efficiencies

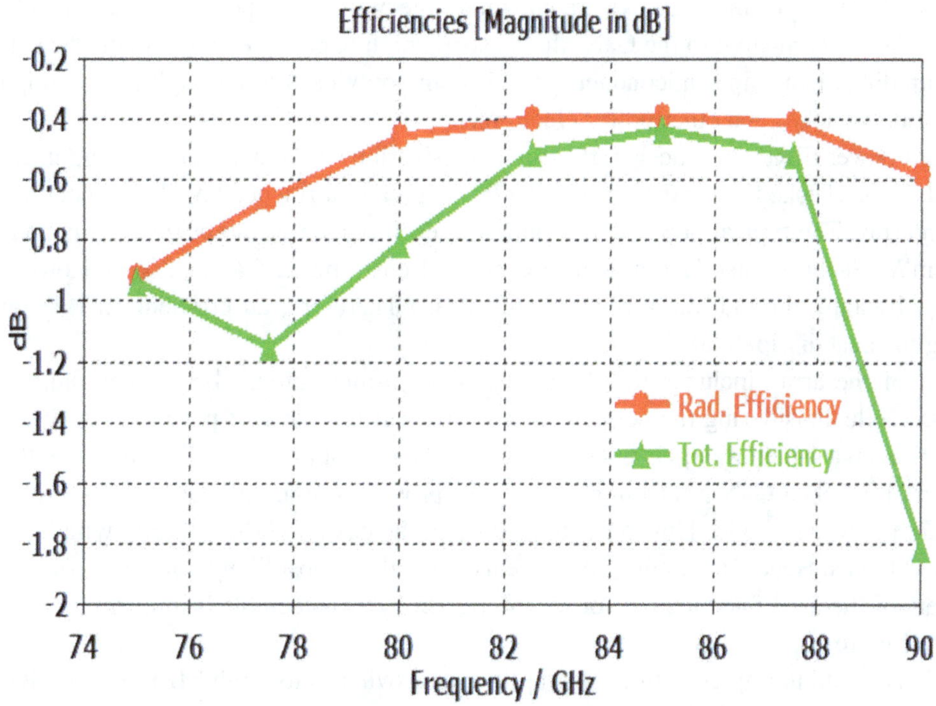

COMPARATIVE ANALYSIS OF SEMICONDUCTOR TECHNOLOGIES

The front-end antenna structure is intensively using large array antenna architecture for the achievements of the high data rate in case of 5G NR standardization and a better range and beamwidth resolution for ACC applications. The motive behind the FE array design is to use high EIRP for achieving a large radial range as well as better bandwidth resolution to avoid the unnecessary incoming signals. The high EIRP can be achieved through the application of array antenna which is directly linked with the beamforming gain.

The average power ($P_{avg\text{-}chan}$) associated with each channel of array contributes towards the antenna gain. For a given value of EIRP, it can be achieved either by using a large array size (say more than 500 antenna elements) with lower, $P_{avg\text{-}chan}$, or by a smaller array size but with a higher average power. The high-power performance can be easily achieved with the use of Gallium Nitride (GaN) technology in comparison with the Silicon Germanium (SiGe) which supports lower-power, because GaN has a higher band gap energy compared to SiGe which helps in attaining performance at higher temperature, 225°C.

The use of GaN results in smaller array size compared to the SiGe. Therefore, given the size constraint for an application, such as in ACC the whole FE design should be confined within the size of a Radome, the GaN suits more than the SiGe. On the other hand, 5G base stations have flexibility in the available size, therefore, SiGe is economically viable for the use while utilizing a lesser average power associated with each RF channel. This helps in achieving a low-cost design for deployments.

The high EIRP requires the use of high power amplifier in the FE. The semiconductor technologies, such as GaN, Gallium Arsenide (GaAs), are known for the high-power performance. The SiGe technology is less efficient compared to the GaN, the reason is high bandgap energy in later one. The GaN is a III/V-group direct bandgap semiconductor and is commonly used in RF high-power amplifiers, low noise amplifier, switches and other power electronics.

With high power EIRP array design, the power dissipation (P_{diss}) in an array design is also an issue which needs to be checked while designing. The P_{diss} is greatly influenced by the number of active elements in an array. The typical power consumption in each branch of the beamforming architecture is about 190 mW (Sadhu, Tousi, Hallin, Sahl, Reynolds, Renstrom, … Garcia, 2017). However, the P_{diss} happens significantly through the array structure itself. Therefore, an efficient AM design helps in minimizing the heat dissipation.

The cost of the array includes the RF components, printed circuit board material, and antenna themselves. While considering the component cost, the current die cost per mm^2 of 150 nm GaN on SiC fabrication on a 4-inch wafers is around 4.5 times the cost of an 8-inch 130 nm SiGe. But, with the introduction of a 6-inch GaN production line, it will provide a high volume, and this would drop the cost up to three times of SiGe. This trend indicates that the cost of GaN will drop with the increase in production volumes. Hence, in coming years the use of GaN is most likely going to grow.

Overview of the use of semiconductor technology in hybrid steerable beamforming architecture is presented in Figure 20.

The GaN is useful in implementing the amplifier and switches for high EIRP because it can sustain high-power performance with the use of a fewer number of active antenna elements. It supports the integrated transmit/receive arrays, power-amplifier, low noise amplifier and integrated transmit/receive switches. Each active RF chain has its own control ICs (phase and amplitude) for the beam steering. Because the control ICs are placed out of the array lattice, this introduces more feeding loss due to routing (insertion loss) and hence it lacks the level of component integration. However, this loss can be

Figure 20. Overview of Semiconductor Technologies

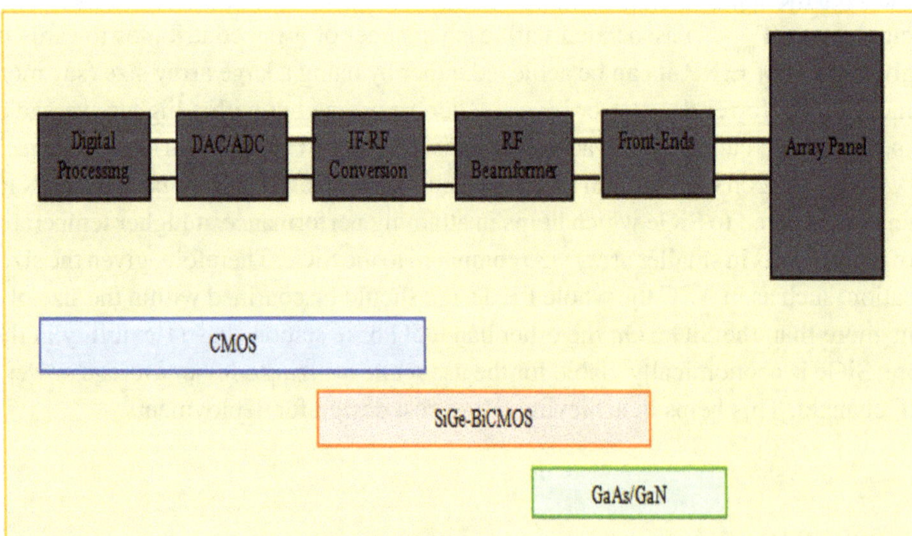

traded off with an oversized array. Although, the GaN technology provides a high EIRP value but its receiver suffers due to extra losses at the feeding network. This technology also needs a DC sequencer, due to depletion-mode which requires dual-supply voltages (positive and negative), which increases the system cost and complexity. Based on GaN technology, Qorvo has designed a front-end transceiver for the beamforming application over the frequency band, 37-40 GHz as presented by Qorvo (2018).

The greatest advantage of SiGe technology is that the beam steering control ICs resides within the same lattice. The control ICs contains the phase control and gain control electronics on a single silicon die. Low feed loss is possible with the placement of die within the lattice of the array antenna. This helps in maximizing the EIRP and receiver gain efficiency. The SiGe provides a high level of integration which results in lower implementation cost. It does not require a DC sequencer which simplifies the system cost and complexity. Typical, the limited transmit power of SiGe, compared to GaN, make the array much larger in size for achieving the same EIRP level as with GaN.

Gallium Nitride was once considered as a sophisticated technology for defense programs such as electronic warfare, however, it's advantages are becoming increasingly acceptable in terms of cost and implementations to fulfill the requirement of 5G and ACC.

FUTURE INDUSTRIAL REVOLUTION: ADDITIVE MANUFACTURING

Perhaps, an industrial revolution starts by accelerating the product development and its mass volume production level, given flexibility in fabricating customized product as per demand, easy setup of manufacturing and assembly line in a small area while giving minimum time and labor as an input. These are all possible with the AM because it is used to realize prototype for research work as well as the same setup would be used for the mass industrial production. It is filling the huge time-gap which exist for transforming a prototype into an industrial product.

For example, steerable mmWave antennas for ACC and 5G will require mass production volumes over the next decades. The same concept of the front-end array antenna architecture is applied in two different applications, 5G NR standardization, and ACC, for two completely different services. In the absence of AM, these two services may have required a completely different production line. Such flexibility and customization of the concept are only possible with the introduction of AM. The cheap and easy manufacturing process, like Additive Manufacturing, is going to leverage a completely new industrial revolution because of its unprecedented scope. Freedom in manufacturing, starting from a very simple three- or two-dimensional object such as a tea-cup or flexible printed antenna to a very complex shape like front-end array antenna design at mmWave, as an integrated system is an unbelievable development in the human history of industrial revolution.

According to IDC forecast, the global data will grow to 163 zettabytes by 2025 (King, 2017). Such tremendous date is mostly generated using mmWave technologies. The unlimited potentials of AM are mainly derived by the goal to put a new benchmark in advanced technologies across the domain. Millimeter Wave component design is one such domain which is getting benefited with it.

CONCLUSION

The 5G antenna network based on steerable beams are bound to offer a much-enhanced wireless network connectivity performance compared to the 4G networks. The auto cruise industries also require hundreds of thousands of such FE design integrated with the radar systems. It is certain that autonomous driving will become reality soon. Therefore, Additive Manufacturing is an ideal prospect for addressing challenges associated the standard and the qualified front-end array antenna fabrications for different promising applications. The main purpose of the additive manufacturing design is to reduce the cost and realization time for the mass production of the qualified RF components as well as complex and integrated systems.

This chapter gives an insight view of the manufacturing capacity based on the Additive Manufacturing concept. The various methods of achieving AM are presented here for better understanding of its potential with the lowest possible input of times, money, space and the freedom of realization and adaptation. These are some key factors which always influence the final products or system implementations. With 5G New Radio intended infrastructure, the flexibility with the component production will certainly help in fast and easy deployments. Likewise, Autonomous Cruise Control industries are pushing for the fast and cheap production of the radar systems. Both applications, 5G and ACC, need a complex front-end array antenna design. The manufacturing of steerable front-end antenna design at millimeter spectrum is a challenging research-work because it comes with complex design, long realization time, and high cost-and-weight. Such complex designs need a completely new level of improvement in manufacturing from the traditional way such as machining technologies. Additive Manufacturing allows an increased flexibility in the manufacturing of complex designs by leveraging design-constraints, allowing it to realize entire designs in one single element that positively impact mass, cost, realization time, assembly quality and RF performance.

This chapter has presented the detail of about additive manufacturing of RF designs such as steerable FE antenna design for adaptive cruise control and 5G network infrastructure supports. As an example, the front-end Array Antenna Architectures and design towards the 5G network supports and the FE radar for adaptive cruise control are discussed. The front-end steerable array antenna and its feeding network designs for the adaptive cruise control over 76-81 GHz are presented for additive manufacturing. Then, a brief analysis of the different cutting-edge semiconductor technologies such as Gallium Nitride (GaN), Silicon Germanium (SiGe) in terms of manufacturing flexibility and choice is also outlined. This chapter put some light about the future market demands about the potential growth of additive manufacturing in context with the steerable front-end design is presented.

Additionally, the reduced weight of the FE antenna will leverage a whole new market in the satellite/space market which may act as Internet-of-LOS (Low Orbit Satellite) network.

REFERENCES

Alexandrea, P. (2017). *3Dnatives*. Retrieved from https://www.3dnatives.com/en/arizton-additive-manufacturing-280920174/

Cadman, D., Zhang, S., & Vardaxoglou, Y. (2016). Fused deposition modeling for microwave circuits and antenna. *International Symposium on Antenna and Propagation (ISAP)*, Okinawa, Japan.

Fezai, F., Menudier, C., Thevenot, M., Arnaud, E., & Mondiere, T. (2015). Reconfigurable Parasitic Element Antenna Using Reflection Phase Shifters. *IEEE Antennas and Wireless Propagation Letters, 14*, 775–778. doi:10.1109/LAWP.2014.2377592

He, Y., Chahal, P., & Papapolymerou, J. (2017). A lego-like reconfigurable cavity using 3-D Polyjet Technology. *International Microwave Symposium*, Honolulu, HI. 10.1109/MWSYM.2017.8058616

Johan, S., William, F., Aurélien, P., Olivier, T., Nicolas, D., Bila, S., ... Gramond, R. P. (2016). Plastic and Metal Additive Manufacturing Technologies for Hyper Frequency Passive Components up to Ka-band. *46th European Microwave Conference*, London, UK.

King, K. (2017). *Data Management Solutions Review*. Retrieved from https://solutionsreview.com/data-management/idc-data-creation-to-reach-163-zettabytes-by-2025/

Kumar, R., Cousin, J. C., & Huyart, B. (2014). 3D Indoor Localization System using Single Reference/Anchor Position based on Multistatic FMCW Radar System. European Microwave Week (EuMW2014), Rome, Italy.

Kumar, R., Louzir, A., & Naour, J. Y. L. (2017). Flexible RFID-Helix Antenna for Sensing the Home Electrical Device Activity. *European Conference on Antenna & Propagation (EuCAP2017)*, Paris, France.

Lacher, R. (2009). *Development of Status of Next Generation Automotive Radar in EU.* ITS Forum, Tokyo, Japan.

Li, Y., Iskander, M. F., Zhang, Z., & Feng, Z. (2013). A new low-cost leaky wave coplanar waveguide continuous transverse stub antenna array using metamaterial-based phase shifters for beam steering. *IEEE Transactions on Antennas and Propagation, 61*(7), 3511–3518. doi:10.1109/TAP.2013.2257649

Liu, B., Gong, X., & Chappell, W. J. (2004). Layer-by-layer polymer stereolithography fabrication for three-dimensional RF components. In *International Microwave Symposium Digest IEEE MTT-S* (vol. 2, pp. 481-484). IEEE.

Manke, I., Kardjilov, N., Strobl, M., Hilger, A., & Banhart, J. (2008). Investigation of the skin effect in the bulk of electrical conductors with spin-polarized neutron radiography. *Journal of Applied Physics, 104*(7), 076109. doi:10.1063/1.2992516

Oscar, A., Peverini, A., Lumia, M., Caliganano, F., Addamo, G., Lorusso, M., … Virone, G. (2017). Selective Laser Melting Manufacturing of Microwave Waveguide Devices. In *Proceeding of the IEEE* (*vol. 105*, pp. 620-631). IEEE. 10.1109/JPROC.2016.2620148

Peterson, B., & Schnaufer, D. (2018). *5G Fixed Wireless Access Array and RF Front-End Trade-Offs. Microwave Journal.*

Qorvo. (2018). *QPF4005 37-40.5 GaN Dual Channel FEM Datasheet.* Retrieved from www.qorvo.com/products/d/da006271

Ramasubramanian, K., & Ginsburg, B. (2017). *AWR1243 sensor: Highly integrated 76-81 GHz radar front-end for emerging ADAS applications*. Retrieved from http://www.ti.com/lit/wp/spyy003/spyy003.pdf

Sadhu, B., Tousi, Y., Hallin, J., Sahl, S., Reynolds, S., Renstrom, O., … Garcia, A. V. (2017). *A 28GHz 32-Element Phased-Array Transceiver IC with Concurrent Dual Polarized Beams and 1.4 Degree Beamsteering Resolution for 5G Communication.* IEEE International Solid-State Circuits Conference (ISSCC), San Francisco, CA.

Sickinger, F., Weissbrodt, E., & Vossiek, M. (2017). 76-81 GHz LTC antenna for an automotive miniature radar frontend. *European Radar Conference EURAD2017*, Nuremberg, Germany.

Tsuchiya, A., & Onodera, H. (2015). Impact of anomalous skin effect on the metal wire for terahertz integrated circuit. *IEEE International Symposium on Radio-Frequency Integration Technology*, Sendai, Japan. 10.1109/RFIT.2015.7377939

Chapter 2
Design Considerations Pertaining to the Application of Complementary Split Ring Resonators in Microstrip Antennas

Sivaranjan Goswami
Gauhati University, India

Kumaresh Sarmah
Gauhati University, India

Angana Sarma
Gauhati University, India

Kandarpa Kumar Sarma
Gauhati University, India

Sunandan Baruah
Assam Don Bosco University, India

ABSTRACT

Metamaterial-based design of microstrip antennas and other microwave structures have gained enormous popularity worldwide among researchers. The complementary split ring resonator (CSRR) is one of the most commonly used metamaterial structures in this direction. The CSRR structure yields a negative value of its effective permittivity at a narrow band near its resonant frequency. CSRR structure was initially proposed as a notch filtering element in microstrip transmission lines because of the negative permittivity. Later, the CSRR structure found its use in antennas and other microwave applications. The CSRR structure is reported to enhance the performance of a microstrip antenna in terms of its gain and bandwidth. In addition, CSRR structure is also used in the design of dual band antennas and antennas with integrated filters. This chapter deals with the practical design aspects relative to these applications of CSRR structures.

DOI: 10.4018/978-1-5225-7539-9.ch002

INTRODUCTION

Microstrip antennas have received wide attention in the past few decades because of its small size and easy installation in portable electronic gadgets. These antennas have made contributions in the miniaturization of modern electronic communication gadgets like mobile phones, bluetooth devices, dongles etc. Theoretically, the length of a rectangular patch antenna is slightly less than half of the resonant wavelength of the antenna (Balanis, 2005). A number of novel approaches have evolved over the past two decades for improving these antennas by reducing the sizes and increasing their gain and bandwidths. A wide range of research is going on all over the world for further enhancement of the performance of the microstrip antennas.

Various approaches have been reported for the improvement of the microstrip antennas in terms of reducing size, improving far field gain, enhancing bandwidth and obtaining resonance at multiple frequencies. Some of the approaches include use of modified geometry (Gao, Li, Leong & Yeo, 2001), multilayered substrate (Lu & Coetzee, 2005), introduction of cut slots of various sizes and shapes on the patch or the ground plane (Khodaei, Nourinia & Ghobadi, 2008; Islam & Samsuzzaman, 2014; Li, Liu, & Gong, 2008), etc. Currently metamaterial based designs of high performance and low size antennas have attracted attention of researchers around the world. Metamaterials are artificial materials that have negative values of permittivity, permeability or both. In (Sigalas, Chan, Ho, & Soukoulis, 1995), authors presented a photonic band-gap structure by arranging thin metallic wires periodically inside a dielectric medium. It has been recognized as a pioneer work in achieving negative value of relative permittivity. Pendry et al. (1999) introduced negative permeability material using split ring resonators (SRR) in (Pendry, Holden, Robbins & Stewart, 1999). Ozbay et al. in (Ozbay, Aydin, Cubukcu & Bayindir, 2003) studied an alternating arrangement of thin wire grid and SRR grid to achieve double negative metamaterial.

The metamaterial unit cell structures as reported in (Pendry, Holden, Robbins & Stewart, 1999) and (Ozbay, Aydin, Cubukcu & Bayindir, 2003) were originally designed for use in electromagnetic metamaterial blocks. However, physical realization of such structures is challenging and expensive. Moreover, such blocks are bulky and hence it is difficult to use them as substrate in microstrip antennas or microstrip lines. The planar versions of these structures later became popular in the design of microstrip antennas and other microwave devices. Nowadays, the SRR structure and its different geometrical variants are extensively used in microstrip antennas. The complement of the SRR structure is known as complementary split ring resonator (CSRR). The CSRR is one of the most popular metamaterial structures reported in the present years. Figure 1 shows the planar form of SRR and the CSRR structures.

Figure 1. Illustrative topology of a planar SRR and CSRR structures

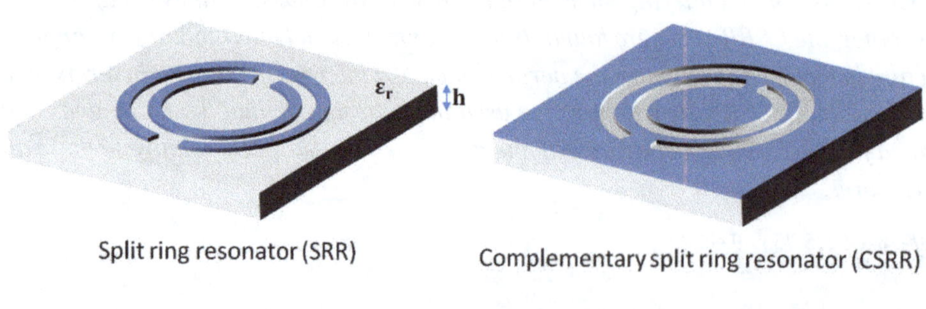

The CSRR structure was originally proposed by Falcone et al. in (Falcone, Lopetegi, Laso, Baena, Bonache, Beruete, Marqués, Martín, & Sorolla, 2004) as a notch filtering element in a microstrip transmission line. The SRR structure provides a negative value of the magnetic permeability (μ) of a metamaterial block. Being a complementary of the SRR structure, the CSRR structure exhibits a negative value of the electric permittivity (ε) at a narrow band near its resonant frequency. Due to this phenomenon, the wave travelling through the microstrip line is blocked at this frequency. That is why it was proposed as a notch filtering element in (Falcone, Lopetegi, Laso, Baena, Bonache, Beruete, Marqués, Martín, & Sorolla, 2004). This work by Falcone et al. opened doors to researchers worldwide. In the recent years, CSRR structure was extensively used by researcher to enhance the performance of microstrip antennas and other microwave devices such as microstrip lines, filters, waveguides, etc.

The primary aim of this chapter is to highlight the effect of the CSRR structure on the performance of microstrip antennas. In recent years, a wide range of research works have been reported where the CSRR structure is used to enhance the performance of the microstrip antennas in a number of ways. Some of these applications of the CSRR structure on microstrip antennas are discussed as follows.

Reduction of Electrical Size of an Antenna

The electrical size of an antenna is expressed as a multiple of the free space wavelength corresponding to its resonant frequency. For example, an antenna with a physical length of 75 mm is considered which resonates at 2 GHz. The free space wavelength corresponding to resonant frequency 2 GHz is 150 mm. Hence, the electrical length of the antenna is said to be λ/2 or 0.5× λ. If another antenna of the same size resonates at 1.5 GHz i.e., free space wavelength of 200mm, the electrical length now becomes 0.375× λ. Thus, the second antenna is electrically smaller than the first antenna. In other words, a much compact antenna is obtained that resonates at 1.5 GHz.

In (Goswami, Sarmah, Sarma, Sarma, & Baruah, 2017) and (Goswami, Sarmah, Sarma, Sarma, & Baruah, 2016), the CSRR structure is shown to reduce the electrical size of microstrip antennas. When the CSRR structures were etched from the ground planes below a rectangular patch antenna, the antenna started resonating at a frequency lower than its original resonant frequency. This phenomenon has two possible explanations. First, when the CSRR structure is etched from the ground plane below a patch, it has a capacitive effect on the antenna. The negative resonant frequency of the CSRR structure does not have a significant role to play in this context. It is the capacitance appearing because of the gaps between the two rings in the CSRR structure that contributes to the reduction of the electrical size of the antenna.

Enhancement of Directive Gain

The directive gain of a transmitting antenna is defined as the ratio of the far field power radiated by the antenna in a given direction to the power radiated by a lossless isotropic antenna in the same direction. Similarly, the directive gain of a receiving antenna is defined as the ratio of the power received from a given direction by that antenna to the power received from the same direction by a lossless isotropic antenna. The ratio of power, after being represented in decibel (*dB*) scale is written with unit dBi. Here the letter ‘i’ denotes that the gain is measured in comparison with an *isotropic* antenna.

In (Goswami, Sarmah, Sarma, Sarma, & Baruah, 2016), (Pandeeswari & Raghavan, 2014) and (Mu, Wang, Jiang, Xu & Huang, 2016) it is reported that when a CSRR structure is etched from the ground plane below a patch antenna, there is an enhancement in the far field gain of the antenna. When CSRR is used, the antenna radiates more power in the main lobe and less power in the side lobes along the unwanted directions. This results in an enhancement of the directive gain of the antenna. Moreover, this also helps in reduction of the radar cross section (RCS) of the antenna (Mu, Wang, Jiang, Xu & Huang, 2016).

CSRR as Integrated Filtering Element in Antennas

The CSRR structure was originally proposed as a notch filtering element in (Falcone, Lopetegi, Laso, Baena, Bonache, Beruete, Marqués, Martín, & Sorolla, 2004) due to the negative value of the effective permittivity that it shows at its resonant frequency. When the permittivity becomes negative, a transmission line ceases to allow electromagnetic waves to pass through it. This phenomenon is utilized in antennas as well. One common application of this phenomenon is in ultra-wide band (UWB) antennas. The UWB antennas are non-resonant antennas. They don't have any specific resonant frequency. These antennas have a low return loss and a high gain at frequencies ranging from 3.1 GHz to 10.6 GHz (Kim, Jo, Jang & Kim, 2011). However, in order to avoid interference, sometime these antennas are intended to notch out some specific frequency bands within this range. For example, the ISM bands of 2.4 GHz, 3.6 GHz and 5 GHz are within the range of the UWB channel. These bands being used in other devices, poses the risk of interference with the UWB channel. It is therefore intended that the UWB antennas are inherently immune to the interference caused by these known ISM band frequencies. This significantly reduces the cost of the UWB communication system as it no longer needs additional filter circuits to notch out these unwanted interfering signals. A number of works have been reported such as (Zhang, Hong, Yu, Kuai, Don, & Zhou, 2008) where CSRR structures of different dimensions are used to notch out these interfering frequencies. In such applications, the dimensions of the CSRR structure are so selected that it resonates at the frequency which is intended to be notched. The UWB antenna is then loaded with CSRR structures of these dimensions. Due to the negative effective permittivity, the antenna ceases to radiate or receive electromagnetic waves at these frequencies.

Another antenna where the CSRR structure is used as an integrated filtering element is reported in (Goswami, Sarmah, Sarma, Sarma, & Baruah, 2017). Here, the ground plane of the antenna is loaded with a CSRR structure to notch out two possible image frequencies when the antenna is used in a superheterodyne receiver. The superheterodyne receiver is a very commonly used circuit where the received signal is first mixed with a locally generated sinusoidal signal of known frequency. After mixing, the carrier of the received signal is translated to a specific frequency known as the intermediate frequency. The demodulator circuit is optimized for its best performance in this intermediate frequency. One common drawback of the superheterodyne receiver is the presence of image frequencies. An image frequency is another frequency which also gets translated to the same intermediate frequency when multiplied with the locally generated sinusoidal wave. This phenomenon leads to interference at the intermediate frequency level. The CSRR structure in this proposed antenna notches two possible image frequencies before and after the resonant frequency of the antenna.

Design of Dual Band Antennas

A dual band antenna is one which resonates at two different frequencies. It should be noted here that the second resonant frequency in this case should not be a harmonic of the fundamental resonant frequency of the antenna. In microstrip antennas fed by a microstrip line or a coaxial cable, there is always a harmonic of the fundamental resonant frequency in which the main lobe in a direction orthogonal to the direction of the main lobe of the antenna at its fundamental resonant frequency (Balanis, 2005). This however is not regarded as a dual band antenna. When an antenna is dual band, the main lobe of the antenna should be in the same direction at both of its resonant frequencies. The shape of the radiation pattern may not be exactly the same at the two resonant frequencies.

The CSRR structure yields a negative value of the effective permittivity in a narrow band near its resonant frequency. At other frequencies, the permittivity is fixed and independent of frequency. However, in this very narrow transition band, the effective permittivity sweeps through a range of positive and negative values and restores itself back to its original value. If the dimensions of the CSRR structure are selected in such a way that this very narrow transition band coincides with the resonant frequency of the microstrip antenna, then the antenna starts resonating at two different nearby frequencies just before and after the resonant frequency of the CSRR structure.

The dual band behavior of the microstrip antennas is reported in numerous works such as (Sarmah, Goswami, Sarma, Sarma & Baruah, 2016; Sarkar, Saurav & Srivastava, 2013) and (Sarmah, Goswami, Sarma & Baruah, 2015). Along with the dual band behavior, it is possible to predict the exact radiation pattern of the antenna from the dimensions of the antenna and the CSRR structure (Sarmah, Goswami, Sarma & Baruah, 2015). In this way, it is possible to convert a single band microstrip antenna to a dual band antenna using CSRR structure.

Arrangement of the Chapter

The key objective of the chapter is to provide a practical insight to the design of CSRR based microstrip antennas. So far a brief introduction to the trends and practices in CSRR based design of microstrip antennas was covered. This is followed by a brief theoretical overview of the CSRR structure along with the analytical and experimental techniques for the measurement of the resonant frequencies of a CSRR structure. The results obtained by the analytical and the experimental techniques are compared. This is followed by the design consideration for compact microstrip antennas with enhanced gain using CSRR structure. The design considerations for reducing the electrical size of an antenna and enhancing its far field gain using CSRR structures is considered with illustrations from published works. After this discussion, the design of antennas with CSRR structures as integrated filtering elements is discussed. In this discussion, the actual working of a CSRR structure is elaborated with examples. Finally, the design considerations for CSRR based dual band antennas are covered with examples. This discussion also includes an insight to the predictability of the far field radiation patterns of a CSRR based microstrip antenna.

AN OVERVIEW OF THE CSRR STRUCTURE

The most common example of a resonant circuit is the LC-tank circuit. Here, the electromagnetic energy is stored as electric field and magnetic field by the capacitor and the inductor respectively. At resonance, the electric energy and the magnetic energy stored must be equal to each other. Another popular resonating device is the cavity resonator. In cavity resonator, a standing wave is generated between the waves reflected from the walls of the cavity. The energy is stored as standing waves of oscillating electric and magnetic fields. Proper phase propagation is important in order to form the standing waves in cavity resonators.

Figure 2. Sample quasi-static resonator. The current that circulates around the ring generates a magnetic field passing through the ring, which effectively behaves as an inductance. The gaps on the ring and the separation between the rings behave as capacitance
(Boybay & Ramahi, 2012)

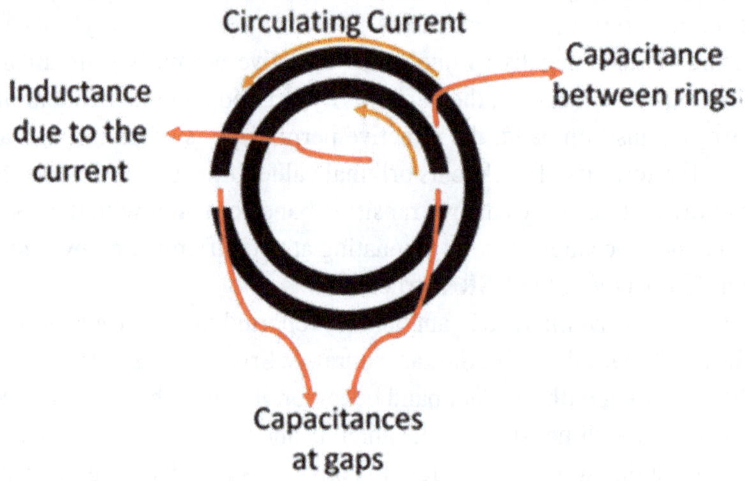

The split ring resonator is a quasi-static resonator where the resonance is not dependent on proper phase propagation. Other examples of other quasi-static resonators include Hilbert curves (Yousefi & Ramahi, 2007), Spiral resonators (Baena, Marqués, Medina & Martel 2004), sinusoidal closed loops (Kabiri & Ramahi, 2009) etc. In quasi-static resonators, a path for electric current flow is generated through a loop as shown in Figure 2. Here, an inductance is generated due to the circulating current through the rings. Capacitance is obtained between gaps and splits of the rings. Hence, the resonance in SRR is analogous to LC resonant circuit. A wide range of works are reported where these structures are analytically modeled as equivalent LC tank circuits.

The complementary structures of these quasi-static resonant geometries also resonate and the impedance of these structures can be obtained using Babinet principle (Falcone, Lopetegi, Laso, Baena, Bonache, Beruete, Marqués, Martín, & Sorolla, 2004). Due to the complementary nature, the inductances in the SRR become capacitances in case of CSRR and the capacitances in SRR become inductances in CSRR. In (Baena, Bonache, Martin, Sillero, Falcone, Lopetegi… Sorolla, 2005), the equivalent LC tank circuits of the SRR and the CSRR are discussed as shown in Figure 3.

A set of analytical expressions for calculating the values of the capacitance, C_C and the inductance, L_C are discussed in (Baena, Bonache, Martin, Sillero, Falcone, Lopetegi… Sorolla, 2005) and (Marques, Mesa, Martel & Medina, 2003) respectively. The value of C_C is given by Equation (1).

$$C_C = \frac{\pi^3 \varepsilon_0}{c^2} \int_0^\infty \frac{[bB(kb) - aB(ka)]^2}{k^2} \left[0.5 \left(1 + \frac{1 + \frac{\varepsilon}{\varepsilon_0} \tanh(kh)}{1 + \frac{\varepsilon_0}{\varepsilon} \tanh(kh)} \right) \right] dk \tag{1}$$

Similarly, the value of L_0 is given by Equation (2).

Figure 3. Topologies of the (a) SRR and (b) CSRR, and their equivalent-circuit models (ohmic losses can be taken into account by including a series resistance in the model). Gray zones represent metalization (Baena et al. 2005)

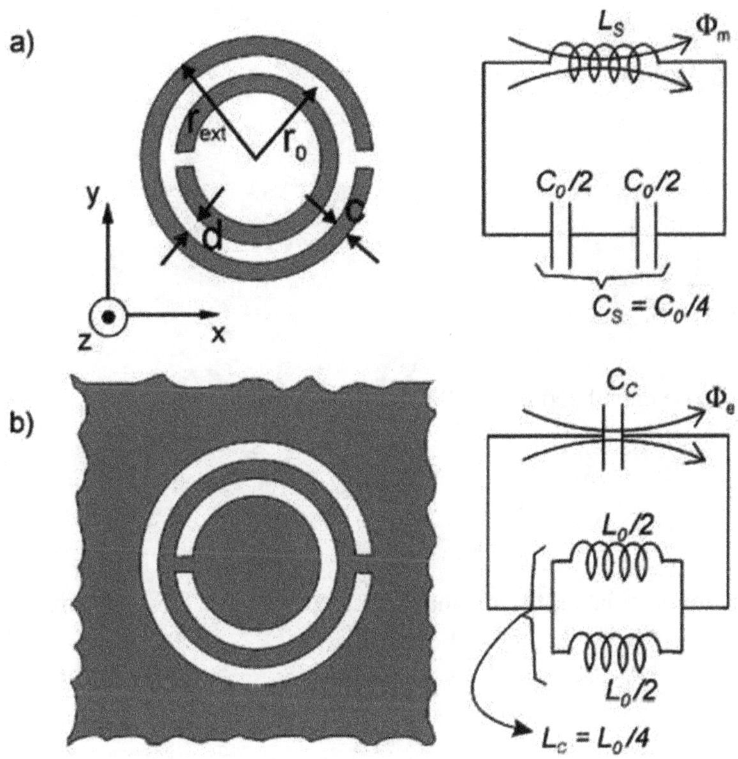

$$L_0 = \frac{\pi^3 \mu_0}{4c^2} \int_0^\infty \frac{[bB(kb) - aB(ka)]^2}{k^2} dk \quad (2)$$

where, μ_0 and c are the magnetic permeability and the velocity of electromagnetic waves respectively in free space. The terms *a*, and *b* are given by Equations (3) and (4) respectively.

$$a = r_0 - \frac{c}{2} \quad (3)$$

$$b = c + a \quad (4)$$

B is given by Equation (5)

$$B(x) = S_0(x)J_1(x) - S_1(x)J_0(x) \quad (5)$$

Here, $S_n(x)$ is the nth order Sturve function of x and $J_n(x)$ is the nth order Bessel function of x given by Equations (6) and (7) respectively.

$$S_n(x) = \sum_{m=0}^{\infty} \frac{(-1)^m}{\Gamma(m+3/2)\Gamma(m+n+3/2)} \left(\frac{x}{2}\right)^{m+1} \tag{6}$$

$$J_n(x) = \left(\frac{x}{2}\right)^n \sum_{m=0}^{\infty} \frac{\left(\frac{-x^2}{4}\right)^m}{m!\Gamma(n+m+1)} \tag{7}$$

The value of L_C can be calculated from L0 using Equation (8).

$$L_C = \frac{L_0}{4} \tag{8}$$

Once, L_C and C_C are calculated, the resonant frequency of the LC-tank circuit can be determined using Equation (9).

$$f_r = \frac{1}{2\pi\sqrt{L_C C_C}} \tag{9}$$

From these expressions, it is evident that the analytical computation of the resonant frequency of the CSRR structure is difficult and prone to errors at various stages of the computation. Moreover, this approach gives only one resonant frequency of the CSRR structure for a particular set of dimensions. However, at times the higher order resonant frequencies of a CSRR structure are also useful for certain applications. These higher order resonant frequencies cannot be determined using this technique.

Experimental Measurement of the Resonant Frequency of CSRR

Due to the limitations of the equivalent LC-tank circuit base analytic model, an experimental model can be used to estimate the resonant frequencies of a CSRR structure. This approach uses the fundamental notch filtering behavior of the CSRR structure, proposed in (Falcone, Lopetegi, Laso, Baena, Bonache, Beruete, Marqués, Martín, & Sorolla, 2004). In this technique, a CSRR structure is etched from the ground plane below a microstrip transmission line. The width of the microstrip line is carefully selected to ascertain that impedance is properly matched between the two ports and the microstrip line. This model is simulated using HFSS® to find the transmission (S21) parameters between the two ports. Due to the negative permittivity of the material at its resonant frequencies, the S21 parameter plot will exhibit a sharp fall in these frequencies as shown in Figure 4. It shows the fundamental and the second order resonant frequencies of the CSRR structure.

Figure 4. Frequency vs. Transmission Parameter (S21) plot that indicates resonant frequency of the CSRR (Sarmah, Goswami, Sarma, Sarma & Baruah, 2016)

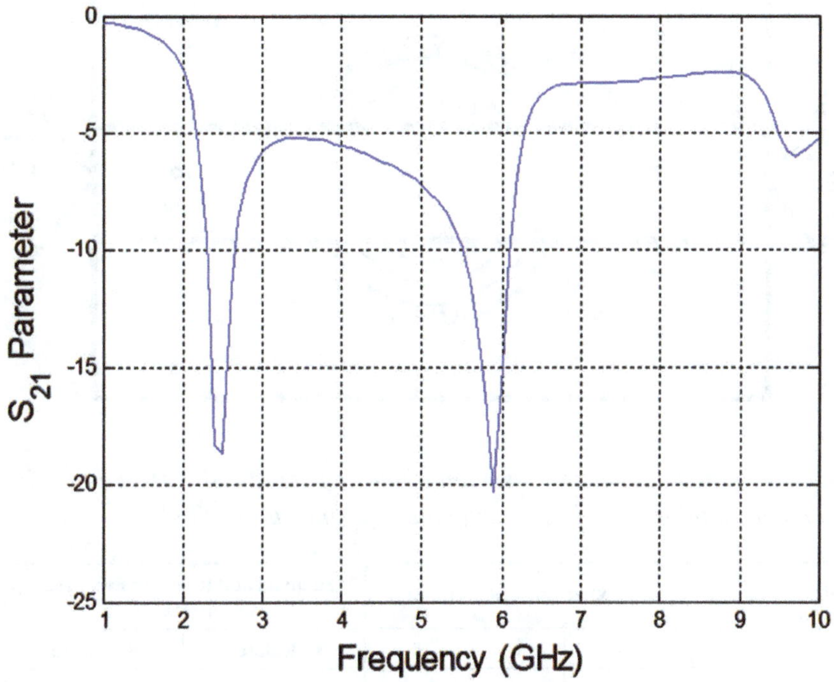

The experimental setup is shown in Figure 5. It is an implementation of the notch filter using CSRR. The frequency response of the notch filter can be observed in Figure 4, where the S21 parameter is plotted against frequency. It is observed that at two frequencies, the wave is notched. These two are the resonant frequencies of the CSRR. This method is reported in works such as (Sarmah, Goswami, Sarma, Sarma & Baruah, 2016) and (Kshetrimayum & Karthikeyan, 2009).

Comparative Analysis of the Two Approaches

Although a transmission line model gives better insights to the design of a microstrip antenna, a full wave solution is always the most reliable and accurate one (Balanis, 2005). HFSS® is a finite element method (FEM) based full wave electromagnetic solver which is accepted by industry as well as academia worldwide. Hence, the experimental approach is more accurate than the analytical approach discussed above.

The analytical approach is complex and requires rigorous numerical computation. However, once a computer program is written to perform all the computational steps of the analytical approach, it gives result faster than the experimental approach.

A computer program is written using an application based software (Matlab®) to perform the computational steps in the analytical approach. The fundamental resonant frequencies are obtained for a number of randomly selected dimensional combinations of the CSRR structure using both of the two approaches. The comparison of the fundamental resonant frequencies found by the two methods is shown in Table 1.

Figure 5. Experimental setup for the measurement of resonant frequency of a CSRR using HFSS®

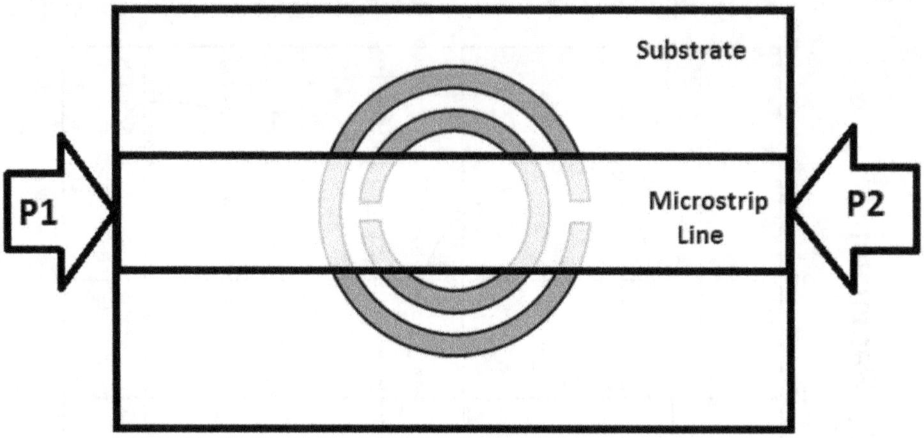

Table 1. Comparison of the fundamental resonant frequency of the CSRR structure calculated from the analytical method and simulated in HFSS® using the experimental method

Serial No.	CSRR Dimensions (mm)			Fundamental Resonant Frequency (GHz)		Percentage Error (%)
	r_0	c	d	Calculated	Simulated	
1	3.50	0.50	2.00	3.13	3.10	0.93
2	3.25	0.25	1.00	3.54	3.50	1.10
3	4.00	0.50	1.00	2.72	2.70	0.73
4	3.75	0.25	1.00	3.10	3.10	0.05
5	3.75	0.75	1.00	2.76	2.90	4.84
6	3.50	1.25	1.00	2.66	2.90	8.42
7	3.50	2.00	1.00	2.42	2.90	16.60
8	3.00	2.00	1.00	2.74	3.30	17.06
9	4.00	2.00	0.50	2.18	2.50	12.79
10	5.00	0.50	1.00	2.04	2.10	2.89

From Table 1 it is seen that the resonant frequencies obtained using the two methods agree. For some dimensions, the two results are much closer than the others. The percentage error can be as low as 0.05% to as high as 17.06%. It may further be inferred from this table that the analytical approach is not highly reliable for practical purposes as the percentage of error is not consistent.

For each of the structures simulated, a second order resonant frequency is also found as shown in Figure 4 which is not shown in this table. Apparently, the second order resonant frequency cannot be directly estimated from the fundamental resonant frequency. This is another drawback of the analytical approach. Hence, the experimental method is chosen for all further investigations related to the authors' analysis of the CSRR structure and its effects on microstrip antennas.

Figure 6. Comparing topologies and dimensions of (a) Circular and (b) Square CSRR

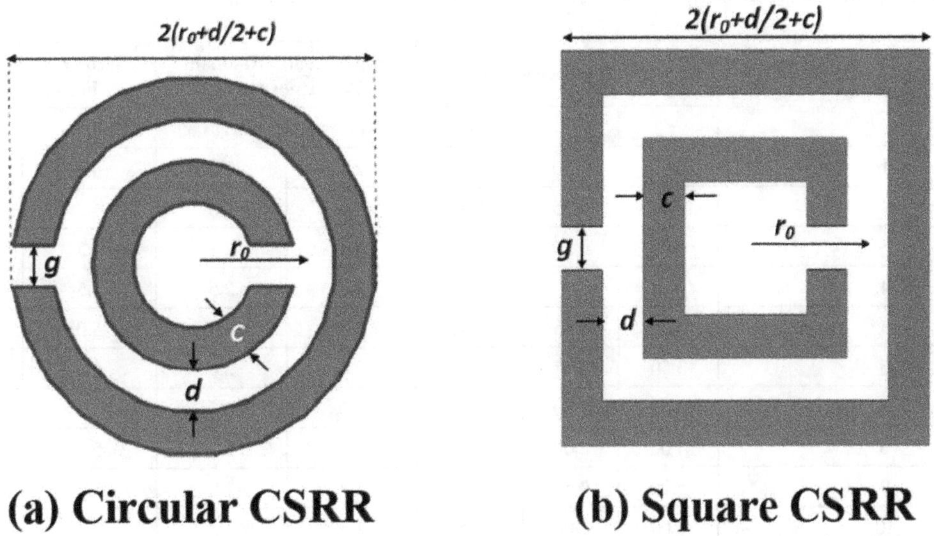

The Square and Circular Complementary Split Ring Resonator

Originally, the SRR and the CSRR structures were designed to be circular in shape. These structures comprises of a pair of concentric split rings separated by a gap. However, when these structures were widely explored by researchers to observe their performance in microstrip antennas, a square version of the SRR and the CSRR structure evolved. A number of works have been reported where the square version of the SRR and the CSRR structures were shown to exhibit properties similar to that of their conventional circular counterparts such as (Pandeeswari & Raghavan, 2014). In this section, an analysis of the square and the circular CSRR structure is presented in terms of their resonant frequencies obtained using the techniques discussed above.

The topologies and the dimensions of the square and the circular form are shown in Figure 6. Here, the widths of the two structures are made identical. A series of experiments is performed to measure the fundamental resonant frequencies of both of the circular and square shape of the CSRR structures. The second order resonant frequencies of these structures are also estimated to compare the two shapes. The comparative analysis is shown in Table 2. It is observed the when the dimensions of the CSRR structure are the same, the square and the circular CSRR structure resonates at the same frequencies irrespective of their shape being square or circular.

From these experimental results, it may be inferred that the behavior of the antenna will not significantly vary based on the shape of the CSRR structure being square or circular as long as the dimensions are maintained to be the same. In the various works discussed in the following sections, both the circular and the square shape of the CSRR structure are used.

Table 2. Comparison of the fundamental and the second order resonant frequencies for the CSRR structures of square and circular shape in HFSS® using the experimental method

Serial No.	CSRR Dimensions (mm)			Fundamental Resonant Frequency (GHz)		Second Order Resonant Frequency (GHz)	
	r_0	c	d	Square	Circular	Square	Circular
1	3.50	0.50	2.00	2.6	2.6	6.1	6.1
2	3.25	0.25	1.00	3.0	3.0	5.3	5.3
3	4.00	0.50	1.00	2.3	2.3	4.5	4.5
4	3.75	0.25	1.00	2.5	2.5	4.5	4.5
5	3.75	0.75	1.00	2.5	2.5	5.0	5.0
6	3.50	1.25	1.00	2.6	2.6	6.0	6.0
7	3.50	2.00	1.00	2.5	2.5	6.8	6.8
8	4.00	2.00	0.50	2.2	2.2	5.3	5.3
9	5.00	0.50	1.00	1.8	1.8	3.5	3.5
10	6.00	2.00	2.00	1.5	1.5	3.4	3.4

DESIGN OF COMPACT ANTENNA WITH ENHANCED GAIN USING CSRR

In this section, the design of antennas with enhanced gain using the CSRR structure is discussed. The CSRR is reported in a number of works to enhance the far field gains of the antennas. In this section, the design of a CSRR structure based antenna with reduced size and enhanced gain is illustrated with two examples reported in (Goswami, Sarmah, Sarma, Sarma, & Baruah, 2016) and (Pandeeswari & Raghavan, 2014).

Slot Loaded Square Patch Antenna With CSRR at Ground Plane

One of the simplest works reported in this direction is (Goswami, Sarmah, Sarma, Sarma, & Baruah, 2016), where a CSRR structure is loaded in the ground plane of a slotted rectangular patch antenna as shown in Figure 7.

The antenna comprises of a rectangular patch antenna on an FR4 epoxy substrate ($\varepsilon_r = 4.4$) of thickness 1.5 mm. The patch is loaded with a rectangular slot and fed by a co-planner microstrip transmission line as shown in Figure 7 (a). On the ground plane, there is a CSRR structure as shown in Figure 7 (b). The dimensions of the antenna are shown in Table 3. Relative permittivity (εr) of the substrate = 4.4

Here, the simulations are separately performed to observe the effects of the CSRR structure and the slot. Figure 8 shows the effect of the CSRR on the resonant frequency of the antenna. Here, the return loss parameter (S11) is used to find the resonant frequency of the antenna. By definition, when the impedance of an antenna is matched at the resonant frequency, the return loss shows a sharp fall. Here, the return loss does not reduce significantly at the resonant frequency. However, the shift in the resonant frequency is clear from the figure. Before the CSRR is applied, the antenna resonates at 5.2 GHz. On loading the antenna with a CSRR structure, the resonant frequency shifts to 4.5 GHz which is significantly less than 4.5 GHz. The length of the antenna is fixed at 12.5 mm. The free space wavelengths (λ_0) corresponding to 5.2 GHz and 4.5 GHz are 57.69 mm and 66.67 mm respectively. Thus, the electrical

Figure 7. Antenna with slot and ground plane for reduction of electrical size and enhancement of gain (Goswami, Sarmah, Sarma, Sarma, & Baruah, 2016)

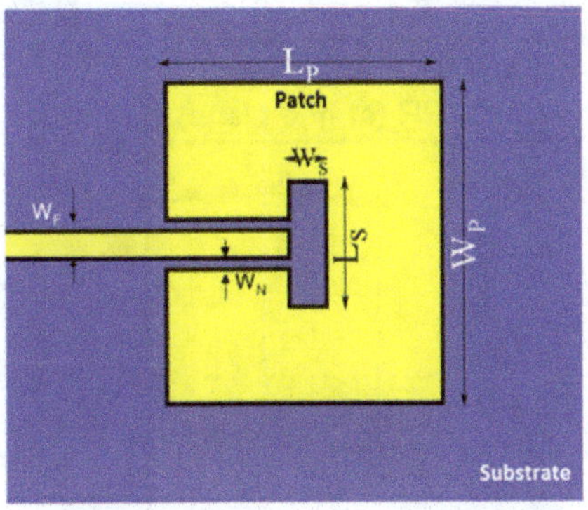

(a) Top view of the slotted patch antenna

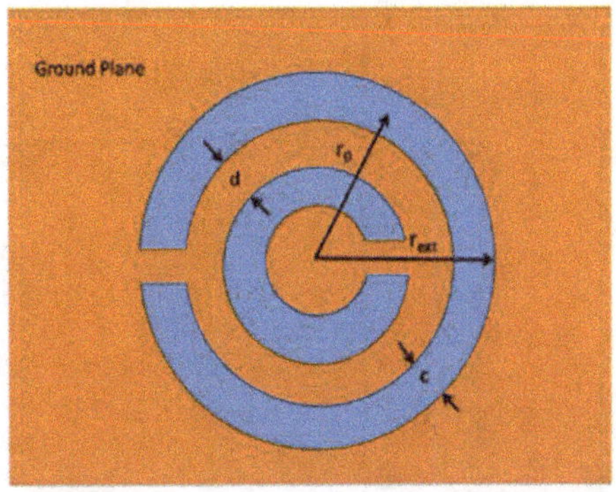

(b) CSRR at the ground plane of the antenna

Table 3. Dimensions of the Antenna Proposed in Goswami, Sarmah, Sarma, Sarma, & Baruah, 2016

Dimension	Value (mm)
Width of the patch (W_p)	12.5
Length of the patch (L_p)	12.5
Height of Substrate (h)	1.5
Length of slot (L_s)	6
Width of slot (W_s)	2
Length of the Notches (L_N)	5.25
Width of the Notches (W_N)	0.5
Width of the Feed Line (W_F)	1
External radius of the CSRR (r_{ext})	5
Internal radius of the CSRR (r_0)	3.5
Width of each complementary ring of the CSRR (c)	2
Gap between the two complementary rings of CSRR (d)	0.05

size of the original patch is $0.22\lambda_0$. On applying the CSRR, the electrical length of the antenna becomes $0.19\ \lambda_{0.}$ It shows that there is a reduction of the electrical length of the antenna by 13.64%.

Figure 9 shows the combined effect of the slot and the CSRR structure on the performance of the antenna. It is observed from the figure that the return loss parameter has significantly reduced when the patch antenna is loaded with slots along with the CSRR structure on the ground plane. It indicates that the slots help in matching the impedance of the antenna.

The frequency response of the far field gain of the antenna along the main lobe is shown in Figure 10 for the patch antenna without and with the CSRR structure at ground plane. Here, it is observed that the gain of the CSRR based antenna with slots is more compared to the original rectangular patch at their respective resonant frequencies. This shows when a CSRR structure is etched from the ground plane below a patch, the far field gain of the antenna is enhanced.

Figure 8. Effect of the CSRR structure on the resonant frequency of the antenna
(Goswami, Sarmah, Sarma, Sarma, & Baruah, 2016)

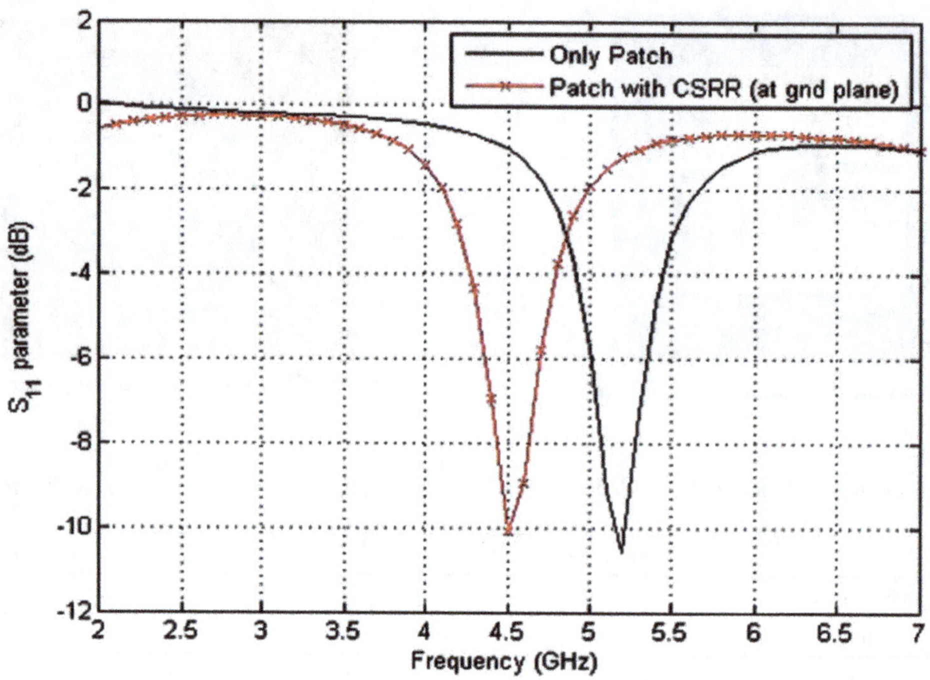

Figure 9. There is an improvement in the return loss (S11) parameter of the antenna when a CSRR structure is etched from the ground plane below a patch
(Goswami, Sarmah, Sarma, Sarma, & Baruah, 2016)

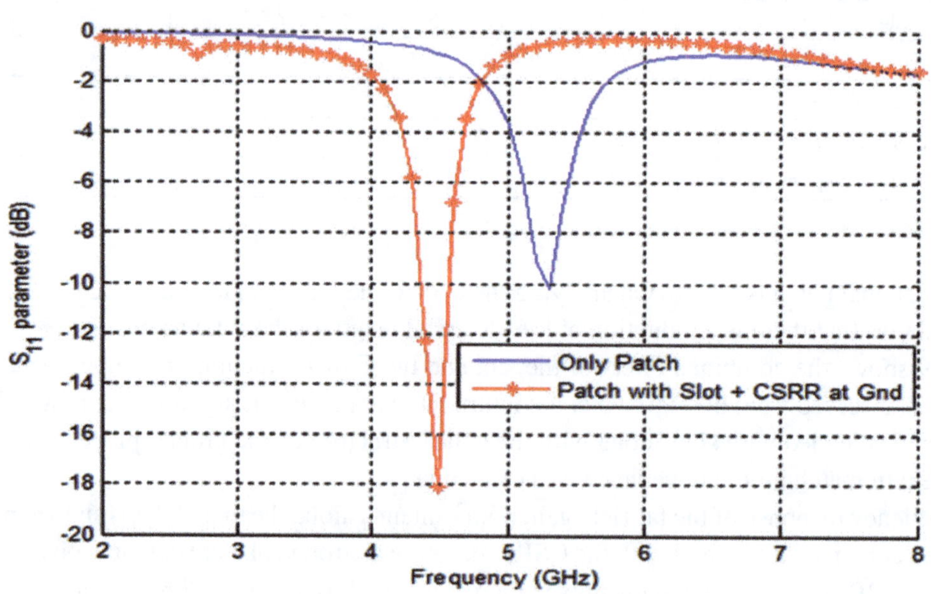

Figure 10. Frequency vs. far field gain plot of the original patch antenna and the proposed slotted patch with CSRR at ground plane along Z-axis ($\theta = 0$ and $\varphi = 0$)
(Goswami, Sarmah, Sarma, Sarma, & Baruah, 2016)

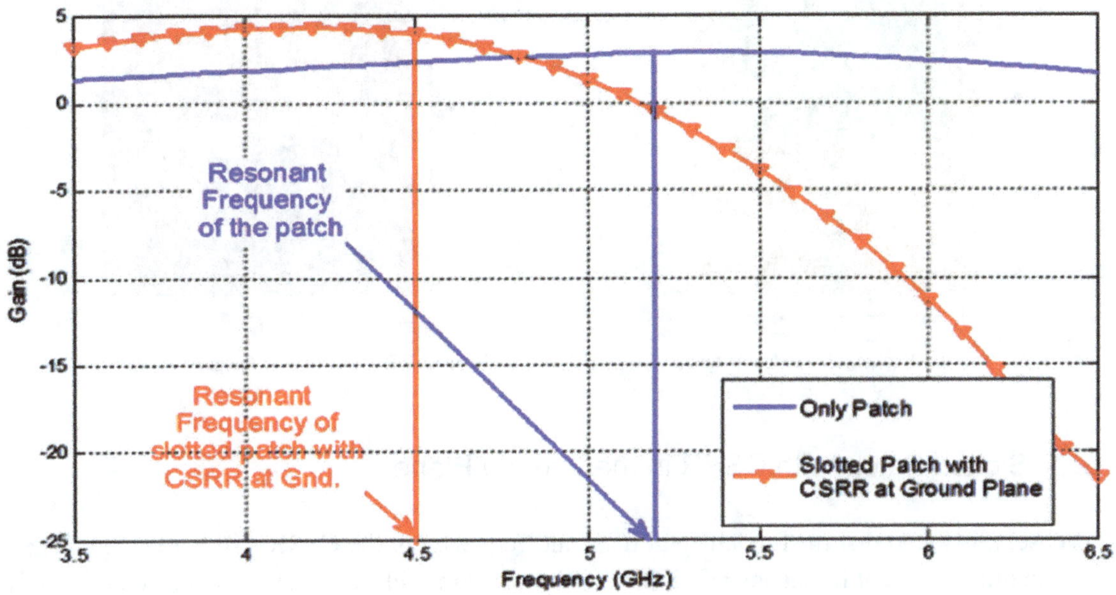

Microstrip Antenna With Complementary Split Ring Resonator Loaded Ground Plane for Gain Enhancement

In (Pandeeswari & Raghavan, 2014), another approach for improvement of the far field gain of the antenna along with a reduction in the electrical size is reported. Here, four CSRR structures are etched from the ground plane below a rectangular patch as shown in Figure 11.

The paper analyzes the various values of the dimensions of the CSRR structure for best performance of the antenna. It is reported that the gain of the antenna is significantly improved after loading the CSRR. Moreover, there is a reduction in the electrical size of the antenna which is evident from the S11 parameter plot given in the paper. The original resonant frequency of the unloaded patch antenna is 2.8 GHz and the resonant frequency of the proposed antenna with four CSRR structures at the ground plane is 2.45 GHz. Hence there is a reduction in the electrical size by about 13%.

Comparative Analysis With the Resonant Frequency of the CSRR Structure

Now, the resonant frequencies of the CSRR structures are observed and are compared with the resonant frequencies of the antennas in both of the above two cases. The resonant frequencies of the CSRR structure are measured using the experimental technique discussed above.

Figure 11. Topology of the antenna reported in Pandeeswari & Raghavan, 2014

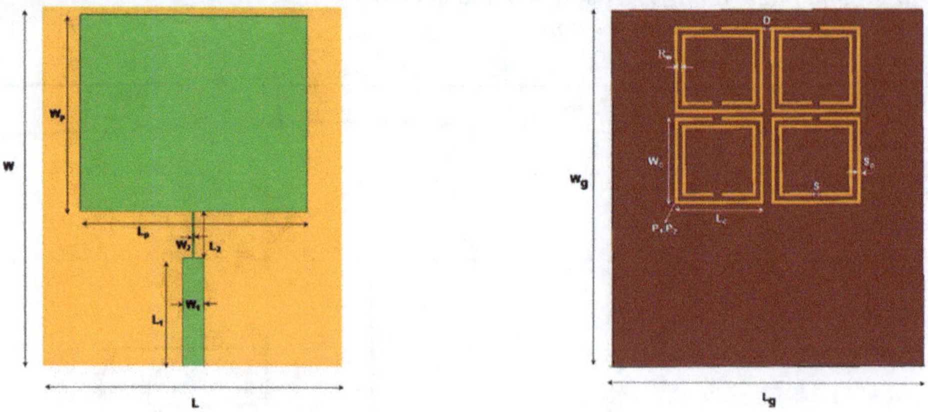

(a) Rectangular Patch Antenna (b) Four CSRR Structures in the ground plane

Case 1: Slotted Patch With CSRR in the Ground Plane

In this case, it is found that the fundamental resonant frequency of the CSRR structure is 2.9 GHz and the second order resonant frequency is 7.2 GHz. None of them is close to the resonant frequency of the antenna which is 4.5 GHz.

Case 2: Rectangular Patch With Four CSRR Structures in the Ground Plane

The four CSRR structures here are identical. The fundamental and the second order resonant frequencies of the CSRR structure in this case are found to be 1.6 GHz and 3.6 GHz respectively. Here also both of these two resonant frequencies of the CSRR structure are far from the resonant frequency of the antenna (2.45 GHz).

From these cases it may be inferred that the negative value of the effective permittivity offered by the CSRR structure has no significant effect on the behavior of the antenna. Thus, the CSRR structure here is simply an implementation of reactive slot that is known to reduce the electrical size of microstrip antennas.

Enhancement of gain in this case is not widely reported in case of simple slot based designs. On the contrary, the gain is reduced after the antenna is loaded with slot. However, the radar cross section (RCS) is also reduced after introduction of the slot. A similar work for reduction of RCS is reported in (Mu, Wang, Jiang, Xu & Huang, 2016) where CSRR structure is used. In this case the far field gain is reportedly increased along with reduction in the RCS due to the use of the CSRR structure.

From this discussion, it may be inferred that the reduction of electrical size of an antenna has nothing to do with the negative permittivity of the CSRR structure, as the resonant frequency of the antenna is well apart from the resonant frequencies of the CSRR structure where the CSRR structure imparts negative values of effective permittivity of the material. The same can be implemented by simple slot based designs as well. However, unlike simple slot based designs, the far field gain of the antenna is enhanced on loading it with CSRR structure.

Moreover, the CSRR structure has many dimensions to tune the antenna for its optimal performance. A simpler rectangular, U-shaped or H-shaped slot usually does not have that many dimensional parameters to tune the antenna in order to achieve an optimal performance.

USING CSRR AS AN INTEGRATED FILTERING ELEMENT IN MICROSTRIP ANTENNAS

In this section, the filtering behavior of the CSRR structure is discussed. The CSRR structure was originally proposed by Falcone et al. in (Falcone, Lopetegi, Laso, Baena, Bonache, Beruete, Marqués, Martín, & Sorolla, 2004) as a filtering element in a microstrip transmission line. Owing to the negative value of its effective permittivity near the resonant frequencies of the CSRR structure, it behaves as a notch filter in the microstrip line. The same phenomenon may be utilized in microstrip antennas to prevent unwanted frequencies from being received by the antenna and avoid possible interference in the system.

This behavior of microstrip antennas is widely utilized in ultra-wide band (UWB) antennas. In this section, we present a detailed analysis of the CSRR structure being utilized to notch out unwanted frequencies in UWB antennas. Further, another example where the CSRR structure is used to design an integrated filter with a rectangular patch that can filter possible image frequencies in a superheterodyne receiver is discussed.

CSRR as Notch Filtering Element in UWB Antennas

An UWB antenna is a non-resonant antenna. There is no definite resonant frequency of a non-resonant antenna. Instead, the antenna has a high value of far field gain and a small value of return loss over the entire UWB band, in the range of 3.1 GHz to 10.6 GHz. Within this band and the nearby bands, there are many other frequencies which are commonly used for other applications. CSRR structures are used to notch out these frequencies which are potential cause of interference to the UWB communication system.

A number of works are available in open literatures where the CSRR structure is used to notch out unwanted frequencies in UWB antennas (Rani & Pandey, 2017; Yadav, Agrawal & Yadav, 2017). In all these works, the UWB antenna is loaded with CSRR structures of different dimensions. Each CSRR structure acts as a notch filter in their resonant frequencies. Here, an UWB antenna is reported which has the inherent capacity to notch four frequency bands, namely 2.4 GHz, 3.5 GHz, 5.5 GHz and 8.2 GHz. These bands are widely used in Wireless Local Area Network (WLAN), World Interoperability for Microwave Access (WiMax), and International Telecommunication Union (ITU). In this publication, the current distribution in a resonating CSRR structure is elaborately discussed which gives a close insight to the working of the CSRR structure at its resonant frequency.

The topology of the antenna proposed is shown in Figure 12. Here, a typical staircase shaped UWB antenna is taken which is fed by a co-planar waveguide (CPW). The antenna is loaded with four CSRR structures. A pair of CSRR structures is symmetrically loaded on the ground area of the CPW feed. Another CSRR structure is loaded near the feed point of the patch. The fourth CSRR structure is loaded at the top of the staircase shaped patch antenna element.

Figure 12. Geometry of the quadruple-band rejection antenna proposed

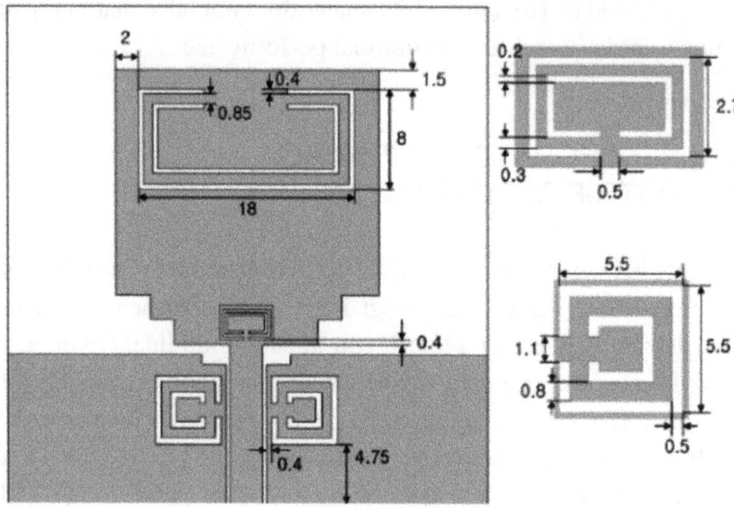

It is to be noted that the CSRR structures used here are co-directional CSRR structures. In this structure, the splits of the two rings are oriented in the same direction. This structure is widely used in UWB antennas. They behave similarly as the conventional CSRR structures and their resonant frequencies can also be measured using the experimental technique discussed above. A conventional CSRR and the co-directional CSRR structure do not usually resonate at the same frequencies even if their dimensions are the same.

Here, the performance of the antenna is evaluated based on voltage standing wave ratio (VSWR) and realized far field gain. In a transmission line, a standing wave is formed between the transmitted wave moving in the forward direction and the reflected wave moving in the reverse direction. The ratio between the amplitudes of these two waves is termed as the VSWR of the transmission line. Thus, the VSWR based analysis is similar to the reflection parameter (S11) based analysis discussed above. When the impedance is properly matched, the amplitude of the transmitted wave should be equal to that of the reflected wave, leading to a VSWR of unity. In practical cases, the VSWR should be less than 2 at the resonant frequency of the antenna. On the other hand, the VSWR exhibits a high value at the frequencies where the impedance is not properly matched and the antenna is not radiating.

The frequency response of the VSWR for this antenna is shown in Figure 13 (a). Here, it is observed that the VSWR is high at the frequencies which are intended to be filtered out. On the other frequencies of the UWB channel, the VSWR is less than 2. Thus, the CSRR behaves as a notch filtering element at these frequencies. The same result is also observed from the realized far field gain plot shown in Figure 13 (b). The gain of the antenna drops at the frequencies where the CSRR acts as a notch filter.

This resonance of the CSRR structure is further explained with the help of the surface current distribution of the antenna at the frequencies which are notched. The surface current distributions at the resonant frequencies of the CSRR structure are shown in Figure 14. Here, the high current density regions are indicated with red color. The low current density regions are indicated with green color.

In Figure 14, the surface current distributions at the four stop bands of the antenna are shown separately. When the CSRR resonates, the surface current is trapped within the close loop of the CSRR structure and it leads to resonance. At the resonant frequency of a particular CSRR structure, the current density

Figure 13. Performance of the antenna proposed based on VSWR and realized far field gain

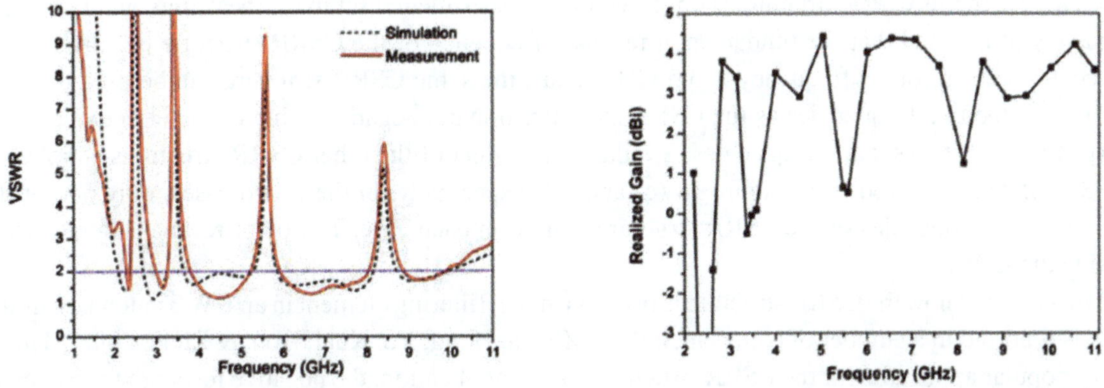

(a) Frequency vs. VSWR (b) Frequency vs. realized far field gain

Figure 14. Surface current distribution of the quadruple-band rejection UWB antenna at (a) 2.4 GHz; (b) 3.5 GHz; (c) 5.5 GHz, and; (d) 8.2 GHz

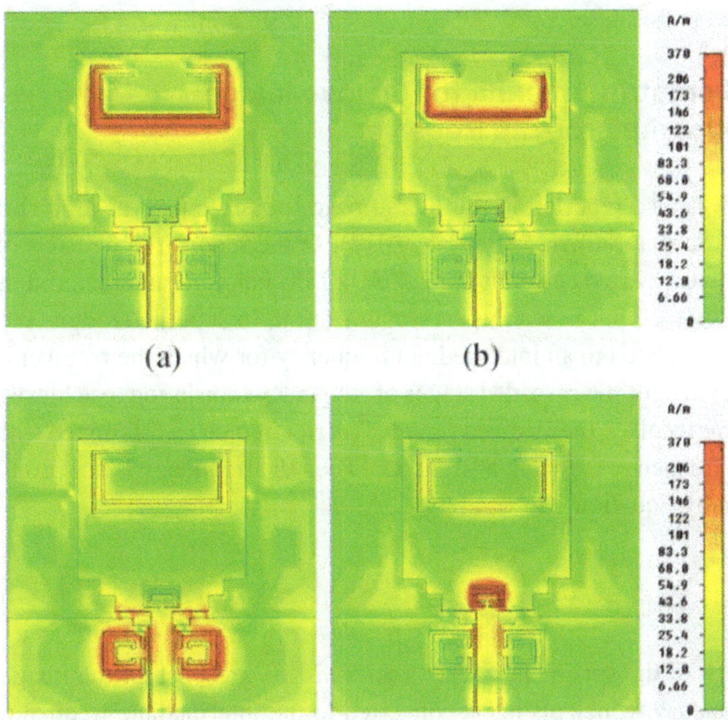

becomes high within the CSRR structure. This phenomenon of current looping in closed path gives rise to the metamaterial behavior of negative permittivity. Pendry et al, in (Pendry, Holden, Robbins & Stewart, 1999), explained how this looping current leads to the effective negative permeability in case of the SRR structure. Similar behavior is observed for the CSRR structure for the electric permittivity.

From Figure 14 (a) and (b), it is observed that the same CSRR structure resonates at 2.4 GHz and 3.5 GHz. When this CSRR structure is observed in the experimental setup for measurement of resonant frequency, it is found that the fundamental resonant frequency of the CSRR structure is 2.4 GHz and its second order resonant frequency is 3.5 GHz. Thus, the same CSRR structure can be used to notch out both of the two frequencies as the CSRR structure imparts negative value of the effective permittivity at both of its resonant frequencies. Similarly, the effect of the other CSRR structures at 5.2 GHz and 8.2 GHz is observed from Figure 14 (c) and (d) respectively. In these two cases, only one of the two resonant frequencies of the CSRR structure is used in each case. The other resonant frequency is outside the UWB range.

We have seen how the CSRR structure is used as a notch filtering element in an UWB antenna. Similar work is reported in a number of papers including (Zhang, Hong, Yu, Kuai, Don, & Zhou, 2008). This is a very popular application of the CSRR structure. In this application, the negative resonant frequency of the CSRR structure is properly utilized. The design procedure for this application of the CSRR structure is straightforward. This includes the selection of the proper dimensions of the CSRR structure in order to make sure that it resonates at the frequencies which are intended to be notched. Then the UWB patch antenna is loaded with the CSRR structure of these dimensions. The antenna will then notch out these frequencies. Both the fundamental and the second order resonant frequencies of the CSRR structure should be considered in this case.

CSRR Based Microstrip Antenna to Remove Image Frequency in Superheterodyne Receiver

The CSRR structure is also reported in the literature for possible uses as a notch filtering element is (Goswami, Sarmah, Sarma, Sarma, & Baruah, 2017). In the report, the CSRR structure is used in an antenna for a superheterodyne receiver at the 2.4 GHz ISM band. A superheterodyne receiver is a very commonly used receiver circuit in wireless communication. In superheterodyne receivers, the original carrier frequency is translated to an intermediate frequency for which the receiver circuit is optimized. In some cases, there are multiple cascaded stages of frequency translation to achieve better performance.

In superheterodyne receiver, the received signal of carrier frequency f_C is mixed with a locally generated sinusoidal signal of frequency f_{LO} using a RF mixer. Then the frequency is converted to an intermediate frequency fIF, given by Equation 10.

$$f_{IF} = | f_C - f_{LO} | \tag{10}$$

A major drawback of the superheterodyne receiver is the presence of the image frequency. It is an unwanted input frequency, which also gets converted to the intermediate frequency after mixing. The image frequency is given by Equation 11.

$$f_{IMG} = \begin{cases} f_C - 2f_{IF} & \text{if } f_{LO} < f_C \\ f_C + 2f_{IF} & \text{if } f_{LO} < f_C \end{cases} \tag{11}$$

In (Goswami, Sarmah, Sarma, Sarma, & Baruah, 2017), the two intermediate frequencies are considered to be 500 MHz and 750 MHz. The local oscillator frequencies are taken to be 1.9 GHz and 3.15 GHz for the intermediate frequencies of 500 MHz and 750 MHz respectively. Accordingly, the two corresponding image frequencies are 1.4 GHz and 3.9 GHz. These two image frequencies are to be rejected by the antenna. The working of the antenna is evaluated with a model of the communication system with superheterodyne receiver using Simulink®. The block diagram of the communication system is show in shown in Figure 15. Here, a noise source is taken that generates the image frequencies. The performance of the antenna is measured in terms of the spectrum analyzer reading before and after passing through the antenna as shown in Figure 15. The error rate at the receiver is also used to evaluate the performance of the antenna.

Figure 15. Block diagram of the communication system for the evaluation of the antenna proposed in (Goswami, Sarmah, Sarma, Sarma, & Baruah, 2017) in a superheterodyne receiver-based communication system

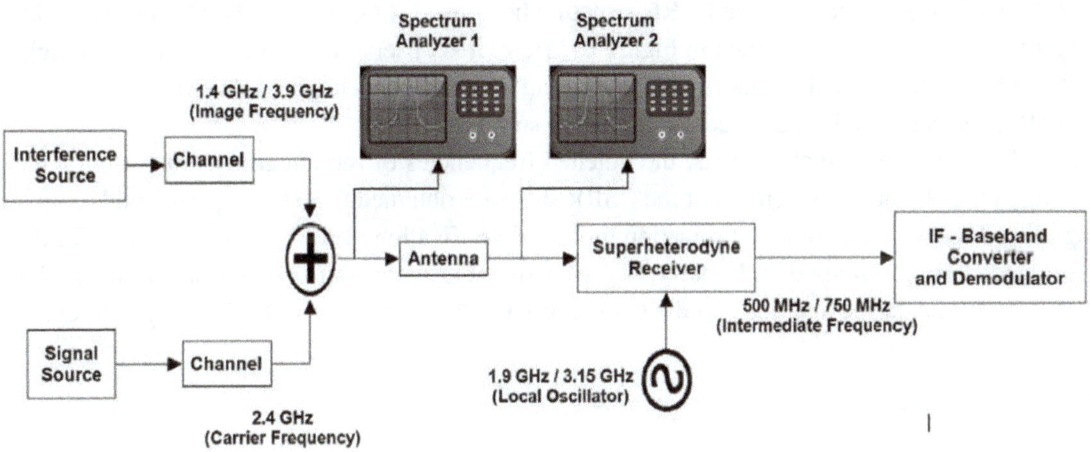

Figure 16. Topology of the image rejection antenna proposed in Goswami, Sarmah, Sarma, Sarma, & Baruah, 2017

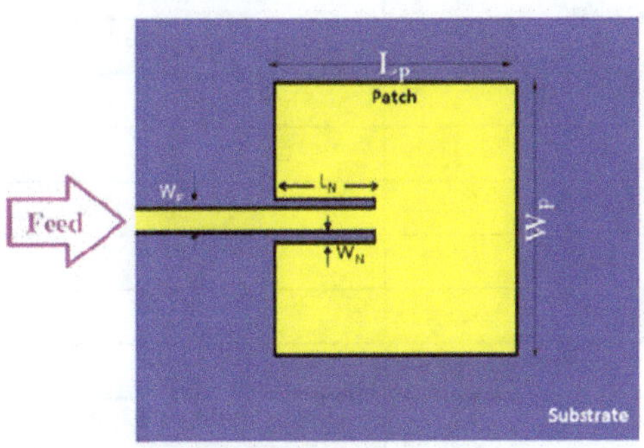

(a) Patch

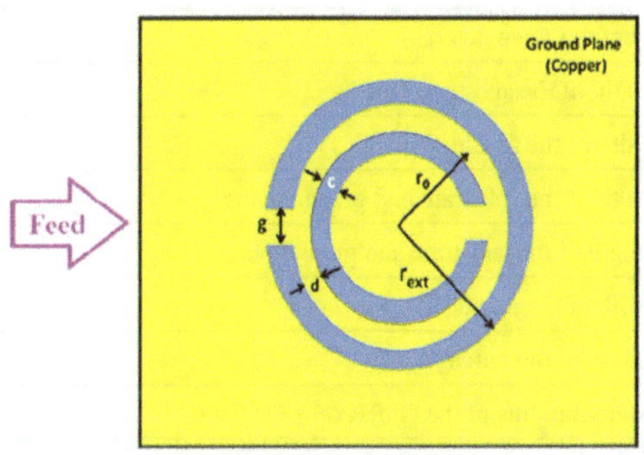

(b) CSRR at Ground Plane

In this work, a rectangular patch antenna on an FR4-Epoxy substrate is used which resonates at 2.4 GHz and a CSRR structure is etched from the ground plane below the patch. The resonant frequencies of the CSRR structure are 1.4 GHz and 3.9 GHz respectively. The topology of this antenna is shown in Figure 16. The various dimensions of the antenna are given in Table 4. Relative permittivity of the substrate (ε_r) is 4.4.

This design is analogous to the CSRR based antennas designed for reduction of electrical size and enhancement of the gain of the antenna. Here also, the resonant frequencies of the CSRR structure are well apart from the resonant frequency of the patch antenna. Therefore, a reduction in the electrical size of the antenna is observed. The dimension of the original patch is 26mm×24.5mm. This antenna has a resonant frequency of 2.8 GHz. On loading the antenna with the CSRR structure at the ground plane, the resonant frequency becomes 2.4 GHz, which results in a reduction of the electrical size of the antenna by 14.3%. The gain of the antenna is also slightly improved after applying the CSRR structure.

As the resonant frequency of the antenna is affected by the CSRR structure, the resonant frequency of the CSRR is also influenced by the antenna. The original resonant frequencies of the CSRR structure were 1.6 GHz and 3.6 GHz. After etching the same CSRR structure from the ground plane of the antenna, the new resonant frequencies of the CSRR structure becomes 1.4 GHz and 3.9 GHz. The far field gain distribution of the antenna is shown in Figure 17. Here, it is observed that the antenna has a high gain at 2.4 GHz. The gain sharply reduces at 1.4 GHz and 3.9 GHz. This happens due to the negative value of the effective permittivity imparted by the CSRR structures.

Therefore, in case of such designs, the notched frequencies of the integrated filter are not exactly the same as the resonant frequencies of the CSRR structure obtained from the experimental setup. Such designs work better with some optimization tools and verification of the designed antenna for desired performance through simulation. In the work reported in (Goswami, Sarmah, Sarma, Sarma, & Baruah, 2017), dimensions of the antenna and the CSRR structure are optimized with the integrated optimistic tool available with Ansys HFSS®.

Table 4. Dimensions of the antenna proposed in Goswami, Sarmah, Sarma, Sarma, & Baruah, 2017

Dimension	Value (mm)
Width of the patch (W_p)	26
Length of the patch (L_p)	24.5
Height of the substrate (h)	1.5
Width of the feeding line (W_F)	2
Width of the substrate and ground plane	50
Length of the substrate and ground plane	50
Width of the notch (W_N)	2.75
Length of the notch (L_N)	7.25
External radius of the CSRR (r_{ext})	9.5
Averaged radius of the CSRR (r_0)	6.5
Width of the rings (c)	2
Width of the splits (g)	2

Figure 17. Frequency vs. far field gain plot for the proposed antenna in Goswami, Sarmah, Sarma, Sarma, & Baruah, 2017

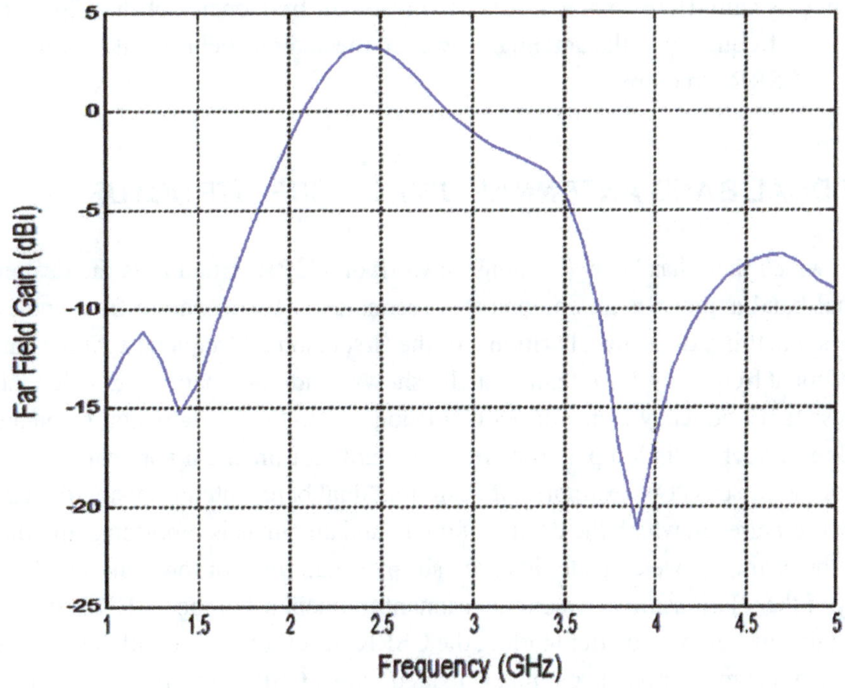

Figure 18. Spectral analysis of the performance of the antenna in a communication system (Goswami, Sarmah, Sarma, Sarma, & Baruah, 2017)

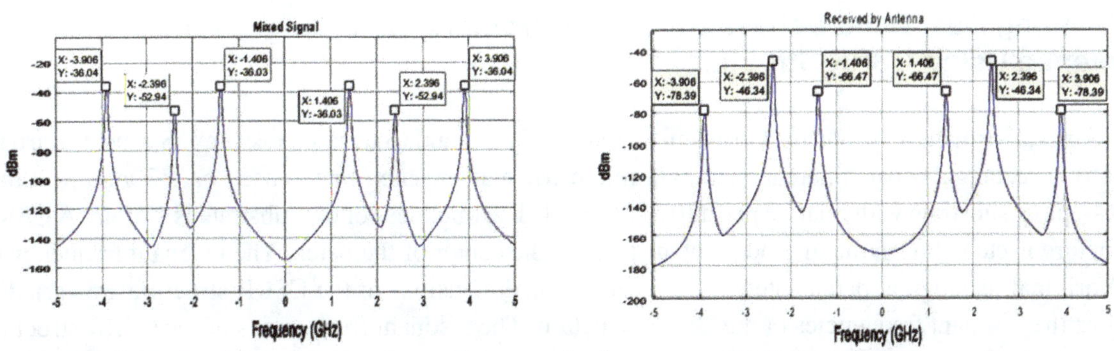

(a) Mixed signal and interference at image frequency in air

(b) Output of the proposed antenna

The performance of the antenna in terms of the spectrum analyzer reading is shown in Figure 18. Here, the communication system is modeled in Simulink® by importing the radiation pattern obtained from Ansys HFSS®. In Figure 18 (a), the spectrum of the desired signal at 2.4 GHz mixed with the interfering signal at 1.4 GHz and 3.9 GHz is shown. Here, the interfering bands have more power than the desired signal of 2.4 GHz. The signal received by the antenna is shown in Figure 18 (b). Due to the effect of the CSRR, the power level is now higher in the 2.4 GHz band compared to the interfering bands of 1.4 GHz and 3.9 GHz.

Another point that is relevant to mention here is that the application of the CSRR structure in this case is analogous to the application where the only intention is to reduce the electrical size of the antenna and enhance its gain. Here also, neither of the resonant frequencies of the CSRR structure is not close to the resonant frequency of the antenna. However, the antenna behaves like a filter at the resonant frequencies of the CSRR structure.

DESIGN OF DUAL BAND ANTENNAS USING CSRR STRUCTURE

Another area in which there has been an extensive work on CSRR structure is the design of dual band antennas. A dual band antenna is an antenna that resonates at two separate frequencies. The second resonant frequency in this case is not a harmonic of the first resonant frequency. At the harmonics of the fundamental resonant frequency, an antenna usually shows a radiation pattern in which the main lobe of the harmonic resonant frequency is in a direction orthogonal to the fundamental resonant frequency or shows grating lobes. Such radiation patterns are not suitable for practical applications.

In this section, we discuss two examples of designing dual band antennas using the CSRR structure. There are many literatures in which the design of dual band antennas is reported using the CSRR structure. However, the examples we consider here are simpler than most of the available designs so that an analytical study of the relationship between the resonant frequencies of the CSRR structure can be made with that of the antenna before and after loading the CSRR structure. First of all, let us consider a simple example in which a rectangular patch antenna is loaded with a CSRR structure at its ground plane. This example is reported in (Sarma, Sarmah, K.K. Sarma, Goswami & Baruah, 2016) and (Sarmah, Goswami, Sarma & Baruah, 2015). The topology of the antenna is exactly the same as the one shown in Figure 16. Just the dimensions and hence the resonant frequency of the CSRR structure is different.

Obtaining Dual Band Behavior From an Ordinary Antenna Using CSRR Structure

In (Sarma, Sarmah, K.K. Sarma, Goswami & Baruah, 2016) and (Sarmah, Goswami, Sarma & Baruah, 2015), a rectangular patch antenna fed by a microstrip transmission line is used on a 30mm × 30mm FR4-epoxy substrate with relative permittivity (ε_r) 4.4. The thickness of the substrate is 1.5mm. A CSRR structure is etched from the ground plane just below the center of the patch. The resonant frequency of the original rectangular patch antenna is 5.1 GHz. The dimensions of the CSRR structure are varied to adjust the resonant frequencies of the CSRR structure. The resonant frequencies of the CSRR structure are measured using the experimental technique discussed above. Now, a comparative analysis of the resonant frequencies of the CSRR structure and the resonant frequencies of the CSRR loaded patch antenna is shown in Table 5. Here, f1 is the lower resonant frequency and f2 is the upper resonant frequency in each case.

From the experimental results shown in Table 5, it may be inferred that a rectangular patch antenna shows resonance at two different frequencies provided the following conditions are met:

1. The fundamental resonant frequency or the lower resonant frequency of the CSRR structure is less than that of the original patch antenna.

Table 5. Comparison of the resonant frequencies of the CSRR and that of the antenna when the CSRR is applied on the ground plane below the patch to obtain dual band characteristics (Sarma, Sarmah, K.K. Sarma, Goswami & Baruah, 2016) and (Sarmah, Goswami, Sarma & Baruah, 2015)

Sl. No.	Resonant Frequency of Patch (GHz)	Resonant Frequency of CSRR (GHz)		Resonant Frequency of Dual Band Antenna (GHz)	
		f_1	f_2	f_1	f_2
1	5.3	3.1	7	4.6	--
2	5.3	2.9	6.8	4.6	--
3	5.3	2.0	4.3	3.7	5.2
4	5.3	2.7	5.2	4.3	5.9
5	5.3	3.1	5.4	4.4	5.9
6	5.3	2.9	5.8	4.4	6.4
7	5.3	2	4.3	3.7	5.2
8	5.1	2.2	5.4	4.0	5.0
9	4.9	3.1	5.4	4.4	5.9
10	4.8	2.7	5.2	4.0	5.7
11	4.5	2.0	4.0	3.5	5.0
12	4.5	2.0	4.3	3.4	5.0

2. The second order resonant frequency or the upper resonant frequency is close to the resonant frequency of the unloaded patch, so that the difference between the second order resonant frequency of the CSRR structure and the resonant frequency of the unloaded patch antenna is less than or equal to 1 GHz.

In (Sarma, Sarmah, Sarma, Goswami & Baruah, 2016), another design is proposed where the shape of the patch is not rectangular. Here, a patch of an arbitrary shape is taken as shown in Figure 19. The dimensions of the patch are labeled in the figure. The resonant frequency of the patch antenna without any CSRR structure is 2.15 GHz. Therefore, a CSRR structure is considered which has a fundamental resonant frequency of 2 GHz and the upper resonant frequency is 4 GHz. Here, it may be noted that the inference derived from Table 5 are slightly twisted. The upper resonant frequency is now greater than the resonant frequency of the unloaded patch and the lower resonant frequency is close to the resonant frequency of the CSRR structure. From the experimental results reported in (Sarma, Sarmah, Sarma, Goswami & Baruah, 2016), it is observed that this combination also yields a dual band behavior in case of a microstrip antenna. The frequency response of the return loss (S11) parameter for this antenna is shown in Figure 20. It is observed that now the antenna has a dual band behavior and it resonates at 1.7 GHz and 2.4 GHz. After fabrication, the measured result shows a further shift in the resonant frequency which may be attributed to manufacturing defects and the external ambient parameters at the time of measurement.

From this result it may be concluded that in order to obtain dual band behavior in any arbitrary shaped patch antenna using CSRR structure, any of the resonant frequencies of the CSRR structure must be close to the resonant frequency of the unloaded patch antenna. The other resonant frequency of the CSRR structure may be either much lesser or much higher than the resonant frequency of the unloaded patch.

Figure 19. Dimensions of the arbitrary shaped antenna for which the dual band behavior of the antenna is tested in Sarma, Sarmah, Sarma, Goswami & Baruah, 2016

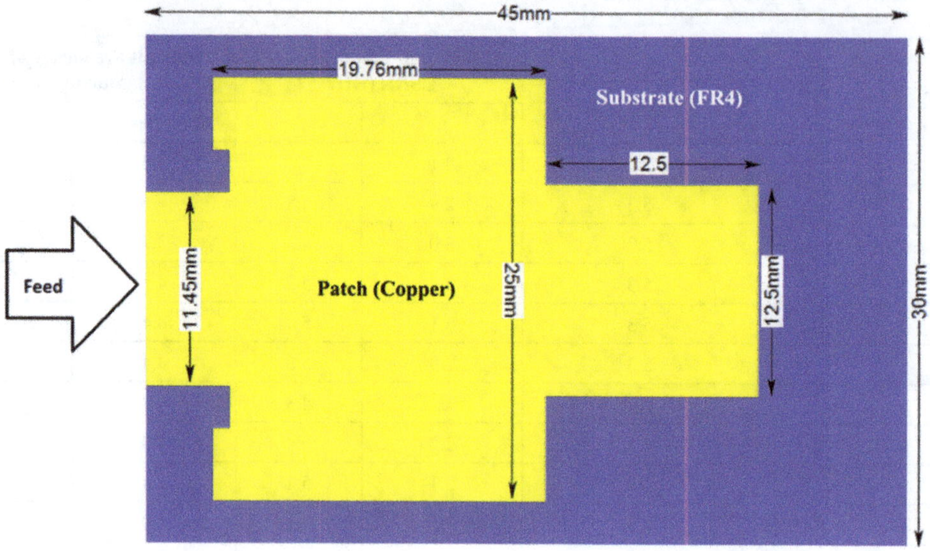

Figure 20. Return loss parameter of the original patch of arbitrary shape studied in (Sarma, Sarmah, Sarma, Goswami & Baruah, 2016) and simulated and measured values of the patch antenna with CSRR at ground plane

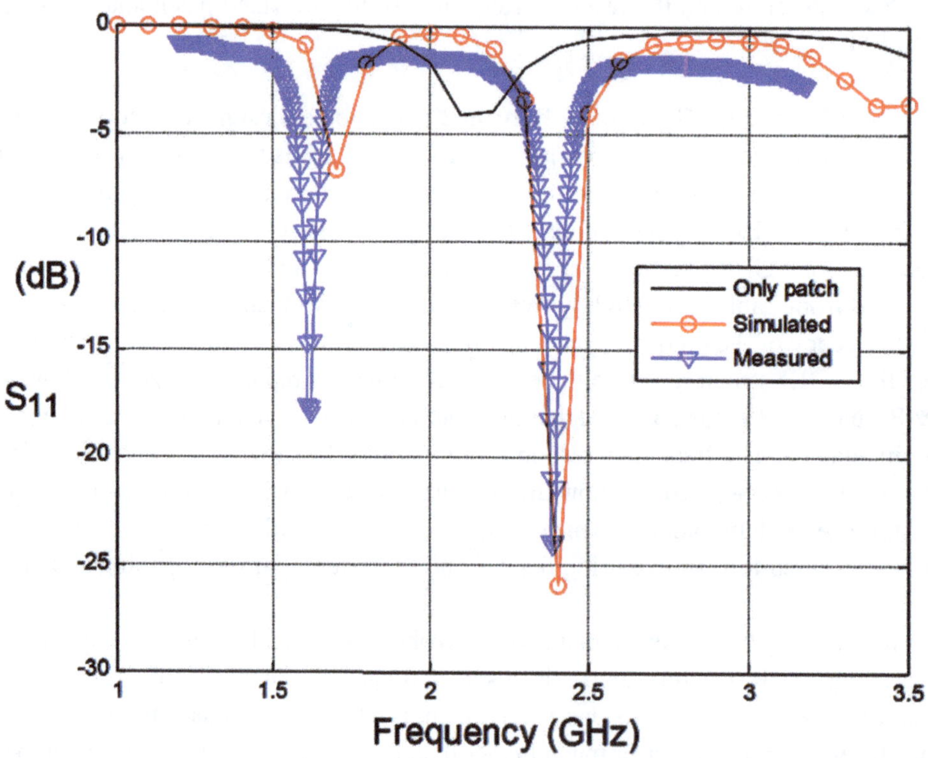

Considerations for Impedance Matching in a Dual Band Antenna

In order to ensure efficient transmission and reception by an antenna, the input impedance of the antenna must match the impedance of the port or connector used to feed the antenna. The impedance of a typical port used for feeding a microstrip antenna is 50 ohm. A sharp fall in the frequency response of the S11 parameter indicates that the impedance is perfectly matched. The lower is the value of the S11 parameter at the resonant frequency, the closer will be the input impedance of the antenna to that of the port used for feeding the antenna. The match in the impedance can also be obtained from the smith chart plot of the antenna.

One of the popular techniques used for matching the impedance in a mcirostrip antenna is the inset feeding technique. In inset feeding technique, the microstrip line used to feed the antenna meets the antenna at an inset etched from the patch antenna as shown in Figures 16 and 19. The theory of the inset feeding technique for a conventional rectangular patch antenna is available in (Balanis, 2005). Another technique for impedance matching is the quarter wavelength transformer. Here, the microstrip feed line is divided into two parts with different widths. The detailed theory of a quarter wavelength transformer can be found in (Pozar, 2011). An example of quarter wavelength transformer used for matching the impedance of a rectangular patch antenna is shown in Figure 11 (a).

Whatever technique is used for matching the impedance in an antenna, it is always difficult to perfectly match the impedance at more than one frequency. Even if the impedance is not matched, the resonant frequency of an antenna can be observed from the frequency response of the S11 parameter. However, there is no significant fall in the value of the S11 parameter. As it may be observed from Figure 20, the resonant frequency of the unloaded patch is 2.15 GHz. However, the value of the S11 parameter at this point is only – 4 dB. This antenna cannot be used for any practical application in its current form. Since the objective here is to obtain a dual band antenna from this patch, no attempts have been made for matching the impedance of the antenna. After the CSRR structure is etched from the ground plane below the patch, the antenna now starts resonating at two different frequencies. The first frequency is 1.7 GHz and the second resonant frequency is 2.4 GHz. The inset length is adjusted to make sure that the impedance is perfectly matched at 2.4 GHz, since it is a useful frequency for WLAN operation. Thus, the simulation result in Figure 20 shows a sharp fall in the value of the S11 parameter to about – 25 dB at 2.4 GHz. However, due to possible manufacturing defects or the effects of the ambient parameters at the time of measurement, the measured result is slightly different from the simulation result.

It is possible to adjust the lengths and widths of the inset, denoted by L_N and W_N respectively in Figure 16 (a), to adjust the return loss parameter at the desired frequency. Here, some optimization technique may be used for this purpose. There are inbuilt optimistic tools in commonly used professional electromagnetic solvers such as Ansys HFSS® and CST Microwave Studio®. There are also instances in the available literatures where soft-computational optimization techniques such as Genetic Algorithm etc. are also used to optimize the performance of different geometrical parameters of microstrip antennas (Sarma, Sarmah, Sarma, Goswami & Baruah, 2016).

Predictable Far Field Radiation Pattern in Dual Band Antennas

In this section, an analysis of the far field radiation patterns of the CSRR based dual band antennas is covered. The far field radiation patterns for three antennas considered in Table 5 are shown in Figure 21. From this observation, it is inferred that the far field radiation pattern of an antenna is predictable from the known resonant frequency of the unloaded patch antenna and that of the CSRR structure loaded at the ground plane.

The radiation patterns considered in Figure 21 (a) correspond to a rectangular patch of resonant frequency 5.3 GHz which is loaded with a CSRR structure of lower and upper resonant frequencies 2 GHz and 4.3 GHz respectively. 4.3 GHz being close to 5.3 GHz, the CSRR loaded antenna resonates at 3.7 GHz and 5.2 GHz. This antenna is mentioned in serial number 7 of Table 5. Similarly, the corresponding frequencies of the radiation patterns shown in Figure 21 (b) and (c) are given in the serial numbers 8 and 10 respectively of Table 5. The observed behavior of the radiation pattern can be summarized as follows (Sarmah, Goswami, Sarma & Baruah, 2015).

- For the lower resonant frequency:
    - There is a back lobe in both E-plane and H-plane which is almost like a mirror image of the major lobe.
    - As the frequency increases, the width of the lobes (beam width) become higher, thus the directivity is reduced.
    - There is a pair of side lobes in the H-plane. The width of the side lobe also increases as the resonant frequency is increased.
- For the upper resonant frequency:
    - The major lobe is very wide.
    - There are minor lobes towards the back of the antenna and also in the sides. However, the behavior of these minor lobes is somewhat random.
    - Some null points are observed at the back and sides of the major lobes.
    - In all cases, the gain along the major lobe is high.

It is seen that when a simple microstrip antenna is converted to a dual band antenna using the CSRR structure, its far field radiation patterns are almost predictable. The beam width of the second resonant frequency is high for both E-plane and H-plane compared to the first resonant frequency.

SUMMARY

In this chapter the design considerations for some trending applications of the CSRR structure are covered. The key notes of the chapter are summarized as follows:

- For design of antennas with reduced electrical size and enhanced gain, the resonant frequency of the CSRR structure should not be near the resonant frequency of the unloaded patch. This separation is usually greater than 1 GHz.
- For using the CSRR structure as an integrated filtering element in an UWB antenna, the resonant frequency of the CSRR must be equal to that of the frequency intended to be notched. The UWB patch antenna is loaded with these CSRR structures.

Figure 21. Far field radiation pattern of some of the obtained resonant frequencies of the dual band antennas
(Sarmah, Goswami, Sarma & Baruah, 2015)

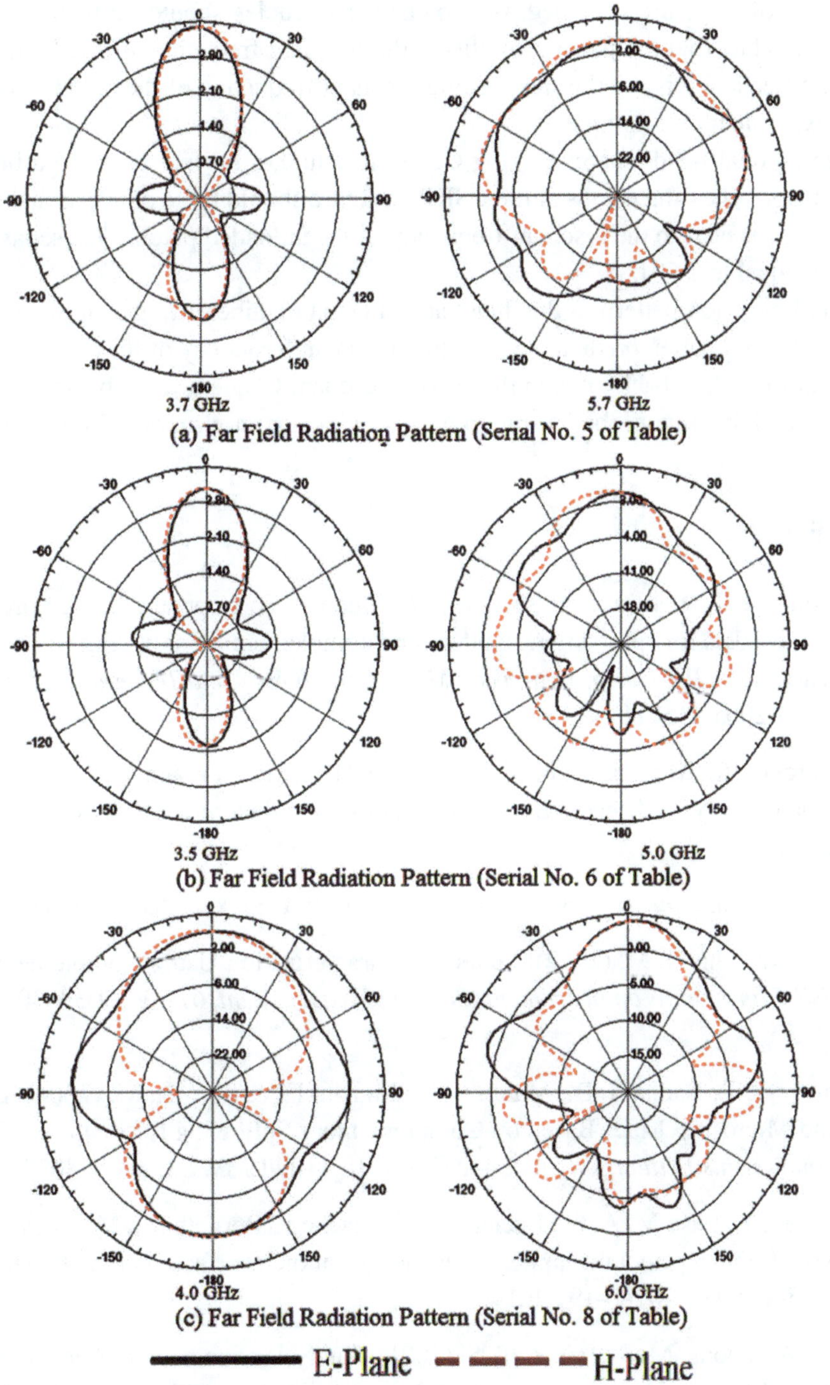

- In order to use a CSRR structure as an integrated filtering element in a resonant antenna, the resonant frequency of the unloaded patch must be slightly greater than the desired resonant frequency. The CSRR will reduce the resonant frequency of the antenna thereby reducing its electrical size. After the CSRR structure is etched, the notched frequencies should be further verified. This is because of the fact that the patch also affects the resonant frequency of the CSRR structure and the notched frequency is not the same as the resonant frequency of the CSRR structure obtained from the experimental technique.
- For designing dual band antennas using CSRR structure, a normal patch of arbitrary geometry should be taken. The dimensions of the CSRR structure should be so adjusted that one of the resonant frequency is near to the resonant frequency of the unloaded patch. This separation should be less than or equal to 1 GHz.
- The far field radiation pattern of the dual band antenna obtained using a single band antenna and a CSRR structure is often predictable. The first resonant frequency of the dual band antenna has a narrow major lobe and back lobe. In the second resonant frequency, the beam-width of the major lobe is high and the minor lobe has random lobes which are not suitable for practical application.

REFERENCES

Baena, J. D., Bonache, J., Martin, F. R., Sillero, M., Falcone, F., Lopetegi, T., ... Sorolla, M. (2005). Equivalent-circuit models for split-ring resonators and complementary split-ring resonators coupled to planar transmission lines. *IEEE Transactions on Microwave Theory and Techniques*, *53*(4), 1451–1461. doi:10.1109/TMTT.2005.845211

Baena, J. D., Marqués, R., Medina, F., & Martel, J. (2004). Artificial magnetic metamaterial design by using spiral resonators. *Physical Review B: Condensed Matter and Materials Physics*, *69*(1). doi:10.1103/PhysRevB.69.014402

Balanis, C. A. (2005). *Antenna Theory: Analysis and Design*. Hoboken, NJ: John Wiley & Sons Inc.

Boybay, M. S., & Ramahi, O. M. (2012). Material Characterization Using Complementary Split-Ring Resonators. *IEEE Transactions on Instrumentation and Measurement*, *61*(11), 3039–3046. doi:10.1109/TIM.2012.2203450

Falcone, F., Lopetegi, T., Baena, J. D., Marques, R., Martin, F., & Sorolla, M. (2004). Effective Negative – ε Stopband Microstrip Lines Based on Complementary Split Ring Resonators. *IEEE Microwave and Wireless Components Letters*, *14*(6), 280–282. doi:10.1109/LMWC.2004.828029

Falcone F., Lopetegi T., Laso M. A. G., Baena J. D., Bonache J., Beruete M., Marqués R., Martín F., & Sorolla M. (2004). Babinet principle applied to the design of metasurfaces and metamaterials. *Physical Review Letters*, *93*(19), 197 401-1–197 401-4.

Gao, S. C., Li, L. W., Leong, M. S., & Yeo, T.-S. (2001). Analysis of an h-shaped patch antenna by using the FDTD method. *Progress in Electromagnetics Research*, *34*, 165–187. doi:10.2528/PIER01060501

Goswami, S., Sarmah, K., Sarma, A., Sarma, K. K., & Baruah, S. (2016, January). *Slot Loaded Square Patch Antenna with CSRR at Ground Plane*. Paper presented at the IEEE International Conference on Microelectronics, Computing and Communication (MicroCom), India. 10.1109/MicroCom.2016.7522558

Goswami, S., Sarmah, K., Sarma, A., Sarma, K. K., & Baruah, S. (2017). Design of a CSRR based Compact Microstrip Antenna for Image Rejection in RF Down-converter based WLAN Receivers. *AEÜ. International Journal of Electronics and Communications*, *74*, 128–134. doi:10.1016/j.aeue.2017.02.004

Islam, M. T., & Samsuzzaman, M. (2014). *Miniaturized Dual Band Multislotted Patch Antenna on Polytetrafluoroethylene Glass Microfiber Reinforced for C/X Band Applications. The Scientific World Journal.*

Kabiri, A., & Ramahi, O. (2009, June). Limitations of artificial magnetic materials with negative permeability. Paper Presented at the *2009 IEEE International Symposium on Antennas and Propagation and USNC-URSI Radio Science Meeting*. 10.1109/APS.2009.5171634

Khodaei, G. F., Nourinia, J., & Ghobadi, C. (2008). A practical miniaturized u-slot patch antenna with enhanced bandwidth. *Progress In Electromagnetics Research B*, *3*, 47–62. doi:10.2528/PIERB07112201

Kim, D. O., Jo, N. I., Jang, H.-A., & Kim, C. Y. (2011). Design of the ultrawideband antenna with a quadruple-band rejection characteristics using a combination of the complementary split ring resonators. *Progress in Electromagnetics Research*, *112*, 93–107. doi:10.2528/PIER10111607

Kshetrimayum R. S., & Karthikeyan S. S. (n.d.). A parametric study on the stopband characteristics of CSRRs. *International Journal of Recent Trends in Engineering*, *1*(3), 274-276.

Li, Y., Liu, Y., & Gong, S. (2008). Microstrip antenna using ground-cut slots for low RCS with size miniaturization techniques. *Progress In Electromagnetics Research Letters*, *1*, 211–220.

Lu, L., & Coetzee, J. C. (2005). Reduced-size microstrip patch antenna for Bluetooth applications. *IEEE Electronics Letters*, *41*(17), 944–945. doi:10.1049/el:20052312

Marques, R., Mesa, F., Martel, J., & Medina, F. (2003). Comparative Analysis of Edge- and Broadside-Coupled Split Ring Resonators for Metamaterial Design—Theory and Experiments. *IEEE Transactions on Antennas and Propagation*, *51*(10), 2572–2581. doi:10.1109/TAP.2003.817562

Mu, J., Wang, H., Jiang, X., Xu, D., & Huang, Y. (2016, October). *Design of Low-RCS and Gain Enhancement Microstrip Antenna Based on Miniaturized Polarisation-Dependent AMC*. Paper presented at the International Symposium on Antennas and Propagation (ISAP), Japan.

Ozbay, E., Aydin, K., Cubukcu, E., & Bayindir, M. (2003). Transmission and reflection properties of composite double negative metamaterials in free space. *IEEE Transactions on Antennas and Propagation*, *51*(10), 2592–2595. doi:10.1109/TAP.2003.817570

Pandeeswari, R., & Raghavan, S. (2014). Microstrip antenna with complementary split ring resonator loaded ground plane for gain enhancement. *Microwave and Optical Technology Letters*, *52*(2), 292–296.

Pendry, J. B., Holden, A. J., Robbins, D. J., & Stewart, W. J. (1999). Magnetism from conductors and enhanced nonlinear phenomena. *IEEE Transactions on Microwave Theory and Techniques*, *47*(11), 2075–2084. doi:10.1109/22.798002

Pozar, D. M. (2011). *Microwave engineering*. John Wiley & Sons Inc.

Rani, R. B., & Pandey, S. K. (2017). A CPW-fed circular patch antenna inspired by reduced ground plane and CSRR slot for UWB applications with notch band. *Microwave and Optical Technology Letters*, *59*(4), 745–749. doi:10.1002/mop.30386

Sarkar, D., Saurav, K., & Srivastava, K. V. (2013, March). *Design of a Novel Dual-band Microstrip Patch Antenna for WLAN/WiMAX Applications Using Complementary Split Ring Resonators and Partially Defected Ground Structure*. Paper Presented at the *Progress in Electromagnetics Research Symposium*, Taipei, Taiwan.

Sarma, A., Sarmah, K., Sarma, K. K., Goswami, S., & Baruah, S. (2016). Soft-computing based Optimization of Low Return Loss Multiband Microstrip Patch Antenna. In J. K. Mandal, S. Mukhopadhyay, & T. Pal (Eds.), *Handbook of Research on Natural Computing for Optimization Problems*. IGI Global. doi:10.4018/978-1-5225-0058-2.ch024

Sarmah, K., Goswami, S., Sarma, A., Sarma, K. K., & Baruah, S. (2016, February). *An Experimental Study on Designing of a Dual Band Antenna using CSRR Structure with a Single Band Antenna*. Paper Presented at the *2nd IEEE International Conference on Computational Electromagnetics (ICCEM)*, China. 10.1109/COMPEM.2016.7588563

Sarmah, K., Goswami, S., Sarma, K. K., & Baruah, S. (2015 December). *Design of a CSRR based Dual Band Rectangular Patch Antenna with Predictable Far Field Radiation Pattern*. Paper Presented at the *6th International Conference on Computers and Devices for Communication (CODEC)*, India.

Sigalas, M. M., Chan, C. T., Ho, K. M., & Soukoulis, C. M. (1995). Metallic photonic band gap materials. *Physical Review B: Condensed Matter and Materials Physics*, *52*(11744). PMID:9980306

Yadav, A., Agrawal, S., & Yadav, R. P. (2017). SRR and S-shape slot loaded triple band notched UWB antenna. *AEÜ. International Journal of Electronics and Communications*, *79*, 192–198. doi:10.1016/j.aeue.2017.06.003

Yousefi, L., & Ramahi, O. M. (2007, March). *New artificial magnetic materials based on fractal Hilbert curves*. Paper Presented at the International Workshop on Antenna Technology (iWAT), UK. 10.1109/IWAT.2007.370119

Zhang, Y., Hong, W., Yu, C., Kuai, Z.-Q., Don, Y.-D., & Zhou, J.-Y. (2008). Planar Ultrawideband Antennas With Multiple Notched Bands Based on Etched Slots on the Patch and/or Split Ring Resonators on the Feed Line. *IEEE Transactions on Antennas and Propagation*, *56*(9), 3063–3068. doi:10.1109/TAP.2008.928815

Chapter 3
Analytical and Calibration Models for Slot Antennas

Kok Yeow You
Universiti Teknologi Malaysia, Malaysia

ABSTRACT

In this chapter, the background, analytical formulations, and calibration routines for slot antennas are briefly reviewed. Performance and operating frequency of the slot antenna are strongly dependent on the dimension and shape of the slot or slotted array on the antenna. Nowadays, most antennas are designed using numerical simulation software for accuracy in analysis. However, analytical formulations still play an important role in the pre-design of the antenna due to the numerical simulation which still requires relatively long period of computation time compared to the analytical calculation. The predicted dimension of the antenna from analytical calculations will only require minor adjustment to optimize its performance in numerical simulation. Hence, the time spent for the antenna design can be shortened. Besides the performance of antenna, the antenna calibration process is crucial as well for releasing systematic errors in the antenna measurements. Some one-port calibration methods are described in detail.

INTRODUCTION

The antenna is an important element for today's technological developments, especially in industry 4.0 practice, Internet of Things (IoT) development, and 5G communication applications. In this chapter, only the types of slot antennas will be introduced and described; since the slot antenna has a simple structure, cost effective, and provide several advantages of usage in certain applications. This chapter focuses more on modeling and calibration of slot antenna. The background principle of slot antenna can be referred to a simple slot antenna, which consists of a narrow slot cut in a thin metal sheet and the slot is excited by a voltage source as shown in Figure 1 (a). The slot radiates waves in a way similar to a dipole antenna (in Figure 1 (b)).

DOI: 10.4018/978-1-5225-7539-9.ch003

Babinet's principle states that the radiation field of the slot is the same as that for the dipole, but the field vectors over the slot are perpendicular to the corresponding field vectors in the case of the dipole (Booker, 1946). It has been shown that the voltage distribution (electric field) in the slot is identical to the current distribution on the dipole antenna. The impedance, Z relationship between slot antenna and dipole antenna can be written as:

$$Z_{slot} Z_{dipole} = \frac{\eta^2}{4} \tag{1}$$

where Z_{slot} and Z_{dipole} are the impedance of the slot antenna and dipole antenna, respectively. While, η ($\approx$ 376.99 Ω) is the intrinsic impedance of free space.

Once the complex value of Z_{dipole} is obtained, the Z_{slot} can be easily found as:

$$Z_{slot} \approx \frac{35530.57}{Z_{dipole}} \tag{2}$$

However, slots cut for the practise slot antenna is not always on the extended thin flat sheet of metal. In the slot antenna design, the slots cut can be either on the wall surface or cross section of the waveguide. Hence, the slot antenna can be categorized into several groups, such as annular slot antenna, waveguide slot antenna, cavity-backed slot antenna, waveguide-fed slot antenna, coaxial-fed slot antenna, and microstrip-fed slot antenna. For instance, the coaxial-fed slot antenna has been widely used as an antenna in underground long tunnels, mines, subway or underground parking spaces, due to the weak propagated radio signal in underground whereby the signal is screened by concrete walls or building. On the other hand, the waveguide-fed slot antenna is usually implemented in radar applications since it is simple, space saving and relatively low cost compared to parabolic reflector and horn antennas. Furthermore, microstrip-fed slot antenna is generally used in mobile phones, which appeared to be wideband, low profile, space and cost savings.

Figure 1. (a) Simple slot antenna. (b) Simple dipole antenna

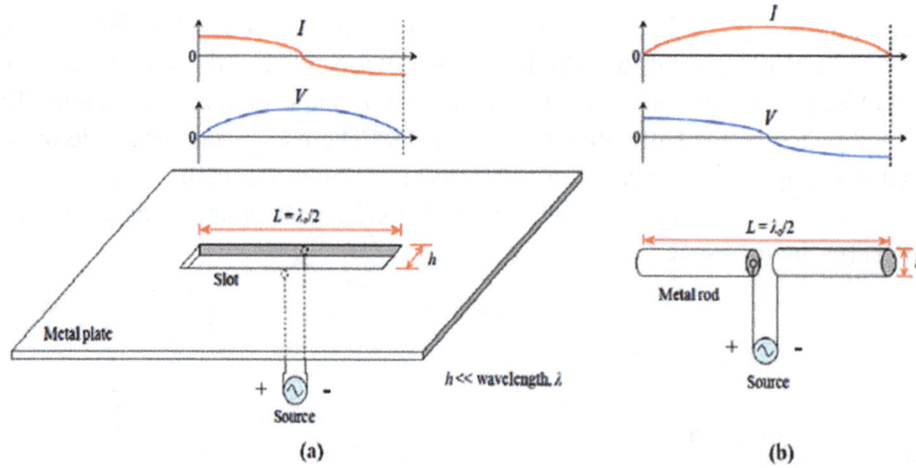

ANNULAR SLOT ANTENNAS

The annular slot antenna is an open-ended coaxial waveguide consisting of inner and outer conductors, with radia a and b respectively, as shown in Figure 2. In general, the medium between the conductors in coaxial line is filled by free space or Teflon. Figure 3 shows another kind of coaxial-type of antenna, namely coaxial slot antenna in which widely used as a applicator in hyperthermia treatment.

ANALYTICAL FORMULATIONS

Characteristic Impedance

In general, commercial RF cables and connectors mostly use a standard 50 Ω coaxial line. Hence, the annular slot antenna is required to have characteristic impedance, Z_o = 50 Ω for impedance matching when connected to the cable or connector. The value of Z_o for the annular slot antenna can be determined based on its dimensions with outer radius of inner conductor, a and inner radius of outer conductor, b, as well as the different conductivity, σ between the inner and outer conductors of the transmission line as (Agilent Technologies, 2010):

$$Z_o = 60\sqrt{\frac{\mu_c}{\varepsilon_c}}\ln\left(\frac{b}{a}\right)+\left(1-j\right)\frac{1}{2\pi}\sqrt{\frac{1}{\pi\varepsilon_c\varepsilon_o f}}\left(\frac{1}{a\sqrt{\sigma_a}}+\frac{1}{b\sqrt{\sigma_b}}\right) \tag{1}$$

where σ_a and σ_b (in S/m) are the conductivities of the inner and outer of conductors, respectively. While, μ_c is the relative permeability for the coaxial line filled material (Non-magnetic material: $\mu_c = 1$). Normally, the characteristic impedance, Z_o can be simplified as:

Figure 2. (a) Annular aperture with outer radius of inner conductor, a and the inner radius of outer conductor, b (You, 2015). (b) Actual photograph of the annular slot antenna

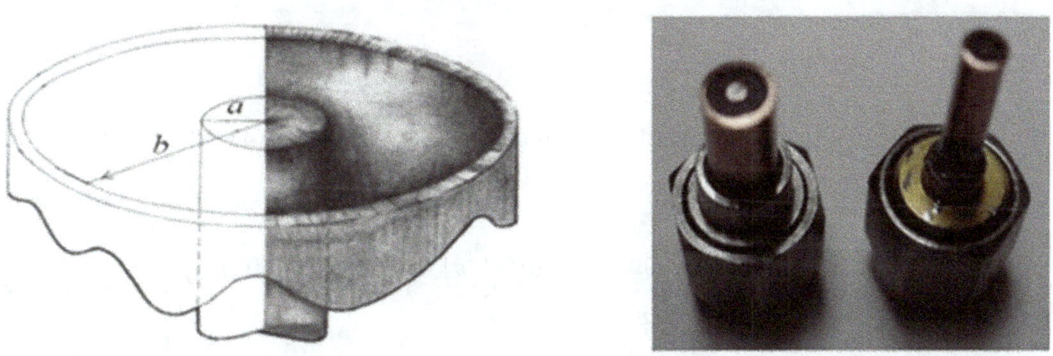

(a) (b)

Figure 3. Coaxial-slot antenna

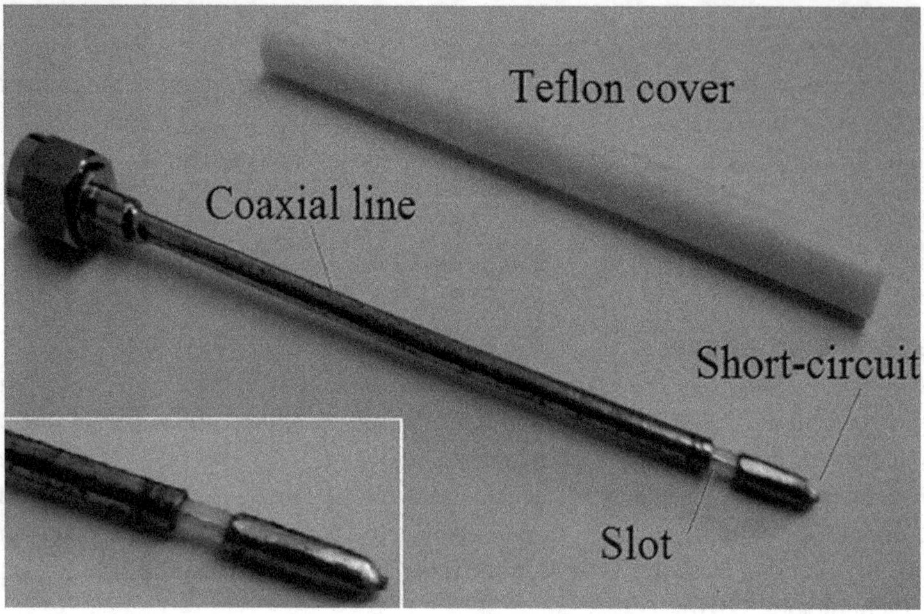

$$Z_o = 60\sqrt{\frac{\mu_c}{\varepsilon_c}}\ln\left(\frac{b}{a}\right) = 50\,\Omega \tag{2}$$

From (1) and (2), the $Z_o = 50\ \Omega$ of transmission characteristic will give $b/a \approx 2.3$ for the coaxial air line and $b/a \approx 3.3$ for the coaxial Teflon-filled line, respectively. However, there are existed various size of annular slot antenna, since dimensions of *a* or *b* can be arbitrary values as listed in Table 1.

Cut-Off Frequency

The operating frequency bandwidth of the antenna is also depended on its dimensions. The maximum limit of the TE_{11} cut-off frequency, f_c propagated in the coaxial transmission line of the antenna determined as:

$$f_c = \frac{c\times\zeta}{2\pi\left(b+a\right)\sqrt{\varepsilon_c\mu_c}} \tag{3}$$

Coefficient value of ζ in (3) is available in Marcuvitz, (1986) (in Figure 4), which can be expressed in empirical polynomial as:

$$\zeta = \begin{Bmatrix} 3.0276859226\times10^{-3}\left(b/a\right)^4 - 3.3123993783\times10^{-2}\left(b/a\right)^3 + 0.1189462449\left(b/a\right)^2 \\ -0.1379387439\left(b/a\right) + 2.0484831228 \end{Bmatrix}, r^2 = 0.999107 \tag{4}$$

The cross sectional visualization of the field distributions in the fundamental TEM mode and higher order TE_{11} mode is shown in Figure 5. Green, (1989) has been investigated the TE_{11} cut-off frequency, f_c based on transverse resonance technique and the closed form expression is written as:

$$f_c = \frac{c}{\pi\left(b+a\right)\sqrt{\varepsilon_c \mu_c}\left[1 - \frac{1}{6}\left(\frac{b-a}{b+a}\right)^2 - \frac{7}{120}\left(\frac{b-a}{b+a}\right)^4 - \ldots\right]} \tag{5}$$

The cut-off frequency, f_c of (3) and (5) can be simplified as:

$$f_c = \frac{c}{\pi\left(b+a\right)\sqrt{\varepsilon_c \mu_c}} \tag{6}$$

Normally, the limit of operation frequency, f_c of the practice coaxial line is always much lower than the TE_{11} cutoff frequency based on the quality and precision of manufacturing. The comparison of calculated cut-off frequency using Equations (3), (5), and (6) are tabulated in Table 1.

Annular Aperture Admittance

The earliest derivation of the integral admittance model by Levine and Papas (1951) concerned the homogeneous case, in which an open-ended air-filled coaxial line with an infinite conducting flange is radiated into the free space as shown in Figure 6. The integral admittance equation was rewritten by Marcuvitz (1967) for arbitrary coaxial filled material and exterior material, in real and imaginary parts expressions of normalized aperture admittance, $\tilde{Y}_{in}$ as:

Figure 4. Coefficient, ζ value for various ratio b/a

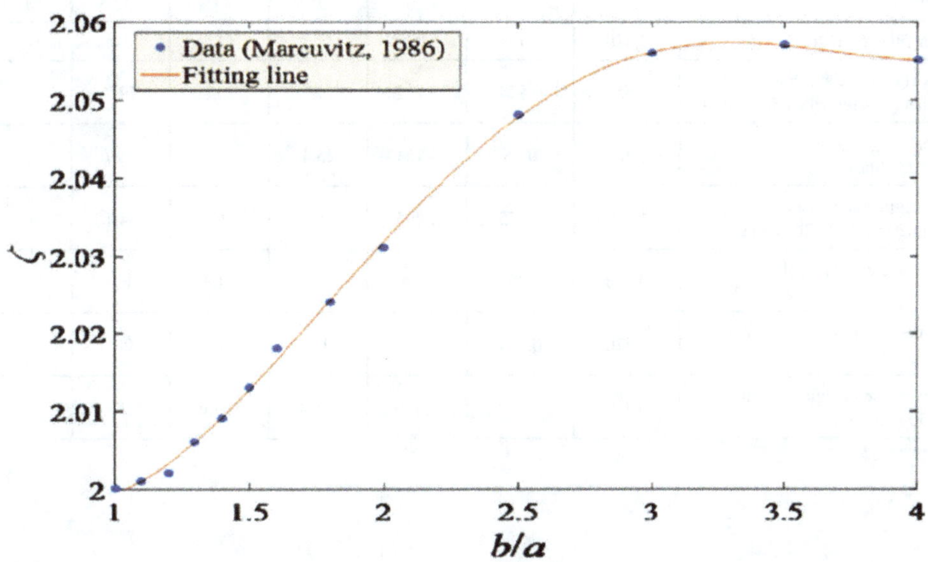

Figure 5. Front sectional view of (a) TEM mode and (b) TE_{11} mode

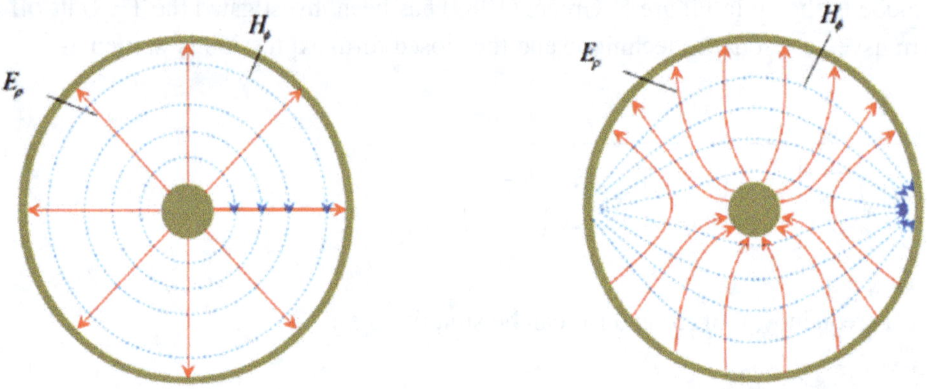

(a) TEM mode (b) TE_{11} mode

Table 1. Dimensions of commercial cable and connectors and its TE_{11} cut-off frequency, f_c

Type of 50 Ω Coaxial Line	b (mm)	a (mm)	b/a	f_c (GHz)			Max Practice Frequency (GHz)
				Eq. (3)	Eq. (5)	Eq. (6)	
14 mm, Air line	7.1437	3.1021	2.303	9.51	9.58	9.31	8.5
7 mm, Air line (N-type connector)	3.500	1.520	2.303	19.4	19.5	19	18
3.5 mm, Air line (SMA-type connector)	1.750	0.760	2.303	38.8	39.1	38	33
2.92 mm, Air line (K-type connector)	1.460	0.634	2.303	46.5	46.9	45.6	40
2.4 mm, Air line (Q-type connector)	1.200	0.521	2.303	56.6	57	55.4	50
1.85 mm, Air line (V-type connector)	0.925	0.4015	2.304	73.5	74	71.9	65
1 mm, Air line (W-type connector)	0.500	0.217	2.304	135.9	136.8	133.1	110
0.8 mm, Air line	0.400	0.1736	2.304	169.9	171.0	166.4	145
0.6 mm, Air line	0.300	0.130	2.308	226.6	228.2	221.9	To be defined
0.4 mm, Air line	0.200	0.087	2.299	339.5	341.8	332.5	To be defined
8.3 mm, Teflon-filled 50 Ω line	3.620	1.124	3.221	14.5	14.8	14	11.7
RG401/U semi-rigid coaxial cable (0.25 inch diameter,Teflon-filled 50 Ω line)	2.655	0.824	3.222	19.7	20.2	19.2	20
SMA connector (Teflon-filled 50 Ω line)	2.05	0.65	3.154	25.4	26	24.7	18
RG402/U semi-rigid coaxial cable (0.141 inch diameter,Teflon-filled 50 Ω line)	1.499	0.455	3.295	35.1	36	34.1	36
RG405/U semi-rigid coaxial cable (0.086 inch diameter,Teflon-filled 50 Ω line)	0.838	0.255	3.286	62.7	64.3	61	40
0.047 inch diameter semi-rigid coaxial cable (Teflon-filled 50 Ω line)	0.470	0.144	3.264	111.7	114.5	108.5	20
0.034 inch diameter semi-rigid coaxial cable (Teflon-filled 50 Ω line)	0.330	0.100	3.300	159.4	163.6	155	40

$$\tilde{Y}_{in} = G_{in} + jB_{in} \tag{7}$$

The real part and imaginary part are called normalized conductance, G_{in} and susceptance, B_{in}, respectively, which can be solved analytically using expansion method as (You, 2015):

$$\begin{aligned} G_{in} &= \frac{\sqrt{\varepsilon_r}}{\sqrt{\varepsilon_c}\ln\left(b/a\right)}\sum_{n=0}^{N} g_n I_n k_2^{2n+4} \\ &= \frac{\sqrt{\varepsilon_r}}{\sqrt{\varepsilon_c}\ln\left(b/a\right)}\left[G_0\left(k_2a\right)^4 + G_1\left(k_2a\right)^6 + G_2\left(k_2a\right)^8 + \ldots + G_N\left(k_2a\right)^{2N+4}\right] \end{aligned} \tag{8a}$$

And:

$$\begin{aligned} B_{in} &= \frac{\sqrt{\varepsilon_r}}{\pi\sqrt{\varepsilon_c}\ln\left(b/a\right)}\sum_{n=0}^{N} s_n k_2^{2n+1} \\ &= \frac{\sqrt{\varepsilon_r}}{\pi\sqrt{\varepsilon_c}\ln\left(b/a\right)}\left[B_0\left(k_2a\right) + B_1\left(k_2a\right)^3 + B_2\left(k_2a\right)^5 + \ldots + B_N\left(k_2a\right)^{2N+1}\right] \end{aligned} \tag{8b}$$

where:

$$I_n = \left[\frac{2^{2(n+1)}\left[\left(n+1\right)!\right]^2}{\left(2n+3\right)!}\right] \tag{9a}$$

$$k_2 = \left(\frac{2\pi f}{c}\sqrt{\varepsilon_r}\right) \tag{9b}$$

where ε_r and ε_c are the relative permittivities for the exterior medium (Air: $\varepsilon_r = 1.0006$) and coaxial line filled material, respectively. For n = even (0, 2, 4,…), the coefficient value in (8a) is given as:

$$g_0 = \left(\frac{a^2 - b^2}{2^2}\right)^2 \tag{10a}$$

$$g_2 = \left(\frac{a^4 - b^4}{2^2 \cdot 4^2}\right)^2 + \left(\frac{a^2 - b^2}{2^2}\right)\left(\frac{a^6 - b^6}{2^2 \cdot 4^2 \cdot 6^2}\right) \tag{10b}$$

$$g_4 = \left(\frac{a^6 - b^6}{2^2 \cdot 4^2 \cdot 6^2}\right) + \left(\frac{a^4 - b^4}{2^2 \cdot 4^2}\right)\left(\frac{a^8 - b^8}{2^2 \cdot 4^2 \cdot 6^2 \cdot 8^2}\right) + \left(\frac{a^2 - b^2}{2^2}\right)\left(\frac{a^{10} - b^{10}}{2^2 \cdot 4^2 \cdot 6^2 \cdot 8^2 \cdot 10^2}\right) \tag{10c}$$

For n = odds (1, 3, 5,…)

$$g_1 = -\left(\frac{a^2 - b^2}{2^2}\right)\left(\frac{a^4 - b^4}{2^2 \cdot 4^2}\right) \tag{10d}$$

$$g_3 = -\left[\left(\frac{a^4 - b^4}{2^2 \cdot 4^2}\right)\left(\frac{a^6 - b^6}{2^2 \cdot 4^2 \cdot 6^2}\right) + \left(\frac{a^2 - b^2}{2^2}\right)\left(\frac{a^8 - b^8}{2^2 \cdot 4^2 \cdot 6^2 \cdot 8^2}\right)\right] \tag{10e}$$

$$g_5 = -\left[\left(\frac{a^6 - b^6}{2^2 \cdot 4^2 \cdot 6^2}\right)\left(\frac{a^8 - b^8}{2^2 \cdot 4^2 \cdot 6^2 \cdot 8^2}\right) + \left(\frac{a^4 - b^4}{2^2 \cdot 4^2}\right)\left(\frac{a^{10} - b^{10}}{2^2 \cdot 4^2 \cdot 6^2 \cdot 8^2 \cdot 10^2}\right) + \left(\frac{a^2 - b^2}{2^2}\right)\left(\frac{a^{12} - b^{12}}{2^2 \cdot 4^2 \cdot 6^2 \cdot 8^2 \cdot 10^2 \cdot 12^2}\right)\right] \tag{10f}$$

The series of (8b) are conveniently re-write in a single expression as:

$$s_n = \left\{\left[\left(-1\right)^n \cdot \int_0^\pi \frac{2 \cdot R^{(2n+1)/2}}{(2n+1) \cdot \left[(2n+1)!\right]} d\phi\right] - \left[\frac{\left(-1\right)^n \cdot \left(2\right)^{2(2n+1)} \left(n!\right)^2}{(2n+1) \cdot \left[(2n+1)!\right]^2}\right]\left(a^{2n+1} + b^{2n+1}\right)\right\} \tag{11}$$

The integration term in (11) can be accurately solved by numerical integral solution. Here, the integration solved by 2nd level of Boole's rule can be written as:

$$\int_0^\pi R^{(2n+1)/2} d\phi = \frac{2h}{45}\left\{\begin{matrix} 7R_1^{(2n+1)/2} + 32R_2^{(2n+1)/2} + 12R_3^{(2n+1)/2} + 32R_4^{(2n+1)/2} + 14R_5^{(2n+1)/2} + 32R_6^{(2n+1)/2} \\ +12R_7^{(2n+1)/2} + 32R_8^{(2n+1)/2} + 7R_9^{(2n+1)/2} \end{matrix}\right\} \tag{12}$$

where $h = \pi / 8$ and;

$$R_1 = a^2 + b^2 - 2ab\cos\left(0\right);\; R_2 = a^2 + b^2 - 2ab\cos\left(\pi/8\right);\; R_3 = a^2 + b^2 - 2ab\cos\left(\pi/4\right);$$

$$R_4 = a^2 + b^2 - 2ab\cos\left(3\pi/8\right);\; R_5 = a^2 + b^2 - 2ab\cos\left(\pi/2\right);\; R_6 = a^2 + b^2 - 2ab\cos\left(5\pi/8\right);$$

$$R_7 = a^2 + b^2 - 2ab\cos\left(3\pi/4\right);\; R_8 = a^2 + b^2 - 2ab\cos\left(7\pi/8\right);\; R_9 = a^2 + b^2 - 2ab\cos\left(\pi\right)$$

The G_n and B_n coefficients (subscript $n = 0, 1, 2,\ldots$) in (8a) and (8b) for $b/a = 2.36$, $b/a = 3.1538$, and $b/a = 3.333$, are respectively listed in Table 2. The comparison of calculated normalized input conductance, G_{in} and susceptance, B_{in} with the numerical simulation results (using COMSOL Multiphysics simulation software) are shown in Figures 7 (a) and (b), respectively. Figure 8 shows the calculated G_{in} and B_{in} using (8a) and (8b) with various sizes of the inner conductor, a. Since the fringing effect at probe aperture has not been taken into account in the (8a) and (8b), thus, the overall calculated values of B_{in} are slightly higher than simulation results.

Annular Aperture Calibration

The complex reflection coefficient, S_{11a_medium} of the medium at the probe aperture (at the *BB'* plane) is should be measured as shown in Figure 9 (b). However, during the measurement process, only the reflection coefficient, S_{11m_medium} at the end of the coaxial line (at the *AA'* plane) is measured. Hence, the measured S_{11m_medium} must be calibrated due to the reflection at the *AA'* plane, which is separated from the external medium (at the *BB'* plane) by a coaxial line. Thus, a de-embedding process should be done to remove the effects of the coaxial line.

In this subsection, a simple open-standard calibration is introduced which standard only requires the probe aperture open to the air as shown in Figure 9 (a). Firstly, the S_{11m_air} for the air is measured. Later, the probe aperture is contacted with external medium and its S_{11m_medium} is measured as shown in Figure 9 (b). The relationship between the S_{11m_air} at the plane *AA'* and S_{11a_air} at the probe aperture *BB'* is expressed in a bilinear equation as:

Figure 6. Spatial behavior of H_ϕ

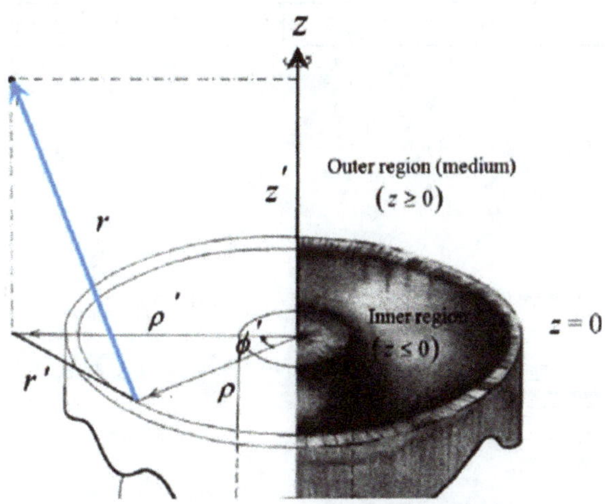

Table 2. Coefficient values of each of the series terms (G_n and B_n) in Equations (8a) and (8b) for b/a = 2.36, b/a = 3.1538, and b/a = 3.333 probes

n	b/a = 2.36		b/a = 3.1538		b/a = 3.333	
	G_n	B_n	G_n	B_n	G_n	B_n
0	0.87005184	2.061739075	3.335177342	3.702084000	4.25977366	4.08456914
1	-0.571589256	1.919124548	-3.650933774	5.739516182	-5.15905921	6.979344183
2	0.178453108	-2.507381026	1.943095250	-12.37093409	3.04859617	-16.61140523
3	-0.034473149	1.076692770	-0.651319483	8.95566282	-1.137464042	13.32960075
4	4.619178022×10^{-3}	$-2.593580866\times10^{-1}$	0.153214914	-3.713265257	0.298311607	-6.14499757
5	$-4.586562712\times10^{-4}$	4.129831004×10^{-2}	-0.026902457	1.032667045	-0.058450164	1.903932961
6	3.525759603×10^{-5}	$-4.732581285\times10^{-3}$	3.672802049×10^{-3}	-0.208601993	8.909134276×10^{-3}	-0.429006044
7	$-2.164577151\times10^{-6}$	4.115921111×10^{-4}	$-4.015805668\times10^{-4}$	0.032160024	$-1.087854273\times10^{-3}$	0.073826565
8	1.087968255×10^{-7}	$-2.818447587\times10^{-5}$	3.597597208×10^{-5}	$-3.916656844\times10^{-3}$	1.088503566×10^{-4}	-0.010039762
9	$-4.558158420\times10^{-9}$	1.561529296×10^{-6}	$-2.689381588\times10^{-6}$	3.866567979×10^{-4}	$-9.089051757\times10^{-6}$	1.106950619×10^{-3}
10	$1.617071267\times10^{-10}$	$-7.149428520\times10^{-8}$	1.703106191×10^{-7}	$-3.157697152\times10^{-5}$	6.429410620×10^{-7}	$-1.009740357\times10^{-4}$
11	$-4.920842165\times10^{-12}$	2.751478771×10^{-9}	$-9.253364969\times10^{-9}$	2.168889397×10^{-6}	$-3.902108837\times10^{-8}$	7.747007819×10^{-6}
12	$1.298646955\times10^{-13}$	$-9.027091809\times10^{-11}$	4.360649538×10^{-9}	$-1.270355636\times10^{-7}$	2.054113497×10^{-9}	$-5.068615039\times10^{-7}$
13	$-3.000571797\times10^{-15}$	$2.554916965\times10^{-12}$	$-1.799247737\times10^{-11}$	6.419955983×10^{-9}	$-9.467587853\times10^{-11}$	2.861331996×10^{-8}
14	-	$-6.302007635\times10^{-14}$	$6.553983989\times10^{-13}$	$-2.827807473\times10^{-10}$	$3.852384357\times10^{-12}$	$-1.407863431\times10^{-9}$
15	-	$1.366785918\times10^{-15}$	$-2.123051402\times10^{-14}$	$1.095234032\times10^{-11}$	$-1.393994108\times10^{-13}$	$6.091064231\times10^{-11}$
16	-	-	$6.155641470\times10^{-16}$	$-3.758972780\times10^{-13}$	$4.514920378\times10^{-15}$	$-2.335239814\times10^{-12}$
17	-	-	$-1.606816187\times10^{-17}$	$1.151119960\times10^{-14}$	$-1.316495561\times10^{-16}$	$7.988388481\times10^{-14}$
18	-	-	$3.795866192\times10^{-19}$	$-3.164642932\times10^{-16}$	$6.927627879\times10^{-19}$	$-2.453238097\times10^{-15}$
19	-	-	$-8.153874958\times10^{-21}$	$7.853690960\times10^{-18}$	$-8.336239280\times10^{-20}$	$6.800880706\times10^{-17}$
20	-	-	$1.599544482\times10^{-22}$	$-1.768185466\times10^{-19}$	$1.826749188\times10^{-21}$	$-1.710392195\times10^{-18}$
21	-	-	$-2.876845308\times10^{-24}$	$3.627833339\times10^{-21}$	$-3.670080498\times10^{-23}$	$3.920050760\times10^{-20}$
22	-	-	$4.760928203\times10^{-26}$	$-6.811131284\times10^{-23}$	$6.784646003\times10^{-25}$	$-8.221292354\times10^{-22}$
23	-	-	$-7.273841916\times10^{-28}$	$1.174570789\times10^{-24}$	$-1.157912920\times10^{-26}$	$1.583713139\times10^{-23}$
24	-	-	-	$-1.866937527\times10^{-26}$	$1.830009548\times10^{-28}$	$-2.811924252\times10^{-25}$
25	-	-	-	$2.743824746\times10^{-28}$	$-2.685918944\times10^{-30}$	$4.616434223\times10^{-27}$

$$S_{11a\_air} = \frac{S_{11m\_air} - e_{00}}{e_{11}S_{11m\_air} + e_{10}e_{01} - e_{00}e_{11}} \tag{13}$$

Similarly, the relationship between measured S_{11m_medium} and S_{11a_medium} is given as:

$$S_{11a\_medium} = \frac{S_{11m\_medium} - e_{00}}{e_{11}S_{11m\_medium} + e_{10}e_{01} - e_{00}e_{11}} \tag{14}$$

Figure 7. Comparison of (a) normalized input conductance, G_{in}, and (b) normalized input susceptance, B_{in} for the Teflon-filled coaxial probe

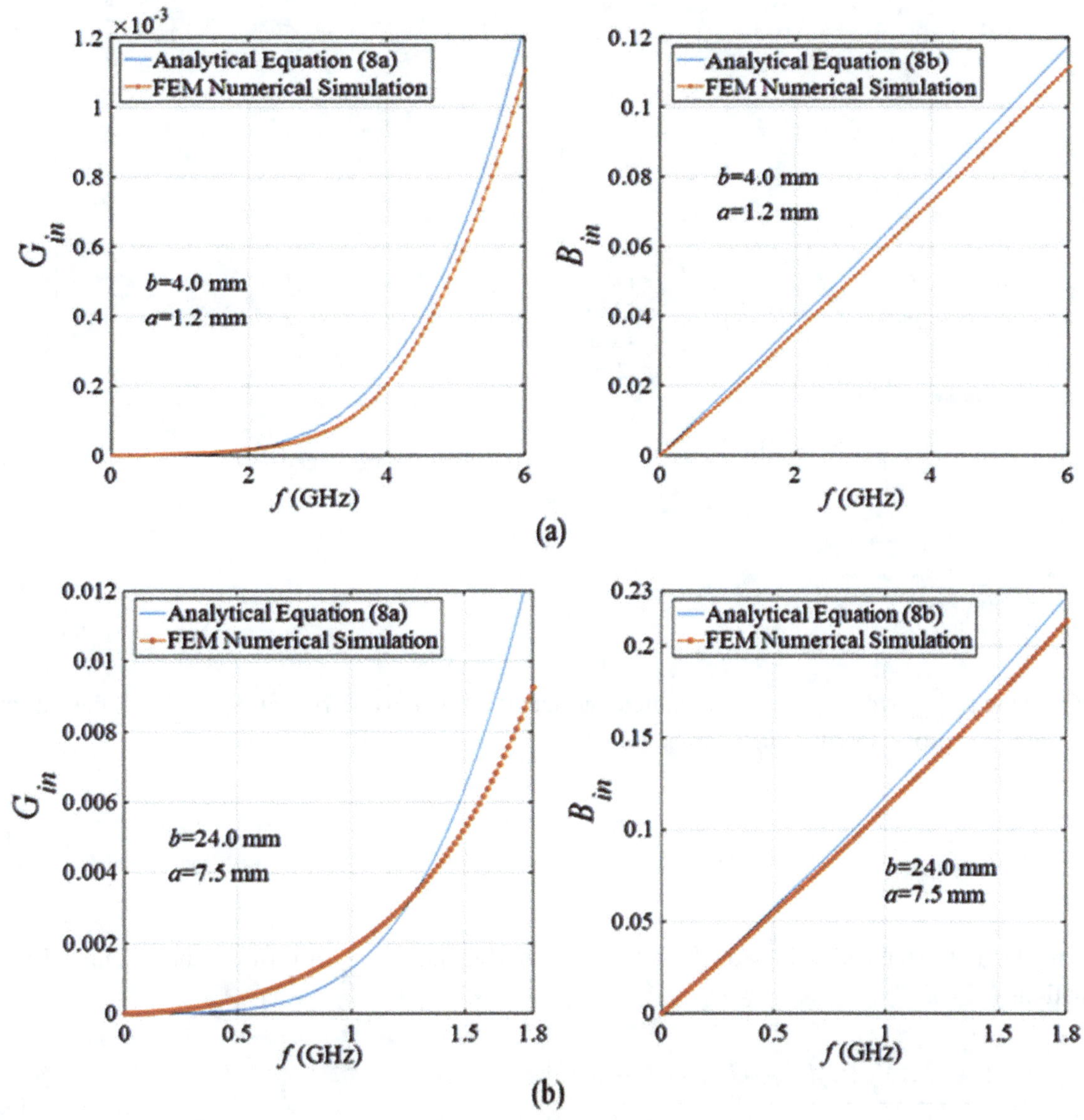

The e_{00}, e_{11} and $e_{10}e_{01}$ are the unknown scattering parameters of the error network for the coaxial line. The e_{00} is the directivity error that causes the failure to receive the measured reflection signal completely from the external medium being tested at plane BB'. The e_{11} is the source match error due to the fact that the impedance of the aperture probe at plane BB' is not exactly the characteristic impedance ($Z_o = 50$ Ω). The $e_{10}e_{01}$ is the frequency tracking imperfections (or phase shift) between plane AA' and medium plane BB'.

For this calibration, the e_{00} and e_{11} terms in (13) and (14) are assumed to be vanished ($e_{00} = e_{11} = 0$). By dividing (13) into (14), yields:

Figure 8. (a) Normalized conductance, G_{in}, and (b) normalized susceptance, B_{in}

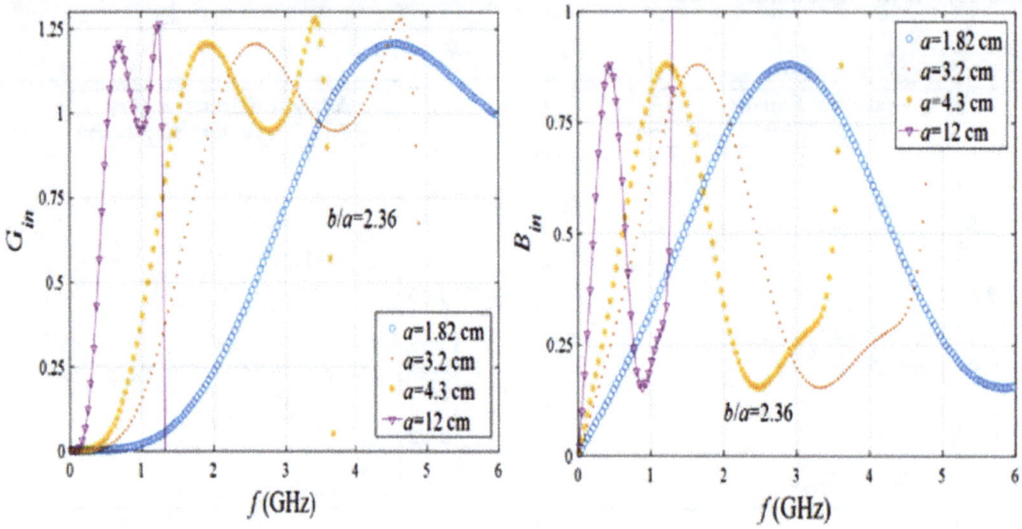

$$\frac{S_{11a\_medium}}{S_{11a\_air}} = \frac{S_{11m\_medium}}{S_{11m\_air}} \tag{15}$$

Once the S_{11m_air} and S_{11m_medium} are obtained, the actual reflection coefficient, S_{11a_medium}, of the external medium at the probe aperture, *BB'* can be found as:

$$S_{11a\_medium} = \frac{S_{11a\_air}}{S_{11m\_air}} \times S_{11m\_medium} \tag{16}$$

The accurate standard values of the reflection coefficient, S_{11a_air} in (16) can be calculated from Equations (17) or (7) .

$$S_{11a\_air} = \frac{1 - j\left(\omega/Y_o\right)\left(C_o f^{-1} + C_1 + C_2 f + C_3 f^2\right)}{1 + j\left(\omega/Y_o\right)\left(C_o f^{-1} + C_1 + C_2 f + C_3 f^2\right)} \tag{17}$$

Symbols $\omega = 2\pi f$ and $Y_o = [(2\pi)/\ln(b/a)]\sqrt{(\varepsilon_o \varepsilon_c/\mu_o \mu_r)}$ are the angular frequency (in rad/s) and characteristic admittance (in siemens), respectively. The complex values of the C_0, C_1, C_2 and C_3 in (17) for air-filled and Teflon-filled coaxial line antenna are given in Tables 3 and 4 (You, 2015).

Figure 9. Error network and finite coaxial line: (a) terminated by air; (b) terminated by a medium

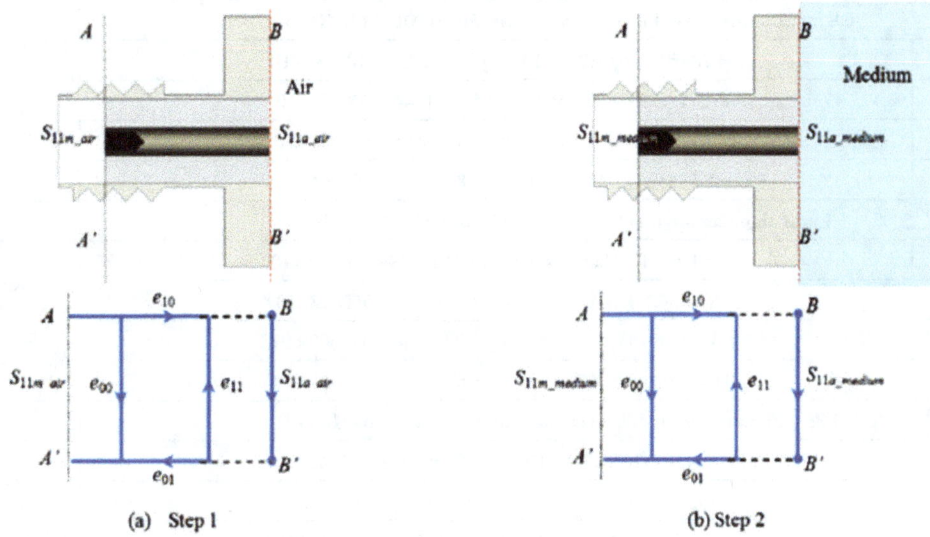

OPEN-ENDED RECTANGULAR SLOT ANTENNAS

The open-ended rectangular slot antenna is a rectangular metal pipe with height of *a* and width of *b*, respectively as shown in Figure 10. This antenna guides high frequency TE_{10} mode wave from one place to slot aperture without significant loss in intensity.

ANALYTICAL FORMULATIONS

Cutoff Frequency

The lower cutoff frequency, f_c and the lower cutoff wavelength, λ_c of the open-ended rectangular slot antenna can be determined as:

$$f_c = \frac{c}{2b} \tag{18a}$$

and;

$$\lambda_c = \frac{c}{f_c} \tag{18b}$$

where c (= 299792458 ms^{-1}) is the velocity of light in the free space and b (in unit meter) is the width of the slot aperture. The effective wavelength, λ_g of propagated wave in the antenna can be calculated as:

Table 3. The complex values of C_0, C_1, C_2, and C_3 in (17) for air-filled coaxial probe

0.8 mm Diameter Air-Filled Coaxial Line (50 Ω) (DC - 145 GHz)		r^2
C_0 (F×Hz)	$-4.789292726763718\times10^{-6}+j\,2.797673463101733\times10^{-7}$	Real part: 0.999091 Imaginary part: 0.986513
C_1 (F)	$4.727152222662021\times10^{-15}-j\,5.416184489706334\times10^{-17}$	
C_2 (F/Hz)	$2.597590226460427\times10^{-27}+j\,2.123350848956879\times10^{-27}$	
C_3 (F/Hz²)	$1.005313944021000\times10^{-38}-j\,2.823225790795064\times10^{-38}$	
1 mm diameter air-filled coaxial line (50 Ω) (DC - 110 GHz)		r^2
C_0 (F×Hz)	$-7.641032127680684\times10^{-6}+j\,1.587314946332905\times10^{-6}$	Real part: 0.999941 Imaginary part: 0.995837
C_1 (F)	$9.878456231235801\times10^{-15}-j\,2.988657030566632\times10^{-16}$	
C_2 (F/Hz)	$1.342783975316819\times10^{-26}+j\,1.238744096419602\times10^{-26}$	
C_3 (F/Hz²)	$3.048176969483076\times10^{-38}-j\,1.885743164943881\times10^{-37}$	
1.85 mm diameter air-filled coaxial line (50 Ω) (DC - 60 GHz)		r^2
C_0 (F×Hz)	$8.854161938791748\times10^{-7}+j\,1.875939931262845\times10^{-6}$	Real part: 0.999941 Imaginary part: 0.995837
C_1 (F)	$1.762402448601485\times10^{-14}-j\,6.060149187981872\times10^{-16}$	
C_2 (F/Hz)	$6.435399746818748\times10^{-26}+j\,4.405321240350318\times10^{-26}$	
C_3 (F/Hz²)	$2.176654611804275\times10^{-38}-j\,1.205196672502096\times10^{-36}$	
2.4 mm diameter air-filled coaxial line (50 Ω) (DC - 50 GHz)		r^2
C_0 (F×Hz)	$2.831224181111995\times10^{-6}+j\,1.740965471884909\times10^{-6}$	Real part: 0.999908 Imaginary part: 0.995119
C_1 (F)	$2.321675171187550\times10^{-14}-j\,7.492628467538283\times10^{-16}$	
C_2 (F/Hz)	$9.427641802250218\times10^{-26}+j\,7.267347365668988\times10^{-26}$	
C_3 (F/Hz²)	$2.310400591396996\times10^{-37}-j\,2.628455410980044\times10^{-36}$	
2.92 mm diameter air-filled coaxial line (50 Ω) (DC - 40 GHz)		r^2
C_0 (F×Hz)	$1.985212051865980\times10^{-6}+j\,1.411848782469896\times10^{-6}$	Real part: 0.999934 Imaginary part: 0.995960
C_1 (F)	$2.854550930878580\times10^{-14}-j\,7.838358138728700\times10^{-16}$	
C_2 (F/Hz)	$1.175070939275526\times10^{-25}+j\,9.786568269710971\times10^{-26}$	
C_3 (F/Hz²)	$8.525043148489875\times10^{-37}-j\,4.546137384996560\times10^{-36}$	
3.5 mm diameter air-filled coaxial line (50 Ω) (DC - 34 GHz)		r^2
C_0 (F×Hz)	$5.454771641548850\times10^{-7}+j\,1.450158120776360\times10^{-6}$	Real part: 0.999883 Imaginary part: 0.988667
C_1 (F)	$3.502775681593356\times10^{-14}-j\,9.547327705077067\times10^{-16}$	
C_2 (F/Hz)	$9.844825448856349\times10^{-26}+j\,1.418823927035124\times10^{-25}$	
C_3 (F/Hz²)	$3.016530462523645\times10^{-36}-j\,7.864274561640535\times10^{-36}$	
7 mm diameter air-filled coaxial line (50 Ω) (DC - 18 GHz)		r^2
C_0 (F×Hz)	$5.443701662884144\times10^{-6}+j\,2.851350015693733\times10^{-6}$	Real part: 0.999830 Imaginary part: 0.995582
C_1 (F)	$6.593688544955866\times10^{-14}-j\,3.057046718186984\times10^{-15}$	
C_2 (F/Hz)	$1.083935327528096\times10^{-24}+j\,7.458506675875836\times10^{-25}$	
C_3 (F/Hz²)	$-5.361028938357830\times10^{-36}-j\,6.993084701733307\times10^{-35}$	

Table 4. The complex values of C_o, C_1, C_2, and C_3 in (17) for Teflon-filled coaxial probe

0.034 Inch Diameter Teflon-Filled Coaxial Line (50 Ω) (DC - 155 GHz)		r^2
C_o (F×Hz)	$-1.330542378303556\times10^{-5} + j\,2.368599791183465\times10^{-6}$	Real part: 0.999183 Imaginary part: 0.996161
C_1 (F)	$8.153818159170725\times10^{-15} - j\,3.190450476884792\times10^{-16}$	
C_2 (F/Hz)	$5.062883997728317\times10^{-27} + j\,9.546523794693330\times10^{-27}$	
C_3 (F/Hz²)	$2.975634480124245\times10^{-38} - j\,1.048718654940674\times10^{-37}$	
0.047 inch diameter Teflon-filled coaxial line (50 Ω) (DC - 108 GHz)		r^2
C_o (F×Hz)	$-8.213515621799459\times10^{-6} + j\,7.737698529183632\times10^{-7}$	Real part: 0.999926 Imaginary part: 0.993230
C_1 (F)	$6.909557679455718\times10^{-15} - j\,1.484097523317373\times10^{-16}$	
C_2 (F/Hz)	$3.502854130485818\times10^{-27} + j\,6.281242909958981\times10^{-27}$	
C_3 (F/Hz²)	$4.022312598828909\times10^{-38} - j\,9.583530205418906\times10^{-38}$	
0.086 inch diameter Teflon-filled coaxial line (50 Ω) (DC - 60 GHz)		r^2
C_o (F×Hz)	$-2.804574843656230\times10^{-6} + j\,4.174304885612845\times10^{-7}$	Real part: 0.999984 Imaginary part: 0.998502
C_1 (F)	$1.186851533636809\times10^{-14} - j\,1.690024435392950\times10^{-16}$	
C_2 (F/Hz)	$1.923475908224855\times10^{-26} + j\,1.498921009393977\times10^{-26}$	
C_3 (F/Hz²)	$1.759844332286992\times10^{-37} - j\,4.750662745684810\times10^{-37}$	
0.141 inch diameter Teflon-filled coaxial line (50 Ω) (DC - 34 GHz)		r^2
C_o (F×Hz)	$-2.922510945357852\times10^{-8} + j\,6.398268867753511\times10^{-7}$	Real part: 0.999987 Imaginary part: 0.996519
C_1 (F)	$2.109812155069884\times10^{-14} - j\,4.062421361810907\times10^{-16}$	
C_2 (F/Hz)	$5.553427141979506\times10^{-26} + j\,5.731100581278211\times10^{-26}$	
C_3 (F/Hz²)	$1.211039943452395\times10^{-36} - j\,2.931027409943640\times10^{-36}$	
4.1 mm diameter Teflon-filled coaxial line (50 Ω) (DC - 24 GHz)		r^2
C_o (F×Hz)	$5.368082994761808\times10^{-7} + j\,2.320071598550666\times10^{-7}$	Real part: 0.999984 Imaginary part: 0.996707
C_1 (F)	$3.002820660256831\times10^{-14} - j\,2.988971515445163\times10^{-16}$	
C_2 (F/Hz)	$1.112989441958266\times10^{-25} + j\,7.730261500907114\times10^{-26}$	
C_3 (F/Hz²)	$3.140652268416283\times10^{-36} - j\,6.786433840933426\times10^{-36}$	
3.5 mm coaxial air line (50 Ω) (DC - 34 GHz)		r^2
C_o (F×Hz)	$5.246383662974980\times10^{-7} + j\,9.979770216145649\times10^{-8}$	Real part: 0.999978 Imaginary part: 0.995279
C_1 (F)	$3.844256020779706\times10^{-14} - j\,2.621453881635401\times10^{-16}$	
C_2 (F/Hz)	$1.423824046186385\times10^{-25} + j\,1.061224893022041\times10^{-25}$	
C_3 (F/Hz²)	$8.056883567329219\times10^{-36} - j\,1.343021766617817\times10^{-35}$	
1 cm diameter Teflon-filled coaxial line (50 Ω) (DC - 8 GHz)		r^2
C_o (F×Hz)	$3.869794378474446\times10^{-7} + j\,1.899111462898054\times10^{-6}$	Real part: 0.999930 Imaginary part: 0.984729
C_1 (F)	$7.046076320174664\times10^{-14} - j\,3.289730164347698\times10^{-15}$	
C_2 (F/Hz)	$3.366341651876158\times10^{-25} + j\,1.205947522968670\times10^{-24}$	
C_3 (F/Hz²)	$6.530096549187823\times10^{-35} - j\,1.506879419314518\times10^{-34}$	

Figure 10. (a)Actual photographs of open-ended rectangular slot antennas; (b) Radiation field from rectangular waveguide aperture

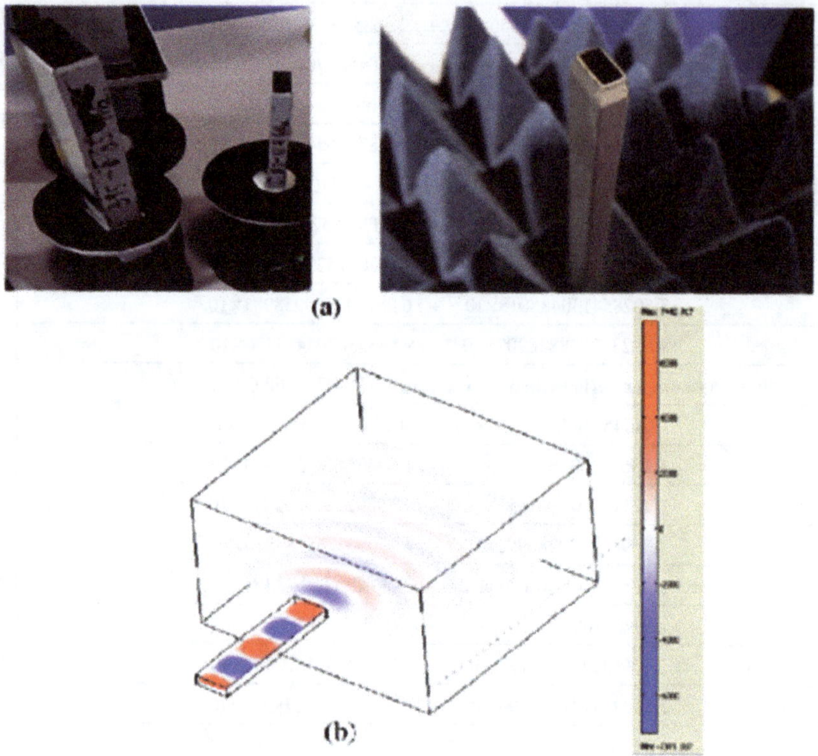

$$\lambda_g = \frac{\lambda_o}{\sqrt{1-\left(\lambda_o/\lambda_c\right)^2}} = \frac{c}{f\sqrt{1-\left\{c/(2bf)\right\}^2}} \tag{19}$$

where f and λ_o (= c/f) are the operating frequency and the wavelength in free space, respectively.

Gain

Stratton, (1941) was initially expressed the gain formulation of the rectangular slot antenna based on the reflection coefficient, S_{11}at slot aperture as:

$$Gain = \frac{\pi k^2 ab}{8\tilde{\beta}\left(1-\left|S_{11}\right|\right)^2}\left|\left[1+\tilde{\beta}+S_{11}\left(1-\tilde{\beta}\right)\right]\left(\frac{2}{\pi}\right)^2\right|^2 \tag{20}$$

where k (= $2\pi/\lambda_o$) is propagation constant in free space and $\tilde{\beta}$ is the normalized propagation constant of the TE_{10} mode which is given as:

$$\tilde{\beta} = \sqrt{1 - \left(\frac{\pi}{kb}\right)^2} \tag{21}$$

Normally, the calculated gain using (20) has lower than the actual value of 1 dB to 2 dB. Yaghjian, (1984) has been improved the (20) by adding a correction term, C_o as:

$$Gain = \frac{\pi k^2 ab}{8\tilde{\beta}\left(1 - \left|S_{11}\right|\right)^2} \left| \left[1 + \tilde{\beta} + S_{11}\left(1 - \tilde{\beta}\right)\right]\left(\frac{2}{\pi}\right)^2 + C_o \right|^2 \tag{22}$$

where C_0 is the positive real constant due to edge current and mutual coupling effects at the slot aperture, which value is depended on the operating wavelength, λ_o and the width of slot aperture, *b*. For far-fields gain estimation, the C_0 in (22) is simply expressed as (Selvan, 1998):

$$C_o \simeq \frac{8}{\pi^2\sqrt{2}}\left(\frac{1}{\tilde{\beta}} - 1\right) \tag{23}$$

Equation (23) can also be interpreted as:

$$C_o \simeq \frac{3}{4\pi}\sqrt{\frac{\lambda}{2b}}\left(\frac{\sin\left(\theta/2\right)}{\cos\theta}\right) \tag{24}$$

where $\theta = \cos^{-1}\left(\tilde{\beta}\right)$.

Directivity

In Collin, (1985), the directivity, *D* expression of the open-ended rectangular slot antennas is given as:

$$D \simeq \frac{64ab}{\lambda_o^3\sqrt{k^2 - \left(\pi/b\right)^2}} \tag{25}$$

Rectangular Aperture Waveguide

The normalized input admittance model of the rectangular aperture waveguide as shown in Figure 11 was derived as:

$$
\begin{aligned}
\tilde{Y}_{in} &= \Re e\left(Y/Y_o\right) + j\Im m\left(Y/Y_o\right) \\
&= \frac{j8b}{k_1 a}\int_0^a \int_0^b \left(a-x\right)\left[D_1\left(b-y\right)\cos\left(\frac{\pi y}{b}\right) + D_2 \sin\left(\frac{\pi y}{b}\right)\right]\frac{e^{-jk_2\sqrt{x^2+y^2}}}{\sqrt{x^2+y^2}}dxdy
\end{aligned}
\tag{26}
$$

The D_1 and D_2 parameters in (26) are given as:

$$
D_1 = \frac{1}{b^2}\left[\frac{k_2^2}{4\pi} - \frac{\pi}{4b^2}\right] \tag{27a}
$$

$$
D_2 = \frac{1}{\pi b}\left[\frac{k_2^2}{4\pi} + \frac{\pi}{4b^2}\right] \tag{27b}
$$

where $k_1 = k_o$ and $k_2 = k_o\sqrt{\varepsilon_r}$ are the propagation constants for rectangular waveguide line, respectively. Symbol ε_r is the relative permittivity for the external medium (outer region).

Rectangular Aperture Calibration

The Open-Short-Load (OSL) technique is rarely used in the one-port rectangular waveguide calibration due to the unavailability commercial open-kit for the rectangular waveguide. In this subsection, the Short-Offset-Offset short (SOO) calibration (Da Silva and McPhun, 1978; Chang *et al.*, 1997) is introduced for waveguide calibration by using waveguide adjustable sliding shorts as shown in Figure 12. The calibration procedures are shown in Figure 13.

In this calibration method, the measured reflection coefficients for one shorted aperture and two different lengths, l of offset short are required. Let S_{11m_1}, S_{11m_2} and S_{11m_3} represent the known measured reflection coefficients at plane *AA'* for the shorted aperture and the two offset shorts at location l_1 and

Figure 11. Rectangular aperture waveguide

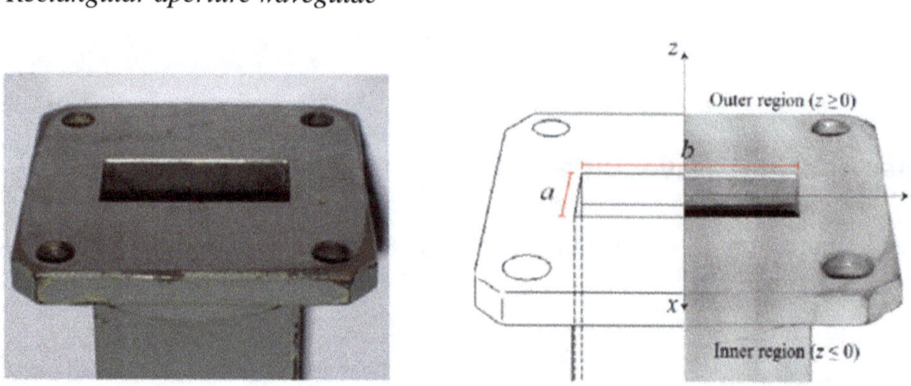

Figure 12. (a) Ku-band and X-band waveguide adjustable sliding shorter; (b) Connection between sliding short with waveguide-to-coaxial adapter

(a) (b)

l_2 from the waveguide aperture, respectively. Before calibration, the selection of the appropriate offset short length, l_1 and l_2 will be an issue. The lengths of the offset shorts can be determined by conditions:

1. The three phase-shift between the S_{11m_1}, S_{11m_2} and S_{11m_3} must not be equal:

$$\angle\left(S_{11m\_2}\right)-\angle\left(S_{11m\_1}\right)\neq\angle\left(S_{11m\_3}\right)-\angle\left(S_{11m\_1}\right)\neq\angle\left(S_{11m\_3}\right)-\angle\left(S_{11m\_2}\right)$$

2. The resolution degree between any three phase-shift must be significant large (> 100°) as shown in Figure 14. In this work, the distance l_1 and l_2 for the offset shorts from the X-band waveguide aperture are equal to 0.007 m and 0.013 m, respectively. In other words, the distance, l_1 and l_2 must be obeyed:

$$l_1 \simeq \frac{\lambda_g}{6} \text{ and } l_2 \simeq \frac{\lambda_g}{3}$$

where λ_g is the wavelength of waveguide (wavelength of the center frequency), which is given in (19).

Once the S_{11m_1}, S_{11m_2}, S_{11m_3}, l_1, and l_2 are obtained, the three unknown complex coefficients (e_{00}, e_{11}, and $e_{10}e_{01}$) values in (14) can be found as:

$$e_{00}=\frac{S_{11m\_1}S_{11m\_2}\left(e^{-2\gamma l_1}-1\right)-S_{11m\_2}S_{11m\_3}\left(e^{2\gamma(l_2-l_1)}-1\right)-S_{11m\_1}S_{11m\_3}\left(e^{-2\gamma l_1}-e^{2\gamma(l_2-l_1)}\right)}{\left(e^{-2\gamma l_1}-1\right)\left(S_{11m\_2}-S_{11m\_3}\right)-\left(e^{2\gamma(l_2-l_1)}-1\right)\left(S_{11m\_2}-S_{11m\_1}\right)} \tag{28a}$$

$$e_{11}=\frac{e^{2\gamma l_1}\left(S_{11m\_2}-e_{00}\right)+e_{00}-S_{11m\_1}}{S_{11m\_1}-S_{11m\_2}} \tag{28b}$$

Figure 13. Calibration procedure of the aperture slot waveguide using three short standards

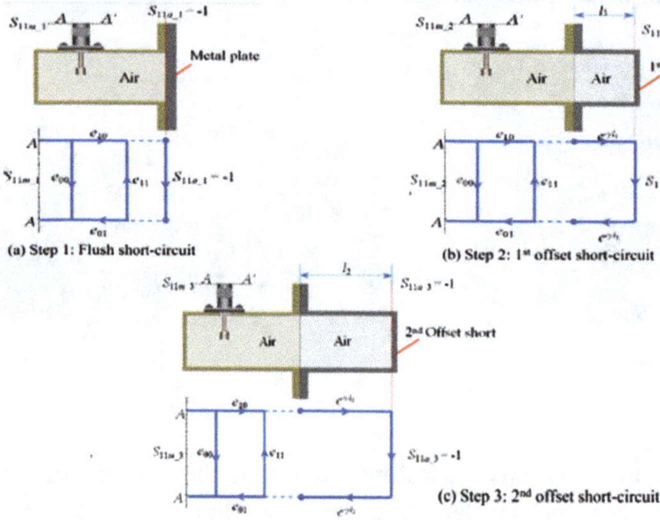

$$e_{10}e_{01} = \left(e_{00} - S_{11m\_1}\right)\left(1 + e_{11}\right) \tag{28c}$$

The complex reflection coefficient, S_{11a_medium}, at the waveguide aperture which is open to the air was measured. Then, the measured S_{11a_medium} was converted to normalized admittance, Y/Y_o parameter by a formula: $Y/Y_o = (1 - S_{11a\_medium})/(1 + S_{11a\_medium})$. The SOO calibration techniques were validated by comparing normalized admittance, Y/Y_o with the literature data (Chang, Chen, & Qian, 1997; Kim, Enkhbayar, Bang, & Ahn, 2010; Hirohide, Serizawa, & Hongo, 2005; Yoshitomi, & Sharobim, 1994; Bodnar & Paris, 1970; Baudrand, Tao, & Atechian, 1988; Compton, 1963) as shown in Figure 15. The real part, $\Re$e(Y/Y_o) and the imaginary part, $\Im$&(Y/Y_o) of admittance results were found to be in good agreement with literature data over the operational range of frequencies.

WAVEGUIDE-FED SLOT ANTENNAS

The waveguide-fed slot antenna (so-called slotted waveguide antenna) is similar to normal rectangular waveguide, except the waveguide has a series of opening slots on the wall surface to allow a small amount of RF signal to radiate out into the surrounding environment as shown in Figure 16.

In waveguide-fed slot antenna design, the foundation of knowledge regarding electromagnetic wave propagation (TE_{10} mode wave) in the rectangular waveguide is indispensable in order to determine the appropriate slot position and layout in the waveguide surface so that the radiate signal properly excited (Gilbert, 2007). Opening slots that are parallel to waveguide wall currents do not radiate. However, when a slot is cut into a waveguide wall and it interrupts the flow of current (narrow slot perpendicular to flow of current), then RF wave is allow to radiate out into the surrounding environment as show in Figure 17.

Figure 14. The three phase-shift of the measured reflection coefficients for the shorted aperture and two offset shorts with l_1 = 0.7 cm and l_2 = 1.3 cm, respectively

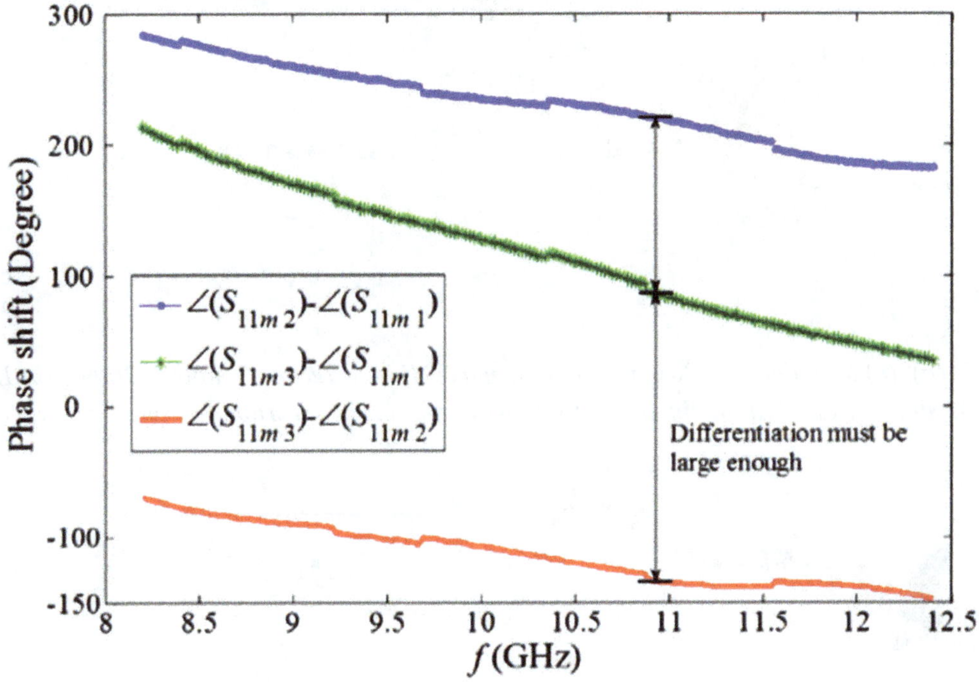

Figure 15. Comparison of real part, Re(Y/Y_o), and imaginary part, I&(Y/Y_o), of the normalized admittance for air

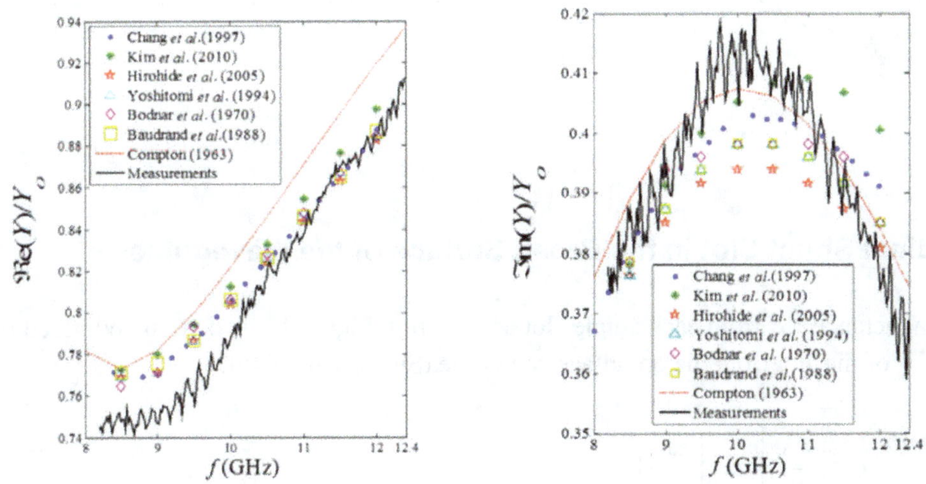

Figure 16. Slots cut in the walls of a rectangular waveguide

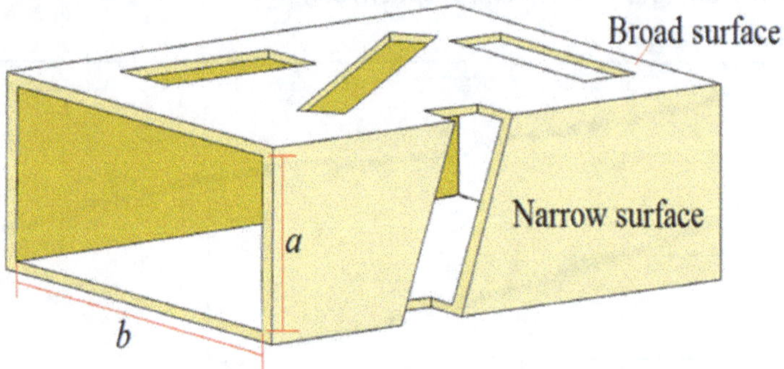

Figure 17. (a), (c) Manufactured X-band slot antennas (Miwave Techno Ventures, 2009); (b) Longitudinal shunt slot array and (d) inclined shunt slot array radiate by interrupting transverse wall currents

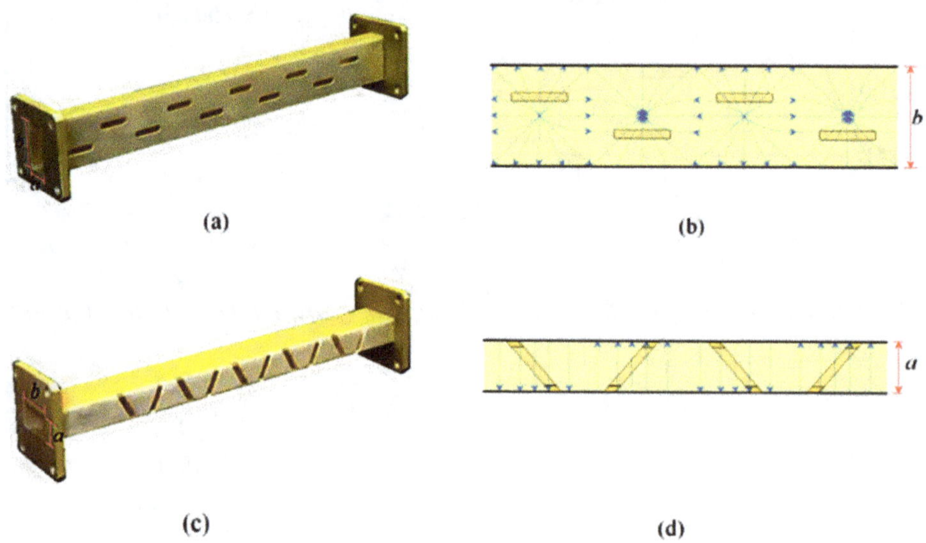

Longitudinal Shunt Slot in the Broad Surface of the Waveguide

Excited conductance, G_n from each single slot as shown in Figure 18, is determined based on the displacement, x_n of the slot from the broadface centerline (Stevenson, 1948):

$$G_n = \frac{2.09\lambda_g b}{\lambda_o a}\cos^2\left(\frac{\pi\lambda_o}{2\lambda_g}\right)\sin^2\left(\frac{\pi x_n}{b}\right) \tag{29}$$

where subscript n in (29) represent the order number of the slot.

Based on (29), the displacement, x_n of the nth slot can be predicted as:

Figure 18. (a) Longitudinal shunt slot in the broad surface of the rectangular waveguide; (b) Equivalent circuit of single longitudinal slot

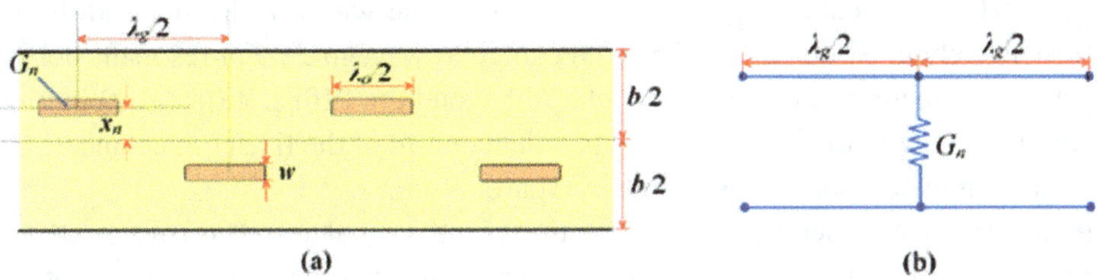

$$x_n = \frac{b}{\pi}\arcsin\left\{\sqrt{\frac{\lambda_o a G_n}{2.09\lambda_g b\cos^2\left(\frac{\pi\lambda_o}{2\lambda_g}\right)}}\right\} \tag{30}$$

where the G_n in (30) can be obtained using trial equation (31) (Collin, 1985):

$$G_n = Ka_n^2 \tag{31}$$

Symbols K and a_n in (31) are the unknown constant and relative excitation amplitudes of the nth slot, respectively. The value of K is determined using Equation (32) and known excitation amplitudes (a_1, a_2, … a_N) of each slot.

$$\sum_{n=1}^{N} G_n = K\sum_{n=1}^{N} a_n^2 = 1 \tag{32}$$

where N is the number of slots. In general, lower side-lobe level (SLLs), broader beam width, and smaller directivity of the slot antenna can be achieved by non-uniform slots displacements, which all slots are at the different distance from the centerline (Collin, 1985; Misilmani, Husseini, & Kabalan, 2015).

For instance, let consider the design of a five-slot resonant array using X-band rectangular waveguide (width, $b = 0.02286$ m and height, $a = 0.01016$ m) with triangular amplitude distribution along the array (relative excitation amplitude levels: $a_1 = 1$, $a_2 = 2$, $a_3 = 3$, $a_4 = 2$, and $a_5 = 1$) as shown in Figure 19 (a). The mostly used of optimized slot width, w in the literature for the X-band waveguide slot is 0.00159 m. Hence, the slot width, w for other operating band of used rectangular waveguide can be determined by referring to the value of 0.00159 m for X-band waveguide as (Misilmani, Husseini, & Kabalan, 2015):

$$\frac{w}{b} = \frac{w_{X-Band}}{b_{X-Band}} \tag{33}$$

and

$$w = 0.0695538 \times b \tag{34}$$

where $w_{X\text{-}Band}$ (= 0.00159 m) and $b_{X\text{-}Band}$ (= 0.02286 m) are the width of the slot and the width of the waveguide, respectively in particular for X-band rectangular waveguide. For resonant slot array design, the short circuit is normally placed a distance of $\lambda_g/4$ beyond the fifth slot (last slot) in order to create an open circuit in shunt with the last slot. The equivalent circuit of the five conductances, G_n connected across a waveguide transmission line is shown in Figure 19 (b).

The normalized input conductance, G_{in} at excitation plane is equal to the sum of all slot conductances, G_n which value is unity at the resonate (or center) frequency (release all input power at center frequency):

$$G_{in} = \sum_{n=1}^{N} G_n = 1 \tag{35}$$

Thus, the reactive field (reactance, X and susceptance, B) of the slot is assumed to be zero at resonant condition. Based on Equations (32) and (35), the value of K can be determined:

$$K\left(1^2 + 2^2 + 3^2 + 2^2 + 1^2\right) = 1 \tag{36}$$

From Equation (36), $K = 0.0526316$. Based on (30) and (31), the slot array with triangular excitation amplitudes distribution will cause the non-uniform slots displacements, x_n from the broadface centerline. The calculated normalized conductances, G_n of each slot [using (31)] and the corresponding slots displacement, x_n [using (30)] are listed in Table 5. It should be noted that slot array design using analytical calculation do not take into account mutual coupling between slots. The slotted waveguide antenna has been simulated using X-band rectangular waveguide and the predicted dimension slots are obtained from Table 5. The simulated far-field radiation patterns of the slotted antenna are shown in Figure 20.

In fact, the performance of slotted waveguide antenna can be optimized using Taylor polynomial series or Chebyshev series that represents the array function of the open slots on the waveguide. By using these series of functions (discrete Taylor distribution instead of triangular distribution), the performance

Figure 19. (a) Longitudinal slot resonant array; (b) Equivalent circuit of the slot array

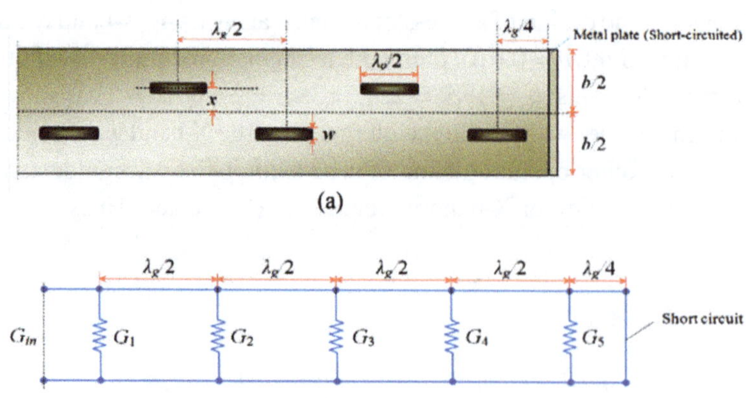

Table 5. List of slot conductances, G_n values and corresponding slots displacements, x_n at 10 GHz

n	a_n	G_n	x_n (m)
1	$a_1 = 1$	$G_1 = 0.0526316$	$x_1 = 0.00180$
2	$a_2 = 2$	$G_2 = 0.2105263$	$x_2 = 0.00372$
3	$a_3 = 3$	$G_3 = 0.4736842$	$x_3 = 0.00600$
4	$a_4 = 2$	$G_4 = 0.2105263$	$x_4 = 0.00372$
5	$a_5 = 1$	$G_5 = 0.0526316$	$x_5 = 0.00180$

Figure 20. Simulated far-field radiation patterns at 10 GHz: (a) Waveguide X-Z cross-section plane; (b) polar plot (at azimuth angle, $\phi = 0^o$); (c) waveguide Z-Y side-section plane, and; (d) 3D view with 12.6 dBi of gain

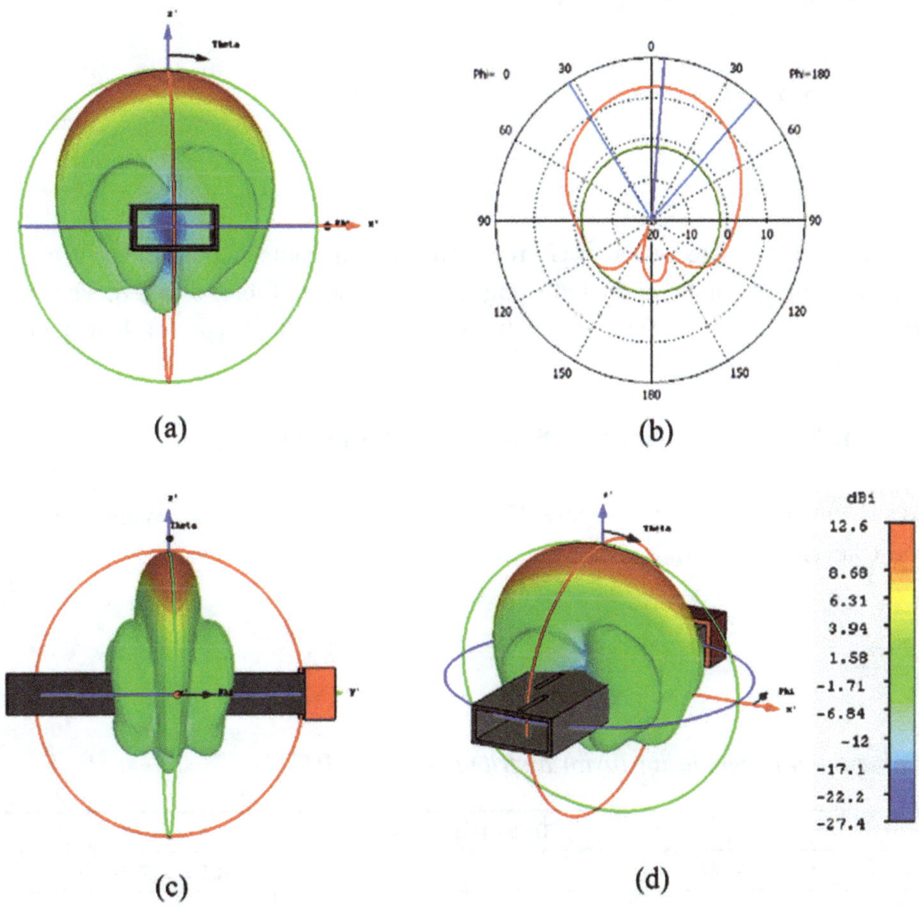

of the antenna, such as gain, side lobe level (SLL), and the number of side lobes can be initially determined. The outputs of the series calculation are the slots' individual excitations, a_n, as listed in Table 6. The detail of discrete Taylor and Chebyshev distribution techniques can be found in Hansen (1992) and Misilmani *et al.* (2015).

Table 6. Comparison of amplitude excitations for five element array

Amplitude Excitation for Each Slot	Triangular Distribution	Taylor Distribution			Chebyshev Distribution		
		Side Lobe Level (SLL)			Side Lobe Level (SLL)		
		-25 dB	-30 dB	-35 dB	-25 dB	-30 dB	-35 dB
a_1	1	0.6234	0.5180657	0.439835	0.3925014	0.3185018	0.2718466
a_2	2	1.159043	1.202881	1.229535	0.7974675	0.7683221	0.744875
a_3	3	1.435113	1.558107	1.66126	1	1	1
a_4	2	1.159043	1.202881	1.229535	0.7974675	0.7683221	0.744875
a_5	1	0.6234	0.5180657	0.439835	0.3925014	0.3185018	0.2718466

Once the amplitude excitation of each slot, a_n is obtained, the evaluation of the normalized conductances, G_n can be done using Equation (37).

$$G_n = \frac{a_n^2}{\sum_{n=1}^{N} a_n^2} \quad n = 1, 2, 3, 4, \text{and } 5 \tag{37}$$

Similarly, from the calculated value of G_n, the corresponding slots displacement, x_n based on Taylor or Chebyshev distribution can be predicted using (30) as listed in Tables 7 and 8. The comparison of simulated return loss, $|S_{11}|$ and far-field gain pattern results using three types of distribution at 10 GHz are shown in Figure 21.

Inclined Shunt Slot in the Narrow Surface of the Waveguide

The inclined slot antenna is shown in Figure 22. Excited conductance, G_n from each single slot can be determined as (Stevenson, 1948):

Table 7. Resultant parameters using Taylor distribution at 10 GHz

Taylor Distribution					
SLL = -30 dB			SLL = -35 dB		
a_n	G_n	x_n (m)	a_n	G_n	x_n (m)
$a_1 = 0.5180657$	$G_1 = 0.045814$	$x_1 = 0.0016753$	$a_1 = 0.439835$	$G_1 = 0.031353$	$x_1 = 0.0013820$
$a_2 = 1.202881$	$G_2 = 0.24699$	$x_2 = 0.0040635$	$a_2 = 1.229535$	$G_2 = 0.24501$	$x_2 = 0.0040453$
$a_3 = 1.558107$	$G_3 = 0.41440$	$x_3 = 0.0055043$	$a_3 = 1.66126$	$G_3 = 0.44728$	$x_3 = 0.0057763$
$a_4 = 1.202881$	$G_4 = 0.24699$	$x_4 = 0.0040635$	$a_4 = 1.229535$	$G_4 = 0.24501$	$x_4 = 0.0040453$
$a_5 = 0.5180657$	$G_5 = 0.045814$	$x_5 = 0.0016753$	$a_5 = 0.439835$	$G_5 = 0.031353$	$x_5 = 0.0013820$

Table 8. Resultant parameters using Chebyshev distribution at 10 GHz

Chebyshev Distribution					
SLL = -30 dB			SLL = -35 dB		
a_n	G_n	x_n (m)	a_n	G_n	x_n (m)
$a_1 = 0.3185018$	$G_1 = 0.04256$	$x_1 = 0.0016137$	$a_1 = 0.2718466$	$G_1 = 0.032736$	$x_1 = 0.0014125$
$a_2 = 0.7683221$	$G_2 = 0.24767$	$x_2 = 0.0040698$	$a_2 = 0.744875$	$G_2 = 0.24578$	$x_2 = 0.0040524$
$a_3 = 1$	$G_3 = 0.41955$	$x_3 = 0.0055469$	$a_3 = 1$	$G_3 = 0.44297$	$x_3 = 0.0057407$
$a_4 = 0.7683221$	$G_4 = 0.24767$	$x_4 = 0.0040698$	$a_4 = 0.744875$	$G_4 = 0.24578$	$x_4 = 0.0040524$
$a_5 = 0.3185018$	$G_5 = 0.04256$	$x_5 = 0.0016137$	$a_5 = 0.2718466$	$G_5 = 0.032736$	$x_5 = 0.0014125$

Figure 21. (a) Simulated return loss, $|S_{11}|$ versus operating frequency.(b) Far-field gain pattern in the altitude plane (at azimuth angle, $\phi = 0°$) at 10 GHz

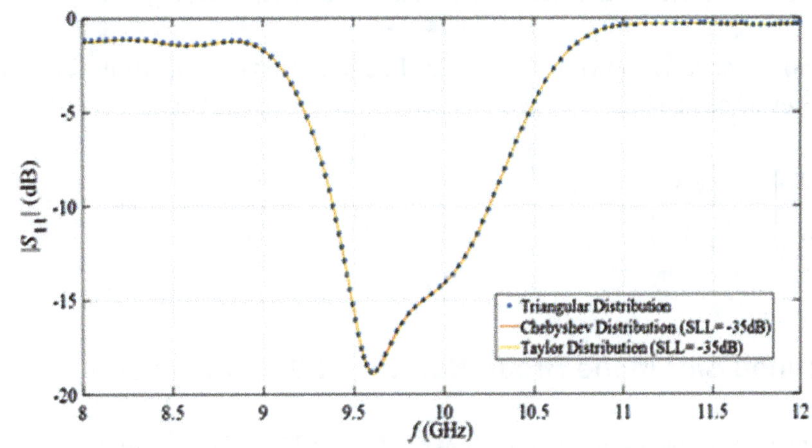

(a)

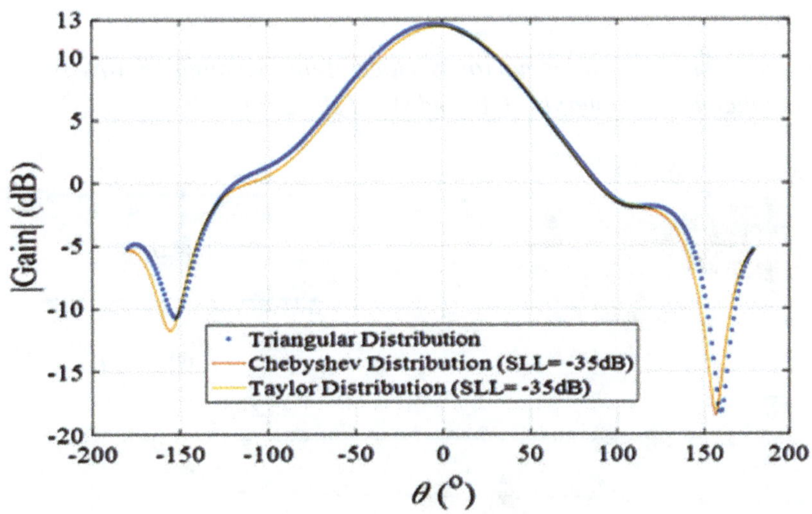

(b)

$$G_n = \frac{30\lambda_g}{73\pi a}\left(\frac{\lambda_o}{b}\right)^3 \left\{\frac{\sin\theta\cos\left[\left(\pi\lambda_o/2\lambda_g\right)\sin\theta\right]}{1-\left(\lambda_o/\lambda_g\right)^2\sin^2\theta}\right\}^2 \tag{38}$$

where θ is the slot inclination angle. The depth of cut d of a slot is given as (Das, Janaswamy Ramakrishna, & Sarap, 1984):

$$d = \begin{cases} 0.21325\lambda_o - \dfrac{a}{2\cos\theta} & for\ \theta < 46.6^\circ \\ 0 & for\ \theta \geq 46.6^\circ \end{cases} \tag{39}$$

Transverse Series Slot in the Broad Surface of the Waveguide

The transverse slot antenna is shown in Figure 23. Excited resistance, R_n from each single slot can be determined as (Stevenson, 1948):

$$R_n = 0.523\left(\frac{\lambda_g^3\lambda_o}{ab}\right)\cos^2\left(\frac{\pi\lambda_o}{4b}\right)\cos^2\left(\frac{\pi x_n}{b}\right) \tag{40}$$

Centered Inclined Slot in the Broad Surface of the Waveguide

Excited resistance, R_n from each single slot as shown in Figure 24 can be determined as (Stevenson, 1948):

Figure 22. (a) Inclined shunt slot in the narrow surface of the rectangular waveguide; (b) Equivalent circuit of single longitudinal slot, and; (c) Inclined shunt slot array

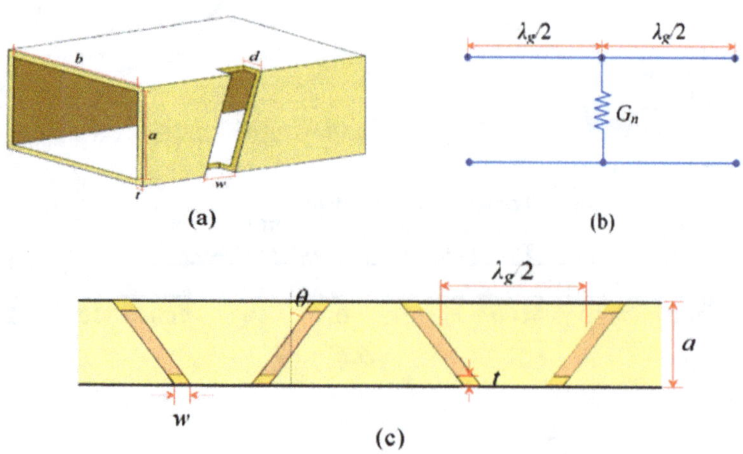

Figure 23. (a) Transverse slot in the broad surface of the rectangular waveguide; (b) Equivalent circuit of single transverse slot

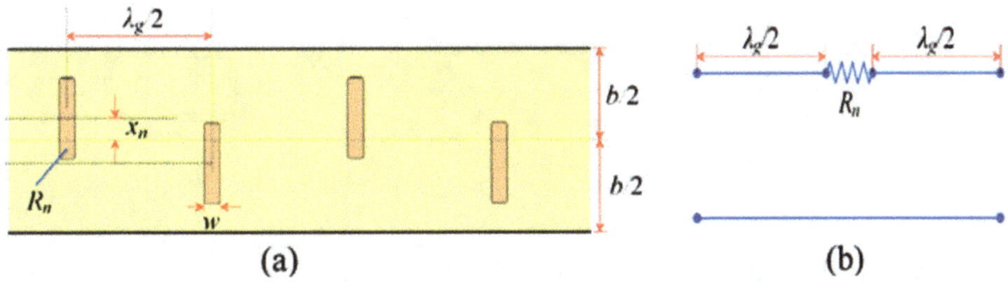

$$R_n = 0.131\left(\frac{\lambda_o^3}{\lambda_g ab}\right)\left\{\left[\frac{\cos\left(\pi\xi/2\right)}{1-\xi^2}+\frac{\cos\left(\pi\kappa/2\right)}{1-\kappa^2}\right]\sin\theta+\frac{\lambda_g}{2b}\left[\frac{\cos\left(\pi\xi/2\right)}{1-\xi^2}-\frac{\cos\left(\pi\kappa/2\right)}{1-\kappa^2}\right]\cos\theta\right\}^2 \tag{41}$$

The parameters ξ and κ in (41) are given as:

$$\xi = \frac{\lambda_o}{\lambda_g}\cos\theta - \frac{\lambda_o}{2b}\sin\theta \;\; \kappa = \frac{\lambda_o}{\lambda_g}\cos\theta + \frac{\lambda_o}{2b}\sin\theta$$

Leaky-Wave Rectangular Waveguide

Leaky rectangular waveguides have simple structure, narrow beamwidth, and frequency scanning capability (in Figure 25), which the angle, ϕ_m of maximum radiation can be approximately determined by:

$$\sin\phi_m = \frac{\beta}{k_o} = \frac{\sqrt{k_o^2-\left(\pi/b\right)^2}}{k_o} \tag{42}$$

Figure 24. (a) Centered inclined slot in the broad surface of the rectangular waveguide. (b) Equivalent circuit of single centered inclined slot

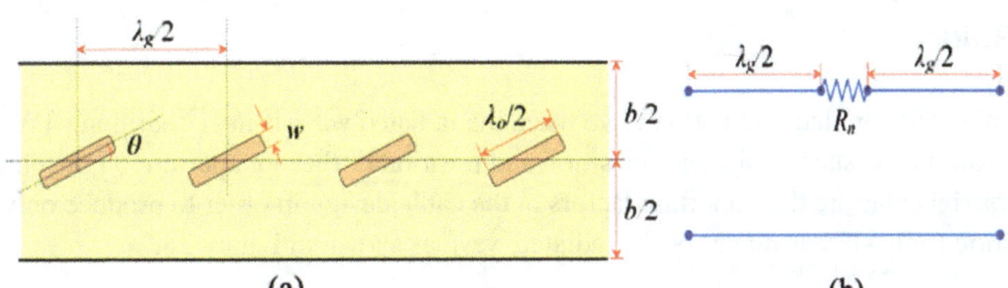

Figure 25. Leaky-wave slotted waveguide

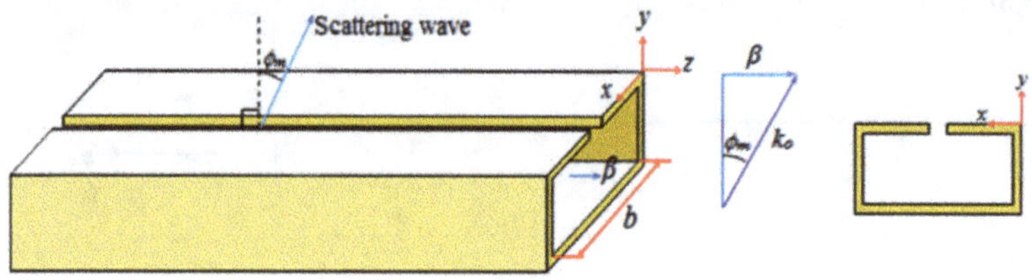

The leaky-wave slotted antennas are implemented in communication-based train control (CBTC) system (Heddebaut, 2009), such as automated subway system to determine the precise position of the train when it is being brought to a halt at a station. Various structure of the leaky-wave antennas, equivalent circuits, and its formulas have been listed in the Table 9.

Besides, there are a lot of slotted rectangular waveguide antennas were studied and proposed in order to improved its gain, power handling capacity, return loss, and operating bandwidth as shown in Figure 26 (Ajioka, Joe, & McFarland, 1984; Lee, Harackiewicz, Jung, & Park, 2005; Tyagi, Mevada, Chakrabarty, & Jyoti, 2017).

LEAKY-WAVE COAXIAL WAVEGUIDE

The coaxial-fed slot antenna (so-called leaky coaxial antenna) is similar to normal coaxial cable as shown in Figure 27 (a), except the antenna has opening slots on the outer conductor surface to allow a small amount of RF signal to radiate out into the surrounding environment from the cable as shown in Figure 27 (b). The dielectric used in the cable is usually the foamed polyethylene with relative permittivity, $\varepsilon_c \approx 1.25$.

In the late 1970's, the leaky coaxial cable has been widely used as an antenna in underground tunnels, mines, subway or underground parking spaces, due to the weak propagated radio signal in underground in which the signal is screened by concrete walls or building. Recently, this kind of radiator has been implemented in indoor building (underground shopping arcades, train stations, long hotel corridors and working office), especially in the blind zone area where the discrete antennas fail to achieve. However, conventional antennas have several issues when it is used to provide wireless access in those areas. Besides, blind zones of the wireless signal will occur, which are out of the antenna coverage areas. A number of antenna units have to be installed every several meter distance in the areas. Recently, the leaky coaxial cables are used to minimize the blind zones and provide a wider wireless access area with one single base station unit and one single frequency channel.

Design Guide

The leaky cable only produces radiating wave when the radiated wave is in -1th harmonic (Wang & Mei, 2001). The parameters such as size, composition and position of the slot aperture on the outer conductor of the coaxial cable are the important factors in the cable design in order to produce only the single mode radiation (only -1th harmonic) of the radiated wave as shown in Figure 28.

Table 9. Antenna Structures and Formulations (Goldstone and Olinert, 1958)

Antenna Structure	Equivalent Circuit	Normalized Admittance
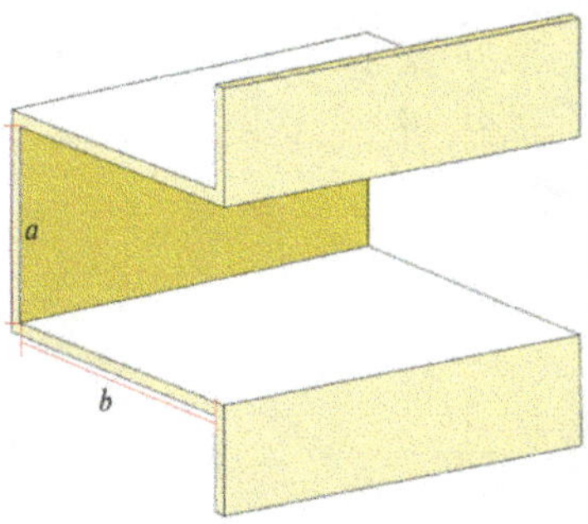		$\frac{G}{Y_o} = \frac{a}{4b}\left[\pi - \frac{a}{b}\ln\left(\frac{3.0522b}{a}\right)\right]$ $\frac{B}{Y_o} = \frac{\kappa_o a}{\pi}\ln\left(\frac{1.5261\pi}{\kappa_o a}\right)$ where $\kappa_o = \frac{\pi}{2b} - \frac{a}{2b^2}\ln\left(\frac{1.5261b}{a}\right)$
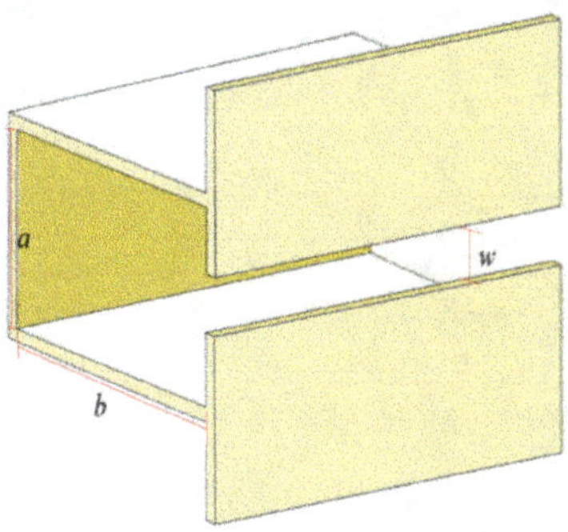		$\frac{G}{Y_o} = \frac{\pi a}{2b}$ $\frac{B}{Y_o} = \frac{a}{b}\ln\left(\frac{1.5261b}{w\sin\left(\pi w/2a\right)}\right)$
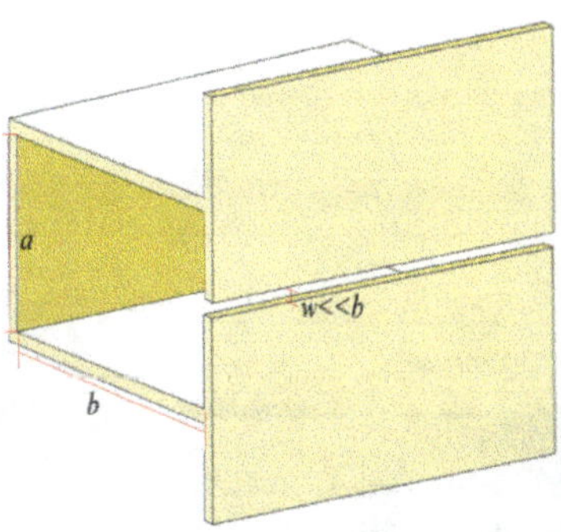		$\frac{G}{Y_o} = 0.57\left\{\frac{\pi^2 a\left[1-\left(w/a\right)^2\right]^2}{16w\cos^2\left(\pi w/2a\right)}\right\}\left(\frac{b}{w}\right)$ $\frac{B}{Y_o} = \left\{\frac{\pi}{\kappa_o a}\cot^2\left(\frac{\pi w}{2a}\right) + 0.312\left\{\frac{\pi^2 a\left(1-\left(w/a\right)^2\right)^2}{16w\cos^2\left(\pi w/2a\right)}\right\}\left(\frac{b}{w}\right)\right\}$
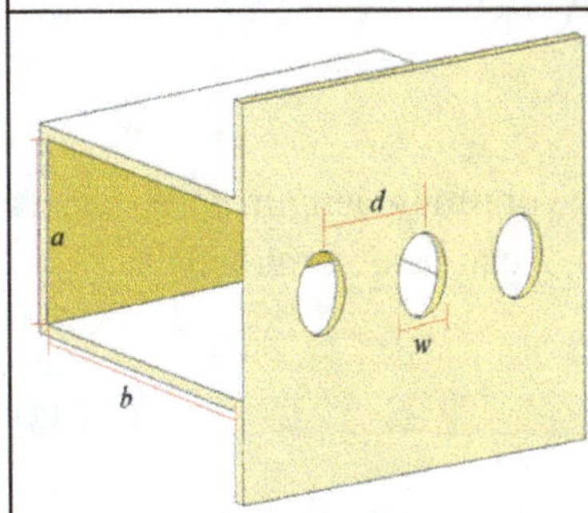		$\frac{G}{Y_o} = \frac{\pi a}{2b}$ $\frac{B}{Y_o} = \frac{6a\,b\,d}{\pi w^3}$

Figure 26. Various proposed rectangular waveguide slot array antennas
(Ajioka, Joe, & McFarland, 1984; Lee, Harackiewicz, Jung, & Park, 2005; Tyagi, Mevada, Chakrabarty, & Jyoti, 2017)

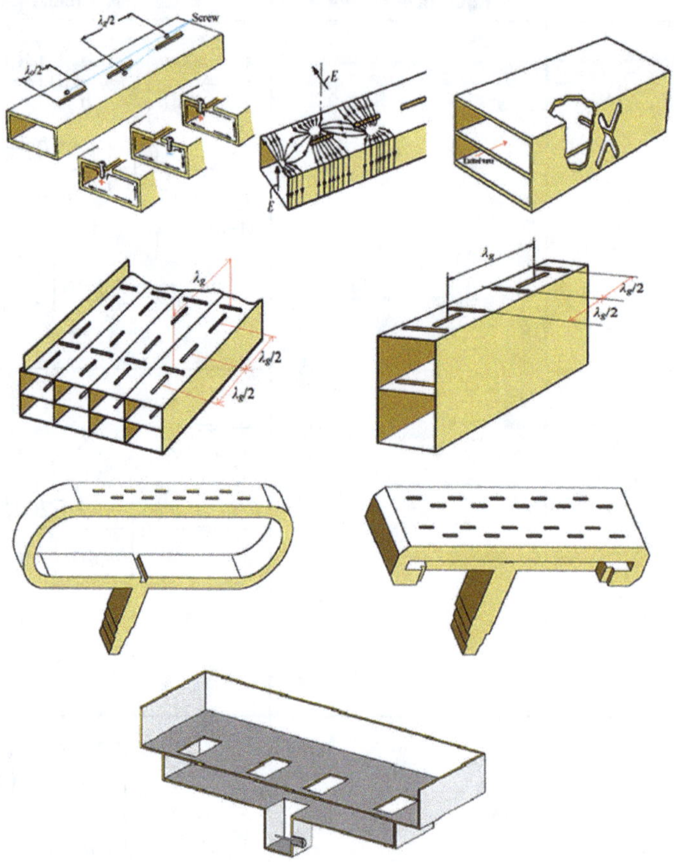

Figure 27. (a) Normal coaxial cable; (b) Leaky coaxial cable

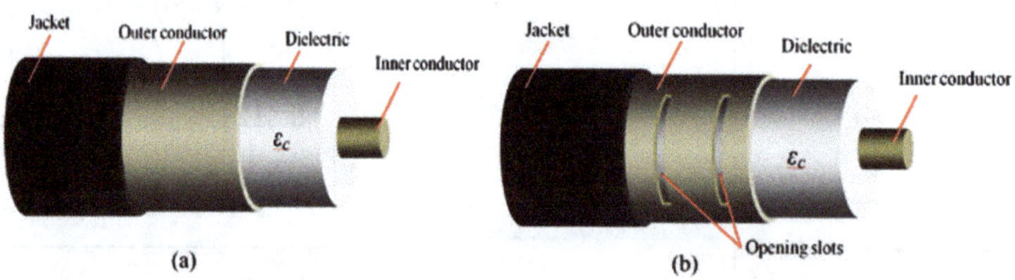

For general leaky cable design, Delogne *et al.* (1980) proposed that the optimum length, d_o (in meter) of each leaky section (in Figure 28) related to the operating wavelength, λ can be determined as:

$$d_o = \frac{\lambda_o}{2\left(\sqrt{\varepsilon_c} - 1\right)} \tag{43}$$

Figure 28. (a) Longitudinal and; (b) zigzag slot aperture of leaky cable

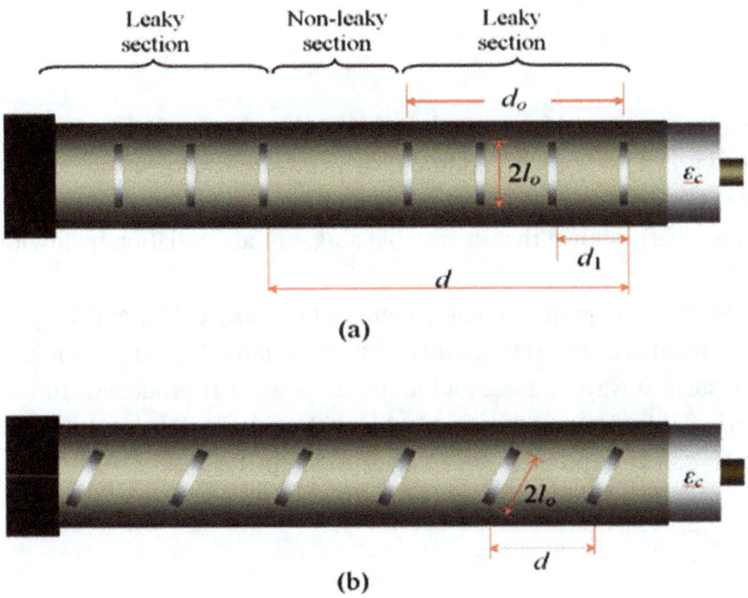

On the other hand, a resonant frequency, f_o within the single mode radiation band is give as (Wang & Mei, 2001):

$$f_o = \frac{c}{d\sqrt{\varepsilon_c}} \text{ (in Hz)} \tag{44}$$

where d is the period distance of slot aperture and c is the velocity of light in free space, respectively. Recently, Li *et al.* (2012) has empirically expressed the resonant frequency, f_o (in GHz) of the leaky cable related to its length of slot aperture, l_o (in meter) as:

$$f_o = \frac{0.0428346}{l_o} + 0.50853 \text{ (in GHz)} \tag{45}$$

When the operating frequency, f increased, higher mode waves (-2th, -3th harmonics and etc) will emerge and mix in the radiated wave. The multiple modes radiated wave may interfere the stability and uniformity of the transmitted electrical fields. The condition for radiation from the mth harmonic in the radial direction is given as (Wang & Mei, 2001):

$$-mf_1 < f < -mf_2 \tag{46}$$

where:

$$f_1 = \frac{c}{d\left(\sqrt{\varepsilon_c} + 1\right)} \text{ and } f_2 = \frac{c}{d\left(\sqrt{\varepsilon_c} - 1\right)}$$

The subscript m is the order of higher-mode radiation harmonics. The operating frequency band of the harmonics is shown in Figure 29. In addition, the measured voltage standing wave ratio (VSWR) data versus normalized frequency, (f / f_o) for the leaky cable, obtained from Wang *et al.* (2001) is shown in Figure 30, which is clearly shown that its useable single mode radiation bandwidth of -1th harmonic is only within f_1 to $2f_1$.

In order to increase the operating frequency band of the leaky cable, normally, the slot aperture on the outer conductor is formed as an array group in a predetermined cycle as shown in Figure 31. For $f_1/f_o < 0.5$ region, only surface wave is exist and no radiated wave is produced from the leaky cable. As known that the electric field, E around the leaky cable with periodically slot aperture on outer conductor as shown in Figure 31, can be expressed as (Wang & Mei, 2001):

$$E = e^{-jk_c z} \sum_{m=-\infty}^{\infty} E_m e^{-j\frac{2m\pi}{d}z} \tag{47}$$

The E_m in (47) is assumed to be expressed as:

$$E_m = \sum_{n=1}^{N} a_n e^{-j\frac{2m\pi}{d}z_n} \tag{48}$$

where the amplitude, a_n in (48) is assumed to be sine function as (Yang, Feng, Zhang, Ren, & Li, 2006):

$$a_n = a_o \sin\left(\frac{2\pi z_n}{d}\right) \tag{49}$$

Figure 29. Operating frequency bands of the harmonics of conventional leaky coaxial cable

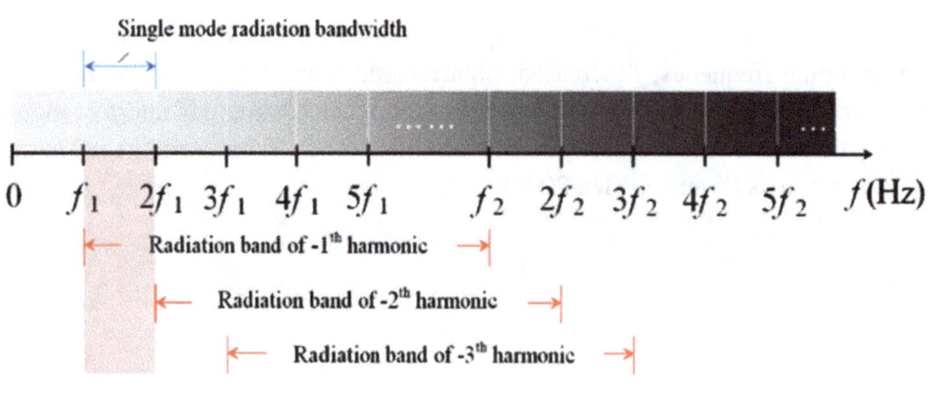

Figure 30. Voltage standing wave ratio (VSWR) of conventional leaky coaxial cable *(Wang & Mei, 2001)*

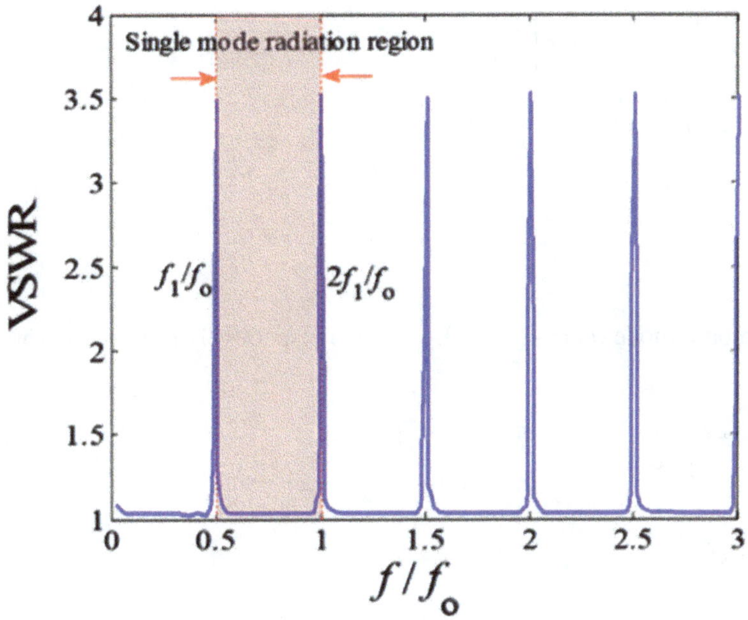

Figure 31. Configurations of leaky coaxial cable with array group of periodic zigzag slot aperture

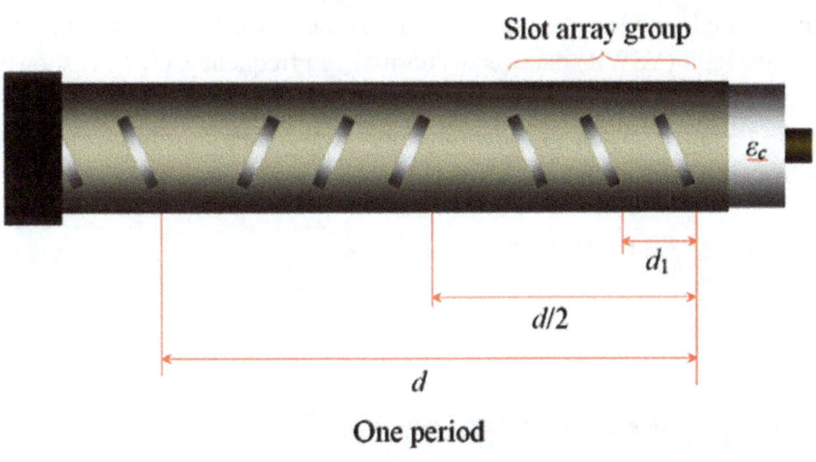

where N is the total number of the slots within one period. Subscript n is the order of slot aperture in the period. The z_n is distance between the nth slot (within one period) and the middle point of a pitch. For zigzag slot leaky cable with $N = 2$ (except the original one), $z_1 = -d/4$ and $z_2 = d/4$, as shown in Figure 32, (48) can be expressed as:

$$\begin{aligned} E_m &= \sum_{n=1}^{2} a_n e^{-j\frac{2m\pi}{d}z_n} \\ &= a_o \sin\left(\frac{2\pi z_1}{d}\right) e^{-j\frac{2m\pi}{d}z_1} + a_o \sin\left(\frac{2\pi z_2}{d}\right) e^{-j\frac{2m\pi}{d}z_2} \\ &= a_o \sin\left(\frac{2\pi}{d}\times\frac{-d}{4}\right) e^{-j\frac{2m\pi}{d}\times\frac{-d}{4}} + a_o \sin\left(\frac{2\pi}{d}\times\frac{d}{4}\right) e^{-j\frac{2m\pi}{d}\times\frac{d}{4}} \\ &= -a_o\left(e^{j\frac{m\pi}{2}} - e^{-j\frac{m\pi}{2}}\right) \end{aligned} \tag{50}$$

Then, each harmonic mode (m = -1, -2, -3, ..., ∞) is respectively replaced in (50), yields

$$\begin{aligned} E_{-1} &= -a_o\left(e^{-j\frac{\pi}{2}} - e^{j\frac{\pi}{2}}\right) \\ &= j2a_o \end{aligned};\quad \begin{aligned} E_{-2} &= -a_o\left(e^{-j\pi} - e^{j\pi}\right) \\ &= 0 \end{aligned};\quad \begin{aligned} E_{-3} &= -a_o\left(e^{-j\frac{3\pi}{2}} - e^{j\frac{3\pi}{2}}\right) \\ &= -j2a_o \end{aligned};$$

$$\begin{aligned} E_{-4} &= -a_o\left(e^{-j2\pi} - e^{j2\pi}\right) \\ &= 0 \end{aligned}$$

From the above calculating results, clearly, the zigzag slot leaky cable with $N = 2$ does not have the radiation wave of even harmonic mode (m = -2, -4, -6, …), but only the odd harmonic mode (m = -1, -5, -7, …). Figure 33 shows that the -2th harmonic is suppressed in the operating frequency band, in which the maximum single mode radiation frequency is increased from $2f_1$ to $3f_1$. The corresponding voltage standing wave ratio (VSWR) data versus normalized frequency, (f/f_o) is shown in Figure 34.

The single mode radiation bandwidth can be further improved using leaky cable with $N = 4$, $z_1 = -d/6$, $z_2 = d/6$, $z_3 = -d/3$ and $z_4 = d/3$, as shown in Figure 35. Using the similar calculation routine, the calculated result from (48) are given as:

$$E_{-1} = j3a_o;\ E_{-2} = 0;\ E_{-3} = 0;\ E_{-4} = 0;\ E_{-5} = -j3a_o$$

Figure 32. Leaky cable with one zigzag slot in a period

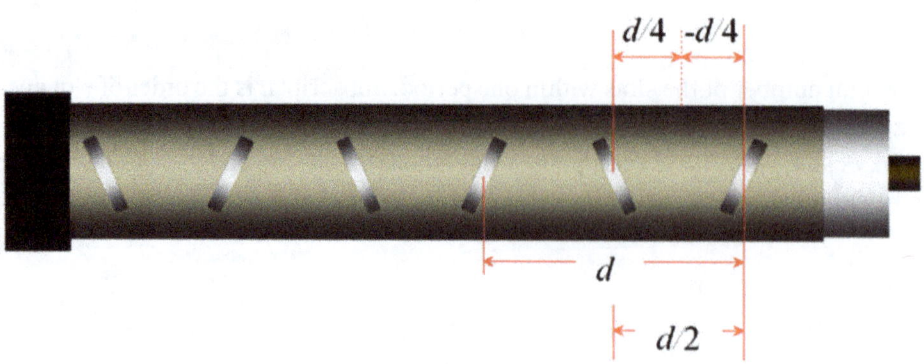

Figure 33. Operating frequency bands of the harmonics of leaky coaxial cable with one zigzag slot in a period

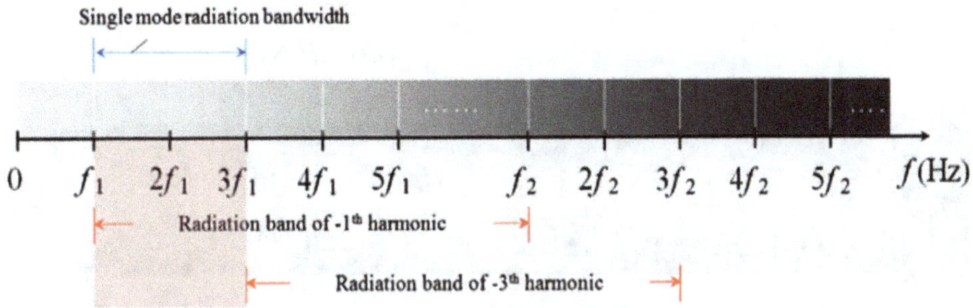

Figure 34. Voltage standing wave ratio (VSWR) of leaky coaxial cable with one zigzag slot in a period (Wang & Mei, 2001)

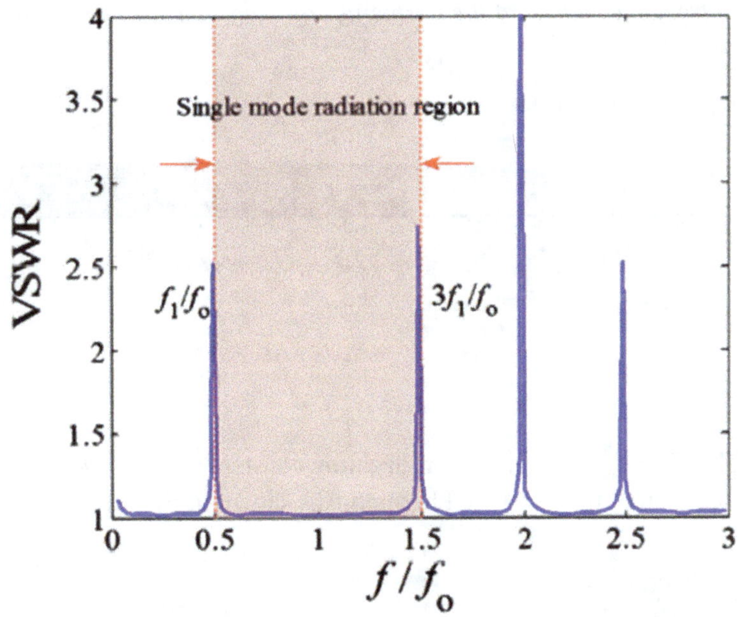

In this case, not only the even harmonic modes (m = -2, -4, -6, …) have been suppressed, but also m = -3. Hence, the limitation of single mode radiation bandwidth of -1th harmonic is expanded to $5f_1$ as shown in Figure 36.

The propagation constant of the mth spatial harmonic in the leaky coaxial cable is given as (Oliner & Jackson, 2007):

$$\beta_m d = \beta_o d + 2m\pi \; (m = -1, -2, \ldots) \tag{51}$$

where d is the period distance of the slot aperture on the leaky cable. On the other hand, the β_o $(= 2\pi\sqrt{\varepsilon_r}/\lambda_o)$ is the wave propagation constant in the coaxial cable (Neglect the slot aperture effects on the outer

Figure 35. Leaky cable with two zigzag slots in a period

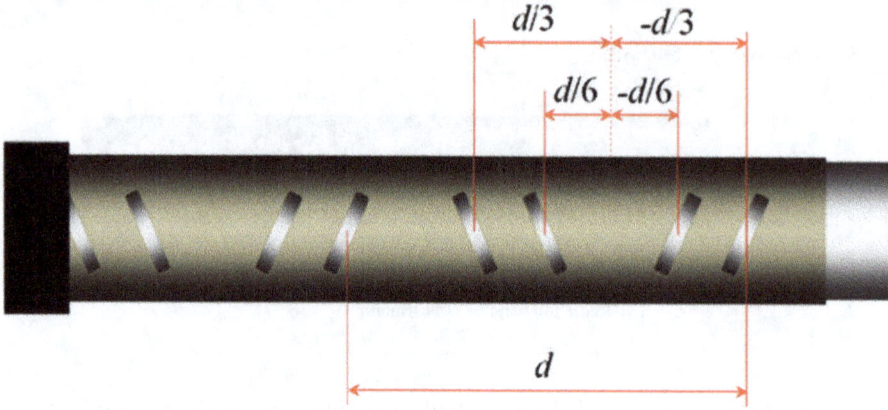

Figure 36. Operating frequency bands of the harmonics of leaky coaxial cable with two zigzag slots in a period

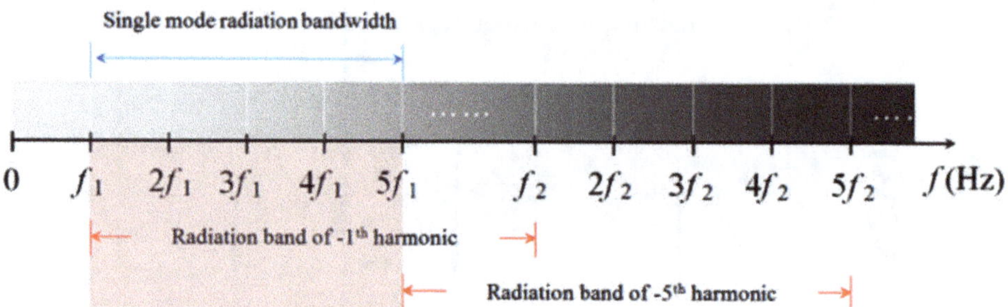

conductor of the cable). The k_o (= ω/c) is the propagation constant of free space. From (51), the angles, ϕ_m of the radiation direction of the mth spatial harmonic as shown in Figure 37, are determined by:

$$\begin{aligned} \sin\phi_m &= \frac{\beta_m d}{k_o d} = \frac{\beta_o d + 2m\pi}{k_o d} \\ &= \frac{\beta_o}{k_o} + \frac{m\lambda_o}{d} \\ &= \sqrt{\varepsilon_r} + \frac{m\lambda_o}{d} \end{aligned} \qquad (52)$$

For a practical radiator, only a single radiated beam is preferred (m= -1), Thus, the angles, ϕ_{-1} of the radiation direction of the -1th spatial harmonic is expressed as:

$$\phi_{-1} = \sin^{-1}\left(\sqrt{\varepsilon_r} - \frac{\lambda_o}{d}\right) \qquad (53)$$

Figure 37. Radiated wave from the slot aperture of the leaky coaxial cable

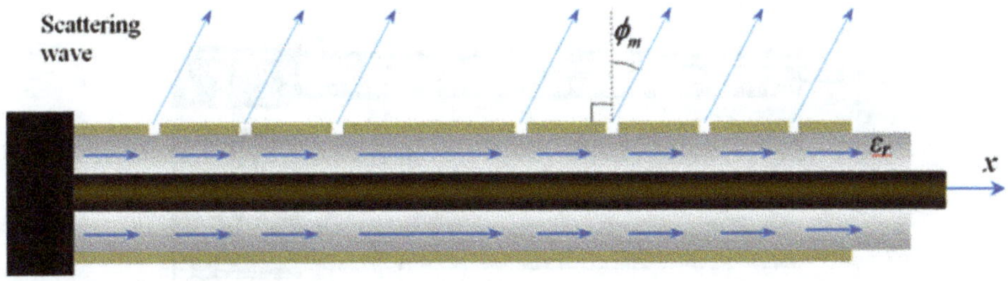

Various Kind of Coaxial-Fed Slot Antennas

In fact, an early use of leaky feeder cable was made by Halstead in 1940. Recently, in the market, most of the leaky coaxial cables are manufactured by RADIAX®. Now, more and more factories produce leaky coaxial cables exist in the market, such as IMEX, Libo, and JTL. There are various types of leaky cables for different applications as shown in Figure 38, and the corresponding features and its implementation are listed in Table 10. Each type of leaky cable has a different shape, size and the arrangement of opening aperture, which may influence the operating wavelength, coupling loss and beam-width of the radiated field (RADIAX®, 2013).

In fact, in research stage, there are a lot of slot array patterns on a coaxial cable were studied and proposed in order to improved its coupling efficiency and operating bandwidth as shown in Figure 39 (Kim, Yun, & Park, 1998; Wang & Mei, 2002; Li & Wang, 2010; Wang, Li, Zhang, & Chen, 2011).

CONCLUSION

This chapter has summarized some of the analytical analysis and calibration techniques of slot antennas. Although slot antenna have been developed since the 1940s, the slot antenna has been widely used and developed in recent years, especially in microstrip-fed slot antenna, which is low profile, relatively low cost, and easy to manufacture. Recent development in the slot antenna have been boosted by the recent developments in the Internet of things (IoT), 5th generation (5G) wireless technology, and fourth industrial revolution (Industry 4.0). In addition, computer numerical control (CNC) machining is increasingly common where it supplies high precision of slot machining in antenna design. At the same

Figure 38. Various kinds of leaky coaxial cables
(RADIAX®, 2013)

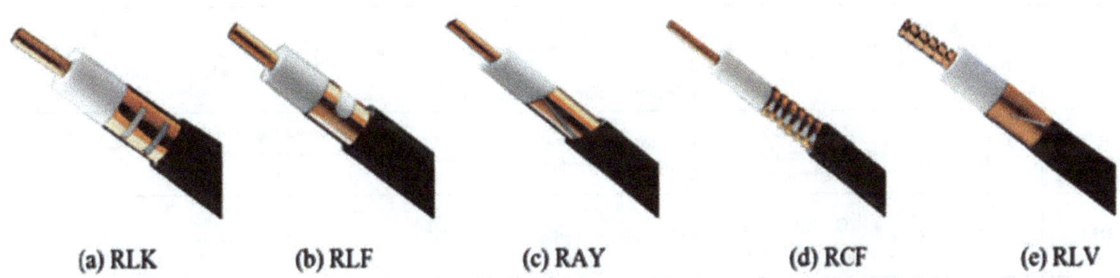

Figure 39. Various proposed slot array patterns
(Kim, Yun, & Park, 1998; Wang & Mei, 2002; Li & Wang, 2010; Wang, Li, Zhang, & Chen, 2011)

Table 10. Features and applications of leaky coaxial cable (RADIAX®, 2013)

Types	Operation Frequency, f	Slot Group Intervals	Features	Use
RLK	0.075 – 0.6 GHz 0.075 – 0.96 GHz 0.075 – 1.9 GHz 0.075 – 2.7 GHz 0.8 – 6 GHz	1 m	• Broadband. • Low coupling loss at 0.9 GHz, 1.8 GHz and 2.2 GHz.	• In-building • Tunnel for areas under 80 m
RLF	0.075 - 0.96 GHz 0.075 - 2.4 GHz	22 m	• Robust. • Low loss. • Broadband. • Negligible influence of dust, salt and moisture accumulation on cable loss.	• Single and multi band systems. • Long tunnels. • Aggressive environments.
RAY	0.075 – 0.96 GHz 0.075 – 2.4 GHz 0.075 – 2.7 GHz 2.4 – 2.5 GHz	≤ 50 cm	• Very low coupling loss. • Optimized for high frequencies and digital transmission. • Optimized for vertical antennas.	• In-building • Dry tunnels
RSF /RCF	0.075 – 2.6 GHz 0.075 – 6 GHz	Milled slots on corrugated outer conductor	• Robust. • Low bending radius.	• Heavy duty applications. • In-building • Mines
RLV	0.15 – 0.96 GHz 0.075 – 2.4 GHz 0.075 – 2.7 GHz	Graded groups of apertures	• Constant attenuation loss along the cable.	• Long tunnels

time, there is a cost-saving method for antenna fabrication, namely the 3D printing method with metal plating. Moreover, the rapid development of electronic design automation (EDA) in RF/microwave field, allowing antenna designs to be done within a short period of time.

Implicitly, analytical analysis has played a very important role in the academic engineering and communication industries. Engineers in telecommunication field are easier to understand the operating background of the slot antennas through analytical analysis, since it is the fundamental knowledge in antenna designs before entering into further numerical simulation. In general, the slot antenna dimensions predicted from analytical calculations only require minor adjustments to optimize their performance in numerical simulations. Therefore, the time spent on slot antenna designs can be shortened.

REFERENCES

Ajioka, J. S., Joe, D. M., & McFarland, J. L. (1984). Slot radiators in septated waveguide. *IEEE Transactions on Antennas and Propagation*, *32*(3), 247–251. doi:10.1109/TAP.1984.1143295

Baudrand, H., Tao, J., & Atechian, J. (1988). Study of radiation properties of open-ended rectangular waveguides. *IEEE Transactions on Antennas and Propagation*, *36*(8), 1071–1077. doi:10.1109/8.7219

Bodnar, D. G., & Paris, D. T. (1970). New variational principle in electromagnetic. *IEEE Transactions on Antennas and Propagation*, *18*(2), 216–223. doi:10.1109/TAP.1970.1139642

Booker, H. G. (1946). Slot aerials and their relation to complementary wire aerials. J. IEE (London), 93, 620–626.

Chang, C. W., Chen, K. M., & Qian, J. (1997). Nondestructive determination of electromagnetic parameters of dielectric materials at X-band frequencies using a waveguide probe system. *IEEE Transactions on Instrumentation and Measurement*, *46*(5), 1084–1092. doi:10.1109/19.676717

Collin, R. E. (1985). *Antennas and Radiowave Propagation*. McGraw-Hill.

Compton, R. T. Jr. (1963). *The aperture admittance of a rectangular waveguide radiating into a lossy half-space. Technical Report, 1691-1*. Columbus, OH: Ohio State University.

Croswell, W. F. (2007). Slot Antennas. In J. L. Volakis (Ed.), *Antenna Engineering Handbook* (pp. 8-1–8-16). McGraw-Hill.

Da Silva, E. F., & McPhun, M. K. (1978). Calibration techniques for one-port measurement. *Microwave Journal*, *21*(6), 97–100.

Das, B. N., Ramakrishna, J., & Sarap, B. (1984). Resonant conductance of inclined slots in the narrow wall of a rectangular waveguide. *IEEE Transactions on Antennas and Propagation*, *32*(7), 759–761. doi:10.1109/TAP.1984.1143401

Delogne, P. P., & Deryck, L. (1980). Underground use of a coaxial cable with leaky sections. *IEEE Transactions on Antennas and Propagation*, *28*(6), 875–883. doi:10.1109/TAP.1980.1142426

Eaton, J. E., Eyces, L. J., & Macfarlane, G. G. (1983). Linear-Array Antennas and Feeds. In S. Silver (Ed.), Microwave Antenna Theory and Design (pp. 257 - 333). The Institution of Engineering and Technology (IET).

Enjiu, R. K., & Perotoni, M. B. (2013). Slotted waveguide antenna design using 3D EM simulation. *Microwave Journal*, *56*(7), 72–84.

Ganchev, S. I., Bakhtiari, S., & Zoughi, R. (1992). A novel numerical technique for dielectric measurement of generally lossy dielectrics. *IEEE Transactions on Instrumentation and Measurement*, *41*(3), 361–365. doi:10.1109/19.153329

Gilbert, R. A. (2007). Waveguide Slot Antenna Arrays. In J. L. Volakis (Ed.), *Antenna Engineering Handbook* (pp. 9-1–9-36). McGraw-Hill.

Goldstone, L. O., & Oliner, A. A. (1959). Leaky-wave antennas I: Rectangular waveguides. *IRE Transactions on Antennas and Propagation*, *7*(4), 307–319. doi:10.1109/TAP.1959.1144702

Green, H. E. (1989). Determination of the cutoff of the first higher order mode in a coaxial line by the transverse resonance technique. *IEEE Transactions on Microwave Theory and Techniques*, *1*(10), 1652–1653. doi:10.1109/22.41018

Hansen, R. C. (1992). Array pattern control and synthesis. *Proceedings of the IEEE*, *80*(1), 141–151. doi:10.1109/5.119573

Heddebaut, M. (2009). Leaky waveguide for train-to-wayside communication-based train control. *IEEE Transactions on Vehicular Technology*, *58*(3), 1068–1076. doi:10.1109/TVT.2008.928635

Hirohide, S. (2005). Radiation for a flanged rectangular waveguide. *IEEE Transactions on Antennas and Propagation*, *53*(12), 3953–3962. doi:10.1109/TAP.2005.859748

Josefsson, L. G. (1991). *A waveguide antenna which includes a slotted hollow waveguide*. U.S. Patent No. EP0545873A2.

Kim, J. H., Enkhbayar, B., Bang, J. H., & Ahn, B. C. (2010). New formulas for the reflection coefficient of an open-ended rectangular waveguide radiating into air including the effect of wall thickness or flange. *Progress In Electromagnetics Research M.*, *12*, 143–153. doi:10.2528/PIERM10033104

Kim, S. T., Yun, G. H., & Park, H. K. (1998). Numerical analysis of the propagation characteristics of multiangle multislot coaxial cable using moment method. *IEEE Transactions on Microwave Theory and Techniques*, *46*(3), 269–279. doi:10.1109/22.661714

Lee, B., Harackiewicz, F., Jung, B., & Park, M. J. (2005). Cavity-backed slot antenna array for the repeater system of a satellite digital multimedia broadcasting service. *IEEE Antennas and Wireless Propagation Letters*, *4*, 389–392. doi:10.1109/LAWP.2005.857967

Levine, H., & Papas, C. H. (1951). Theory of the circular diffraction antenna. *Journal of Applied Physics*, *22*(1), 29–43. doi:10.1063/1.1699816

Li, Y., & Wang, J. H. (2010). Polarization property of leaky coaxial cable with overlapped triangle slots. *IEEE Antennas and Wireless Propagation Letters*, *9*, 1049–1052. doi:10.1109/LAWP.2010.2090332

Li, Y., Wang, J. H., Zhang, Z., & Chen, M. (2012). Empirical formula for the electric field calculation in rectangular slots on leaky coaxial cable. *IEEE Transactions on Antennas and Propagation*, *60*(12), 5825–5833. doi:10.1109/TAP.2012.2208929

Marcuvitz, N. (1986). Waveguide Handbook. The Institution of Engineering and Technology (IET). doi:10.1049/PBEW021E

Misilmani, H. M. E., Husseini, M. A., & Kabalan, K. Y. (2015). Design of slotted waveguide antennas with low sidelobes for high power microwave applications. *Progress In Electromagnetics Research C.*, *56*, 15–28. doi:10.2528/PIERC14121903

Miwave Techno Ventures. (2009). *Slotted Broad Wall Microwave Antenna.* Retrieved from https://www.indiamart.com/proddetail/slotted-broad-wall-microwave-antenna-3736329988.html

Oline, A. A., & Jackson, D. R. (2007). Leaky-Wave Antennas. In J. L. Volakis (Ed.), *Antenna Engineering Handbook* (pp. 11-1–11-56). McGraw-Hill.

RADIAFLEX. (2013). *The Optimal Confined-Coverage Solution. Radio Frequency Systems.* RFS.

Sage, G. P. L. (2016). 3D printed waveguide slot array antennas. *IEEE Access: Practical Innovations, Open Solutions*, *4*, 1258–1265. doi:10.1109/ACCESS.2016.2544278

Selvan, K. T. (1998). Simple formulas for the gain and far-field of open-ended rectangular waveguides. *IEE Proceedings. Microwaves, Antennas and Propagation*, *145*(1), 80–84. doi:10.1049/ip-map:19981552

Stevenson, A. F. (1948). Theory of slots in rectangular wave-guides. *Journal of Applied Physics*, *19*(24), 24–38. doi:10.1063/1.1697868

Stratton, J. A. (1941). *Electromagnetic Theory*. New York: McGraw-Hill.

Technologies, A. (2010). *Specifying calibration standards and kits for Agilent vector network analyzers: application note 1287-11.* Agilent Technologies Inc.

Tyagi, Y., Mevada, P., Chakrabarty, S., & Jyoti, R. (2017). High-efficiency broadband slotted waveguide array antenna. *IET Microwaves, Antennas & Propagation*, *11*(10), 1401–1408. doi:10.1049/iet-map.2016.0682

Wang, J. H., Li, Y. J, Zhang, Z., Chen, M. (2011). Radiation mechanism and polarization properties of leaky coaxial cables. *Radio Science, 46*(RS2009), 1-12.

Wang, J. H., & Mei, K. K. (2001). Theory and analysis of leaky coaxial cables with periodic slots. *IEEE Transactions on Antennas and Propagation*, *49*(12), 1723–1732. doi:10.1109/8.982452

Wang, J. H., & Mei, K. K. (2002). Design of leaky coaxial cables with periodic slots. *Radio Science*, *37*(5), 1–9. doi:10.1029/2000RS002534

Yaghjian, A. D. (1984). Approximate formulas for the far fields and gain of open ended rectangular waveguide. *IEEE Transactions on Antennas and Propagation*, *32*(4), 378–384. doi:10.1109/TAP.1984.1143332

Yang, X. D., Feng, S., Zhang, X., Ren, X. Y., & Li, W. X. (2006). Research of single mode radiation leaky coaxial cable used for ultra wide-band communication. *Proceedings, 2006 International Conference on Communications, Circuits and Systems* (vol. 2, pp. 1261-1264). Academic Press. 10.1109/ICCCAS.2006.284874

Yoshitomi, K., & Sharobim, H. R. (1994). Radiation from a rectangular waveguide with a lossy flange. *IEEE Transactions on Antennas and Propagation*, *42*(10), 1398–1403. doi:10.1109/8.320746

You, K. Y. (2015). *RF Coaxial Slot Radiators: Modeling, Measurements, and Applications*. Artech House.

You, K. Y. (2017). Materials Characterization Using Microwave Waveguide System. In S. K. Goudos (Ed.), *Microwave Systems and Applications* (pp. 341–358). InTech. doi:10.5772/66230

You, K. Y., Abbas, Z., Malek, M. F. A., & Cheng, E. M. (2014). Non-destructive dielectric measurements and calibration for thin materials using waveguide-coaxial adaptors. *Measurement Science Review*, *14*(1), 16–24. doi:10.2478/msr-2014-0003

Chapter 4
Miniaturization of Printed Microstrip Antennas Array by Using Defected Ground Structure Technique

Otman Oulhaj
University Abdelmalek Essaadi, Morocco

Amar Touhami Naima
University Abdelmalek Essaadi, Morocco

Aghoutante Mohamed
University Abdelmalek Essaadi, Morocco

ABSTRACT

In this chapter, the authors present different techniques used to miniature microstrip antennas, particularly planar antennas array, for different applications demanding small dimensions. This will cover DGS, slot technique, and metamaterials. After the presentation of these techniques based on theoretical studies, the second part of this chapter will be about the authors' contribution in the miniaturization of microstrip antennas arrays. This part will include the presentation of some miniature antennas array which they have validated into simulation and measurement by using DGS techniques. The different structures were validated into simulation by using tow electromagnetic solvers ADS (advanced design system) and CST-MW (computer simulation technology) which permit one to validate and to verify the different performances of antennas arrays as radiation pattern, matching input impedance and small dimensions.

DOI: 10.4018/978-1-5225-7539-9.ch004

INTRODUCTION

The rapid development of telecommunications systems has allowed the creation and innovation of several technologies as well as led to a wave of new wireless devices and systems to meet the demands of multimedia applications. Multi-frequency and multi-mode devices such as cell phones, wireless local area networks (WLANs) and wireless personal area networks (WPANs) place multiple requests on the antennas. Mainly, antennas must have a high gain, a small physical size, wide bandwidth, versatility, integrated installation, etc. In particular, the bandwidths for input impedance, polarization or axial ratio, radiation patterns and gain are becoming the most important factors affecting the application of antennas in contemporary fields and future wireless communication systems. On the one hand, there is a tendency towards the miniaturization of components linked to mobile devices, and on the other hand there is an increasing demand for fast data transfer, which in turn requires broadband components and multiband behavior. These two contradictory constraints must be raised with inexpensive solutions, and providing a high yield.

Since the design of the microstrip antenna half a century ago, its application in telecommunication systems has increased rapidly, especially over the past two decades. Due to the many unique and interesting properties of the microstrip antenna, there seems to be little doubt that it will continue to find many applications in the future. These properties include a low profile, lightweight, compact and compliant with the mounting structure, easy to manufacture and to integrate with solid-state devices. The results of these properties have contributed to the success of microwave antennas not only in military applications such as aircraft, missiles, rockets and spacecraft, but also in commercial areas such as mobile satellite communications, direct broadcast satellite (DBS) system, global positioning system (GPS), remote sensing and hyperthermia. Although the microstrip antenna is generally known for its bandwidth weakness, recent technological advances have improved its bandwidth by a few percent to tens of percent.

In recent years, the reduction in the size of components has attracted a great deal of interest among researchers in the telecommunications world. The race to this miniaturization is motivated by the integration of microstrip antennas in the architecture of mobile terminals to reduce their size as much as possible.

An advantage inherent to patch antennas is the ability to have polarization diversity. Patch antennas can easily be designed to have vertical, horizontal or circular polarizations, using multiple feed points, or a single feed point with asymmetric patch structures. This unique property allows patch antennas to be used in many types of communications links that may have varied requirements.

The importance of circular polarization is that we can have the same magnitude with phase difference of 90 degrees. This kind of polarization produces two orthogonal components $E\theta$ and $E\varphi$ with 90 degrees out of phase.

The use of antenna array is one of the possible solutions permitting to get higher directivity and a pattern adjustment. The choice and shape of the radiating element depends on the application and on certain factors: the resonance frequency, the radiated power, the polarization, the gain as well as the antenna bandwidth. One of the disadvantages of antenna array is the array area because of the association of many elements and the different microstrip lines network assuring the feeding of each antenna element. To resolve this problem many methods and techniques can be used as mentioned in (Sidhu, 2014). The networking of several antennas makes it possible to compensate for the limitations of an antenna alone. An antennas array is conventionally constituted by separately excited or sub-array radiating elements in order to increase the gain and radiation performances (for example, to increase the directivity in a direction of space). One of the main advantages of antennas arrays is their ability to scan the beam in certain directions of space. The method includes feeding each source with electronically controlled amplitude and phase. The advantage is, unlike reflector antennas, there isn't a heavy mechanical rotation system whose large inertia prevents the rapid exploration of space.

Several methods have been proposed recently to miniaturize microstrip antennas saving their radiation performances such as DGS technique. This method includes the use of high dielectric-constant material, the introduction of slots in the resonating patch the integrating of a single-layer patch into a two-layer structure, the inverted-F configuration, the quarter-wave-patch approach, the genetic algorithm, and the use of photonic band-gap materials.

When DGS is introduced in a microstrip antenna, the defect geometry etched in the ground plane disturbs its current distribution. This disturbance affects the transmission line characteristics such as the line capacitance and inductance. In other words, introducing DGS in a microstrip antenna can result in an increase of the effective capacitance and inductance which influence the input impedance and current flow of the antenna and thus reducing its size with respect to a given resonance frequency.

MICROSTRIP ANTENNAS ARRAY

Patch Antenna

Definition

An antenna with printed radiating elements, commonly called a patch antenna, is a microstrip antenna of particular shape that consists of a ground plane and a dielectric substrate whose surface carries one or more metal elements (Figure 1).

The printed antenna is not a new concept. The first publications appeared in 1953 with G. A. Deschamps and in 1955 with the filing of a French patent by H. Gutton and G. Baissinot. However, it is only from 1970 that there had been a great interest for this technology because of the strong need of the military and space industry for thin antennas. Indeed, at first this type of antenna has mainly become more common. Microstrip antennas are widely used in the microwave frequency region because of their simplicity and mainly for communications applications.

The prodigious development of these printed structures is linked to the considerable progress made in the 1980s in the field of miniaturization, the integration of electronic circuits and especially low-loss dielectric substrates. It should be that emphasized their low weight, reduced size, low cost and a planar configuration are compatible with integrated circuits and possibly conformable. Thus, studies and achievements intensify and target many uses in the civil, military and medical fields.

Several studies of printed structures have been collected in a special issue of D.M. Pozar (Pozar, 1992) and in two books, that of I. J. Bahl and P. Bhartia (Bahl, 1980) and that of J. R. James et al. (Hall, 1981). The use of plated antennas has become almost universal in all mobile communication systems. These antennas are light, compact and inexpensive. They are manufactured by photolithographic engraving of printed circuits, which have reached very high level of precision that facilitates mass production.

The microstrip antennas, in their basic configuration, consist of a conductive metal plate (usually between 17.5 to 35 microns thick in microwave and 9 microns in millimeter (Grenier, 2016) of specific shape constituting the radiating element profiled on the upper face of a dielectric substrate. Figure (1) illustrates an example of a microstrip antenna.

Figure 1. Microstrip patch antenna geometry
(Hall, 1981)

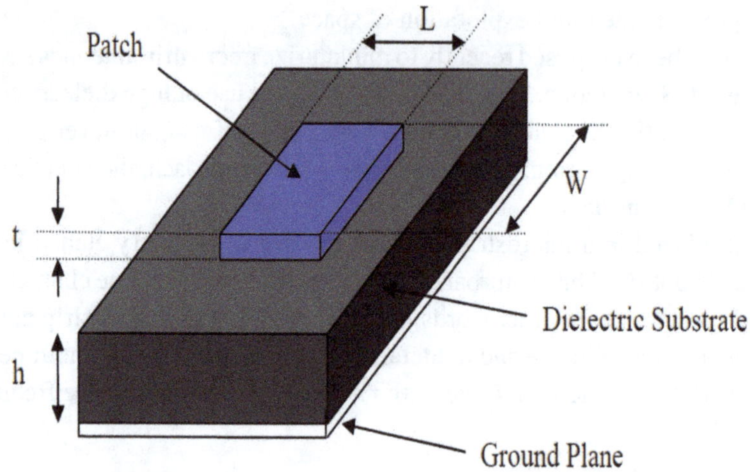

The shape of the conductive patch can be varied, but it influences the modes that are likely to be excited in the antenna, and therefore the nature of the radiation (Grenier, 2016). However, to simplify the analysis and prediction of the performance of the antenna, the plate is generally square, rectangular, circular or elliptical.

More elaborate geometries as shown in Figure (2), however, are used to respond to specific constraints on the antenna and in particular to generate circular polarization.

Among all these forms, the rectangular element is unquestionably the easiest to analyze for the understanding of the radiation mechanisms of microstrip antennas. Indeed, although the analysis of radiation is widespread for all forms of plates, it will be limited here to conventional plates including rectangular shape to simplify the analysis and for the prediction of performance.

The length L of the antenna is often between (Pozar, 1992):

Figure 2. Different geometries of the patch
(Hall, 1981)

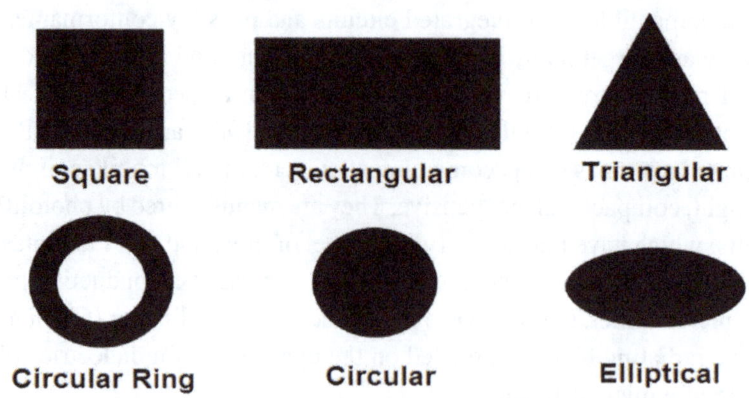

$$0.333»_0 < L < 0.5»_0 \quad (1)$$

The thickness of the substrate is always increased by (Grenier, 2016):

$$0.003»_0 < h < 0.05»_0 \quad (2)$$

The choice of the values of these parameters generally conditions the type of the desired application.

Radiation Characteristics

Different models can be used to predict the resonant frequency, bandwidth, input impedance, and radiation characteristics of the microstrip antenna, namely the radiation pattern, the quality factor, and the efficiency. Indeed, there are three ways to see the radiation of a microstrip antenna (Pozar, 1989):

- The resonant cavity model.
- The transmission line model (Bahl, 1980).
- The full-wave model indicating the integral equations and the moments method.

These methods are based on the nature of the physical phenomena that occur within the antenna. Given the complexity of such an analysis, a set of approximations are put in place to simplify the model developed. The most popular models that provide sufficiently concise results are the model of the transmission line and the model of the cavity.

The model of the transmission line is the simplest; it gives a good physical interpretation; however, it is less precise. For its part, the model of the cavity is distinguished by good accuracy and at the same time has a significant complexity.

The rectangular patch is the most common form of patch antennas and simple to analyze by the model of the lines and the cavity.

In the rest of the study, the transmission line model will be presented for the case of a rectangular shape.

Transmission Line Model

It has already been pointed out that this model is simple but less precise and lacks flexibility. Basically, the line model interprets the microstrip antenna as two slots separated by a transmission line of length L and low characteristic impedance Zc shown in the Figure 3.

Our goal is to produce the radiation; indeed, part of the electrical flow is in the substrate and the other part in the air. As a result, the dominant mode of propagation would be the quasi-TEM mode. Then, use the effective dielectric constant (McGraw-Hill Book Co, 1961) to represent the overflow and propagation of the wave across the line.

The value of the effective dielectric constant is given by (McGraw-Hill Book Co, 1961):

Figure 3. Microstrip antenna represented as two radiating slots
(Schaubert, 1989)

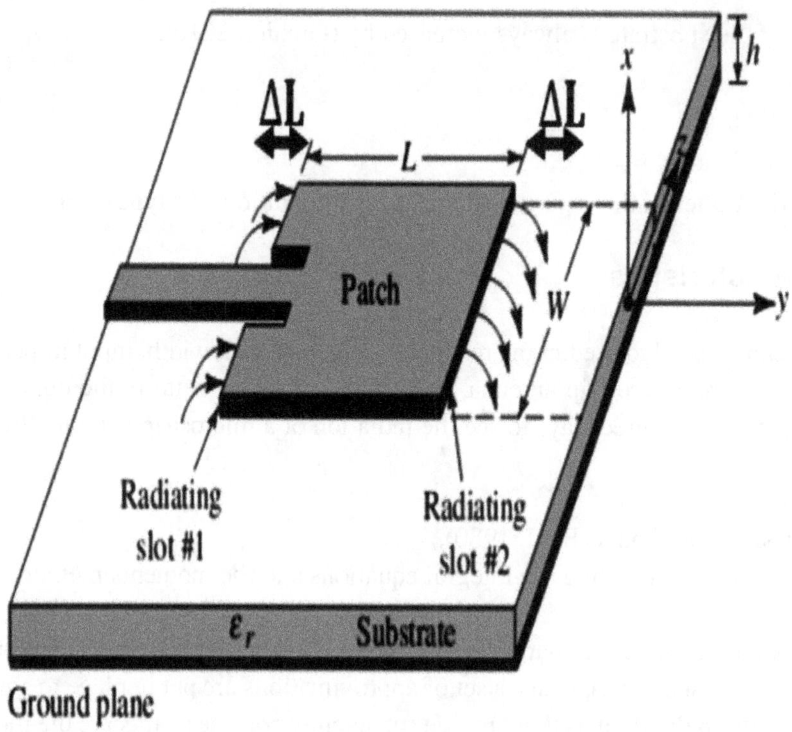

$$\varepsilon_{eff} = \frac{1}{2}\left(\varepsilon_r + 1\right) + \frac{1}{2}\left(\varepsilon_r - 1\right)\left(1 + 12\frac{h}{w}\right)^{-\frac{1}{2}}, w/h \geq 1 \tag{3}$$

Or:

μ_r .is the dielectric constant of the substrate.

h is the height of the substrate.

W is the width of the plate.

It is already clear that the fields of the edges make the dimensions of the patch are electrically superior to its physical dimensions. This is well shown in Figure 3 where the length of the patch is extended by an amount ΔL. An approximation of this quantity of length is given by (Joshi, 1982):

$$\frac{\Delta L}{h} = 0.412 \frac{\left(\varepsilon_{eff} + 0.3\right)\left(\frac{w}{h} + 0.264\right)}{\left(\varepsilon_{eff} - 0.258\right)\left(\frac{w}{h} + 0.8\right)} \tag{4}$$

The effective length L_{eff} will be:

$$L_{eff} = L + 2\Delta L \tag{5}$$

Each radiating gap is represented by an equivalent parallel admittance Y ((I. J. Bahl, 1980) (conductance G and susceptance B), and therefore the equivalent circuit will be as follows (Figure 4)

Hence the input admittance of the patch antenna based on this equivalent circuit is easily obtained as:

$$Y_{in} = G + jB + Y_0 \frac{G + j\left(B + Y_0 tan\beta L\right)}{Y_0 + j\left(G + jB\right) tan\beta L} \tag{6}$$

Figure 4. The equivalent Circuit for a Microstrip Radiating Element
(Bahl, 1980)

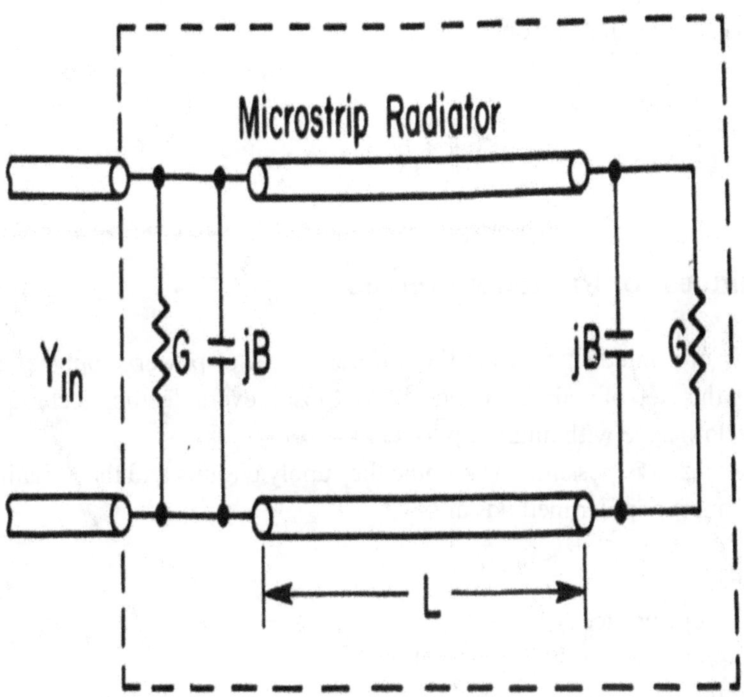

where $G = \frac{1}{R_r}$, β .s the propagation constant= $\frac{2\pi\sqrt{\varepsilon_{eff}}}{\lambda_0}$ and $B = C.\omega$, the susceptance of the slot, is represented by a capacity of expression following:

$$C = \frac{\Delta L\sqrt{\varepsilon_{eff}}}{cZ_0} \tag{7}$$

And so

$$B = k_0 \frac{\Delta L\sqrt{\varepsilon_{eff}}}{Z_0} \tag{8}$$

where $k_0 = \frac{2\pi}{\lambda_0}$ and

R_r is the radiation resistance per unit length of the slot $= \dfrac{120\lambda_0}{1 - \frac{(k_0 h)^2}{24}}$ (9)

since at resonance, the imaginary part of the admittance Y_{in} .is zero, the resonant frequency of the radiating element can be calculated from:

$$tan\beta L = \frac{2Y_0 B}{B^2 + G^2 - Y_0^2} \tag{10}$$

Feeding Techniques for Printed Antennas

Given the sensitivity of printed antennas to the different types of power supply, a very particular care must be reserved for this step of realization. For this purpose, several feeding techniques are highlighted to achieve better performance with minimum losses.

The theory shows that it is possible to combine the supply circuits and the radiating elements on the same substrate. The most popular methods are:

1. Microstrip Line.
2. Coaxial Probe (coplanar feed).
3. Proximity Coupling.
4. Aperture Coupling.

Microstrip Line Feed

This technique involves placing a microstrip line (usually 50Ω) In contact with the radiating element (Figure 5).

Coaxial Feed

Coaxial feeding or probe feeding as shown in Figure (6), used to power plated antennas, is a well-known technique. The inner conductor of the coaxial connector extends through the dielectric and is soldered to the patch while the outer conductor is connected to the ground plane.

Figure 5. Microstrip line feed
(Pozar, 1992)

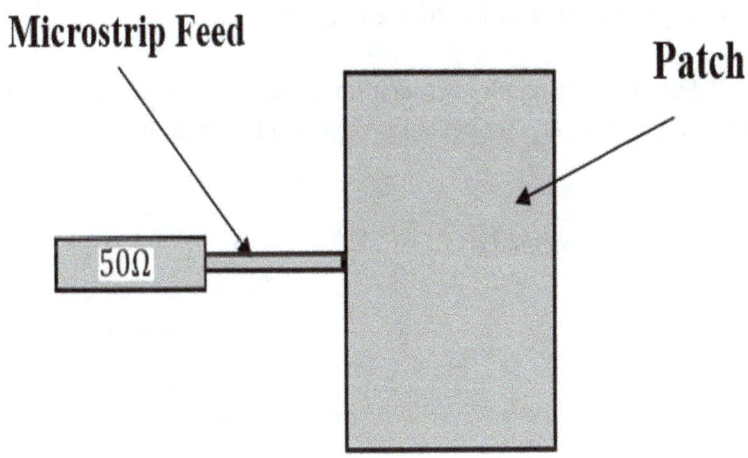

Figure 6. Coaxial line feed
(Pozar, 1992)

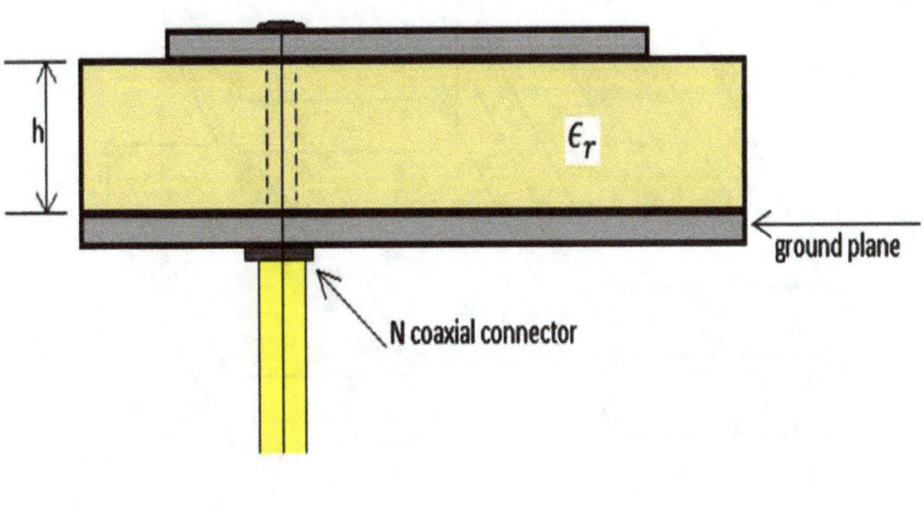

Proximity Coupled Feed

In this type of power supply, the radiating patch and the power strip line are separated by the ground plane as shown in Figure (7). The conjunction between the patch and the power line is provided by a slot or an opening in the ground plane (Katehi, 1984).

Aperture Coupled Feed

This feeding technique is also known as electromagnetic coupling. Two dielectric substrates are employed such that the feed line is between the two substrates and the patch is printed on the upper substrate as shown in Figure (8).

The main advantage of this feeding technique is that it eliminates parasitic radiation due to the power supply and provides a very high bandwidth (above 13%), due to the overall increase in the thickness of the microstrip antenna.

Antennas Array Configuration and Feed Networks

An antenna array is by definition the regular association of identical antennas to create a particular form of radiation. The radiated power is greater because we multiply the number of radiating elements. The

Figure 7. Proximity coupling feed method
(Katehi, 1984)

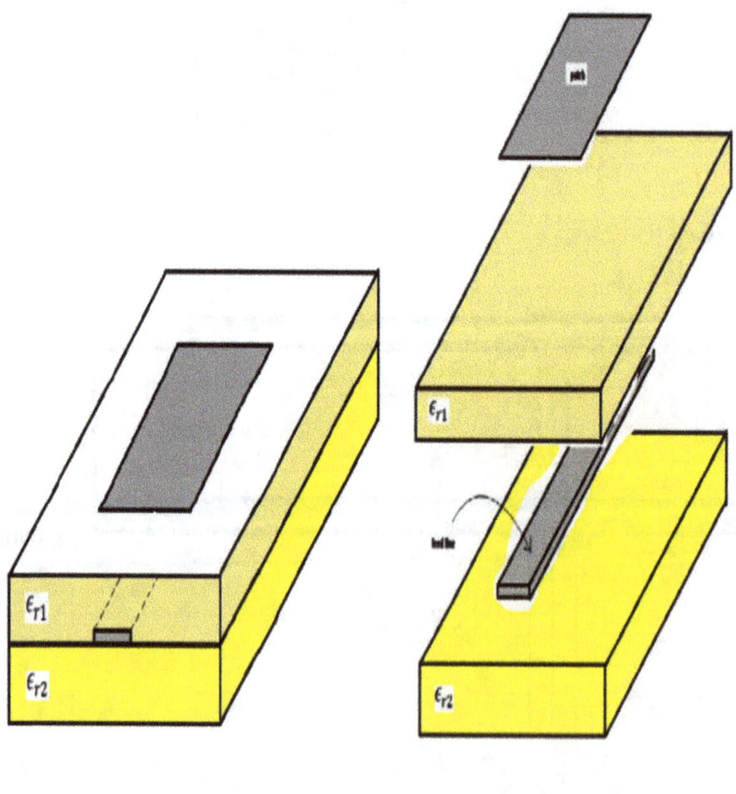

Figure 8. Aperture coupling feed method
(Gronau, 1986)

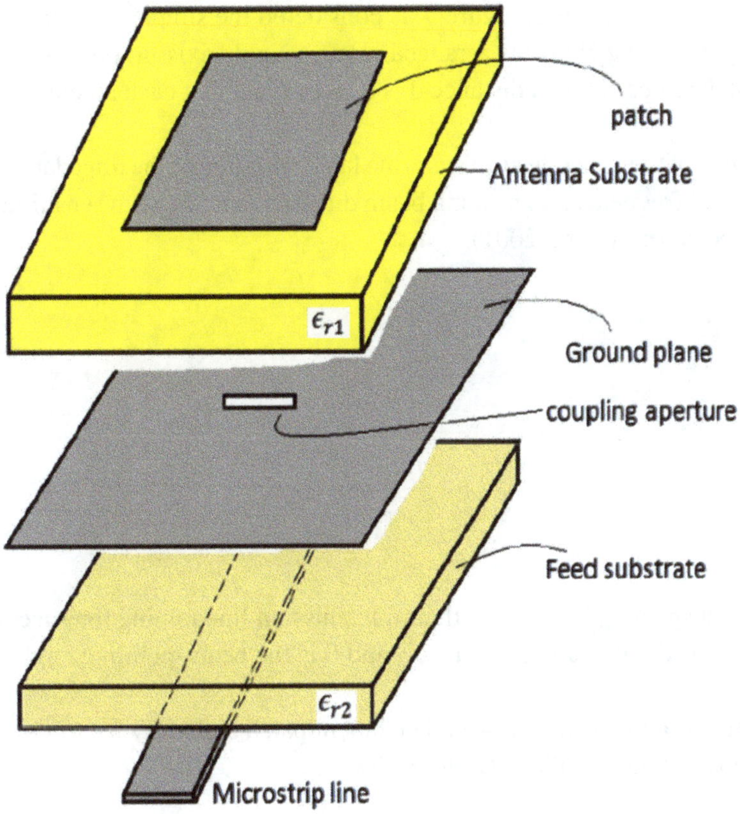

radiation results from the in-phase addition of fields from each element. The possible combinations are therefore numerous and lead to great flexibility in the design of networks.

The applications of the antenna arrays are numerous and use all types of elements: cornets, wired antennas, printed antennas, etc.

Different Topologies of Antenna Arrays

Microstrip antennas are widely used as a single element but also very popular in the design of antenna arrays. The main limitation of the microstrip is that it effectively radiates only over a narrow band of frequencies and can not operate at the high-power levels of the waveguide, coaxial line or even the strip-line. Antenna arrays are used to scan the beam of an antenna system, to increase directivity, to gain and improve various other functions that would be difficult with a single element antenna. In the microstrip antenna array, array elements can be fed by a single or multiple line in a feed network arrangement. Depending on their feeding method, the feed arrays are classified as:

- Series-feed network
- Corporate-feed network
- Combination of series and corporate feed network toolbar

Microstrip Series-Feed Array Antenna

The linear antenna array, as shown in Figure 9, is considered the simplest of the antenna arrays. It is a linear assembly of N radiating elements arranged along a single axis. In order to simplify its analysis, we will assume that the inter-element distance d is constant and that each antenna radiates equivalently in space.

The main limitation of the series-feed arrays is the large variation of the impedance and beam-pointing direction over a band of frequencies. The main beam direction and the scan sensitivity can be calculated from the following equations (Garg. 2001).

$$d\,sin\theta + \sqrt{\varepsilon}l = \lambda = \frac{c}{f} \tag{11}$$

$$\frac{\partial\theta}{\partial f} = -\frac{c}{df^{2}cos\theta} \tag{12}$$

where d is the element spacing, l is the length of transmission line joining the successive elements, c is the velocity of light, f is the operating frequency, and θ is the beam-pointing angle measured from the broadside direction.

The radiated field of the E-plane for a single element patch and array factor can be expressed using the following formulas (Balanis, 2005; Mailloux, 2007).

Figure 9. (a) Generic diagram of a linear array of antennas (b) Series-feed array (four elements, f=2.45Ghz) (Garg. 2001)

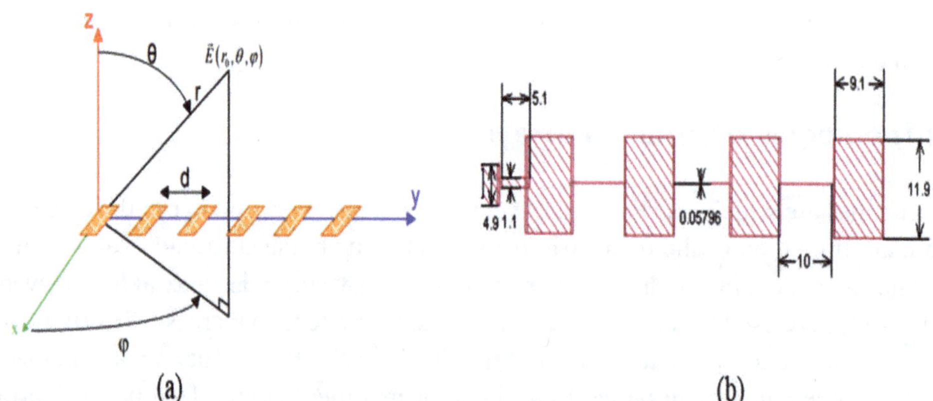

$$E = j\frac{k_0 W V_0 e^{-jk_0 r}}{r\pi}\left\{\frac{\sin\left(\frac{k_0 h}{2}cos\varphi\right)}{\frac{k_0 h}{2}cos\varphi}\right\}cos\left(\frac{k_0 L_{eff}}{2}sin\varphi\right) \tag{13}$$

Here W is the width of the patch antenna, Le is the extended length, $V_0=hE_0$ is the voltage across radiating slot of the patch, h is the substrate height and r is the far field distance from the antenna.

$$FA = \frac{sin\left(N\pi dx\left(u-u_0\right)\right)}{sin\left(\pi dx\left(u-u_0\right)\right)} \tag{14}$$

Here $u=\sin\theta$, $u_0=\sin\theta_0$, dx is the element spacing and N is the number of elements.

Microstrip Parallel-Feed Array Antenna

There is another popular method to feeding an microstrip antenna array which is the corporate feeding. For the parallel supply, 2^n radiating elements are feeding in parallel by n stages of 2^{m-1} power dividers at each stage m. This is accomplished by using either tapered lines or using quarter wavelength impedance transformers. Here, in the Fig. 10, the patch elements are connected by using the quarter wavelength impedance transformer method.

The radiated field formula that is given in Equation (13) it is similar to that of the series feed array and the array factor as given in (Stutzman, 1998) as:

Figure 10. Corporate-feed array (four element)
(Md. Tanvir, 2011)

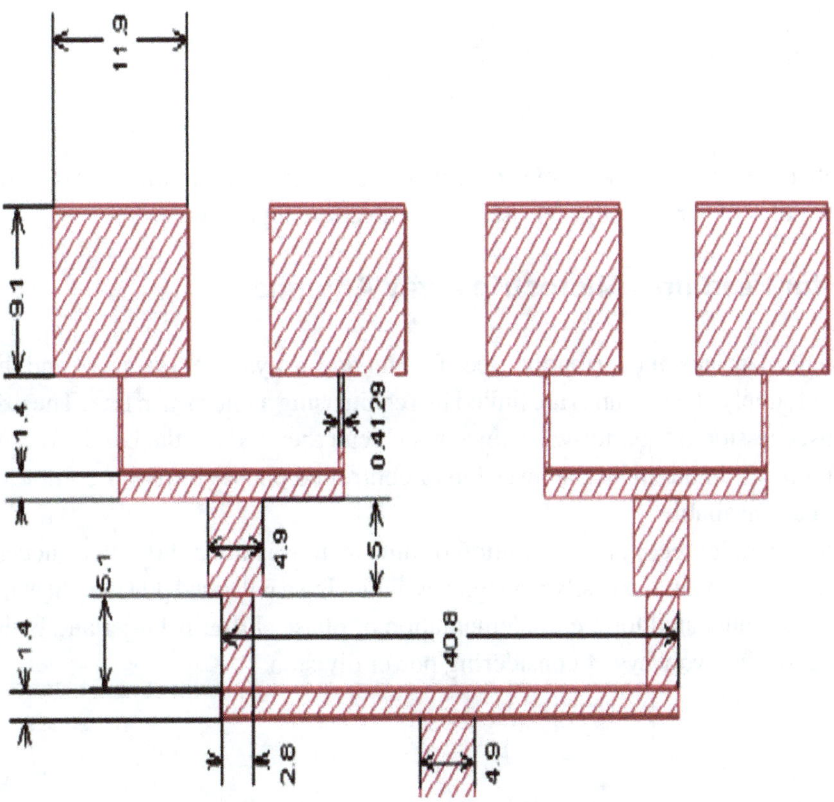

$$FA = \frac{sin^2\left(N\pi\left(\frac{dx}{\lambda}\right)sin\theta\right)}{N^2 sin^2\left(\pi\left(\frac{dx}{\lambda}\right)sin\theta\right)} \tag{15}$$

We can obtain the normalized power radiation pattern of antenna arrays by combining the element radiation pattern and the array factor.

Microstrip Parallel-Series Feed Network

It is also possible to realize an antenna array by combining of series feed and corporate feed. An 8-elemants array antenna can be constructed by using this method as shown in Figure 11. The same series or corporate feeding techniques can be applied to the two-dimensional array of antennas. It is a two-dimensional rectangular planar array whose aperture illumination can be separated into two orthogonal planes such as the horizontal and vertical planes and the radiation pattern may then be written as the product the radiation patterns in these two planes.

The array factor of this antenna with element spacing in the x and y direction of dx and dy respectively as given in (Visser, 2005) is:

$$FA = \frac{sin^2\left(N\pi\left(\frac{dx}{\lambda}\right)sin\theta_a\right) \times sin^2\left(M\pi\left(\frac{dy}{\lambda}\right)sin\theta_a\right)}{\left(N^2 sin^2\left(\pi\left(\frac{dx}{\lambda}\right)sin\theta_a\right)\right) \times \left(M^2 sin^2\left(\pi\left(\frac{dy}{\lambda}\right)sin\theta_a\right)\right)} \tag{16}$$

where N = number of vertical elements of array that gives rise to the azimuth angle, θa and M = number of horizontal elements of array that gives rise to the elevation angle, θe.

Techniques for Feeding a Microstrip Array Antenna

There are two major families of distribution trees for antenna arrays: the series tree and the parallel tree. For the series feed supply, the antennas are linked in a chain using a microstrip line. The excitation is then performed by a succession of quarter-wave lines which can then reduce the bandwidth of the antennas shown in the Figure 12. In addition, the insertion of controlled phase shifters is complex because there is only a small space available.

In the case of a parallel feed, the distribution of information is done through a succession of power dividers. Optimization can then be easily managed by achieving a balanced power split at different stages. In addition, the space allocated for the implementation of phase shifter is larger and is therefore easier. We will briefly describe two ways of considering power division:

Figure 11. 8-Elements microstrip array antenna, combination of series and corporate feed
(Md. Tanvir, 2011)

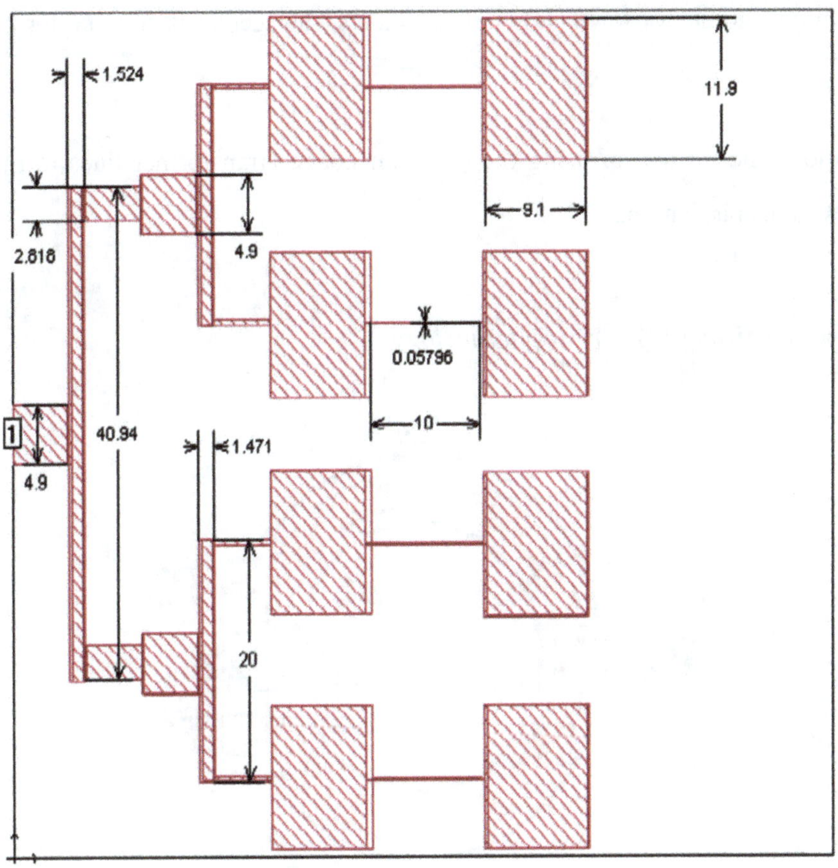

Figure 12. Series-feed network
(Pozar, 1992)

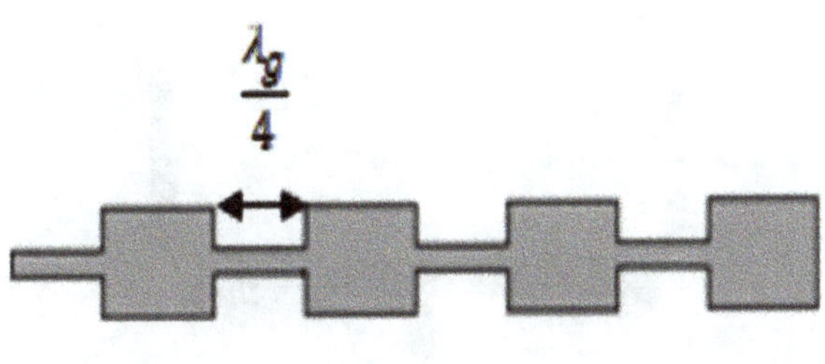

- **Wilkinson Divider (Figure 13):** The Wilkinson power divider is widely used in microwave applications. This divider is composed of an impedance input Z_0 and two quarter-wave lines ($\frac{\lambda_g}{4}$) with characteristic impedance $Z = Z_0\sqrt{2}$ followed by two impedance outputs Z_0. A resistor $R = 2Z_0$ is located between the two outputs and allows the maintenance of the adaptation in the case of an output power imbalance.

- **Impedance Divider (Figure 14):** It is a power divider based on impedance transformation by quarter wave lines or soft impedance transition lines known as tapered lines. While the first solution is more suitable for low bandwidth structures, the second is suitable for high bandwidth structures.

Figure 15 shows the method of using the ($\frac{\lambda_g}{4}$) impedance transformer lines to match the 100Ω patches to a 50Ω transmission line.

Figure 13. Wilkinson divider in microstrip technology
(Wikipedia, 2018)

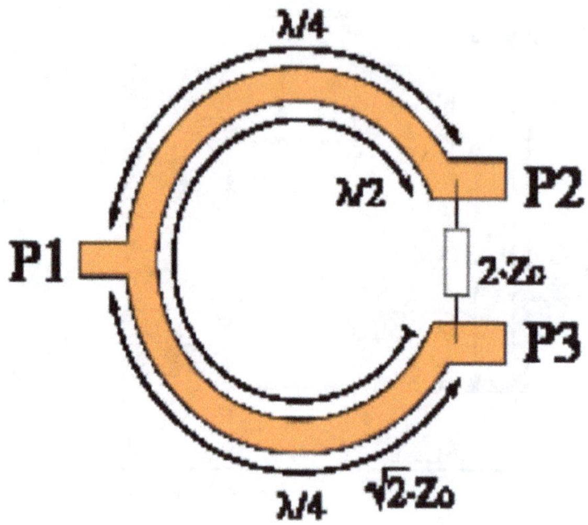

Figure 14. Impedance divider in microstrip technology
(Balanis, 2005)

Figure 15. $\frac{\lambda_g}{4}$ *Impedance transformer matching*
(Pozar, 1992)

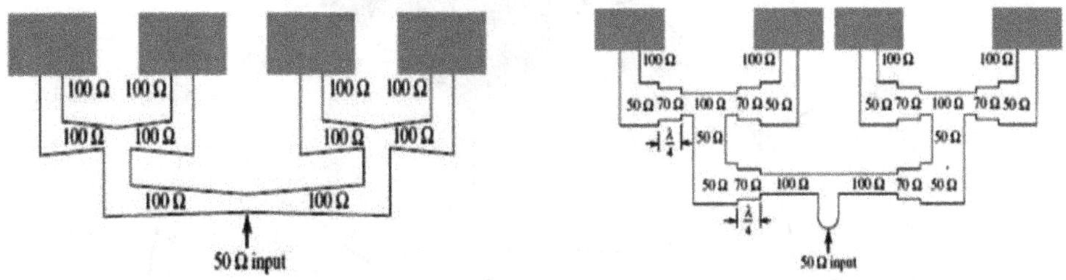

(a)Tapered lines (b) $\frac{\lambda_g}{4}$ Transformers

MINIATURIZATION OF MICROSTRIP ANTENNAS

Several antenna miniaturization techniques exist, each subject to well-identified physical limitations, whose purpose is to artificially affect the dimensions of the resonant circuit. Among these techniques, we find the electrical elongation of the path of the current by modifying the geometry of the antenna, the integration of localized elements, the addition of planes reflectors, short circuits or the use of particular materials. In the following, we will present a synthesis of some miniaturization techniques existing in the literature.

Miniature Antennas

The term miniature antenna encompasses several major categories of antennas. The main category concerns electrically small antennas. These antennas have very small dimensions compared to the wavelength.

The size of an antenna is relatively always expressed to the wavelength associated with its operating frequency. To define the domain boundary of electrically small antennas, it is necessary to define the sphere of Chu (Chaurasia, 2016).

There are several ways to define an electrically small antenna. The most widespread is that given by Wheeler in 1947. For Wheeler, an antenna is said to be electrically small if the sphere enclosing it satisfies the following condition: $a < \frac{\lambda_0}{2\pi}$ where λ_0 is the wavelength in vacuum.

This condition is expressed as ka < 1 Where a is the radius of the sphere and K is the wave number in the propagation medium such that $K = \frac{2\pi}{\lambda_0}$. The minimum sphere encompassing the antenna will be named the sphere of Chu as shown in Figure (16).

The second category groups the small antennas by physical constraints on one dimension. In this case, the size of the antenna structure is extremely small in one dimension.

The third category concerns antennas that are physically small, that is, whose dimensions are small in the strict sense. And finally, the last category is functionally small antennas. These are antenna sys-

Figure 16. Equivalent sphere of an antenna
(Chaurasia, 2016)

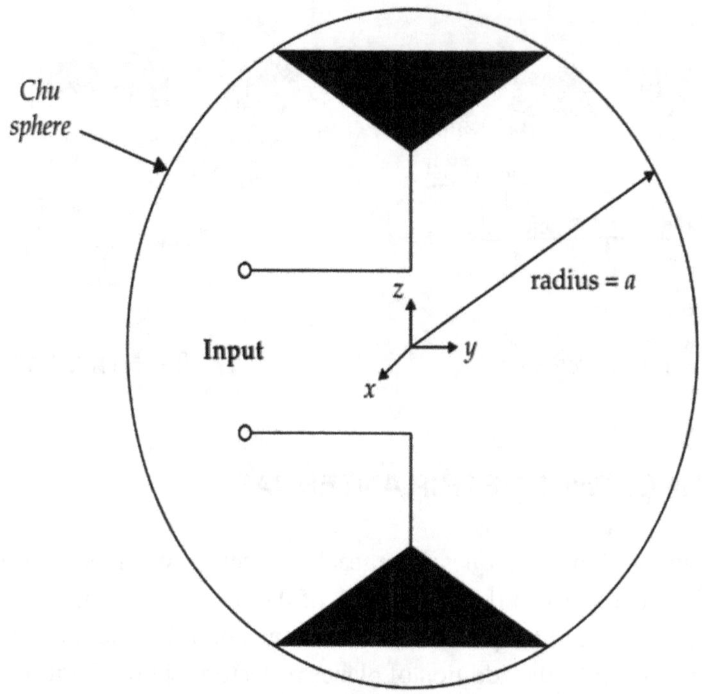

tems incorporating additional functions (active antenna) but do not necessarily require an increase in the size of the antenna.

To conclude, it is necessary to add to the list of antennas presented magnetic loops and slots. However, these antennas may appear to be complementary to dipole antennas.

MINITURIZATION TECHNIQUES FOR PRINTED ANTENNAS

Modification of Geometry

This miniaturization technique is currently the most widespread and concerns planar antennas. We list in this section the different miniature antenna structures based on the modification of the geometry.

Meandering Antennas

The meandering antenna makes it possible to reduce the physical length of an antenna without reducing its electrical length. In general, there are meanders with lines that exist in wired and printed technology (Figure 17-a) and slit meanders in printed technology (Figure 17-b). The meanders with printed lines consist of a radiating element folded in the shape of a meander. Slotted meanders, dual geometries of line meanders, are obtained by creating a meander-shaped slit in a metal plane. Figure 17-a shows an example of a meander monopole with a length of L = 137.6cm and a resonance frequency of Fr =

91.5MHz. An example of a slotted meander antenna is given in Figure 17-b which resonates at a frequency Fr = 2.65GHz.

Fractal Antennas

In the case of fractal antennas, the radiating element is a fractal shape according to the Hilbert, Koch or Sierpinski geometry. These geometries are actually mathematical curves optimized to best fill a square, triangular or any surface from an elementary cell of the same shape. An example of a four-iteration Koch fractal construct is given in Figure 18.

Figure 18 gives an example of a printed fractal antenna obtained from theVon Koch geometry (Baliarda, 2000). The antenna is printed on a Roger Duroid 5880 substrate (εr = 2.2), 0.8mm thick. It resonates at the frequency FR = 825MHz.

Adding Slots

One of the techniques used to lengthen the electrical path of the currents is the introduction of slots in the radiating structure. The principle used is not only to force the currents to bypass the inscribed slots and thus lengthen their path, but also to induce capacitive and inductive effects modifying the input impedance of the antenna. Figure 20 corresponds to the structure studied by Nguyen et al. (Nguyen, 2005). By changing the characteristics of the slot, the resonance frequency is changed. Indeed, the longer the slot, the lower the resonance frequency. We find the same results concerning the width of the slot. There are also folded U-shaped, H-shaped, butterfly-shaped but also curvilinear.

Figure 17. Example of meandering antennas:(a) antenna meander in wired technology and; (b) slotted meander antenna in microstrip technology
(Best, 2002)

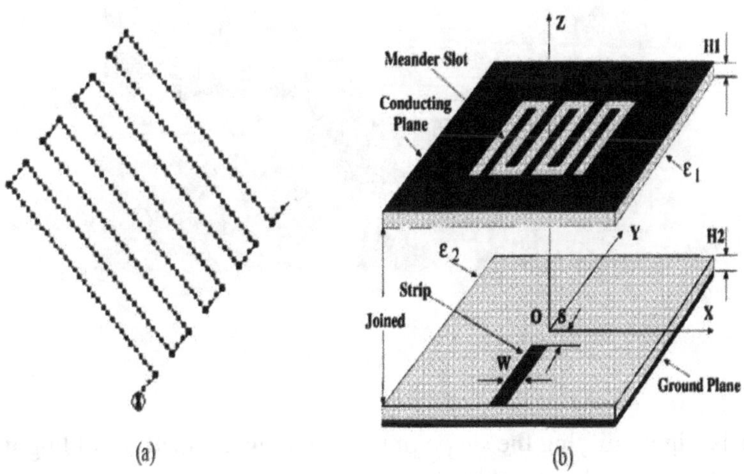

Figure 18. Procedure for designing a fractal according to Koch geometry geometry
(Baliarda, 2000)

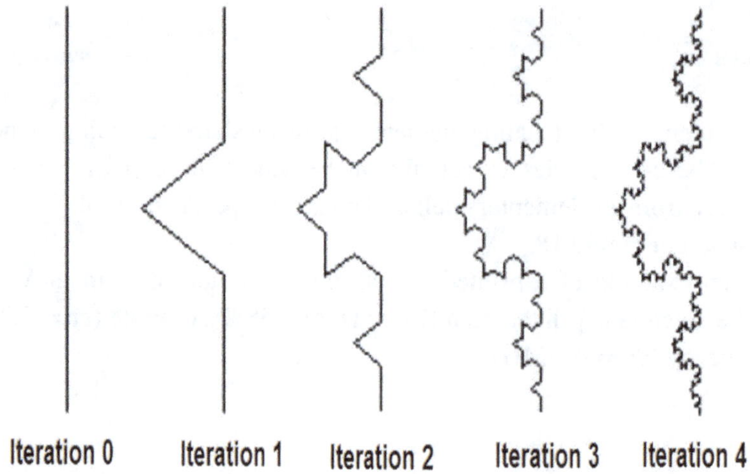

Figure 19. Antenna printed with Koch fractals of order 3 geometry
(Baliarda, 2000)

Folding

This technique consists in modifying the shape of the radiating element or folding it on itself out. This operation naturally leads to an element that physically occupies less space than an unfolded structure. Figure (21) shows an example of a folded antenna.

Figure 20. Extension of current path by adding slots
(Nguyen, 2005)

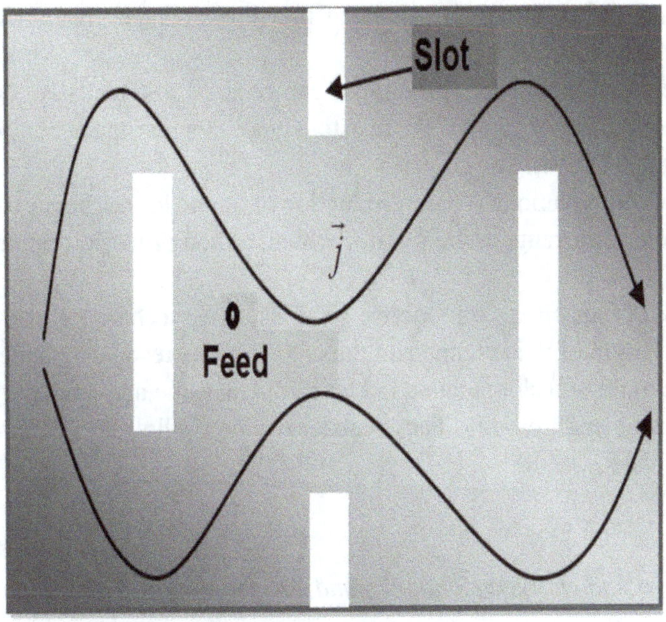

Figure 21. Antenna folded (a) bow-tie; (b) vertical folding
(Foltz, 1998)

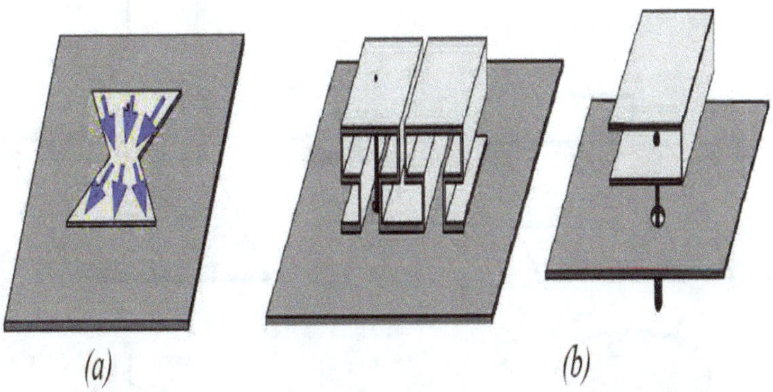

The main counterpart of this type of operation resides in the increase of the ohmic losses and thus a decrease of the efficiency and the gain. The case presented in Figure 21(a) is also intended to broaden the bandwidth. In the second case of folding in Figure 21(b), the inconvenience of this technique lies in the significant increase in the height of the antenna.

Using Localized Elements

The techniques mentioned previously concerning the modification of the geometry of the antenna are equivalent to a localized charge addition that goes from the discrete element to the conductive line section. Indeed, a folded monopole is equivalent to a monopoly charged by an inductor as can be seen in Figure 22 (a). In the same way, a monopoly charged by a metal roof can be seen as a monopoly loaded by a capacitance (Figure 22 (b). Microstrip antennas in which slots have been made may also be represented by equivalent LC circuits.

Regarding the discrete components, three types can be used, the resistive, capacitive or inductive loads. In practice, a discrete component of the resistive type (low resistance of 1 to 2 ohms) can replace the short-circuit tongue of a PIFA.

The disadvantage of this technique is the introduction of ohmic losses, hence a decrease in efficiency due to the concentration of currents in the resistor which, although weak, remains more resistive than a short circuit.

Concerning the use of capacitive or inductive sections, these techniques also aim at artificially increasing the electrical length of the antenna and thus reducing its resonant frequency.

We illustrate an example of a slot antenna fed by a coplanar line and loaded by capacitors placed at the folded ends of the slot (Figure I-11). These capacitive loads allowed a 22% reduction in the initial resonant frequency (K. Luk,, 1998).

Figure 22. Monopoly (a) and charged monopoly, and; (b) their equivalent circuit
(Luk, 1998)

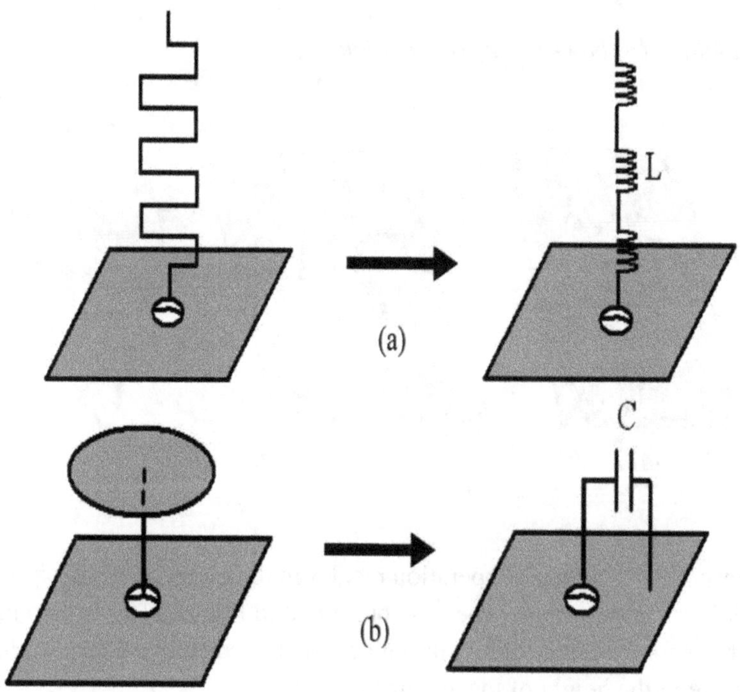

Figure 23. Slotted antenna with capacitive load
(Scardelletti, 2008)

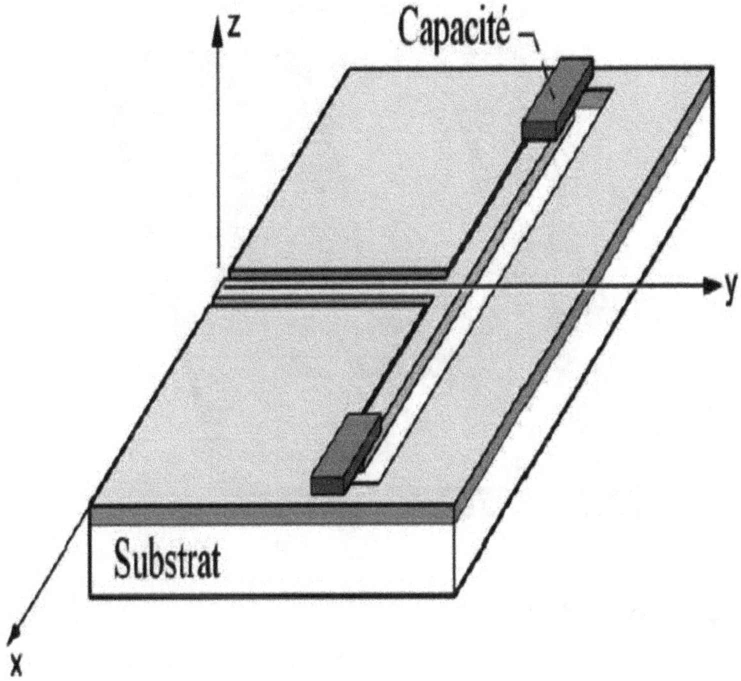

ADDITION OF SHORT CIRCUIT

PIFA Antenna (Planar Inverted-F Antenna)

One of the most used methods in the field of printed antennas is to introduce one or more short circuits between the radiating element and the ground plane. By observing the distribution of the electric field of the first resonant mode or fundamental mode, TM_{100}, in the cavity virtually delimited by electric and magnetic walls as previously shown, it is found that the field vanishes in the middle of the resonant length. This means that by positioning a perfect electrical wall in this vertical plane, the distribution of the equipotential lines is not affected. Thus, the dimensions of the structure can be reduced by half. This structure is called quarter wave printed antenna. The PIFA antenna is shown in Figure (24).

This type of antenna has no dielectric substrate which makes its radiation very interesting for mobile communications since it is very uniform in all planes. When the patch is bypassed with a metal wall, it results in an antenna of longer physical length, indeed, the overall effective inductance of the antenna increases and current paths on the radiating element are naturally lengthened which results in a decrease in the resonant frequency, equivalent to a reduction in the size of the antenna. The two major drawbacks of this miniaturization technique are the degradation of antenna gain and polarization purity.

Figure 24. PIFA antenna structure
(Huynh, 2003)

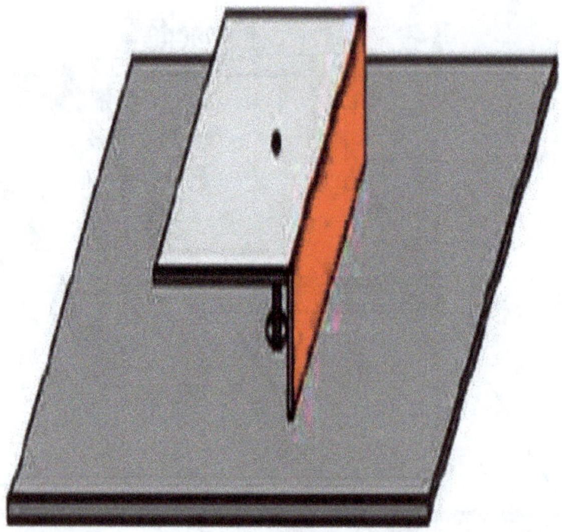

Antenna Wire Monopolar Plate

By adding one or more wires, whose role is to create a parallel resonance, between the patch and the ground plane of the PIFA antenna we obtain a monopolar plate wire antenna as shown in Figure 25.

The size of the patch can be decreased up to $\frac{\lambda}{4}$.

Miniaturization Using Materials

Printed antennas are widely used antennas because of their ease of design, integration and low cost. However, given the context of miniaturization and their relatively large initial sizes, these antennas are often subject to electrical size reduction by the use of materials. We will detail several types of materials used so far to miniaturize the antennas.

Figure 25. Monopolar wire-plate antenna structure

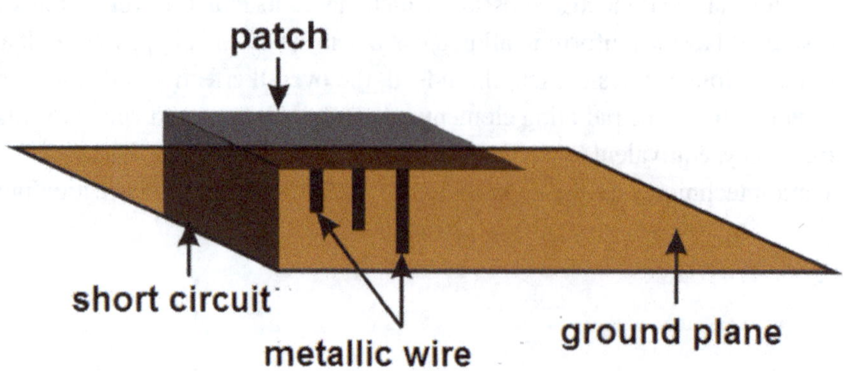

Use of Dielectric Materials

A simple miniaturization technique consists in modifying the radioelectric characteristics of the environment close to the antenna. For reasons of simplicity of implementation, it is preferable, in the case of printed antennas, to modify the substrate that supports the radiating element. Thus, the field lines are concentrated under the antenna. This phenomenon can be explained by the fact that an electric field applied in a highly primitive medium affects the organization of electrical charges, especially on the displacement of charges and the reorientation of electric dipoles. It is well known that high-contrast materials reduce the physical size of the antenna due to a shorter wavelength in the material. Using a dielectric with a high relative permittivity, ε_r typically of the order of 10 to 20, the guided wavelength λ_g decreases, as shown in Equation (17).

$$\lambda_g = \frac{\lambda_0}{\sqrt{\varepsilon_r \mu_r}} \tag{17}$$

Or

λ_0:is the wavelength in free space.

ε_r is the relative permittivity of the material.

μ_r is the relative permeability of the material.

This type of loading leads to a reduction of the bandwidth BW because the quality factor Q is increased.

$$BW = \frac{f}{Q\sqrt{2}} \tag{18}$$

with:

$$Q = \frac{1}{Q_{rad}} + \frac{1}{Q_{cond}} + \frac{1}{Q_{diel}} \tag{19}$$

Q_{rad} Q_{cond} and Q_{diel} are the quality factors of the radiated power of the power loss in the conductor and the power loss in the dielectric respectively.

This is due to a concentration of the electric field in the areas of high permittivity, which makes it difficult to progressively move from propagation in a guided medium to propagation in free space. Indeed, this method results in the storage of a large amount of electromagnetic energy in the near-field region due to the existence of a strong impedance contrast between the dielectric material and the surrounding region in free space.

As a result, for a given operating frequency, it is possible to realize an antenna of reduced dimensions with respect to an antenna which would be printed on a substrate with a lower permittivity. For example, for a microstrip antenna, its loading by a dielectric can be carried out either on the entire surface of the radiating element or in a localized manner or in the entire volume of the cavity. The dielectric materials available for this type of use generally have losses, $tan\delta$ in the order of 0.01 to 0.03, and the confinement of the electric field lines in the dielectric substrate generates a strong capacitive coupling between them. antenna and the ground plane leading to a drop in performance, including a drop in efficiency and thus gain, and the strongly resonant behavior of this type of antennal structure allows the impedance matching only on a very narrow band, and because of this strong resonance, the antenna efficiency becomes lower, regardless of the dielectric losses, so it is a compromise between the desired performance and the reduction in the size of the antenna. antenna.

In practice, the miniaturization of the antennas by using high-contrast materials has become attractive because of the low loss characteristics of the recently available ceramics (ε_r can reach 80 while maintaining and $tan\delta$ of the order of 0.001).

The advantages are therefore their low dielectric losses and excellent temperature stability. Antenna efficiency can be expressed in terms of quality factors (Kula, 2004):

$$\eta = \frac{1}{1 + \frac{Q_{rad}}{Q_{cond}} + \frac{Q_{rad}}{Q_{diel}}} \tag{20}$$

This implies that Q_{rad} must be small compared to Q_{diel} and Q_{cond} to propose efficiencies of nearly one.

For cavity type antennas, such as the microstrip antenna, miniaturization occurs rather easily because the actual mode is trapped in the material placed between the upper metal pellet and the ground plane. In order to illustrate the use of ceramic substrates, Figure 26 shows both the physical length and the radiation efficiency of a $\frac{\lambda_g}{2}$ microstrip antenna at 900 MHz as a function of the permitivity of the ceramic material.

From Figure 26 we find, as one might expect, that the length of the antenna decreases as the permittivity increases. However, efficiency is also a decreasing function of permittivity. This is partly due to the reflection of the wave at the interface between the ceramic substrate and the air, which increases at the same time as ε_r. We can also discuss the basic laws governing the compromises on miniaturization as an explanation. Indeed, the increase in permittivity causes a reduction in size which also contributes to the fall in efficiency. For reasonable values of efficiency (> 0.7) only materials with $\varepsilon_r < 30$ can be used.

Use of Magnetic Materials

Materials with high permeability μ_r. are called “magnetic” materials. They have the advantage of improving both the bandwidth and the radiating device adaptation. Work has shown the possibility of miniaturizing antennas using magnetic materials. The technique used is based on the same principle of

Figure 26. Evolution of the efficiency according to permittivity
(Ghosh, 2001)

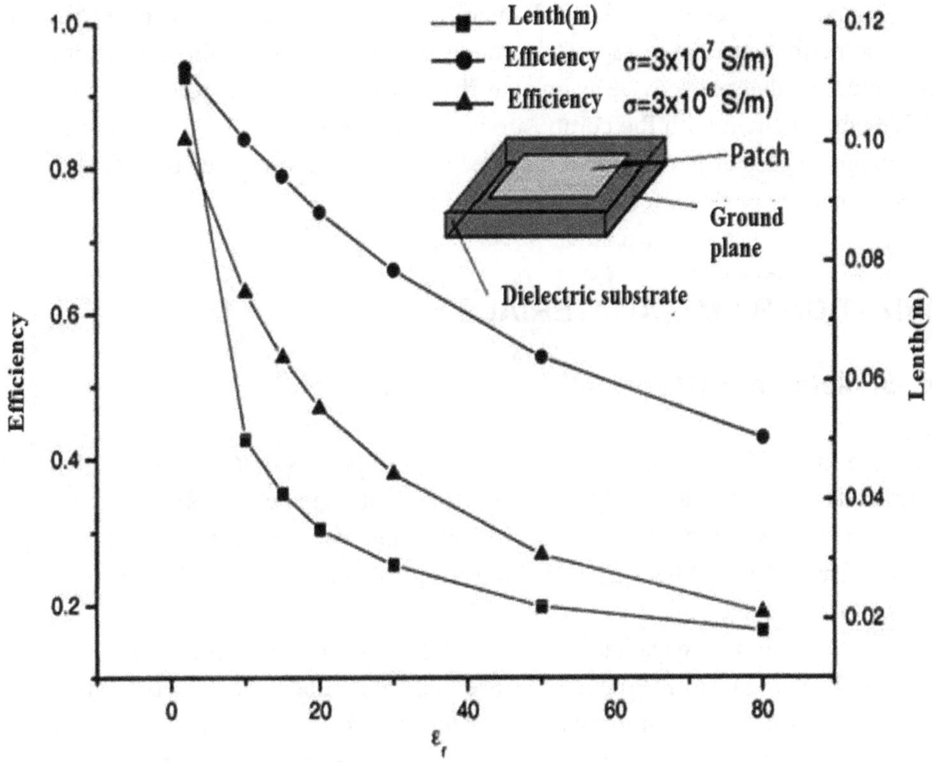

dielectric materials which consists in reducing the electrical length and therefore the size of the antennal structure according to the preceding Equation (17).

Use of Natural Magneto-Dielectric Materials

Another commonly used antenna miniaturization technique is to charge the antenna using magneto-dielectric materials whose permittivity and permeability are both greater than 1. This technique makes it possible to miniaturize an antenna without degrading its performance in terms of bandwidth. According to the work of Hansen and Burke (Hansen, 2000), the bandwidth of a square microstrip antenna is expressed by:

$$BW = \frac{96\sqrt{\frac{\mu_r}{\varepsilon_r}}\frac{h}{\lambda_0}}{\sqrt{2}\left(4+17\sqrt{\varepsilon_r\mu_r}\right)} \tag{21}$$

Where h denotes the height of the substrate and λ_0 the wavelength in the vacuum.

Sarabandi et al. have developed a material consisting of a stack of dielectric and magneto-dielectric layers to miniaturize a printed antenna.

An in-depth study was conducted by P. Ikonen et al. (Ikonen, 2006) on the impact of magneto-dielectrics on the performance of a microstrip antenna. They found that the bandwidth obtained with the magneto-dielectric material is twice as great as that of the dielectric material. On the other hand, the radiation efficiency obtained with the composite is about half that obtained with the dielectric. Thus, the use of the magneto-dielectric material is a source of bandwidth increase but it can lead to losses that can reduce the efficiency of the antenna.

MINIATURIZATION BY METAMATERIALS

Definition of Metamaterials

Metamaterials are materials artificially designed to have properties that are not readily available naturally. Metamaterials have simultaneously negative permittivity and permeability values. Materials with a single negative permittivity are called ϵ-negative(ENG) and materials with only negative permeability are called μ-negatives (MNG). Materials with both negative permittivity and permeability are referred to as double negative (DNG).

The response of a system to an excitation can be largely determined by the properties of the materials involved. In the case of an electromagnetic response, the most important properties are the permittivity ε_r and the permeability μ_r. The permittivity of a material describes the measurement of the ability to withstand the formation of an electric field in itself, while the permeability of a material is a similar response, but for the support of magnetic fields.

The materials in general can then be classified according to the values of ε and μ as shown in the Figure 27 (Mahmoud, 2004).

Use of Metamaterials in the Miniaturization of Printed Antennas

In order to overcome the problems of availability of natural magnetic materials operating at low losses at SHF frequencies, artificial magnetic materials have emerged as a promising solution.

Recently, several studies have been conducted to use these new materials in antennal structures to miniaturize them (Figure 28).

Hao, et al. (Karkkainen, 2005) studied theoretically and experimentally the influence of a metamaterial on the resonance frequency of a patch antenna (Figure 29). In fact, the uniform ground plane is replaced by a periodically etched ground plane of Complementary Split Ring Resonators (CSRRs), as shown in Figure 18. This antenna was manufactured for two different values of the permittivity of the dielectric substrate εr = 3 and εr = 4.7. The measurement results confirm that the resonance frequency of the antenna incorporating CSRRs is significantly lower than that on conventional substrate.

Maslovski and Ikonen recently developed a pattern called metal-solenoid that artificially creates magnetic behavior with a particular arrangement. Karkkainen and Ikonen used a dense and compact network of SRR as an artificial magnetic substrate to reduce the resonance frequency of a microstrip antenna. The digitally studied square antenna has dimensions 60x60mm^2 and is feeding by a coaxial cable placed 1mm from the edge (Figure 30). The distance between the radiating element and the ground plane is H

Figure 27. Permittivity, permeability and refractive index diagram
(Mahmoud, 2004)

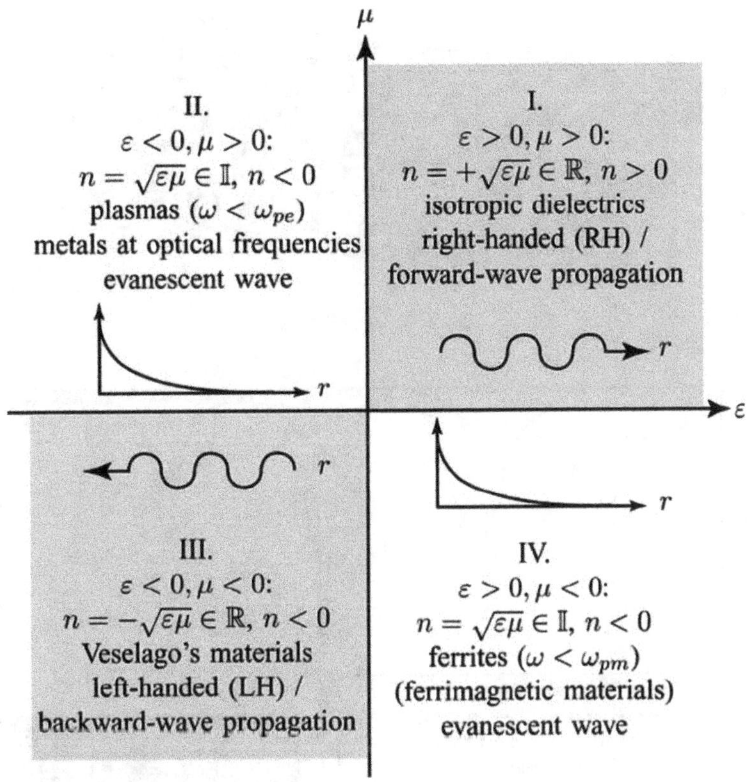

Figure 28. (a)Estimated material parameters for the implemented magnetodielectric substrate and the manufactured prototype antenna. Solid lines for the real parts, dashed lines for the imaginary parts, and; (b) Measured reflection coefficient with different material fillings
(Ikonen, 2006)

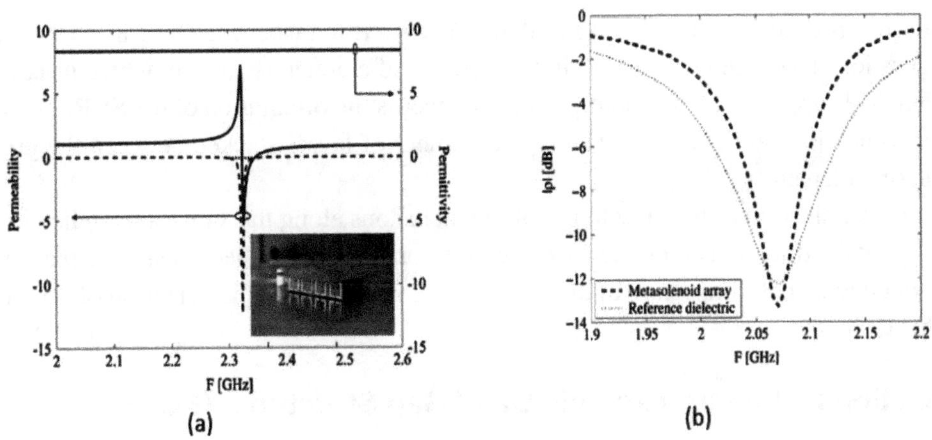

Figure 29. Geometry of the microstrip antenna charged by CSRR and; (b) Coefficient of reflection (Karkkainen, 2005)

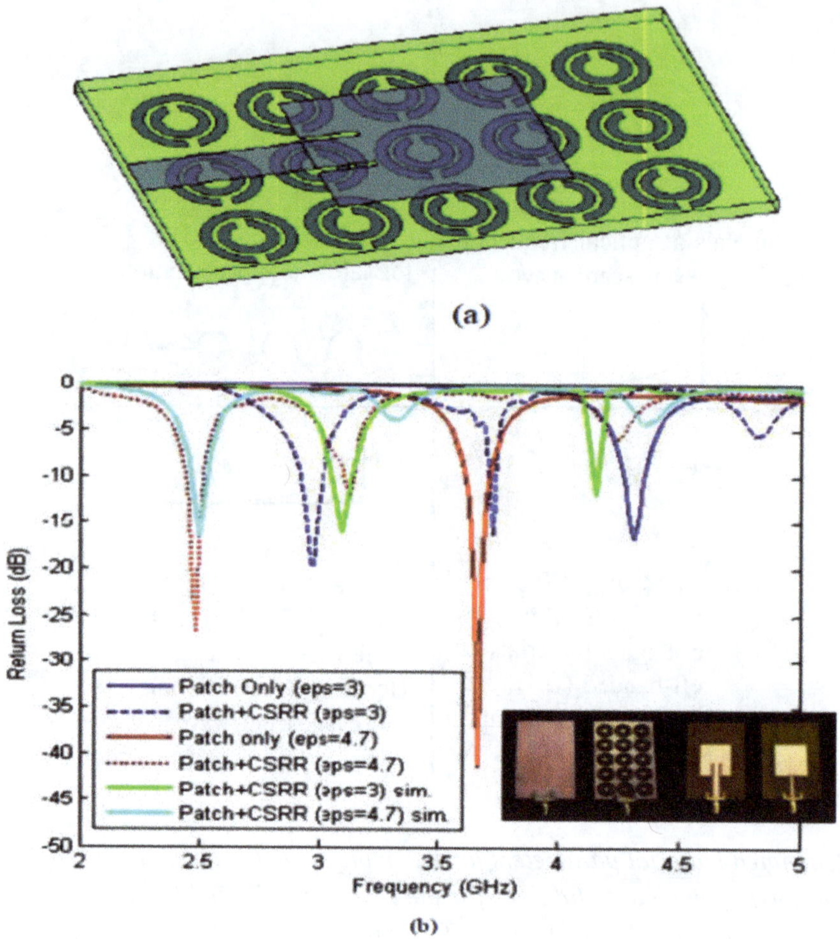

= 10 mm. In the absence of SRR, FDTD simulation gives a resonance frequency at 2.18 GHz. Various ideal configurations have been studied, all in the absence of a dielectric matrix which usually plays the role of mechanical support. The first configuration concerns the orientation of the SRRs and the second the height at which they are placed as well as the number of layers stacked between the ground plane and the radiating element.

Indeed, the examination of the orientation of the inclusions along the x or y axis indicates, as shown in Figure 30, that the direction along the x axis seems more interesting because the displacement of the resonant frequency is more important in this case. As we can see, a miniaturization of 20% is obtained thanks to this technique.

Miniaturization by Electromagnetic Band Gap Structures (EBG)

EBG structures have been of great interest in the electromagnetic community because of the electromagnetic properties that characterize them. They are used extensively in the field of antenna engineering.

Figure 30. (a) A magnetic patch antenna. Axis of the SRRs to x-direction;(b) The input return loss of the magnetic patch antenna
(Simovski, 2004)

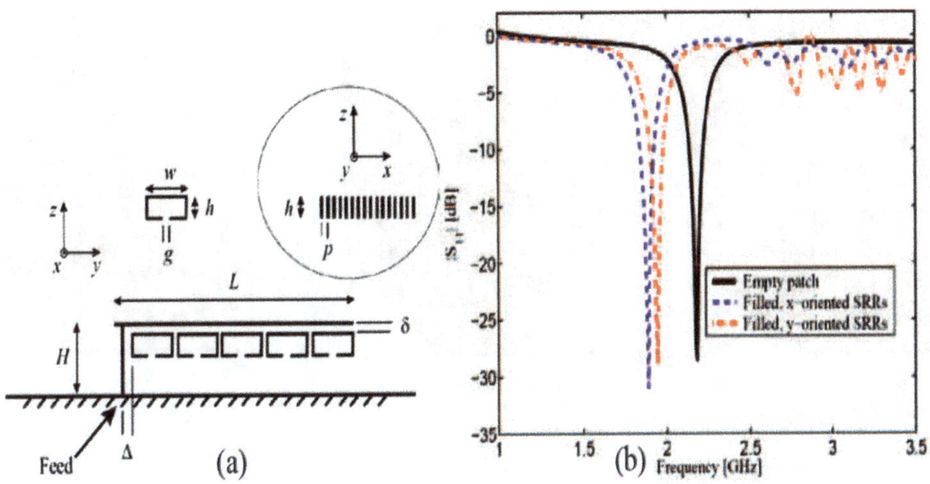

EBG structures are periodic arrangements of the dielectric or metallic elements. They prevent the passage of the electromagnetic wave to certain nails of incidence at certain frequencies called partial bands gap. At specific frequencies, these EBGs prevent the passage of waves in all directions and this frequency region is called a complete band gap. EBG structures are feasible in one, two or three dimensions (Figure 31).

These periodic structures can cause several forbidden bands; however, it should be noted that these bands are not only due to the periodicity of the structure, but also to the individual resonance of each element of the structure.

The tape stop property of EBG structures makes it easy to use multiple applications using this principle. Indeed, at these frequencies, the waves will be reflected and the structure reacts as a mirror. At other frequencies, it will be considered a transparent medium. This concept is illustrated in Figure (32).

Therefore, the use of EBG structures is the best solution that helps to reduce the size of the antenna by eliminating surface waves while keeping its best performance in terms of gain, efficiency, bandwidth ... etc.

Although EBG structures have been used by many researchers in a wide variety of applications (filters, antennas, amplifiers, waveguides, etc.), they have a fundamental disadvantage: the difficulty of modeling due to too many parameters to consider. like the pattern period, shape, size and location. This disadvantage has helped the evolution of Machined Mass Plan Structures (DGS).

Defected Ground Structures (DGS)

Defected Ground Structure (DGS) is a form of pattern engraved on the ground plane. The applications of the defects in the plane of mass (DGS) also find many advantages in term of miniaturization and rejection of the side lobes, and the rejection of the bands of frequency. They have been extensively used in the design of miniature microstrip antennas. The DGS-related disturbance of the current line distribution in the ground plane modifies the characteristics of the propagation line. These DGS, equivalent

Figure 31. Examples of different EBG, (a) 1D, (b) 2D and (c) 3D configurations. The red and yellow areas represent regions of different refractive indices
(Dalia, 2012)

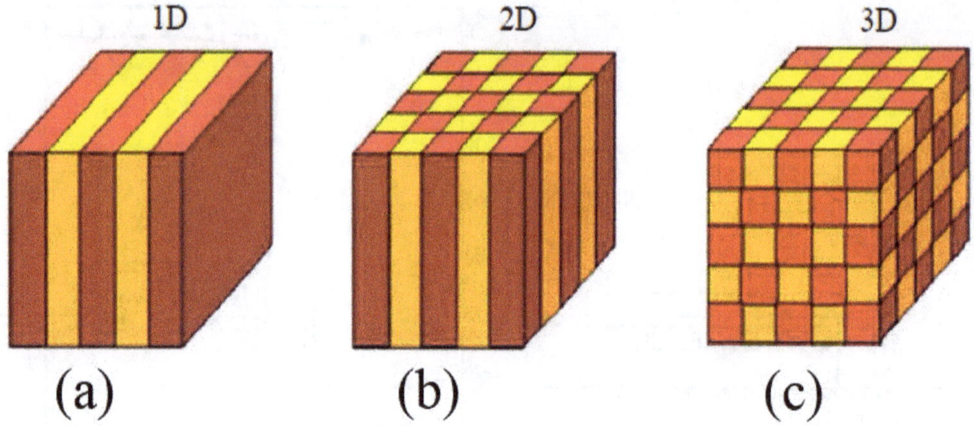

Figure 32. Diagram illustrating the application of EBG as a mirror and its comparison with a metal reflector
(Dalia, 2012)

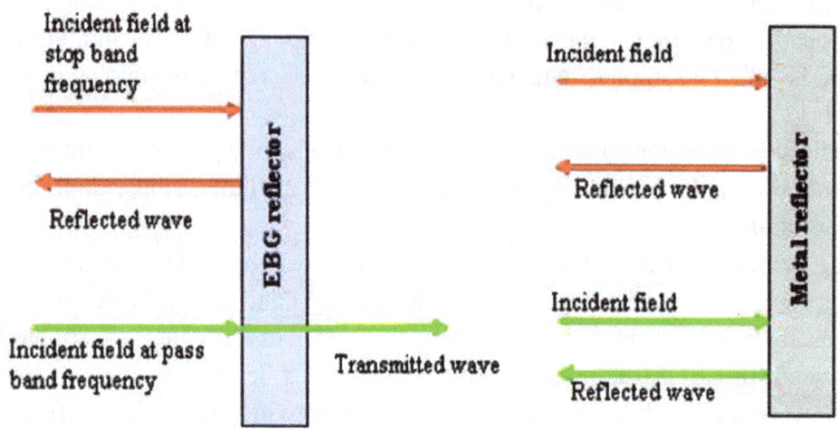

to resonators, impose transmission zeros, hence a frequency behavior that has interesting properties. In other words, any defect etched in the ground plane under the microstrip line changes the effective capacitance and inductance of microstrip line by adding slot resistance, capacitance, and inductance.

DGS are EBG structures, which are characterized by their excellent property of rejection of frequency bands. They are widely used in several passive or active devices namely filters, amplifiers, resonators, couplers and antennas.

Since EBG-based structures are difficult to model because of their many design parameters, DGS-based designs are attracting increasing interest from many researchers. Indeed, these can offer the benefits of EBG based structures while keeping a simplified modeling.

DGSs are arbitrary structures that are etched into the ground plane of microstrip antennas and can take many forms as shown in Figure (33) (Dalia, 2012).

Figure 33. Different DGS structures
(Dalia, 2012)

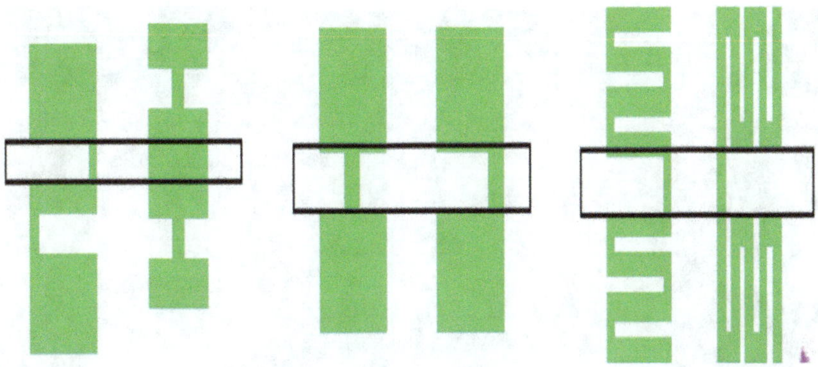

Other more complex 2D defects engraved in the ground plane have also been proposed, including meander lines, slit ring resonators CSSR and fractal lines.

Through the use of several types of disturbances in the ground plane, the authors report respectively 34%, 42% and 70% reduction in size of a microstrip antenna compared to conventional structures referred to.

The last two proposed antenna configurations have been designed as shown in Figure 34 using photolithographic techniques. The reduction of the resonance frequency of the antenna is due to the increase of the electrical and magnetic coupling between the mass and the radiating antenna plate. The measured and simulated results are shown in Figure 35. The second proposed antenna was also fabricated and multi-band operation and antenna size reduction were performed, and the comparison between simulation and measurement of the This reflection is presented in Figure 36.

In addition, there are other structures called DMS, these structures are based on the same principle as the DGS. They can take the same forms mentioned before for the DGS and they are also characterized by their simplicity of extraction of the equivalent circuit and they are used for the same applications mentioned before. The only difference that differs from the DGS is that they are engraved on the ground plane antennas. However, DMS are etched on the antenna transmission line.

High Impedance Electromagnetic Surface (HIES)

Unlike conventional conductors, these surfaces prevent the propagation of surface waves and they reflect electromagnetic waves without reverse phase. Its geometry consists of a metal plate whose texture is in the form of tiles as shown in Figure (37). They are connected with the ground plane by vertical vias. This structure acts as a filter to prevent the propagation of electric current and consequently the elimination of surface waves.

This structure can be modeled by the following equivalent circuit (Figure 38).

The impedance of the equivalent circuit is given by the following equation:

$$Z = \frac{jL\omega}{1-\omega^2 LC} \tag{17}$$

Figure 34. The fabricated two antennas
(Dalia, 2012)

Figure 35. Measured and simulated reflection coefficient of antenna without and with one and two gasket unit cell
(Dalia, 2012)

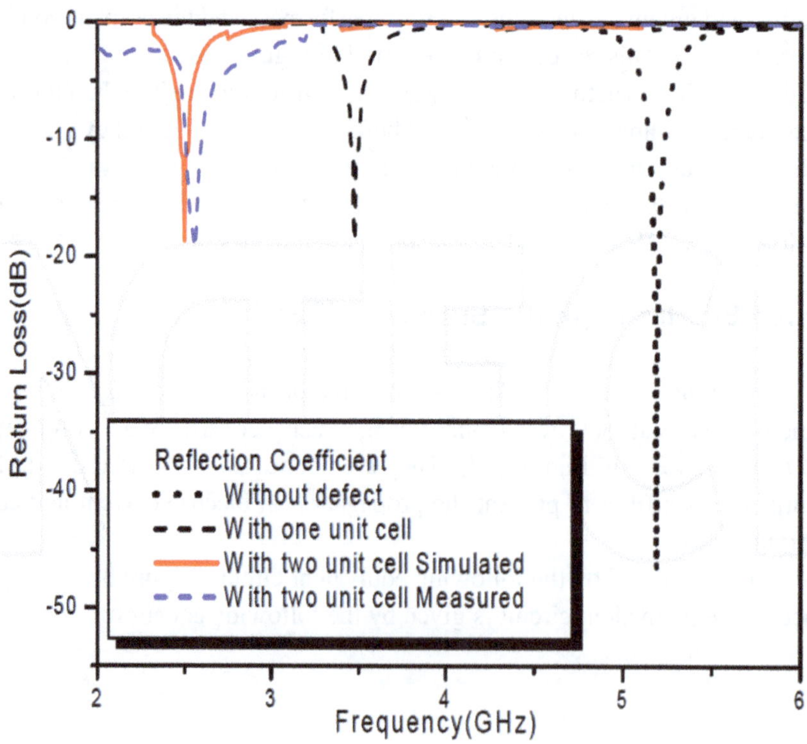

Figure 36. Measured and simulated reflection coefficient of antenna without and with two-unit cell of spiral DGS simulated and measured
(Dalia, 2012)

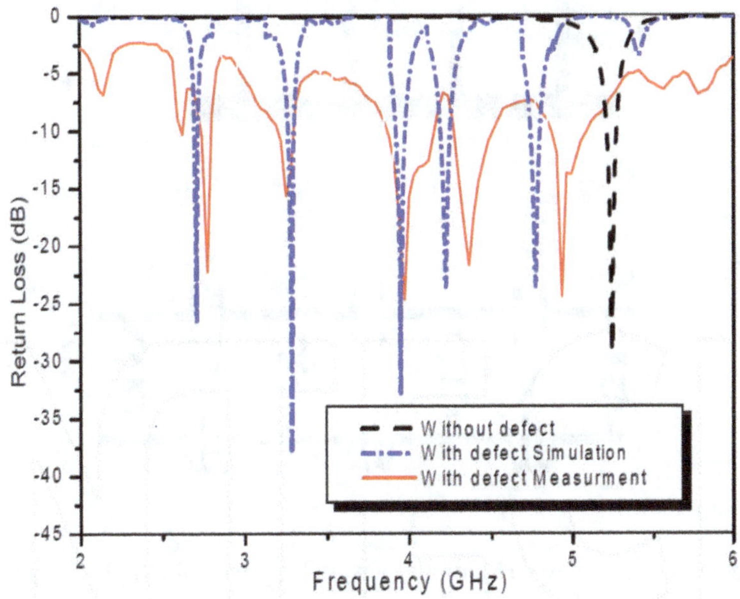

Figure 37. Top view of the high-impedance surface

Figure 38. Equivalent circuit model for the high-impedance surface
(Dalia, 2012)

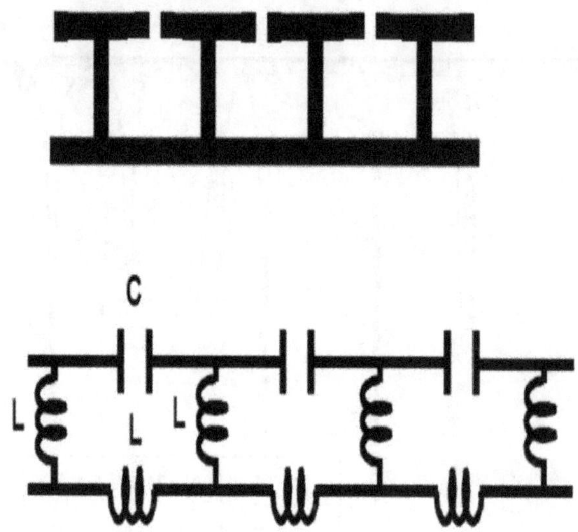

For low frequencies we will have TM mode waves. However, for high frequencies, there will be waves of the TE mode.

NEW DESIGN EXAMPLES OF MINIATUREMICROSTRIPANTENNAS ARRAY FOR WIRELESS COMMUNICATIONS APPLICATIONS

The following sections will discuss and present our contributions in this research field of miniaturization of antennas array by introducing some miniature Antennas array based on the use of DGS technique and other methods.

A Miniature Microstrip Patch Antenna Array With Defected Ground Structure (Oulhaj, 2016)

The proposed microstrip patch antenna array is shown in Figure 39. The two truncated corners are used to obtain circular polarization. This antenna is built on an FR4 substrate with 1.6mm height, the dielectric constant is 4.4 and the loss tangent is 0.021. The antenna is designed and optimized by using CST Microwave Studio tool. We introduce the DGS in order to shift the resonance frequency of the microstrip antenna array without DGS. We study its effect on the antenna properties, specially the resonance frequency. CST Microwave Studio has carried out the simulations. In the absence of DGS, we find that the resonance frequency of our microstrip patch antenna array is at 10 GHz.

Shown in Figure 39 is the size of the fabricated DGS microstrip antenna array that operates at 7.5 GHz. In order to measure the scattering parameters of the proposed antenna, we have been employing a Rohde and Schwarz ZVB 20 vector network analyzer whose frequency range is limited to 20 GHz.

Figure 39. Prototype of DGS antenna array: (a) top view; (b) bottom view

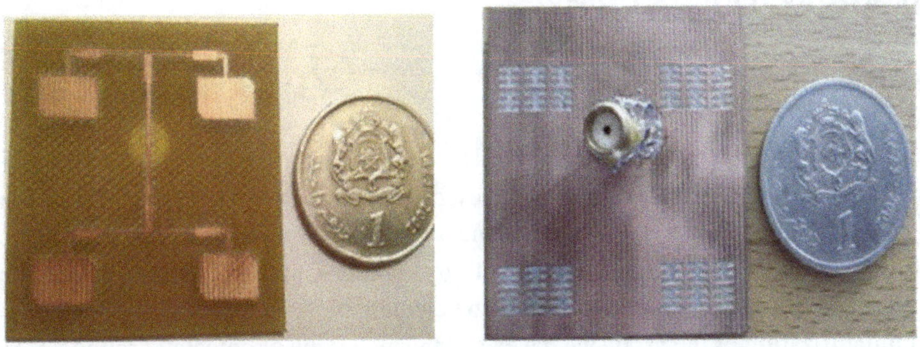

(a) (b)

The designed antenna structure is fabricated and tested; the measurement and simulation results are in good agreement as shown in Figure 40. The final circuit has a return loss of -17dB at resonance frequency equal to 7.5GHz.

In this study we have carried out the design of a miniaturized antenna array with DGS structure. This antenna has good performances in terms of matching input impedance with radiation pattern. The resonance frequency of the initial antenna (without DGS) has been shifted from 10 GHz to 7.5 GHz after introducing DGS. Thus, by introducing DGS we have been able to reduce the antenna size up to 43.75% as compared with the antenna without DGS.

Figure 40. Measurement and simulation results for the antenna array with DGS resonating at 7.5 GHz

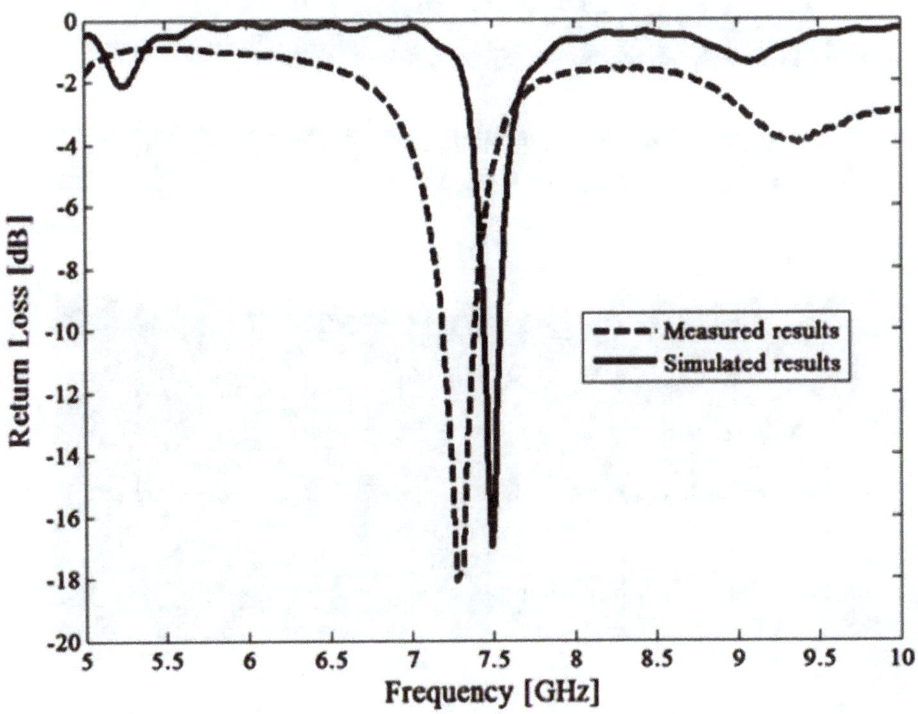

A NEW DESIGN OF A WIDEBAND MINIATURE ANTENNA ARRAY (OULHAJ, 2017)

Antenna Array Design Considerations

After the validation of the patch element at 10 GHz, the idea was to construct the array by using this element and in the same time to shift the frequency band to a low frequency in order to miniature the final antenna array and to let the same performances in terms of matching input impedance, circular polarization with an increasing of gain and directivity. We have started to achieve the array by using microstrip lines and T-junction to feed each antenna element.

For the miniaturization of the array, we have used the technique of defected ground. In order to tune and to change the high bandwidth by a low one, we have used the optimization techniques integrated in CST-MW. The whole circuit is optimized and simulated taking into account a high meshing density. After many series of optimization and sweep we have reached the goal fixed by validating the proposed final antenna array structure depicted in Figure 42. This figure shows the top view of the array and the back view presenting the final structure of the optimized defected ground.

The obtained measured and simulated results are illustrated in Figure 42.

Figure 41. A photograph of the fabricated antenna array: (a) Top Face; (b) Back Face

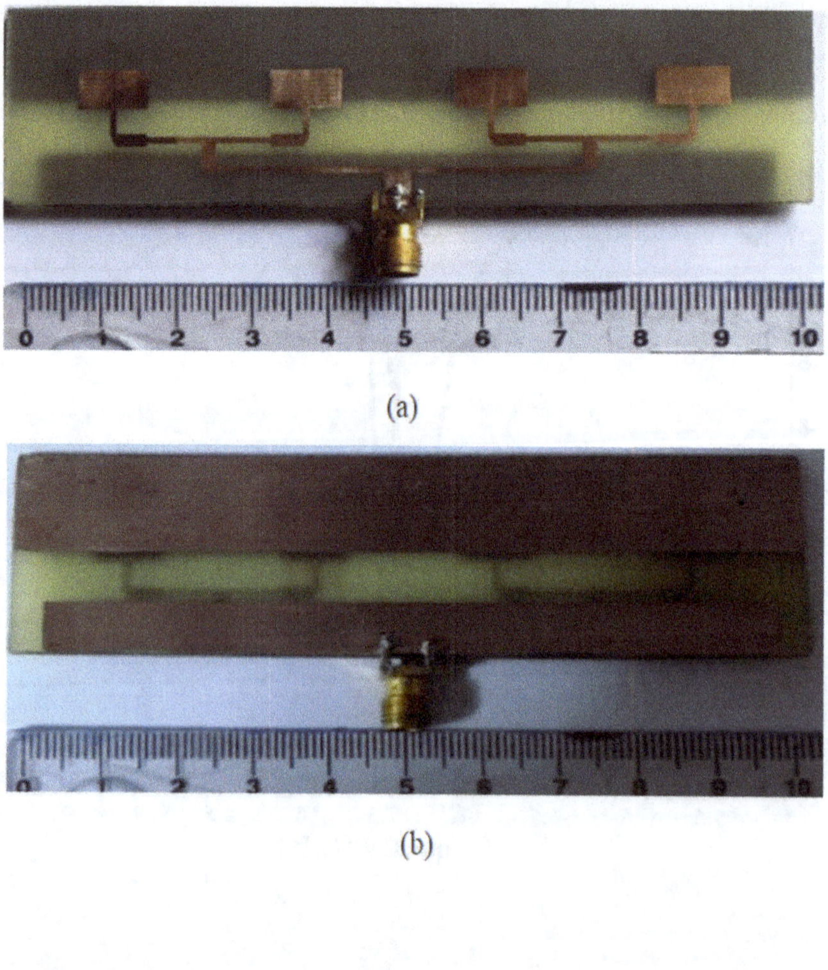

(a)

(b)

Figure 42. Measured and simulated results vs frequency

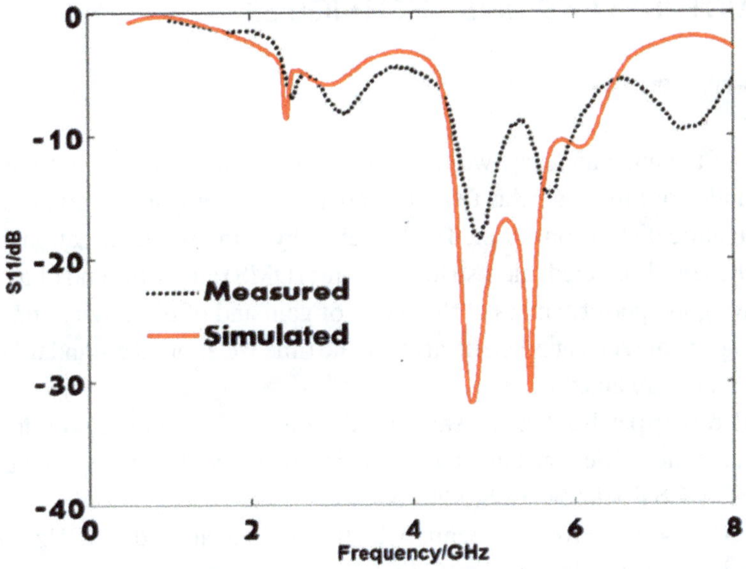

The measured results permit to obtain an antenna array which functions between 4.53GHz and 5.98GHz, with a reflection coefficient which is not under -10 dB at 5.32GHz. Which is due to the characterization of the substrate which can present a difference concerning the permittivity and conditions of fabrication. The final circuit validates the idea of using DGS to miniature the antenna array which functions in a low frequency band having dimensions of an antenna validated at 10 GHz.

Table 1 presents a comparison between the proposed antenna array study and other configuration in literature.

From this table we can conclude that the proposed and validated antenna array presents good performances in comparison with other structures validated in literature, in terms of dimensions, gain and relative bandwidth.

Table 1. Performance of the proposed antenna and other structures

Bandwith(Ghz)	Demensions(mm^2)	Gain (dB) Relative Bandwidth
Proposed antenna array 4-6 Structure 1 9.5-10.6	102.48 - 31.39 114-61.2	5 37.59% 4.25 11%
Structure 2 24.05-24.25	Not plane, length 160 mm	1.6 0.8%
Structure 3 2.32-2.58	Not plane, length 325 mm	3 10.6%

ANTENNA ARRAY MINIATURIZATION UTILIZING DGS (DEFECTED GROUND STRUCTURE) AND DMS TECHNIQUES

Antenna Element Design

To start the design of the antenna array, we have to define the antenna element which will be used in the array. The main idea of this study was to realize an antenna element operating in a given frequency in order to have miniature dimensions of the final antenna by using miniaturization techniques defected ground structure (DGS) and defected microstrip structures (DMS) and at the end of inserting this antenna into the array to have good performances at the levels of gain and of directivity etc.

We chose to design an antenna at 5.8GHz, at the same time we applied a standard technique allowing the realization of a miniature antenna.

The proposed microstrip patch antenna is shown in Figure 43. This antenna is built on an FR4 substrate with 1.6mm height, the dielectric constant is 4.4 and the loss tangent is 0.021. The antenna is designed and optimized by using CST Microwave Studio tool.

As shown in Figure 44, we have an antenna which functions around 5.8GHz with good matching input impedance and having a reflection coefficient under -26.46dB.

In addition, after doing a study of the conventional antenna at the frequency5.8 GHz, Figure 45 shows the difference in size between the conventional antenna and the DGS and DMS antenna at the same resonance frequency of 5.8 GHz.

Indeed, thanks to the introduction of DGS and DMS cells, we have noticed a miniaturization rate of 60%. As such, the use of these structures is considered one of the best solutions to save space in the design and manufacture of microwave components.

Figure 43. The proposed patch antenna element:(a) top view (b) bottom view (L=7 mm, W=15.55mm, l=7.54mm, w=0.52mm, W1=0.3mm, L1=19.54mm, W2=0.83mm, L2=5.67mm, L3=8.6mm, L4=10.94mm)

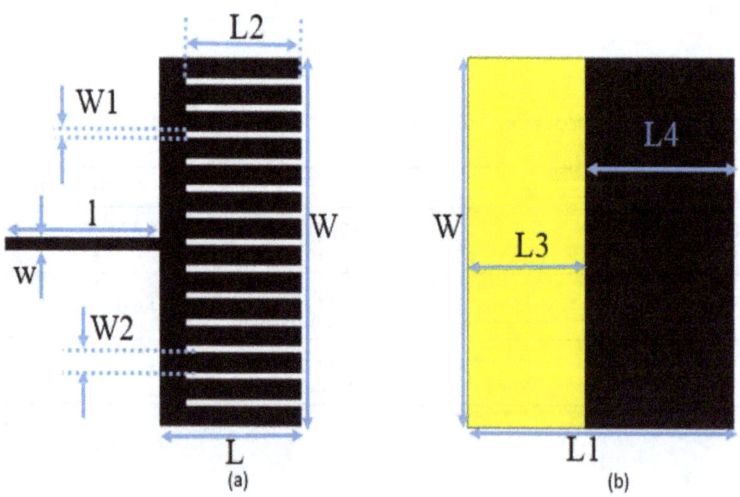

Figure 44. Reflection coefficient versus frequency

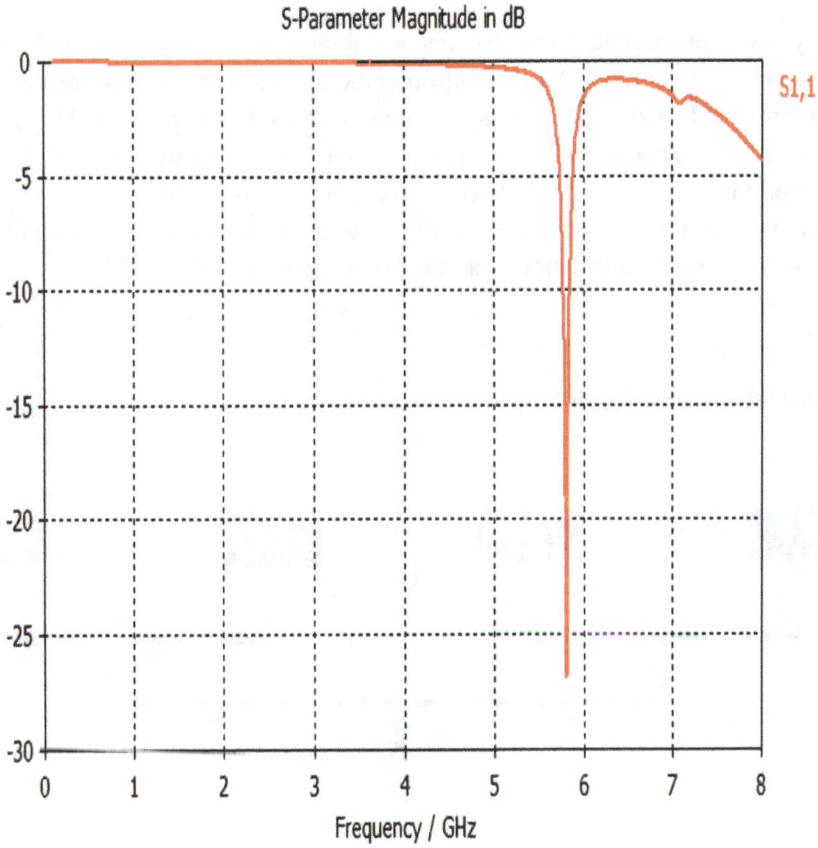

Figure 45. Conventional Antenna Compared to the DGS and DMS Antenna

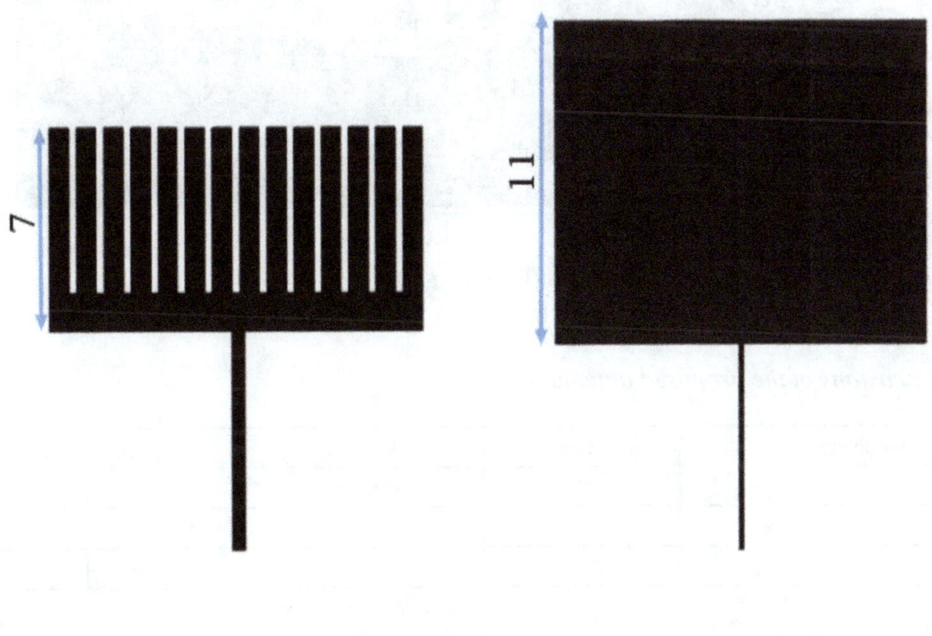

Microstrip Patch Antenna Array

The antenna array based on previous microstrip patch antenna associated with the calculated microstrip lines and power dividers. After many series of optimization and sweep, we have reached the goal fixed by validating the proposed final antenna array structure depicted in Figure 46. This figure shows the top view of the array and the back view presenting the final structure of the optimized defected ground.

The different optimized parameters of the final circuit are listed in Table 2.

The final array has been validated in simulation to have a good input impedance matching the frequency 5.8GHz, with a reflection coefficient as shown in Figure 47, under -10dB.

Figure 46. The proposed antenna array

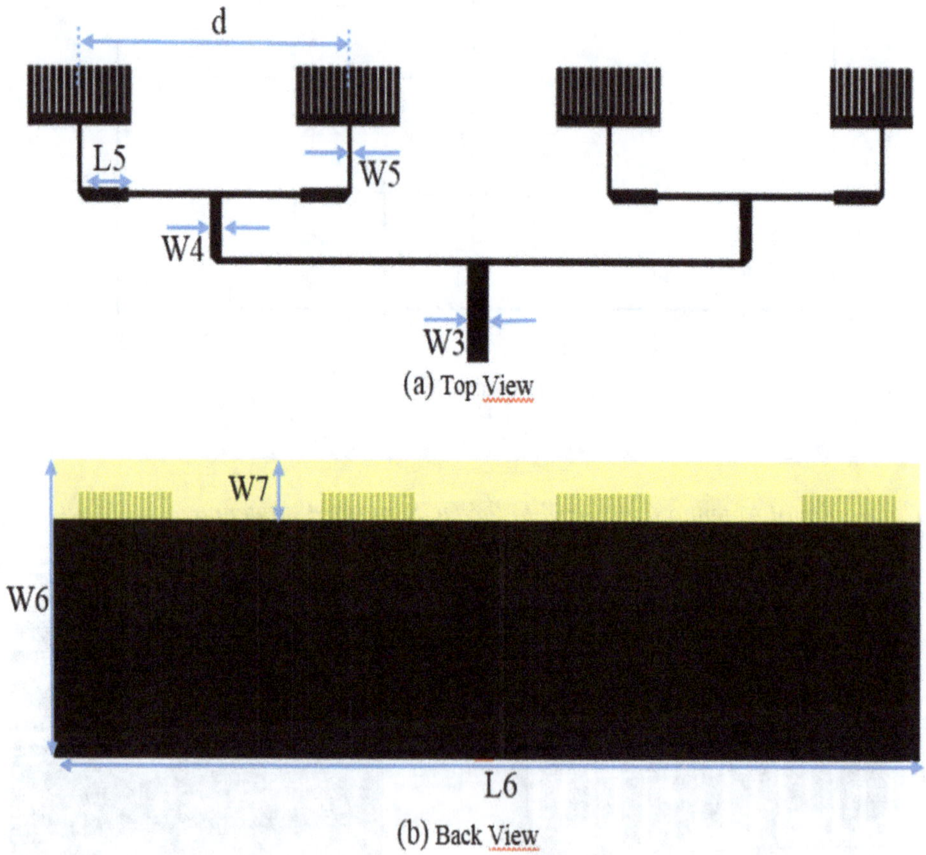

Table 2. Dimensions of the proposed antenna array

Parameter	L5	L6	W3	W4
Value(mm)	7	145.09	3.15	1.66
Parameter	W5	W6	W7	d
Value(mm)	0.52	39.54	8	41

Figure 47. Variation of reflection coefficient versus frequency

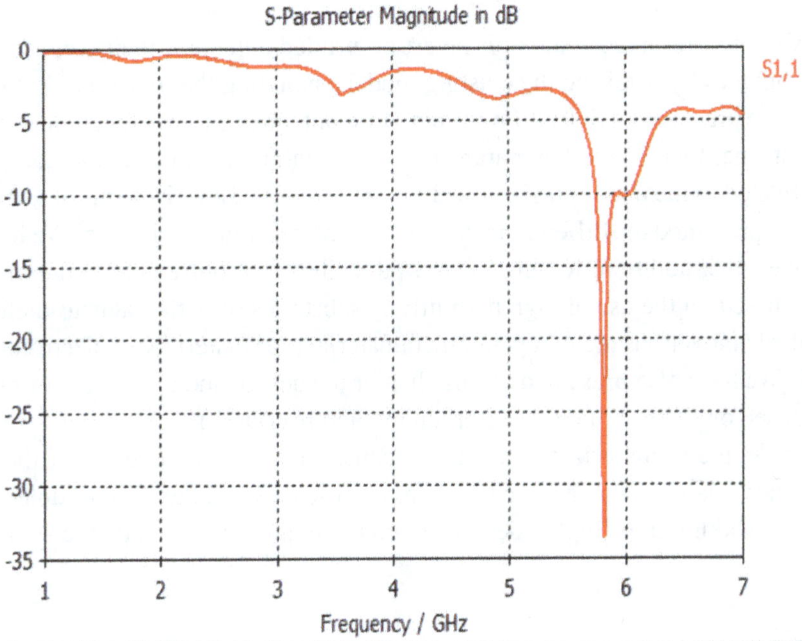

Figure. 48 shows, respectively, the patch antenna array radiation pattern in the E-plane and the H-plane using the DGS and DMS cells etched on the ground plane and on the antenna at the 5.8 GHz resonance frequency.

For the radiation behavior of the antenna array, we can analyze the array radiation patterns in the E and H planes, as shown in Figure 48. As we have seen, we have a multi-beam directional.

Figure 48. The radiation pattern in the E and H plane at 5.8GHz

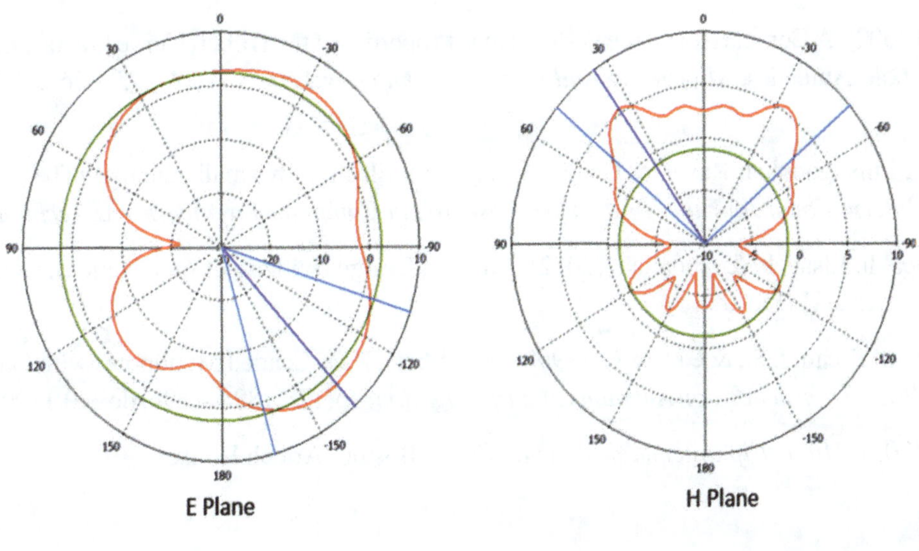

CONCLUSION

In this chapter, we first presented a state of the art on printed antennas, their characteristics and their different feed techniques by detailing the existing models including the transmission line model.

Subsequently, we presented a definition of miniature antennas and the fundamental physical limits of this type of antenna. One of the key parameters of miniature antennas is the quality factor whose minimum value related to the maximum bandwidth has been and still is the focus of interest of researchers. In addition, we presented the different techniques of antenna miniaturization. We have reviewed the classical techniques (slot addition, folding...), as well as the work presented on the miniaturization of printed antennas, based on the use of high permittivity substrates or with magneto-dielectric materials, or with the technical innovations including new metamaterial type materials, or electromagnetic bandgap structures (EBG).We have also presented in this chapter the design and realization of miniature printed antenna arrays, based on the use of defected ground structures (DGS).

As a general rule, the miniaturization of the antennas results in a reduction of the bandwidth and the efficiency of the radiator. To obtain the best compromise between the area occupied by the antenna, its gain and its bandwidth will be highly dependent on the requirements set by the suitable application.

REFERENCES

Alam. (n.d.). Design and performance analysis of microstrip array antenna. *Session 5AP*.

Bahl, I. J., & Bhartia, P. (1980). *Microstrip antennas*. Dedham, MA: Artech House.

Balanis, C. A. (2005). Antenna theory: analysis and design (3rd ed.). Academic Press.

Baliarda, C. P., Romeu, J., & Cardama, A. (2000, November). The Koch monopole: A small fractal antenna. *Antennas and Propagation, IEEE Transactions*, *48*(11), 1773–1781. doi:10.1109/8.900236

Barlevy, A. S., & Rahmat-Samii, Y. (2001, June). Characterization of electromagnetic band-gaps composed of multiple periodic tripods with interconnecting vias concept, analysis, and design. *IEEE Transactions on Antennas and Propagation*, *49*(3), 343–353. doi:10.1109/8.918607

Best, S. R. (2002). A Comparison of the Performance Properties of the Hilbert Curve Fractal and Meander Line Monopole Antennas. *Microwave and Optical Technology Letters*, *35*(4), 258–262. doi:10.1002/mop.10576

Chaurasia, Pimpalgaonkar, Raval, & Upadhyaya. (2016). Electrically small antennas: The art of miniaturization. *International Journal of Advanced Research in Computer Engineering & Technology*, *5*(2).

Dalia, Elsheakh, Elsadek, & Abdallah. (2012). Antenna Designs with Electromagnetic Band Gap Structures Metamaterial. InTech.

Foltz, H. D., McLean, J. S., & Crook, G. (1998, December). Disk-Loaded Monopoles with Parallel Strip Elements. *IEEE Transactions on Antennas and Propagation*, *46*(12), 1894–1896. doi:10.1109/8.743844

Garg, R. (2001). *Microstrip antenna design handbook*. Boston: Artech House.

Ghosh, I. S., Hilgers, A., Schlenker, T., & Porath, R. (2001). T.shlenker, R.Porah, "Ceramic microwave antennas for mobile applications. *Journal of the European Ceramic Society*, *21*(15), 2621–2628. doi:10.1016/S0955-2219(01)00327-2

Grenier, D. (2016). *Antennes et Propagation radio (PhD thesis)*. Université Laval Québec Canada.

Gronau, G., & Wolff, I. (1986, May). Aperture-coupling of a rectangular microstrip resonator. *Electronics Letters*, *22*(10), 554–556. doi:10.1049/el:19860378

Hall, P. S., Wood, C., & James, J. R. (1981). *Microstrip antenna theory and design*. London: IEE Press.

Hansen, R. C., & Burke, M. (2000). Antennas with magneto-dielectrics. *Microwave and Optical Technology Letters*, *26*(2), 75–78. doi:10.1002/1098-2760(20000720)26:2<75::AID-MOP3>3.0.CO;2-W

Huque, Hosain, Islam, & Chowdhury. (2011). Design and Performance Analysis of Microstrip Array Antennas with Optimum Parameters for X-band Applications. *International Journal of Advanced Computer Science and Applications*, *2*(4).

Huynh, M.-C., & Stutzman, W. (2003). Ground plane effects on planar inverted-f antenna (pifa) performance. *Microwaves, Antennas and Propagation, IEE Proceedings*, *150*(4), 209–213. 10.1049/ip-map:20030551

Ikonen, M., & Tretyakov. (2006). Antennas with Artificial Magneto-Dielectric substrates: The Dominant Role of Frequency Dispersion. In *Antennas and Propagation Society International Symposium*. IEEE.

Ikonen, P. M. T., Rozanov, K. N., Osipov, A. V., Alitalo, P., & Tretyakov, S. A. (2006). Magnetodielectric Substrates in Antenna Miniaturization: Potential and Limitations. *Antennas and Propagation, IEEE Transactions*, *54*(11), 3391–3399. doi:10.1109/TAP.2006.884303

Joshi & Das. (1982). Impedance of a radiating slot in the ground plane of a microstripline. IEEE Trans. Antenna and Propagation, AP-30.

Karkkainen, M., & Ikonen, P. (2005). Patch Antenna with Stacked Split-Ring Resonators as Artificial Magneto Dielectric Substrate. *Microwave and Optical Technology Letters*, *46*(6), 554–556. doi:10.1002/mop.21048

Katehi, P. B., & Alexopoulos, N. G. (1984, November). On the modeling of electromagnetically coupled microstrip antennas-the printed strip dipole. *IEEE Transactions on Antennas and Propagation*, *AP-32*(11), 1179–1186. doi:10.1109/TAP.1984.1143235

Kula, Psychoudakis, & Liao, Chen, Volakis, & Halloran. (2006). Patch Antenna Miniaturization Using Thick Truncated Textured Ceramic Substrates. *IEEE Antennas & Propagation Magazine*, *48*(6).

Luk, K., Chair, R., & Lee, K. (1998). Small rectangular patch antenna. *Electronics Letters*, *34*(25), 2366–2367. doi:10.1049/el:19981643

Mahmoud, S. F. (2004). A New Miniaturized Annular Ring Patch Resonator Partially Loaded by A Metamaterial Ring with Negative Permeability and Permittivity. *IEEE Antennas and Wireless Propagation Letters*, *3*, 19–22. doi:10.1109/LAWP.2004.825092

Mailloux, R. J. (2007). Electronically Scanned Arrays. *Synthesis Lectures on Antennas*, *2*(1), 1–82. doi:10.2200/S00081ED1V01Y200612ANT006

Marques, R. (2007). Metamaterials with Negative Parameters Theory, Design, and Microwave Applications. Academic Press.

Necibi, Ferchichi, Vuong, & Gharsallah. (2014). Miniaturized CSRR TAG Antennas for 60GHz Applications. *International Journal of Electrical and Computer Engineering, 4*(1), 64-74.

Nguyen, H. T., Noghanian, S., & Shafai, L. (2005). Microstrip patch antenna miniaturization by slots loading. *IEEE Antennas and Propagation Society International Symposium*, 215-218. 10.1109/APS.2005.1551525

Oulhaj, Touhami, Aghoutane, & Tazon. (2016). A Miniature Microstrip Patch Antenna Array with Defected Ground Structure. *IJMOT, 11*(1).

Oulhaj, O., Touhami, N. A., Aghoutane, M., & Kchairi, A. B. (2017, August). A New Design of a Wideband Miniature Antenna Array. *Iranian Journal of Electrical and Computer Engineering, 7*(3), 1850–1857.

Pozar. (1992, January). Microstrip antennas. *IEEE Trans. Antennas Propagation.*

Pozar, D. M. (1992, January). Microstrip antennas. *Proceedings of the IEEE, 80*(1), 79–81. doi:10.1109/5.119568

Pozar, Schaubert, & Adrian. (1989, June). Effect of microstrip antenna substrate thickness and permittivity: Comparison of theories and experiment. *IEEE Transactions on Antennas and Propagation, AP-37*(6), 677–682.

Scardelletti, M. C., Ponchak, G. E., Merritt, S., Minor, J. S., & Zorman, C. A. (2008). Electrically small folded slot antenna utilizing capacitive loaded slot lines. *Radio and Wireless Symposium, 2008 IEEE*, 731-734. 10.1109/RWS.2008.4463596

Schaubert, D. H., Pozar, D. M., & Adrian, A. (1989, June). Effect of Microstrip Antenna Substrate Thickniss and Permittivity: Comparision of Theories and Experiment. *IEEE Transactions on Antennas and Propagation, AP-37*(6), 677–682. doi:10.1109/8.29353

Simovski, Sochava, & Tretyakov. (2004). New compact and wide-band high impedance surface. In *Antennas and propagation Society International Symposium*. IEEE.

Stutzman, W. L. (1998). Estimating directivity and gain of antennas. *Antennas and Propagation Magazine, IEEE, 40*(4), 7–11. doi:10.1109/74.730532

Time-Harmonic Electromagnetic Fields. (1961). *McGraw-Hill Book Co.*

Visser, H. J. (n.d.). Array and Phased Array Antenna Basics. John Wiley & Sons. doi:10.1002/0470871199

Chapter 5
1D Electromagnetic Band Gap Analysis and Applications

Abdelmoumen Kaabal
Abdelmalek Essaâdi University, Morocco

Mustapha El Halaoui
Abdelmalek Essaâdi University, Morocco

Saida Ahyoud
Abdelmalek Essaâdi University, Morocco

Adel Asselman
Abdelmalek Essaâdi University, Morocco

ABSTRACT

In this chapter, a detailed study of the one-dimension electromagnetic band gap (1D-EBG) structures and their application in a directional antenna design are presented. To improve the ability and analyze and understand the behavior of 1D-EBG, three techniques of analysis are developed. The results show that the periodicity of the dielectric permittivity makes it possible to stop the waves propagation in certain frequency bands. A comparison between the different methods shows an excellent agreement. An evolution of the transmission coefficient of a structure consisting of six layers with a cavity of thickness equal a wavelength in the middle of the structure, shows that there is a peak of transmission which is formed at the center frequency of the band gap and reflects a resonance phenomenon. This phenomenon of frequency filtering is exploited for the design of a directive EBG antenna by introducing an excitation to the cavity center.

DOI: 10.4018/978-1-5225-7539-9.ch005

INTRODUCTION

Electromagnetic band gap (EBG) structures have attracted the attention of several researches in recent years, the EBG structures are regular arrays of materials with different refractive indices and are classified mainly into three categories, that is, one-dimensional (1D) (Joannopoulos, Johnson, Winn,, & Meade, 2011; Moghaddam, Entezar, & Adl, 2013), two-dimensional (2D) (Baccarelli, Paulotto, & Di Nallo, 2007), and three-dimensional (3D) (Mizeikis, Seet, Juodkazis, & Misawa, 2004; Johnson, & Joannopoulos, 2000) structures according to the dimensionality of the stack. The EBG structures that work in the microwave and far-infrared regions are relatively easy to fabricate. Those that work in the visible region, especially 3D ones are difficult to fabricate because of their small lattice constants. Into design a 3D EBG structures appropriately, there appears a frequency range where no electromagnetic eigenmode exists. Frequency ranges of this kind are called band gaps, since they correspond to band gaps of electronic eigenstates in ordinary crystals (Yablonovitch, 1993; John, 1987). In order to gain an intuitive understanding of the photonic bands and band gaps, the 1D EBG structures well be studied in more detail, then their applications in a directional antenna design. The simplest case in which two materials different are stacked alternately. Actually, many basic ideas are common and they will be utilized to build fundamental theories of the EBG structures (Sakoda, 2004; Kushwaha, & Kumar, 2014; Winn, Fink, Fan, & Joannopoulos, 1998; Subodha Mishra, & Satpathy, 2003), as will be shown.

The one-dimensional electromagnetic band gap structure consists in stacking two layers of materials with different permittivities whose periodicity in only one direction (Choubani, Choubani, Gharsallah, David, & Mastorakis, 2010; Joannopoulos, Johnson, Winn, & Meade, 2011). This arrangement is not a new idea. Lord Rayleigh (1887) published one of the first analyses of the optical properties of multilayer films (Rayleigh, 1888). As will be seen, this type of photonic crystal can act as a mirror (Bragg mirror) for light with a frequency within a specified range, and it can localize light modes if there are any defects in its structure. These concepts are commonly used in dielectric mirrors and optical filters (Baumeister, 1958; Fattal, Li, Peng, Fiorentino, & Beausoleil, 2010).

Taking the x-axis a direction perpendicular to the surface of the dielectric layers, and the y-axis a direction of the polarization. Only the electromagnetic waves propagated in the x direction and polarized linearly are deal here. The band gaps are found if the thicknesses of each layer are equal to a quarter of the guided wavelength.

In this chapter, the authors have developed the technique of the 1D EBE analysis and applications. these techniques has notably contributed to improving the ability to analyze and understand the behavior of this type of structure. The chapter is divided into four major parts.

In the first part, a study of the propagation of electromagnetic waves in periodic structures that consists of solving Maxwell's equations taking into account the Bloch-Floquet property, this method leads to the formulation of the dispersion equation to determine the bandgaps frequencies (Collin, 1960; Busch, Von Freymann, Linden, Mingaleev, Tkeshelashvili, & Wegener, 2007; Cheney, & Kincaid, 2012).

The eigenvalue problems for the radiation field in the EBG structures will be formulated in second part. An Hermitian operator related to the electric field will be introduced, and the properties of its eigenfunctions will be studied in detail. A straight forward solution to the exact eigenvalue equation for the one-dimensional periodic problem is obtained through the use of the Fourier series. The eigenvalue problem of the wave equation of the radiation field in the EBG structure will be formulated in the first step, and a general numerical method to solve it as will be given (Zeidler, 1995; Björck, 2016; Rylander, Ingelström, & Bondeson, 2012).

The S parameters are the most desirable in a reception and transmission system, will be given in the part 3. In order to determine the electromagnetic band gaps, it is necessary to calculate the reflection and the transmission coefficients using the simplest method; it is the transfer matrix method. The transfer matrix method is very familiar in light transmission processing in stratified media (Li, & Lin, 2003; Winn, Fink, Fan, & Joannopoulos, 1998; Choubani, Choubani, Gharsallah, David, & Mastorakis, 2010). Assuming a cell composed of N layers along the z direction, so that the refractive index in each layer can be considered a constant. For each of the two modes TE or TM, the electromagnetic fields on the left and right of the layer are linked by the transfer matrix of 2x2. So the problem to be solved and to determine the transfer matrix of each cell.

It is shown that the periodicity of the dielectric permittivity makes it possible to stop the wave propagation in a certain frequency bands. This notion of forbidden band can be extended in two and three dimensions, but the nature of Maxwell's equations becomes very complicated to solve analytically, which makes the use of digital tools indispensable. In part four, a comparison between the results found by the analytical methods for determining electromagnetic band gaps in a 1D periodic structure, those found by the Transmission Line Matrix (TLM) method and those extracted from High Frequency Structural Simulator (HFSS) simulation. An excellent agreement between the calculated eigenfrequencies and the numerical method will be shown.

Finally, an application of the EBG structures in the electromagnetic waves control. For example, they are utilized as high-reflection mirrors and antireflection coatings. However, most investigations related to the control of the radiation field in recent years have been concentrated on 2D and 3D EBG structures, because they offer a rich variety of new physics of the radiation field (Hajj, Rodes, Serhal, Chantalat, Thevenot, Monediere, & Jecko, 2007; Sievenpiper, 2008; Tam, 2007). The analysis of the results make it possible to produce such a periodic structure capable of concentrating the radiation in particular direction at the given frequency. The dispersion diagram and the S parameters are so necessary to determine the forbidden bands. The EBG structures are widely used especially for applications that concerns the realization of directional antennas; that is called EBG-antenna (Thevenot, Cheype, Reineix, & Jecko, 1999; Cheype, Serier, Thèvenot, Monédière, Reineix, & Jecko, 2002; Yeo, & Kim, 2009; Kaabal, Ahyoud, Asselman, & Faize, 2013), as will be detailed in the last part of this chapter. The EBG-antenna is generally consists of two interfaces and an excitation source. The upper interface conventionally consists of one or more dielectric layers; a surface of metal rods or dielectric, and the lower interface explained as a ground plane.

ELECTROMAGNETIC STUDY OF THE DISPERSION DIAGRAM OF A ONE-DIMENSIONAL PERIODIC STRUCTURE

The one-dimensional electromagnetic band gap structure consists of stacking two layers of materials with different permittivities with periodicity in one direction. There are forbidden bands if the thicknesses of each layer are equal to a quarter of the guided wavelength λ_g. The photonic structure have been dealt with consists of alternating plates of thickness a and dielectric permittivity ε_a with others of thickness b and dielectric permittivity ε_b, we define the network period $p = a + b$. This structure is illustrated in Figure 1.

Figure 1. Infinite EBG structure

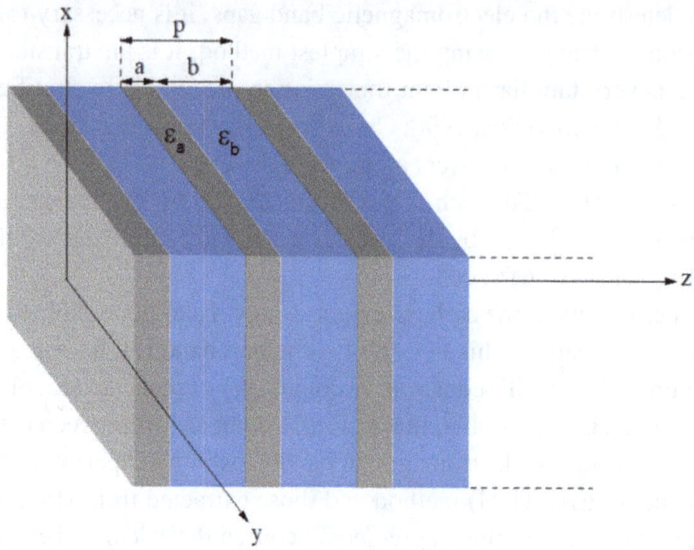

To simplify the study of the resolution of *Maxwell* equations, the configuration of the one-dimensional EBG structure consists of alternating layers of the dielectric materials with others of air (see Figure 2). This structure will interact with an external electromagnetic aggression, it exists an interaction with a plane wave propagating in incidence normal at the resonance frequency f_0, it is exactly the case of the antennas applications. Figure 2 shows the electromagnetic wave diffraction incident with the EBG structure in both directions of propagation; one in the positive sense of (oz) and the other in the negative direction.

The *Maxwell* equations for an electromagnetic wave propagating through a dielectric medium are written as follows:

- Magnetic flux conservation:

$$div\vec{B} = 0 \tag{1}$$

- Maxwell-Faraday:

$$\overrightarrow{rot}\vec{E} = -\frac{\partial \overrightarrow{B}}{\partial t} \tag{2}$$

- Maxwell-Gauss:

$$div\vec{D} = \rho_{charges} \tag{3}$$

- Maxwell-Ampère:

$$\overrightarrow{rot}\vec{H} = \vec{J} + \frac{\overrightarrow{\partial D}}{\partial t} \tag{4}$$

where,

E: The electric field $(V.m^{-1})$
D: The electric induction $(C.m^{-2})$
B: The magnetic induction (T)
H: The magnetic field $(A.m^{-1})$
$\rho_{charges}$: The density of free charges $(C.m^{-3})$
J: The density of the free current $(A.m^{-2})$

For linear[1] and isotropic[2] medium, $\vec{D}$ and $\vec{B}$ are simply connected respectively to $\vec{E}$ and $\vec{H}$ via the permittivity and permeability ε and μ.

$$\vec{D} = \varepsilon\vec{E} \tag{5}$$

$$\vec{B} = \mu\vec{H} \tag{6}$$

Taking into account $\varepsilon = \varepsilon_0\varepsilon_r$ and $\mu = \mu_0\mu_r$, with ε_r and μ_r are respectively the relative permittivity and permeability of the medium while ε_0 and μ_0 are those of vacuum.

In a perfect medium, the permanent densities of charges and the current are zero (neutral and insulating material).

$$\rho = 0 \quad and \quad \vec{J} = \vec{0} \tag{7}$$

Figure 2. 1D EBG structure interacting with plane waves

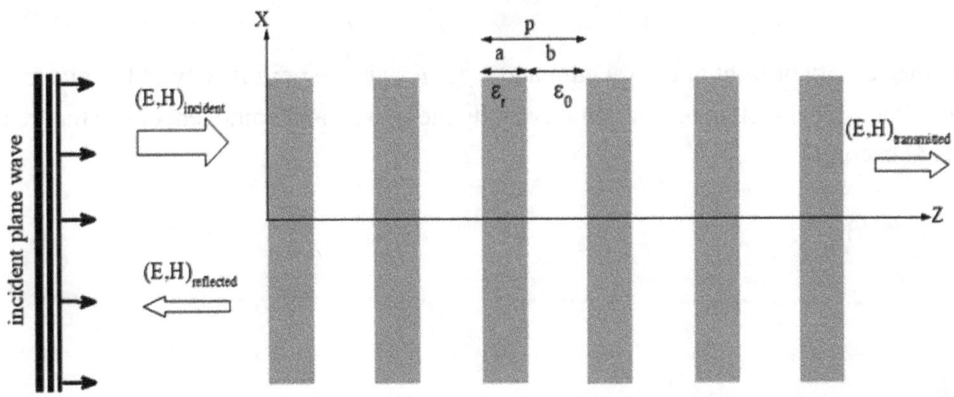

In a homogeneous medium, the dielectric permittivity ε and the magnetic permeability μ are isotropic; we obtain the new expressions of Maxwell's equations as follows:

$$div\vec{E} = 0 \tag{8}$$

$$div\vec{H} = 0 \tag{9}$$

$$\overrightarrow{rot}\vec{E} = -\mu\frac{\overrightarrow{\partial H}}{\partial t} \tag{10}$$

$$\overrightarrow{rot}\vec{H} = \varepsilon\frac{\overrightarrow{\partial E}}{\partial t} \tag{11}$$

And for a harmonic regime, the electromagnetic field can be written in the following equation:

$$\vec{E}(\vec{r},t) = \vec{E}(\vec{r})e^{-j\omega t}$$

and

$$\vec{B}(\vec{r},t) = \vec{B}(\vec{r})e^{-j\omega t} \tag{12}$$

with $\omega = 2\pi f$ is the angular frequency.

The second-order differential equation for the electric field $\vec{E}$; satisfying the wave equation (13) in each dielectric layer, is obtained by combining the equations of Maxwell (Equations 8 and 11):

$$\Delta\vec{E} + \frac{\omega^2.\varepsilon_r(x,y,z)}{c^2}\vec{E} = \vec{0} \tag{13}$$

where c is the celerity of light in the vacuum and $\varepsilon_r(x,y,z)$ is the permittivity of the dielectric layer.

In a one-dimensional system of a periodicity along the z-axis and homogeneous in the xy plane, the Equation (13) becomes:

$$\frac{\partial^2 E(z)}{\partial z^2} + \frac{\omega^2.\varepsilon_r(z)}{c^2}E(z) = 0 \tag{14}$$

As the permittivity is periodic, its period is p and its form is described in Figure 3. Thus, the Equation (14) can be solved easily by considering a one-dimensional periodic lattice.

- If $0 < z < a$ then:

$$\frac{\partial^2 E_a(z)}{\partial z^2} + \frac{\omega^2.\varepsilon_r}{c^2} E_a(z) = 0 \tag{15}$$

- If $a < z < a + b$ then:

$$\frac{\partial^2 E_b(z)}{\partial z^2} + \frac{\omega^2}{c^2} E_b(z) = 0 \tag{16}$$

The solutions of the differential Equations (15 and 16) are respectively:

$$E_a(z) = A\sin(\alpha z) + B\cos(\alpha z) \tag{17}$$

and:

$$E_b(z) = C\sin(\beta z) + D\cos(\beta z) \tag{18}$$

where $\alpha = \dfrac{\omega\sqrt{\varepsilon_r}}{c}$ and $\beta = \dfrac{\omega}{c}$.

By using the fact at the interface, at the point M for example, the function $E(z)$ and its derivative $E'(z)$ are continuous. Thus, according to the theorem of *Bloch-Floquet*, any solution $E(z)$ that satisfies the wave Equation (14) in a periodic structure is written as follows:

Figure 3. Periodic dielectric constant $\varepsilon_r(z + p) = \varepsilon_r(z)$

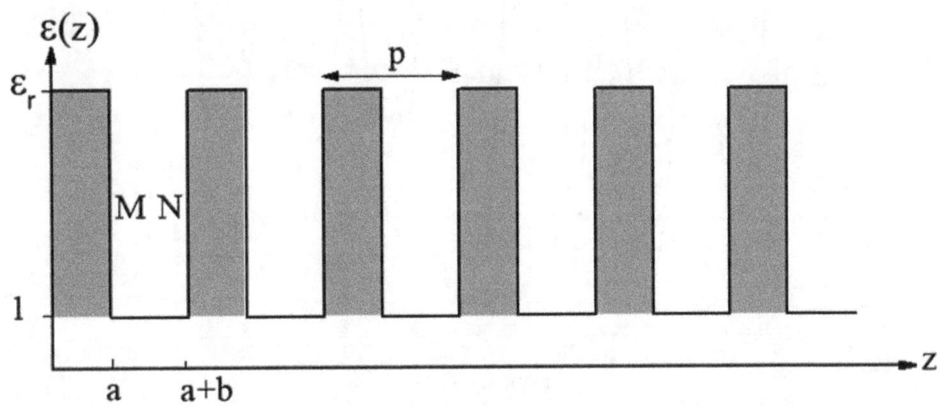

$$E(z) = u(z)e^{jkz} \tag{19}$$

where $u(z)$ is a periodic function with the same period as the distribution of the permittivity, i.e. $u(z+p) = u(z)$, with $k = 2\pi / \lambda$ is a wave constant.

It is shown that the dispersion relation can be written as in Equation (20):

$$\cos(\alpha a)\cos(\beta b) - \sin(\alpha a)\sin(\beta b)\frac{\varepsilon_r + 1}{2\sqrt{\varepsilon_r}} = \cos[k(a+b)] \tag{20}$$

From the dispersion equation (20), the wave constant is written as follows:

$$k = \frac{1}{a+b}\arccos\left[\cos(\alpha a)\cos(\beta b) - \sin(\alpha a)\sin(\beta b)\frac{\varepsilon_r + 1}{2\sqrt{\varepsilon_r}}\right] \tag{21}$$

The Figure 4 represents the variation of the first member of the equation (20) as a function of the frequency, it's observed that the first member of the equality can be less than -1 or greater than +1, but the second member is always between -1 and $+1$.

Figure 4 shows that, if the first member of the dispersion equation exceeds ±1, there are the frequency bands where the reduced wave constant k_r ($k_r = \frac{(a+b).k}{\pi}$) is undefined as shown in Figure 5, i.e. no wave can propagate in these frequency bands, called band gaps. Consequently, the one-dimen-

Figure 4. Dispersion diagram for $b = \frac{\lambda_{10GHz}}{4}$, $a = 3.b$ *and* $\varepsilon_r = 9$

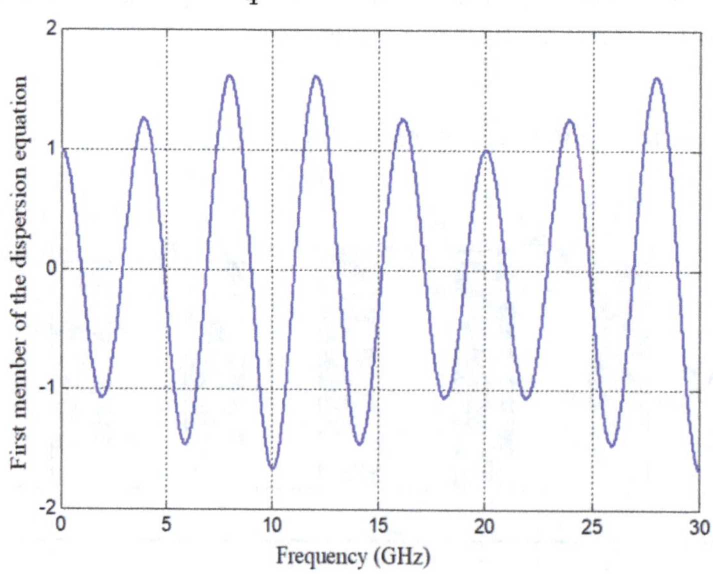

Figure 5. The dispersion diagram representation of 1D periodic structure: $b = \frac{\lambda_{10GHz}}{4}$ *and* $a = 3.b$ *for* $\varepsilon_r = 9$

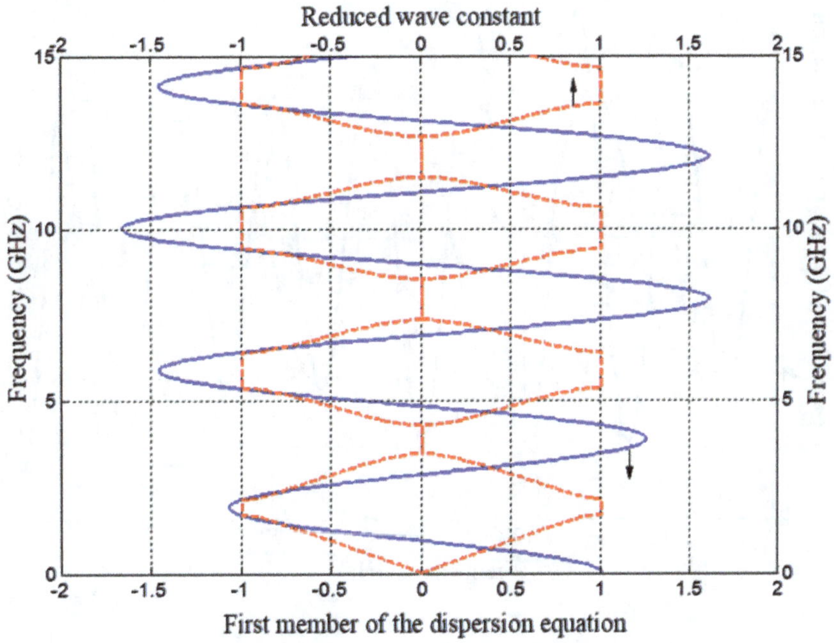

sional periodic structure prevents the propagation of electromagnetic waves to propagating into these frequency bands.

Figure 6 shows that the width of the band gaps depends on the permittivity of the dielectric material used.

For an application of the frequency f_0 given, in the antennas design for example. It is necessary for the first forbidden band to be centered on the frequency f_0, i.e. the first member of the dispersion equation has an extreme value in this frequency. For this reason, Equation (22) is used:

$$\frac{d(1^{er}\,membre)}{df} /_{f=f_0} = 0 \tag{22}$$

So,

$$(\varepsilon_r + 3)\sin(\alpha a)\cos(\beta b) + (2\sqrt{\varepsilon_r} + \varepsilon_r + 1)\cos(\alpha a)\sin(\beta b) = 0 \tag{23}$$

This condition is achievable if:

$$\alpha a = \frac{\pi}{2}$$

Figure 6. Dispersion diagram: $b = \frac{\lambda_{10GHz}}{4}$ *and* $a = 3.b$

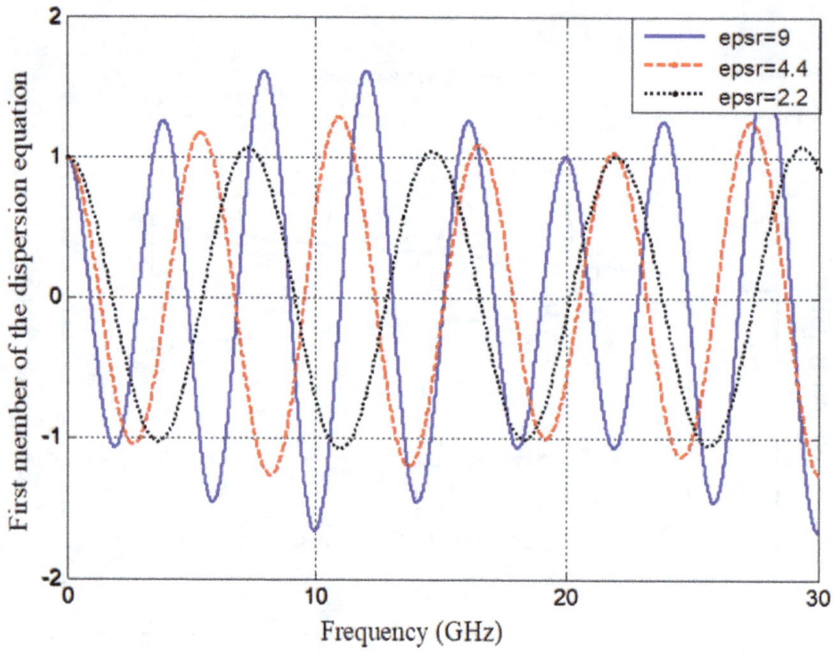

and

$$\beta b = \frac{\pi}{2} \tag{24}$$

Using the expressions $\alpha = \frac{\omega\sqrt{\varepsilon_r}}{c}$ and $\beta = \frac{\omega}{c}$, the following equation is found:

$$a = \frac{c}{4f_0\sqrt{\varepsilon_r}} = \frac{\lambda_g}{4} \quad \text{and } b = \frac{c}{4f_0} = \frac{\lambda_0}{4} \tag{25}$$

where ε_r the relative permittivity of the dielectric material, λ_0 is the wavelength in the vacuum corresponding to the central frequency f_0 of the forbidden band whereas λ_g in the dielectric.

To obtain the forbidden bands around the frequency f_0 or $(2n+1)f_0$ with $n \in \mathbb{N}$ (see Figure 7), it is necessary that the thickness of the layers is equal to $\frac{\lambda}{4}$ in order to obtain a destructive interference of the electromagnetic waves.

To build the periodic structure of the Figure 2, the thicknesses of the air layer and the dielectric layer are:

Figure 7. Dispersion diagram: $b = \frac{\lambda_{10GHz}}{4}$ *and* $a = \frac{\lambda_{10GHz}}{4\sqrt{\varepsilon_r}}$ *for* $\varepsilon_r = 9$

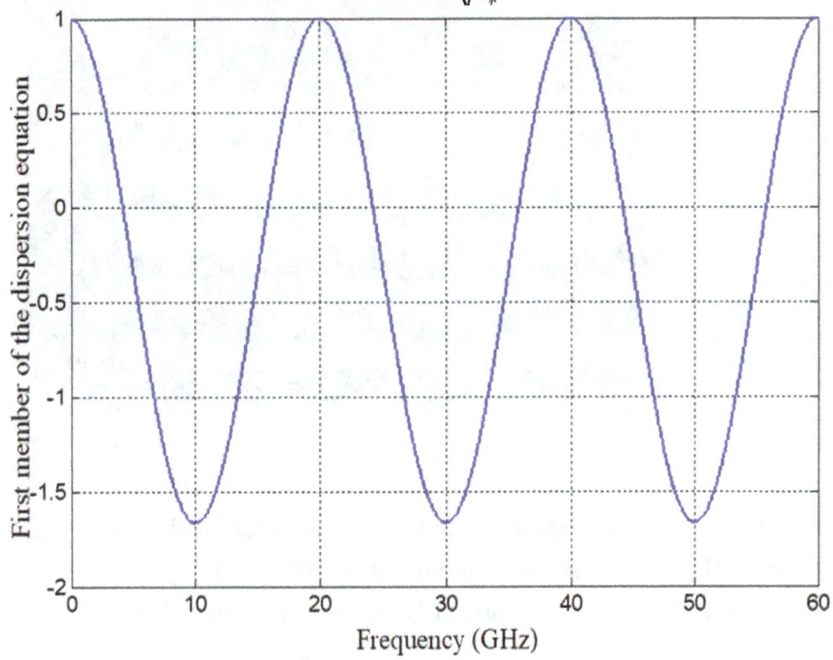

$$a = \frac{\lambda_g}{4} = \frac{c}{4f_0\sqrt{\varepsilon_r}}$$

$$b = \frac{\lambda_0}{4} = \frac{c}{4f_0} \tag{26}$$

PLANE WAVE DECOMPOSITION METHOD

In order to acquire an intuitive understanding of the electromagnetic forbidden bands, the 1D periodic structure is examined by the plane wave decomposition method. Consider the z-axis is perpendicular to the surface of the dielectric layers, as shown in Figure 8. Only the normal incidence of electromagnetic waves is studied, which propagate in *z* direction and linearly polarized along x-axis to assimilate their practical application in the antennas. The structure is periodic according to *z* direction with a spatial period p, and homogeneous in x and y directions.

For convenience, the electric field of the propagating wave is denoted by the function $\vec{E}(x, y, z, t)$. If the electric field has no component in *z* direction (the direction of propagation), the propagating mode is noted TE_z. If the magnetic field has no component in *z* direction, the propagating mode is noted TM_z.

Figure 8. 1D Bragg network: the periodicity is only following z direction with a spatial period p

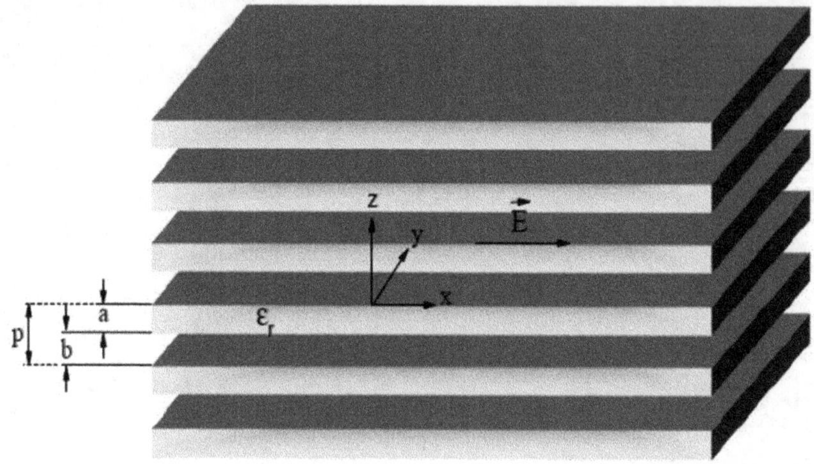

In the case under study, the wave is propagating in normal incidence, which means that the two modes are equivalent. It is enough to calculate one of them, TE_z mode for example.

In the harmonic regime, *Helmholtz* Equation (13) can be written as below:

$$\frac{1}{\varepsilon(z)}.\Delta\vec{E}(x,y,z)+k_0^2.\vec{E}(x,y,z)=0 \tag{27}$$

where $k_0 = \omega^2\varepsilon_0\mu$.

To obtain the frequencies allowed to propagate in the EBG structure, searching for the possible eigenvalues k_0^2 of the equation (27) is required. The electric field is extended as a periodic plane wave function along *z* direction with the period p. According to the theorem of *Bloch-Floquet*:

$$\vec{E}(x,y,z)=E_p(z).e^{jk_z.z}.\vec{x} \tag{28}$$

Where $E_p(z)$ is the periodic electric field along the *z* direction; i.e. $E_p(z)=E_p(z+p)$, and k_z is a propagation constant following the *z* direction.

According to the theorem of *Fourier* (Cheney, & Kincaid, 2012), the periodic electromagnetic field of a period p can decompose into Fourier series according to Equation (29):

$$E_p(z)=\sum_n A_n e^{j\frac{2\pi n}{p}z} \tag{29}$$

Replacing the expression of $E_p(z)$ in Equation (28), it is found that:

$$\vec{E}(x,y,z) = \sum_n A_n e^{j\frac{2\pi n}{p}z}.e^{jk_z.z}.\vec{x} \tag{30}$$

Since the dielectric constant $\varepsilon(z)$ is periodic, $\dfrac{1}{\varepsilon(z)}$ is also periodic and it can be broken down into Fourier series:

$$\frac{1}{\varepsilon(z)} = \sum_m B_m e^{j\frac{2\pi m}{p}z} \tag{31}$$

As the electric field must satisfy the wave Equation (27), the eigenvalue equation to be solved becomes:

$$\sum_m B_m e^{j\frac{2\pi m}{p}z} \Delta\left(\sum_n A_n e^{j\frac{2\pi n}{p}z}.e^{jk_z.z}.\vec{x}\right) + k_0^2 \sum_n A_n e^{j\frac{2\pi n}{p}z}.e^{jk_z.z}.\vec{x} = 0 \tag{32}$$

So:

$$\sum_m B_m e^{j\frac{2\pi m}{p}z} \left\{-\sum_n A_n \left[\left(\frac{2\pi n}{p} + k_z\right)^2\right] e^{j\frac{2\pi n}{p}z}.e^{jk_z.z}.\vec{x}\right\} + k_0^2 \sum_n A_n e^{j\frac{2\pi n}{p}z}.e^{jk_z.z}.\vec{x} = 0 \tag{33}$$

By multiplying the equation (32) by a base vector $\left| e^{j\frac{2\pi q}{p}z} \right>$ in a *Hilbert* space (Zeidler, 1995), then we integrate over the period $\int_{-p/2}^{p/2} dz$ using the property of the Equation (34):

$$\int_{-p/2}^{p/2} \left< e^{j\frac{2\pi n}{p}z} \middle| e^{j\frac{2\pi q}{p}z} \right> dz = p.\delta(n-q) \tag{34}$$

where:

$$\delta(n-q) = \begin{cases} 1 & if \quad n = q \\ 0 & if \quad n \neq q \end{cases}$$

It is found that:

$$\sum_m B_m . \sum_n A_n \left[\left(\frac{2\pi n}{p} + k_z\right)^2\right] \int_{-p/2}^{p/2} \left< e^{j\frac{2\pi(m+n)}{p}z} \middle| e^{-j\frac{2\pi q}{p}z} \right> dz = k_0^2 \sum_n A_n \int_{-p/2}^{p/2} \left< e^{j\frac{2\pi n}{p}z} \middle| e^{-j\frac{2\pi q}{p}z} \right> dz \tag{35}$$

According to Equation (34):

$$\sum_m B_m \cdot \sum_n A_n \Big[\big(\frac{2\pi n}{p} + k_z\big)^2 \Big] p.\delta[(m+n)-q] = k_0^2 \sum_n A_n p.\delta(n-q) \tag{36}$$

Finally, the eigenvalue equation to be solved is:

$$\sum_n A_n B_{q-n} \Big[\big(\frac{2\pi n}{p} + k_z\big)^2 \Big] = k_0^2 A_q \tag{37}$$

The calculation of the B_{q-n} coefficient is obtained from the Equation (31). According to Fourier's theorem, B_m is written this way:

$$B_m = \frac{1}{p} \int_{-p/2}^{p/2} \frac{1}{\varepsilon(z)} e^{-j\frac{2\pi m}{p}z} dz \tag{38}$$

i.e:

$$B_m = \frac{1}{p} \Big\{ \int_{-p/2}^{p/2} \big(\frac{1}{\varepsilon(z)} - 1 \big) e^{-j\frac{2\pi m}{p}z} dz + \int_{-p/2}^{p/2} 1 e^{-j\frac{2\pi m}{p}z} dz \Big\} \tag{39}$$

where:

$$\varepsilon(z) = \begin{cases} \varepsilon_r & si \quad -\frac{a}{2} < z < \frac{a}{2} \\ 1 & si \quad -\frac{p}{2} < z < -\frac{a}{2} \quad or \quad \frac{a}{2} < z < \frac{p}{2} \end{cases} \quad \text{(see Figure 9)}$$

By the analogy to the Equation (39), B_{q-n} is written as follows:

$$B_{q-n} = \frac{1}{p} \Big\{ \int_{-p/2}^{p/2} \big(\frac{1}{\varepsilon(z)} - 1 \big) e^{-j\frac{2\pi(q-n)}{p}z} dz + \int_{-p/2}^{p/2} 1 e^{-j\frac{2\pi(q-n)}{p}z} dz \Big\} \tag{40}$$

By integrating this equation in all regions of the cell in Figure 9:

Figure 9. The periodic structure Cell

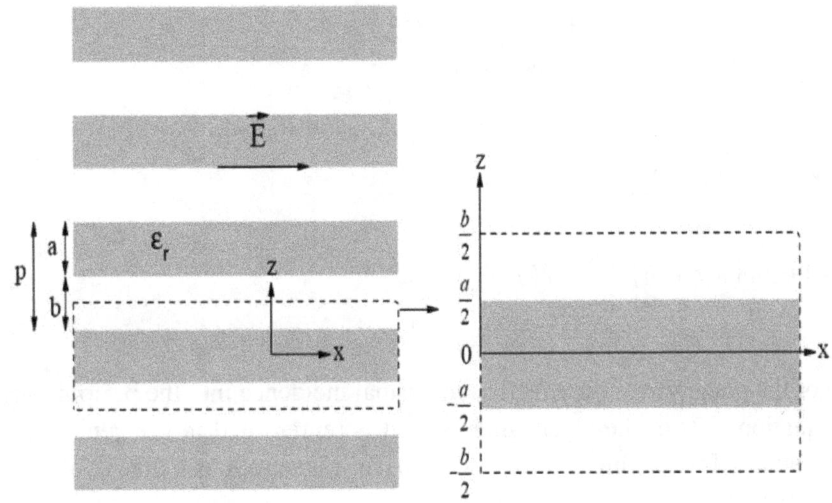

$$B_{q-n} = \frac{1}{p}\{\int_{-p/2}^{-a/2}(\frac{1}{\varepsilon(z)}-1)e^{-j\frac{2\pi(q-n)}{p}z}dz + \int_{-a/2}^{a/2}(\frac{1}{\varepsilon(z)}-1)e^{-j\frac{2\pi(q-n)}{p}z}dz$$
$$+ \int_{a/2}^{p/2}(\frac{1}{\varepsilon(z)}-1)e^{-j\frac{2\pi(q-n)}{p}z}dz + \int_{-p/2}^{p/2} e^{-j\frac{2\pi(q-n)}{p}z}dz\} \quad (41)$$

So,

$$B_{q-n} = \frac{1}{p}\{\frac{p}{-j2\pi(q-n)}(\frac{1}{\varepsilon_r}-1)\left[e^{-j\frac{2\pi(q-n)}{p}z}\right]_{-a/2}^{a/2} + p.\delta(q-n)\} \quad (42)$$

Therefore,

$$B_{q-n} = \frac{1}{p}\{\frac{p}{j2\pi(q-n)}(\frac{1}{\varepsilon_r}-1)\left[2j\sin\left(\pi(q-n).\frac{a}{p}\right)\right] + p.\delta(q-n)\} \quad (43)$$

By simplifying the Equation (43), it is found that:

$$B_{q-n} = (\frac{1}{\varepsilon_r}-1)\frac{\sin\left(\pi(q-n).\frac{a}{p}\right)}{\pi(q-n)} + \delta(q-n) \quad (44)$$

Or,

$$B_{q-n} = \frac{a}{p}\left(\frac{1}{\varepsilon_r} - 1\right)\frac{sin\left(\pi(q-n).\frac{a}{p}\right)}{\pi(q-n)\frac{a}{p}} + \delta(q-n) \tag{45}$$

Finally, the expression of B_{q-n} is:

$$B_{q-n} = \frac{a}{p}\left(\frac{1}{\varepsilon_r} - 1\right)sinc\left(\pi(q-n).\frac{a}{p}\right) + \delta(q-n) \tag{46}$$

The problem of the plane waves propagation in normal incidence into the periodic structure consists of solving the Equation (37) of the eigenvalues which is represented in the general case as $Ax = \lambda$ (Björck, 2016; Rylander, Ingelström, & Bondeson, 2012); such an A is a matrix of $n \times n$, λ is an eigenvalue and $x \neq 0$ is the corresponding eigenvector.

The Equation (36) is solved by constructing a matrix for the left-side terms and then determining its eigenvalues k_0^2 and corresponding eigenvectors A_n. The eigenvalues corresponding to the modal solutions are modes that can be propagated in the periodic structure. Using Matlab, The eigenvalue equation can be easily solved. The results of the eigenvalues are presented in Figure 10.

Figure 10. Eigenvalues representation of 1D periodic structure: $b = \frac{\lambda_{10GHz}}{4}$ *and* $a = 3.b$ *for* $\varepsilon_r = 9$

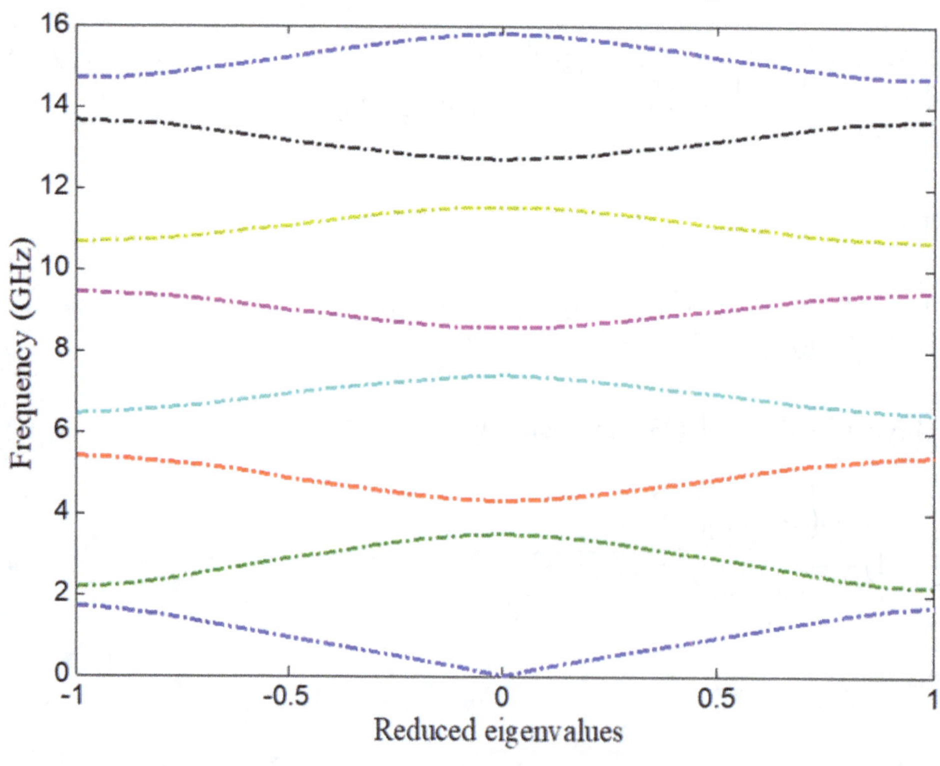

From Figure 10, the eigenvalue equation no longer has solutions in some frequency bands. These are then the electromagnetic band gaps.

Figure 11 shows that the results found with the plane wave decomposition method are similar to those found with the previous method (First part), which are represented in Figure 5.

TRANSFER MATRIX METHOD

The *S* parameters are most desirable in a receiving and transmitting system. To determine the electromagnetic band gap, the reflection coefficient Γ and the transmission coefficient *T* must be calculated using the simplest method, the transfer matrix method. The transfer matrix method is very familiar to the light transmission processing in stratified media. Supplying a cell composed of N layers along the *z* direction, so that the refractive index in each layer can be considered a constant. For each of the two modes TE and TM, the electromagnetic fields on the left and right of the layer are linked by the transfer matrix M_i of 2×2. So the problem to solve is to determine the transfer matrix of each cell.

Consider the sinusoidal uniform plane wave $\vec{E}_i = E_0 e^{j(\omega t - \vec{k}.\vec{z})} \vec{e}_x$ polarized rectilinear along the *x*-axis and propagating in the vacuum along z-axis; within normal incidence to the surface of the dielectric layer (see Figure 12).where,

$(\vec{E}_{in}, \vec{H}_{in})$ The incident electromagnetic field at the dielectric-air interface.

Figure 11. Comparison of two methods

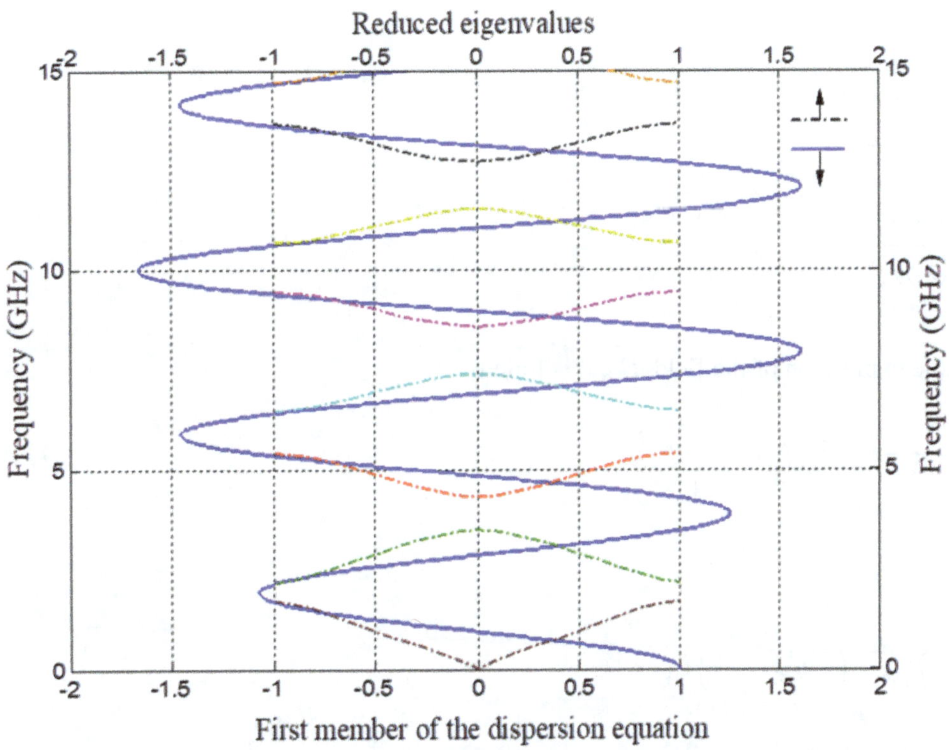

Figure 12. Normal incidence of plane wave across a discontinuity

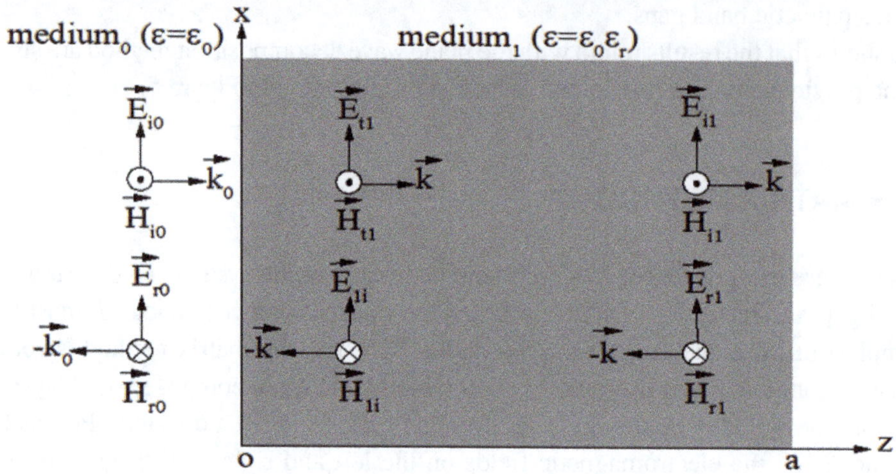

$(\vec{E}_{rn}, \vec{H}_{rn})$ The electromagnetic field reflected at the dielectric-air interface.
$(\vec{E}_{tn}, \vec{H}_{tn})$ The electromagnetic field transmitted to the dielectric-air interface.
$(\vec{E}_{ni}, \vec{H}_{ni})$ The incident electromagnetic field at the dielectric-air interface in the negative propagation direction.

The incident magnetic field is deduced by using the plane wave properties:

$$\frac{\partial \ldots}{\partial t} = j\omega \ldots \tag{47}$$

From Equation (10),

$$\overrightarrow{H_i} = -\frac{\vec{\nabla} \times \overrightarrow{E_i}}{j\omega\mu} \tag{48}$$

Using the expression of the field $\overrightarrow{E_i}$, it is found:

$$\overrightarrow{H_i} = \frac{1}{j\omega\mu} E_0 \vec{e}_x \times \overrightarrow{grad}\left(e^{j(\omega t - \vec{k}\vec{z})}\right) \tag{49}$$

i.e.:

$$\overrightarrow{H_i} = \frac{1}{j\omega\mu} E_0 \vec{e}_x \times \left(-j\vec{k} e^{j(\omega t - \vec{k}\vec{z})}\right) \tag{50}$$

So,

$$\overrightarrow{H_i} = \frac{1}{\omega\mu}\vec{k} \times E_0 e^{j(\omega t - \vec{k}\vec{z})}\vec{e}_x \tag{51}$$

Therefore, the expression $\overrightarrow{H_i}$ is given by the following equation:

$$\overrightarrow{H_i} = \frac{1}{\omega\mu}\vec{k} \times \vec{E}_i \tag{52}$$

where $\vec{k} = k.\vec{e}_z = \omega\sqrt{\varepsilon\mu}\vec{e}_z$

Using the notation of the characteristic impedance of the medium $\eta = \sqrt{\frac{\mu}{\varepsilon}}$, the field $\overrightarrow{H_i}$ can be written as:

$$\overrightarrow{H_i} = \frac{1}{\eta}\vec{e}_z \times \vec{E}_i \tag{53}$$

Since, $\vec{E}_i$ is directed along the x-axis:

$$\overrightarrow{H_i} = \frac{1}{\eta}E_i\vec{e}_y \tag{54}$$

With the same reasoning, the reflected field (where $\vec{k} = -k.\vec{e}_z$) is written as follows:

$$\overrightarrow{H_r} = -\frac{1}{\eta}E_r\vec{e}_y \tag{55}$$

At the interface $z = 0$, the tangential component of the electric and magnetic fields is continuous:

$$\begin{cases} E_{0,z=0} & = E_{1,z=0} \\ H_{0,z=0} & = H_{1,z=0} \end{cases} \tag{56}$$

Note that, $E_{0,z=0} = E_0$, $H_{0,z=0} = H_0$, $E_{1,z=0} = E_{10}$ and $H_{1,z=0} = H_{10}$

So,

$$\begin{aligned} E_0 &= E_{10} \\ H_0 &= H_{10} \end{aligned} \tag{57}$$

where:

$$\begin{aligned} E_0 &= E_{i0} + E_{r0} \\ H_0 &= H_{i0} + H_{r0} \end{aligned} \tag{58}$$

From Equations (54) and (55):

$$\begin{aligned} E_0 &= E_{i0} + E_{r0} \\ H_0 &= \frac{1}{\eta_0}(E_{i0} - E_{r0}) \end{aligned} \tag{59}$$

Similarly, the expression of the fields in the other side of the interface at $z = 0$ is:

$$\begin{aligned} E_{10} &= E_{1i} + E_{t1} \\ H_{10} &= \frac{1}{\eta}(E_{t1} - E_{1i}) \end{aligned} \tag{60}$$

In the same way, the expression of the electromagnetic field at the interface in $z = a$ is:

$$\begin{cases} E_{1,z=a} = E_1 = E_{i1} + E_{r1} \\ H_{1,z=a} = H_1 = \dfrac{1}{\eta}(E_{i1} - E_{r1}) \end{cases} \tag{61}$$

Using the plane wave propagation theory, E_{i1} and E_{r1} differs respectively of E_{t1} and E_{1i} only by a phase difference due to the propagation through the dielectric layer:

$$\begin{aligned} E_{i1} &= E_{t1}e^{-jka} \\ E_{r1} &= E_{1i}e^{jka} \end{aligned} \tag{62}$$

By this way, leads to the equations system giving the expression of the electromagnetic field at the interface in $z = a$ following:

$$\begin{cases} E_1 = E_{t1}e^{-jka} + E_{1i}e^{jka} \\ H_1 = \dfrac{1}{\eta}(E_{t1}e^{-jka} - E_{1i}e^{jka}) \end{cases} \tag{63}$$

The expressions of E_{t1} and E_{1i} extract from Equation (63) will be replaced in Equation (60):

$$\begin{cases} E_{01} & = \frac{1}{2}(E_1 - \eta H_1)e^{-jka} + \frac{1}{2}(E_1 + \eta H_1)e^{jka} \\ E_{01} & = \frac{1}{\eta}\left\{\frac{1}{2}(E_1 + \eta H_1)e^{jka} - \frac{1}{2}(E_1 - \eta H_1)e^{-jka}\right\} \end{cases} \tag{64}$$

By grouping the electromagnetic field E_1 and H_1 at the interface ($z = a$), it is found that:

$$\begin{cases} E_{01} & = \frac{1}{2}E_1\left(e^{jka} + e^{-jka}\right) + \frac{\eta}{2}H_1\left(e^{jka} - e^{-jka}\right) \\ H_{01} & = \frac{1}{\eta}\left\{\frac{1}{2}E_1\left(e^{jka} - e^{-jka}\right) + \frac{\eta}{2}H_1\left(e^{jka} + e^{-jka}\right)\right\} \end{cases} \tag{65}$$

So,

$$\begin{cases} E_{01} & = cos(ka).E_1 + j\eta sin(ka).H_1 \\ H_{01} & = j\frac{1}{\eta} sin(ka).E_1 + cos(ka).H_1 \end{cases} \tag{66}$$

Using the fact that the tangential components of the electromagnetic field are continuous, according to Equation (57):

$$\begin{cases} E_0 & = cos(ka).E_1 + j\eta sin(ka).H_1 \\ H_0 & = j\frac{1}{\eta} sin(ka).E_1 + cos(ka).H_1 \end{cases} \tag{67}$$

This equation can be written in the following matrix:

$$\begin{bmatrix} E_0 \\ H_0 \end{bmatrix} = \begin{bmatrix} cos(ka) & j\eta sin(ka) \\ j\frac{1}{\eta} sin(ka) & cos(ka) \end{bmatrix} \begin{bmatrix} E_1 \\ H_1 \end{bmatrix} \tag{68}$$

For a linear medium, according to the Equations (5 and 6) it is found that:

$$\begin{bmatrix} E_0 \\ B_0 \end{bmatrix} = \begin{bmatrix} cos(ka) & j\frac{sin(ka)}{\gamma_1} \\ j\gamma_1 sin(ka) & cos(ka) \end{bmatrix} \begin{bmatrix} E_1 \\ B_1 \end{bmatrix} \tag{69}$$

Generally, for the ith layer:

$$\begin{bmatrix} E_{i-1} \\ B_{i-1} \end{bmatrix} = \begin{bmatrix} cos(\delta_i) & j\dfrac{sin(\delta_i)}{\gamma_i} \\ j\gamma_i sin(\delta_i) & cos(\delta_i) \end{bmatrix} \begin{bmatrix} E_i \\ B_i \end{bmatrix} \tag{70}$$

Thus,

$$\begin{bmatrix} E_{i-1} \\ B_{i-1} \end{bmatrix} = M_i \begin{bmatrix} E_i \\ B_i \end{bmatrix} \tag{71}$$

$$M_i = \begin{bmatrix} cos(\delta_i) & j\dfrac{sin(\delta_i)}{\gamma_i} \\ j\gamma_i sin(\delta_i) & cos(\delta_i) \end{bmatrix} \tag{72}$$

where $\delta_i = ak_i$ and $\gamma_i = \dfrac{k_i}{\omega}$.

In the case of the normal incidence $k_i = (\dfrac{\omega}{c})n_i$, where $n_i = \sqrt{\varepsilon_{r,i}}$ is the refraction index of the ith layer, a is its thickness and c is the celerity of the light in the vacuum.

The global transfer matrix corresponding to the entire structure (an *N*-layers stack (see Figure 13)) is the product of the particular transfer matrices M_i:

$$M = \begin{bmatrix} m_{11} & m_{12} \\ m_{21} & m_{22} \end{bmatrix} = \prod_{i=1}^{n} M_i \tag{73}$$

In this stack:

$$\begin{bmatrix} E_0 \\ B_0 \end{bmatrix} = \begin{bmatrix} m_{11} & m_{12} \\ m_{21} & m_{22} \end{bmatrix} \begin{bmatrix} E_n \\ B_n \end{bmatrix} \tag{74}$$

The tangential components of the electromagnetic field are continuous in the last interface of the periodic structure. Therefore,

Figure 13. N-layer stack

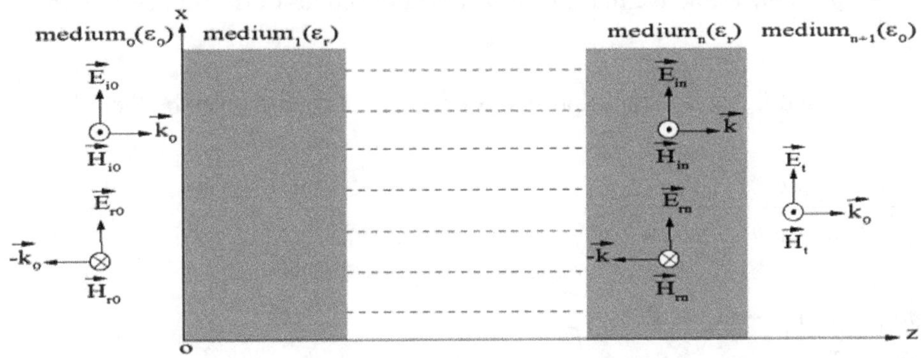

$$\begin{cases} E_n = E_{n+1} \\ H_n = H_{n+1} = \dfrac{1}{\eta_{n+1}} E_t \end{cases} \tag{75}$$

Since the structure is excited only from the left, the field after the last interface will contain only the transmitted component (E_t, H_t).

$$\begin{cases} E_{n+1} = E_t \\ H_{n+1} = H_t = \dfrac{1}{\eta_{n+1}} E_t = \dfrac{1}{\eta_0} E_t \end{cases} \tag{76}$$

The matrix (74) becomes,

$$\begin{bmatrix} E_0 \\ B_0 \end{bmatrix} = \begin{bmatrix} m_{11} & m_{12} \\ m_{21} & m_{22} \end{bmatrix} \begin{bmatrix} E_t \\ B_t \end{bmatrix} \tag{77}$$

According to the Equations (59 and 79), the matrix (77) is written as follows:

$$\begin{bmatrix} E_{i0} + E_{r0} \\ \dfrac{\mu_0}{\eta_0}(E_{i0} + E_{r0}) \end{bmatrix} = \begin{bmatrix} m_{11} & m_{12} \\ m_{21} & m_{22} \end{bmatrix} \begin{bmatrix} E_t \\ \dfrac{\mu_0}{\eta_0} E_t \end{bmatrix} \tag{78}$$

Finally, it is found that:

$$\begin{bmatrix} E_{i0} + E_{r0} \\ \Gamma_0(E_{i0} + E_{r0}) \end{bmatrix} = \begin{bmatrix} m_{11} & m_{12} \\ m_{21} & m_{22} \end{bmatrix} \begin{bmatrix} E_t \\ \Gamma_0 E_t \end{bmatrix} \tag{79}$$

where $\Gamma_0 = \frac{n_0}{c}$, n_0 is the refractive index of medium at the edges of the unit cell. In the studied case, $n_0 = 1$.

Noted $E_i = E_{i0}$ and $E_r = E_{r0}$, the expressions of E_{i0} and E_{r0} are given by the following equation:

$$\begin{cases} E_t = \dfrac{2\Gamma_0}{m_{11} + \Gamma_0^2 m_{12} + m_{21} + \Gamma_0 m_{22}} E_i \\ E_r = \dfrac{\Gamma_0 m_{11} + \Gamma_0^2 m_{12} - m_{21} + \Gamma_0 m_{22}}{\Gamma_0 m_{11} + \Gamma_0^2 m_{12} + m_{21} + \Gamma_0 m_{22}} E_i \end{cases} \tag{80}$$

The periodic structure is interacted with a plane wave propagating in incident normal to the resonance frequency f_0. The diffracted waves are characterized by the transmission and the reflection coefficients. The Figure 14 illustrates the method used to calculate the transmission coefficient T and the reflection coefficient Γ.

To calculate the transmission coefficient T, the observation point is placed in the area of the transmitted field behind the periodic structure. In addition, to obtain the reflection coefficient Γ, the observation point is placed in the area of the field reflected between the excitation and the structure as shown in Figure 14.

The transmission coefficient T of a periodic structure corresponds to the ratio between the amplitude of the transmitted and the incident waves:

$$T = \frac{E_{transmitted}}{E_{incident}} \tag{81}$$

And the reflection coefficient Γ is the ratio between the amplitude of the reflected and the incident waves:

Figure 14. Position of observation points to calculate the transmission and reflection coefficients

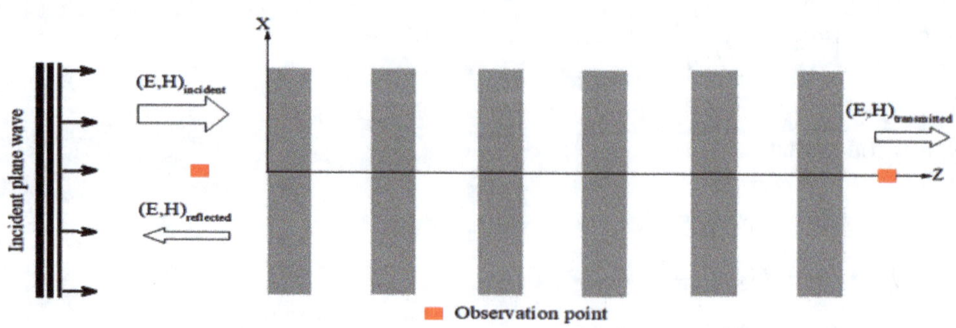

$$\Gamma = \frac{E_{reflected}}{E_{incident}} \tag{82}$$

According to (80), the transmission and the reflection coefficients for the entire stack of layers are written:

$$T = \frac{2\Gamma_0}{\Gamma_0 m_{11} + \Gamma_0^2 m_{12} + m_{21} + \Gamma_0 m_{22}} \tag{83}$$

$$\Gamma = \frac{\Gamma_0 m_{11} + \Gamma_0^2 m_{12} - m_{21} + \Gamma_0 m_{22}}{\Gamma_0 m_{11} + \Gamma_0^2 m_{12} + m_{21} + \Gamma_0 m_{22}} \tag{84}$$

RESULTS AND DISCUSSIONS

The 1D-EBG studied consists of six dielectric plates whose dielectric constant $\varepsilon_r = 9$ and thickness $a = \frac{\lambda_g}{4}$ corresponding to the frequency $f_0 = 10GHz$, these dielectric layers are separated by air layers whose thickness $b = \frac{\lambda_0}{4}$ (see Figure 8). From Figure 15, it can be seen that the periodic structure makes it possible to realize a total reflection for certain frequency bands, where the transmission coefficient (Equation 84) is almost zero ($\ll -20dB$) and the reflection coefficient (Equation 83) is equal to 1 (or $0dB$) in these bands.

To explain this phenomenon, we represent the transmission coefficients and the dispersion diagram in the same figure (see Figure 16). The results obtained are similar, so the band gaps can be determined from the transmission coefficient when it is less than $-20dB$, where the electromagnetic waves are considered totally reflected due to multiple reflections in the dielectric layers. This property is determined using the dispersion diagram, where in the band gaps no wave vector can satisfy the Helmholtz Equation (13).

The depth and the width of the band gap depend on the periods number of the EBG structure and the relative permittivity ε_r of the dielectric material used. Figure 17 shows the influence of the periods number of the structure for the relative permittivity of the layers is $\varepsilon_r = 9$. It is found that, if the periods number increases, the transmission well becomes deeper; this is due to multiple reflections on the dielectric layers.

Figure 18 shows the evolution of the band gaps for the structure of six-period as a function of the dielectric permittivity of the layers. When the permittivity increases, the well becomes deeper and wider too.

In this section, it is shown that the periodicity of the dielectric permittivity makes it possible to stop the wave propagation in a certain frequency band. This notion of forbidden band can be extended in

Figure 15. S parameters in normal incident: a=2.5mm and b=7.5mm

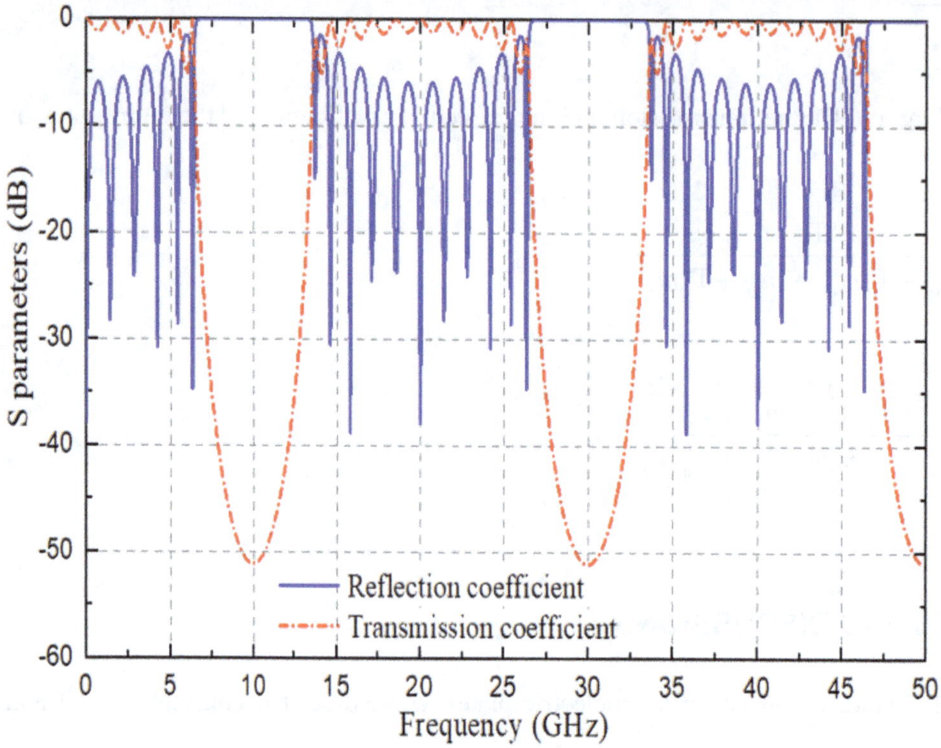

Figure 16. Superposition of dispersion diagram and transmission coefficient

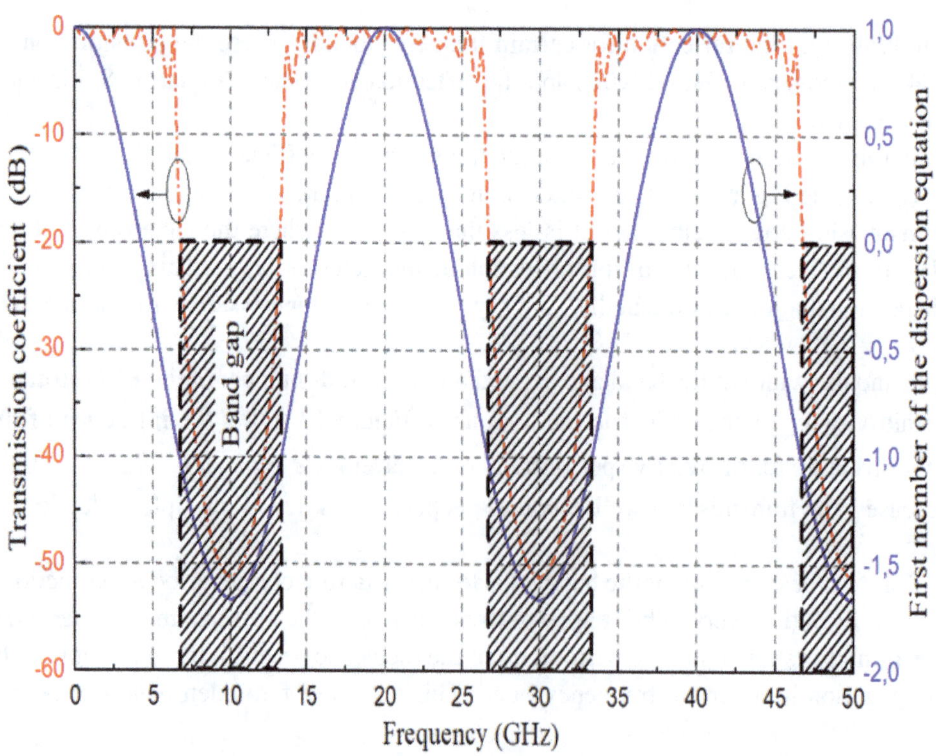

Figure 17. Influence of the periods number on the transmission coefficient of the EBG structure for $\varepsilon_r = 9$

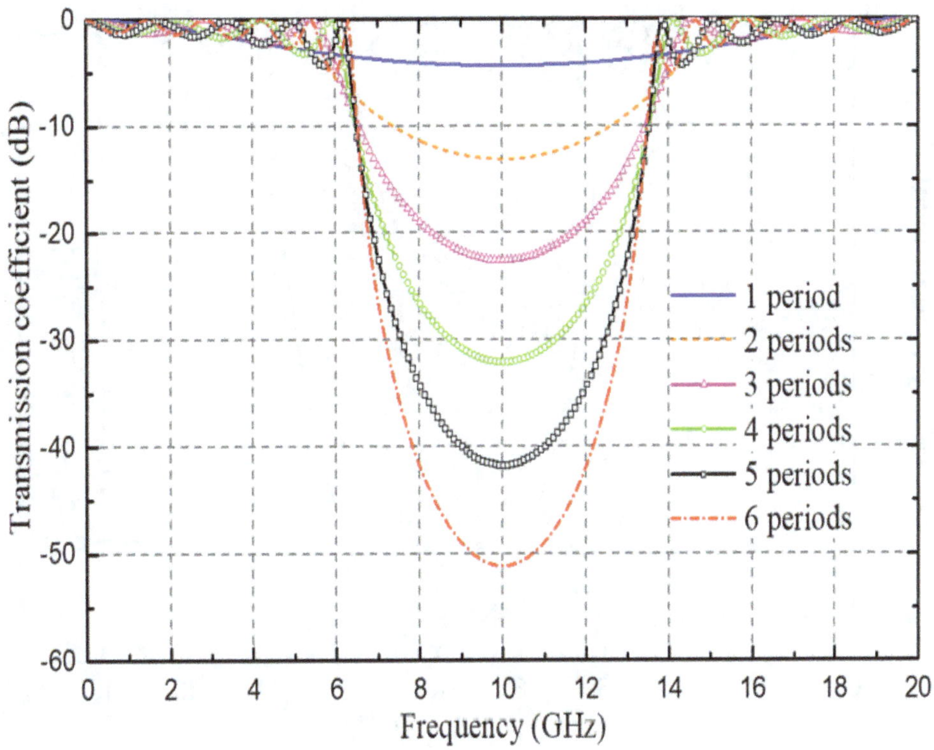

two and three dimensions, but the nature of Maxwell's equations becomes very complicated to solve analytically, which makes the use of digital tools indispensable.

Figure 19 illustrates a comparison between three different methods for determining the electromagnetic band gaps of the 1D-EBG structure consisting of 6-periods with $\varepsilon_r = 9$; the analytical method, the transmission line matrix (TLM) and the simulation by HFSS (High Frequency Structural Simulator).

From Figure 19, it is observed that there is a good agreement between the results found by the analytical method, those found by the TLM method and those extracted from HFSS. This comparison validates the choice of using HFSS in the modeling and design of EBG antennas thereafter.

HFSS software is one of the commercial tools most used for the complex RF circuit design. To obtain the dispersion diagram using periodic conditions between two parallel planes (Master / Slave) (see Figure 19).

For the TLM method, the Absorbing Boundary Conditions (ABC) are used (Morente, Porti, & Khalladi, 1992). At higher frequencies, there is a dispersion of the method, which makes Perfectly Matched Layers (PML) necessary.

Figure 18. Influence of the relative permittivity of materials on the transmission coefficient for 6-periods EBG structure

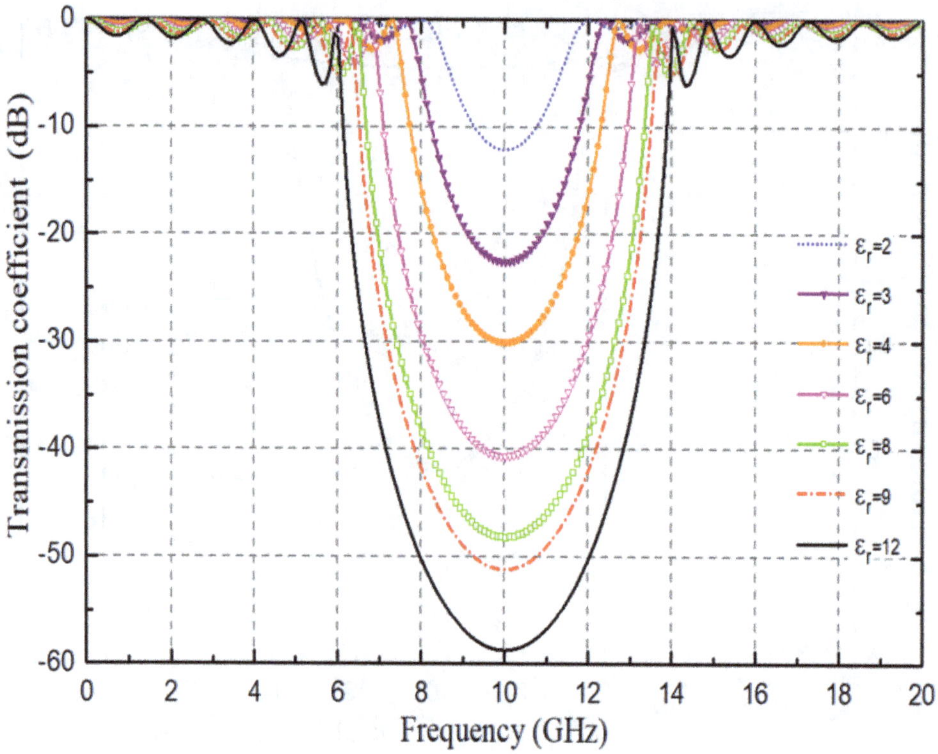

1D EBG STRUCTURE FOR DIRECTIVE ANTENNA APPLICATIONS

To apply the EBG structure in antennas, its band gaps properties are used. The first step of the design is the construction of a cavity (or a defect) in the center of the EBG structure corresponding to the wavelength which transmission in this structure is prohibited. However, several types of defects may be used, which may include lacunars defects, substitution defects, or additive defects (Lo, 2012). In general, the nature of the defect modifies the characteristics of the permitted band (center frequency f_0, bandwidth Δf) (Toubeh, 2011).

Figure 21 shows the evolution of the transmission coefficient of structure having six layers with a thickness cavity λ_0 in the middle of the structure. According to this figure, a transmission peak is formed at the center frequency $f_0 = 10GHz$ of the forbidden band, which is reflects a resonance phenomenon. This phenomenon of frequency filtering is exploited for the design of the EBG directive antenna by introducing an excitation at the cavity center.

Figure 22 shows that, by constructing the cavity in the center of the structure, the mapping of the tangential electric field has a plane of symmetry at the resonant frequency f_0. This plane of symmetry can be replaced by a metallic plane (or a ground plane) (Hashmi, Zeb, & Esselle, 2014), because the mapping of the electric field shows a cancellation of the tangential E field component at this plane of symmetry. Thus, by applying the theory of electrical images, the behavior of the half-structure above

Figure 19. Transmission coefficient T found with three different methods

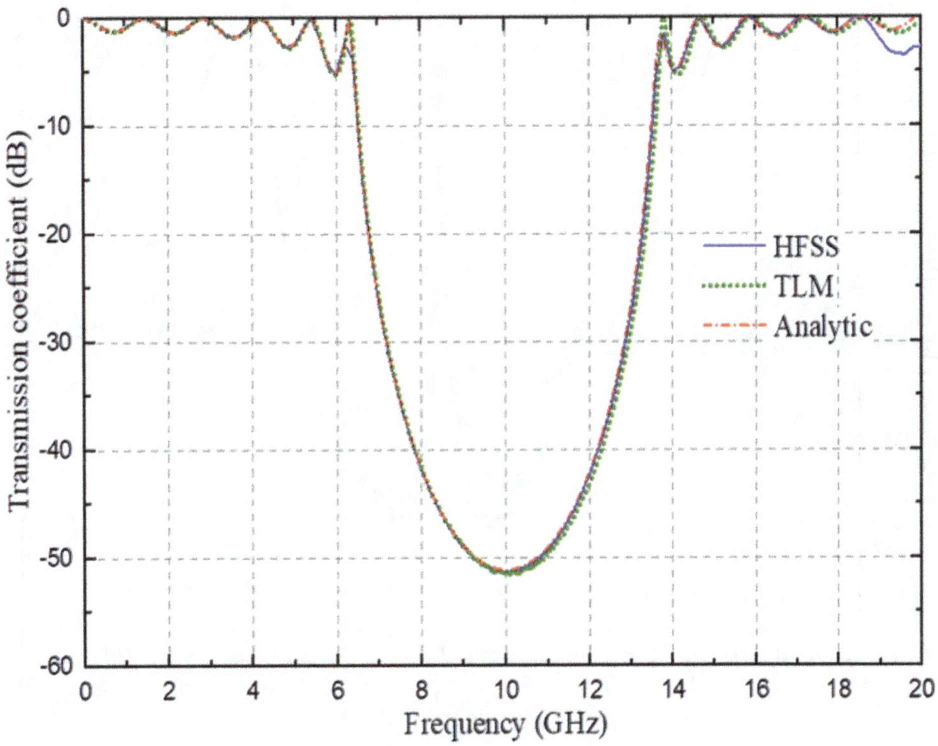

Figure 20. HFSS model

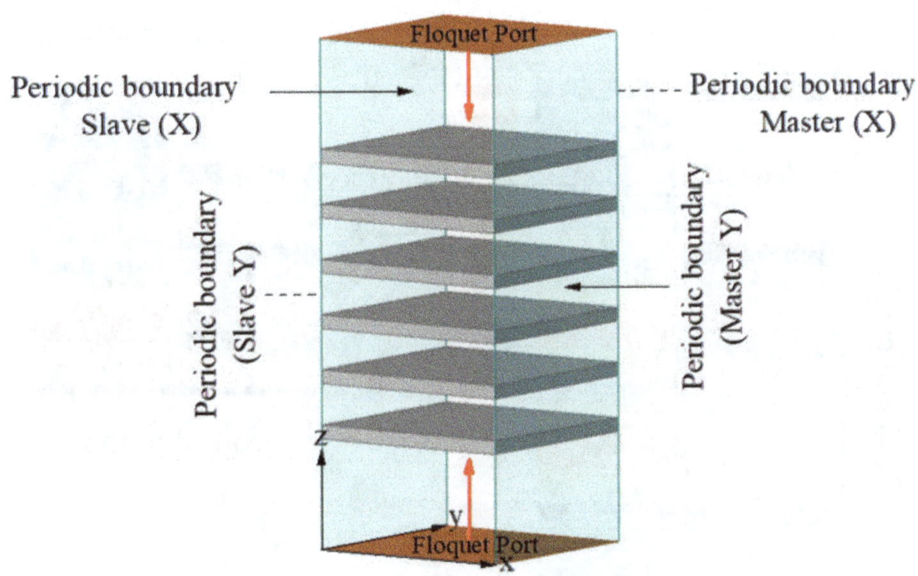

Figure 21. Transmission coefficient of the 1D-EBG structure with a defect

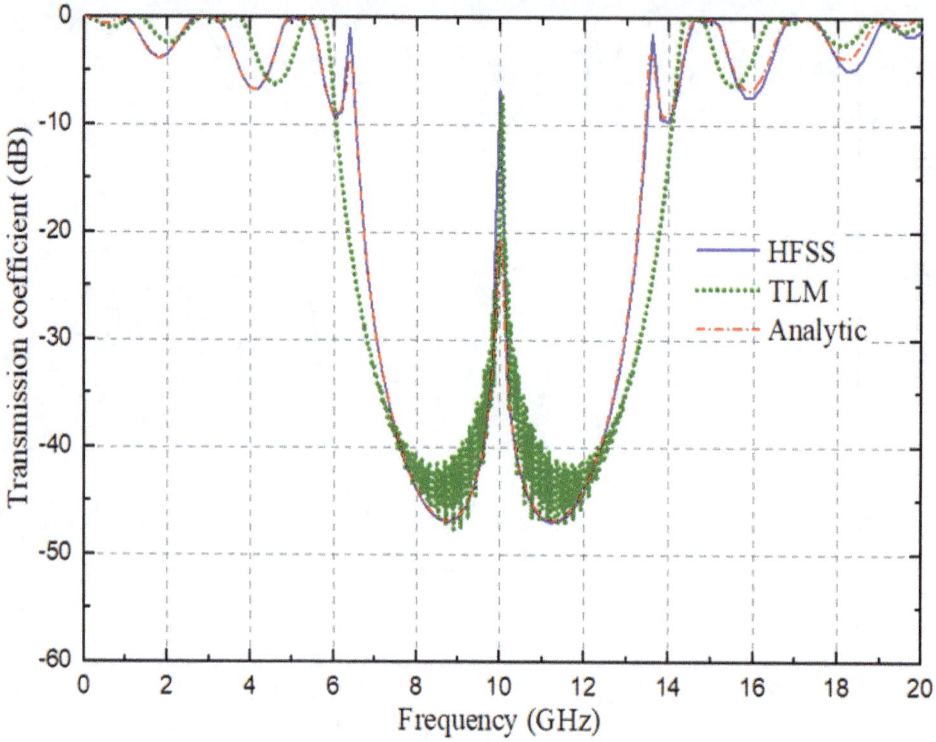

Figure 22. Tangential electric field mapping in EBG structure

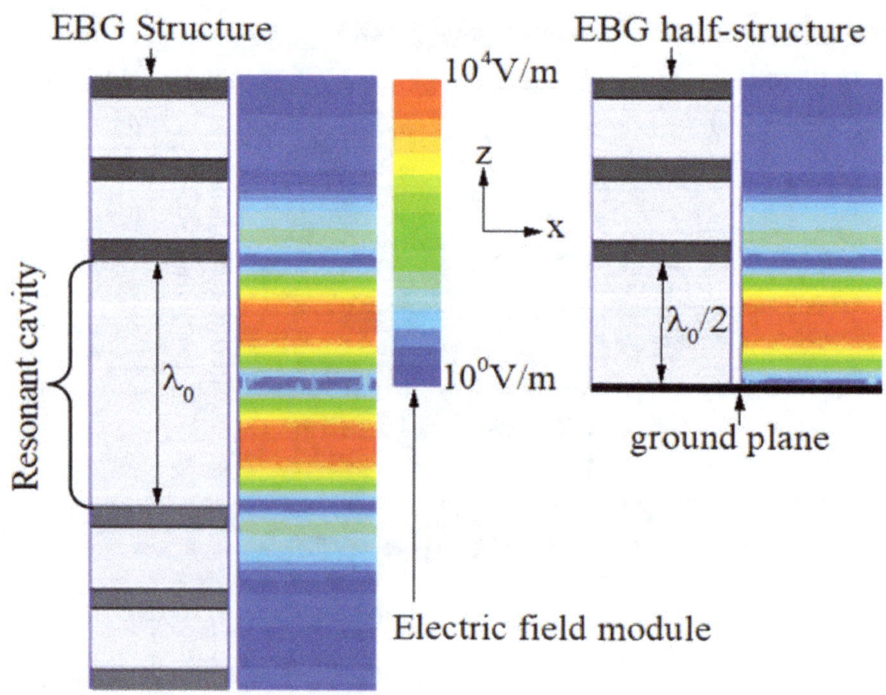

the ground plane is equivalent to that of the EBG structure with defect (Thevenot, Cheype, Reineix, & Jecko, 1999). At the level of the ground plane, a source of excitation is placed; that can be one or more patches, dipoles, slots, a horn or also a waveguide.

The planar EBG antenna behaves as a resonator composed of two mirrors (see Figure 23). The first mirror, being a metal ground plane; it is totally reflective and has a reflection phase φ_{inf} equal to π. The second mirror is generally constituted by a stack of partially reflective dielectric layers and have a reflection phase φ_{sup} whose profile is linearly decreasing and value close to π; this mirror is located at the height h_{cavity}.

$$h_{cavity} = \frac{\lambda_0}{2} \tag{85}$$

One of the most developed ways for the EBG application to microwave frequencies in the antennas domain is their effects on the antenna gain, directivity, and radiation (Leger, 2004; Xu, Zhao, & Luo, 2008; Arand, & Bazrkar, 2015; Yang, Alexopoulos, & Yablonovitch, 1997). Other research has shown that the EBG structure behaves like a perfect reflector in the forbidden band, and that the radiation is mainly directed in the air rather than in the backward substrate (Zeb, Ge, & Esselle, 2012; Leger, Monediere, & Jecko, 2005; Li, 2009), which suggests a significant increase in the EBG structure efficiency of and reduces the level of the sidelobes (Brown, Parker, & Yablonovitch, 1993).

A good functioning of the EBG antenna imposes certain conditions on the excitatory source:

1. The excitation element must be a radiating system capable of transforming guided energy (antenna power supply) into radiated energy.
2. The polarization of the energy must be in agreement with the operating polarization of EBG structure, i.e. electric and magnetic fields are tangential to dielectric layers.
3. The optimal operation is obtained by favoring the coupling between the excitation and the structure.

Figure 23. EBG antenna design

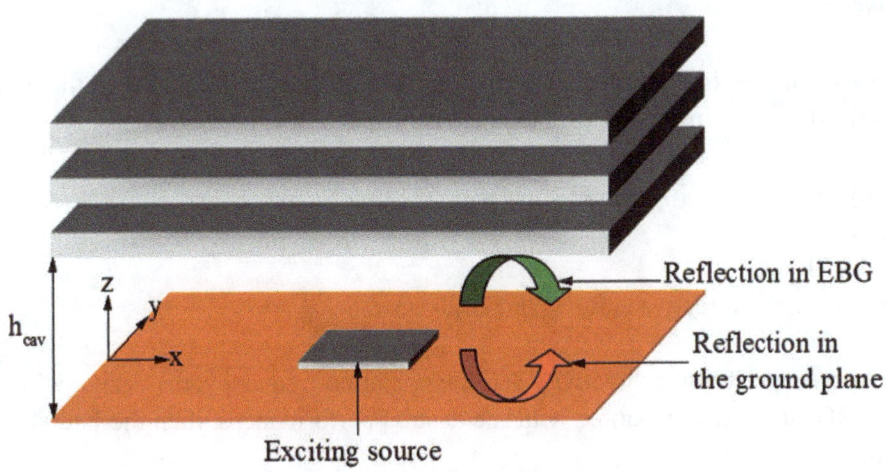

Figure 24. Patch antenna operation

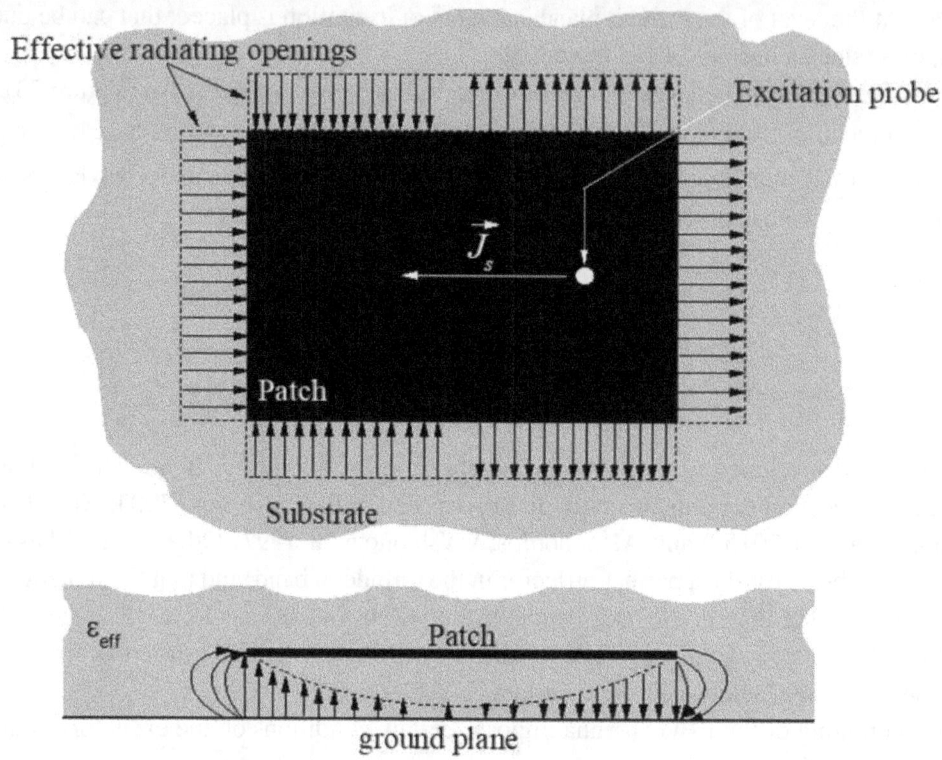

The choice of patch antenna is motivated by its simplicity of modeling and production. This radiating element can be of any shape: square, rectangular, circular, polygon etc. When the radiating element is traversed by a surface electric current $\vec{J}_s$, the electric field will be vertical and located between the ground plane and the patch. The magnetic field constitutes a loop around the radiating element. The polarization of the magnetic field excited by the patch is identical to that found in the cavity of the EBG structure, which allows a good coupling between the patch and the EBG structure.

1. **Objective**

The purpose of this design is primarily to validate the method and concept of the directive EBG-antenna. For this, the following specifications must be respected:

- Directivity: $20dB$.
- Frequency: $5.8GHz$.
- Polarization: Linear.
- Radiation pattern is directive in the E and H planes.

For this purpose, specifying a directivity and the operation frequency, the design will be started by the choice of EBG structure compatible with the desired performances, then the lateral dimensioning of the antenna.

2. Choice of EBG Structure

The first step is to determine the quality coefficient Q necessary to obtain the desired directivity; this is represented on a chart in (Leger, 2004). From the Figure 2, the quality coefficient necessary to obtain a directivity of $20dB$ is about 50. Figure 27 illustrate another chart that shows the relationship between the quality coefficient and the relative permittivity, as well as the number of dielectric layers required to obtain the desired directivity (see Figure 25).

The parameters of the EBG structure concerning the relative permittivity of dielectrics and the number of layers needed to obtain a quality coefficient of 50 are determined from Figure 27.

For the quality coefficient equal 50, there are two possibilities:

- Three layers with $\varepsilon_r = 2.6$. Two layers with $\varepsilon_r = 5$. To determine the size of the antenna, the first case ($\varepsilon_r = 2.6$) is chosen:

3. EBG-Antenna Geometry

a. Lateral dimension

The first important dimension of the antenna is the lateral side L. To have a desired directivity (around 20 dB), a relation deduced from the equation (antenna gain equation) makes it possible to write (Leger, 2004):

$$L = \sqrt{\frac{10^{\frac{D_{dB}}{10}} . \lambda_0^2}{0.8\pi^2}} \tag{86}$$

where:

- L :. The lateral side of the antenna (m).

Figure 25. Number of layers

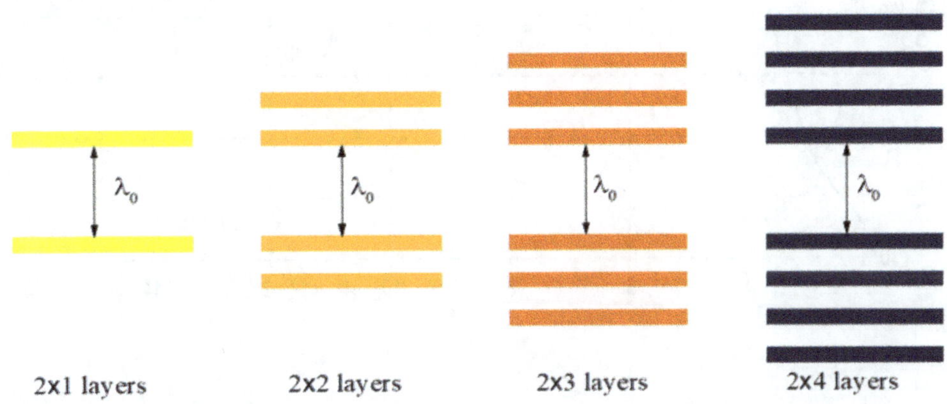

Figure 26. Maximum Directivity according to the coefficient of quality

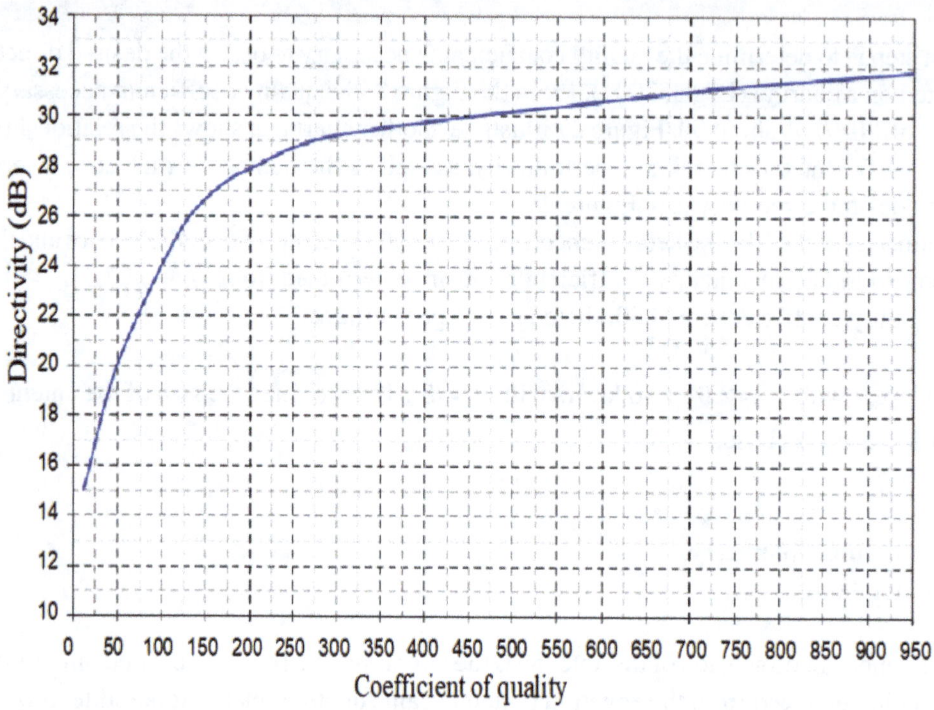

Figure 27. Coefficient of quality according to the number of layers

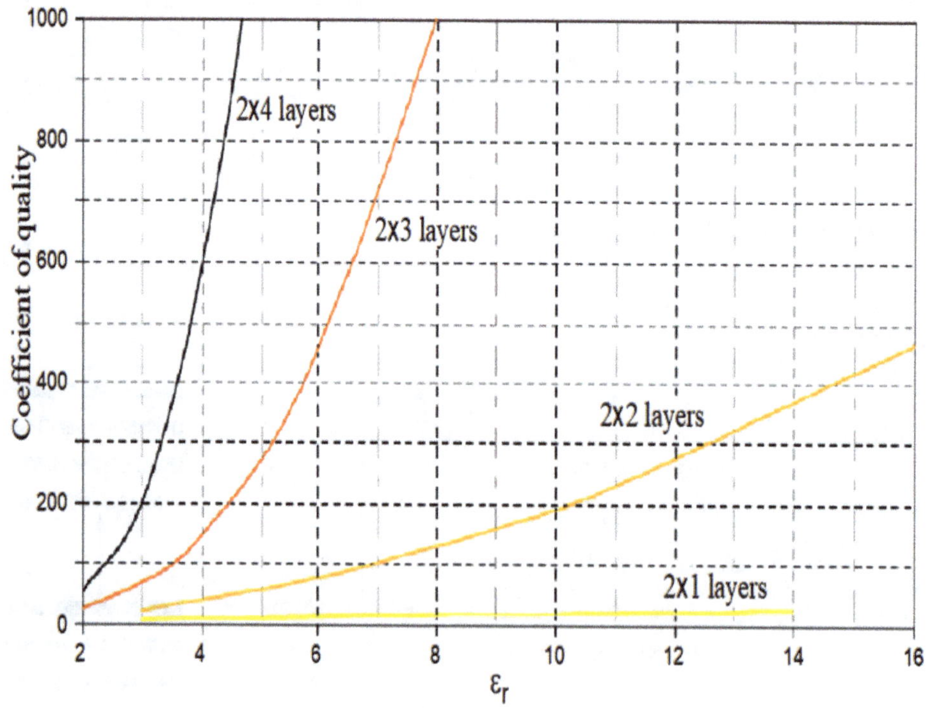

- D_{dB}. The desirable directivity of the antenna (dB).
- λ_0. The wavelength in vacuum corresponds to the resonance frequency $5.8GHz$.

For a directivity $D_{dB}=20dB$. it is found that $L=184mm$.

b. Dimensions of the cavity and the EBG structure.

Figure 28 illustrate the EBG antenna design. The cavity height to have a transmission peak at the center of the band gap has been defined in the Equation (85). However, during the passage in finite structure, the cavity height is readjusted in order to design the chosen operating frequency, which can be estimated by the following equation:

$$f_r = \frac{c}{2}\sqrt{\left(\frac{1}{l}\right)^2 + \left(\frac{1}{L}\right)^2 + \left(\frac{1}{h_{cavity}}\right)^2} \quad (87)$$

where,

- l, L: The lateral dimensions of the antenna.
- h_{cavity}.: The height of the air cavity.
- c: The speed of light in the vacuum.
- f_r: The operating frequency (5.8GHz).

Replacing l and L by the infinite value in Equation (87), it is obtained again the value of the equation (85) for the cavity of the EBG resonator. For the reasons of simplicity, the antenna is a square geometry, that is to say $l = L = 184mm$. According to Equation (87), it is obtained the height of the resonator cavity $h_{cavity} = 26.4mm.$. The thickness of the dielectric layers as well as the inter-plate air spaces are calculated by applying Equation (26), which gives respectively $h_{diel} = a = 8mm$ and $h_{air} = b = 13mm.$

Figure 28. The dimensions of the 5.8GHz EBG-antenna

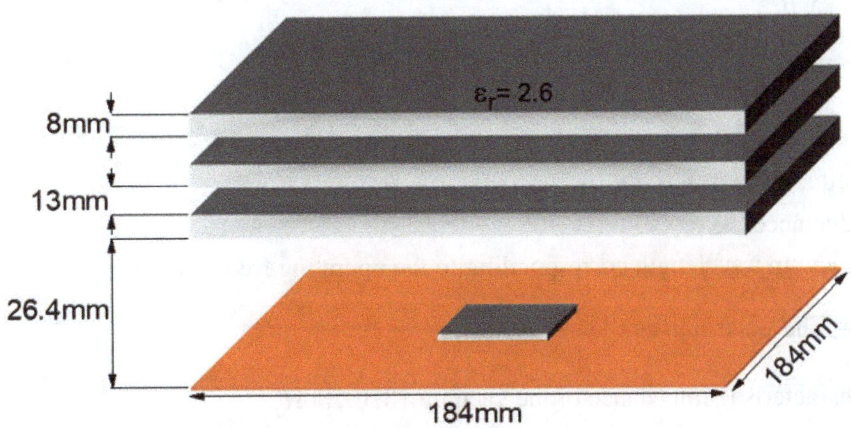

4. Exciting Source

Because of simplicity for the realization, we will choose a simple patch antenna as an exciting source. The radiating element of the patch antenna is etched on a substrate FR4 of height 1.57mm and permittivity $\varepsilon_r = 4.4$. To calculate the dimensions of the patch, the formulas (88) are used (Constantine, 2005, Chapter 14). For the resonance frequency equal to 5.8GHz, it is found that $W_{patch} = 15.7mm$ and $L_{patch} = 11.8mm$..

$$\begin{cases} \dfrac{\Delta L}{h} & = 0.412 \dfrac{\left(\varepsilon_{eff} + 0.300\right)\left(\frac{w_{patch}}{h} + 0.264\right)}{\left(\varepsilon_{eff} - 0.258\right)\left(\frac{w_{patch}}{h} + 0.800\right)} \\ \varepsilon_{eff} & = \dfrac{\varepsilon_r + 1}{2} + \dfrac{\varepsilon_r - 1}{2}\left(1 + 12\dfrac{h}{w_{patch}}\right)^{-1/2} \\ w_{patch} & = \dfrac{c}{2f_r}\sqrt{\dfrac{2}{\varepsilon_r + 1}} \\ l_{patch} & = \dfrac{c}{2f_r\sqrt{\varepsilon_{eff}}} - 2\Delta L \end{cases} \tag{88}$$

$$\begin{cases} G = \dfrac{\pi w_{patch}}{\eta \lambda_0}\left[1 - \dfrac{\left(hk_0\right)^2}{24}\right] \\ R_e = \dfrac{1}{2G} \\ x = \dfrac{l_{patch}}{\pi} sin^{-1}\sqrt{\dfrac{Z_0}{R_e}} \end{cases} \tag{89}$$

where,

- ε_{eff} The effective permittivity.
- G Conductance (S).
- λ_0 The vacuum wavelength corresponding to the operating frequency f_r (m).
- $K_0 = \dfrac{2\pi}{\lambda_0}$ The wave constant $\left(m^{-1}\right)$.
- η The characteristic impedance of the vacuum $\left(376.730\Omega\right)$.

- R_e The resistance at the input of the patch antenna (Ω).
- Z_0 The characteristic impedance of the SMA connector $\left(50\Omega\right)$.

Figure 29 illustrates the dimensions of the patch antenna used to excite the EBG cavity. The dimensions of the patch and the position of the probe have been optimized to achieve the best adaptation possible. The optimized dimensions are W_(patch)=16mm and L_(patch)=12.2mm. The patch is feed by a SMA coaxial probe, whose internal conductor diameter equals and the outer conductor diameter is For a desirable adaptation, the excitation must locate at the distance from the patch center, this parameter is calculated using the formulas (89) given in (Milligan, 2005), it is found that.

5. The Reflection Coefficient

The reflection coefficient of the antenna with and without EBG structure is shown in Figure 30. In both cases (with and without EBG), the reflection coefficient is less than $-10dB$ .around 5.8GHz. It can be noted that there is a slight shift at the presence of EBG between the results found with HFSS and CST, which can be attributed to the use of two different methods. Both software are generally popular for antenna design because of the ease in simulations, the High Frequency Structure Simulator (HFSS) is based on the finite element method (FEM) (Ansys Inc, 2011). While Computer Simulation Technology (CST) Microwave Studio is based on the finite integral method (MIF) and on the method Finite Difference Time Domain (FDTD) (CST, 2011).

Figure 29. Excitation source dimensions

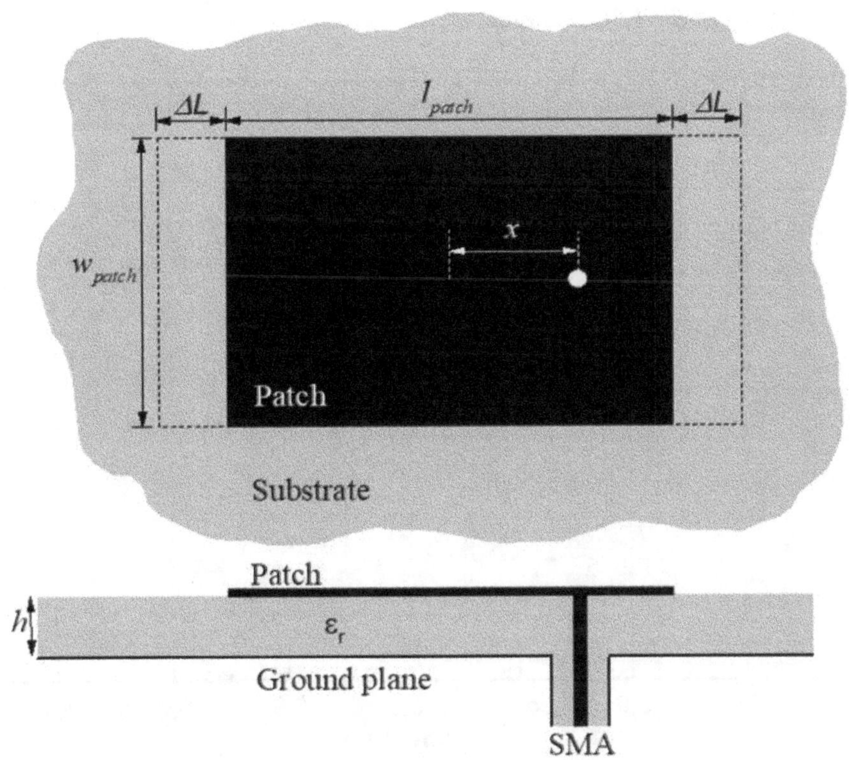

Figure 30. Reflection coefficient of the antenna

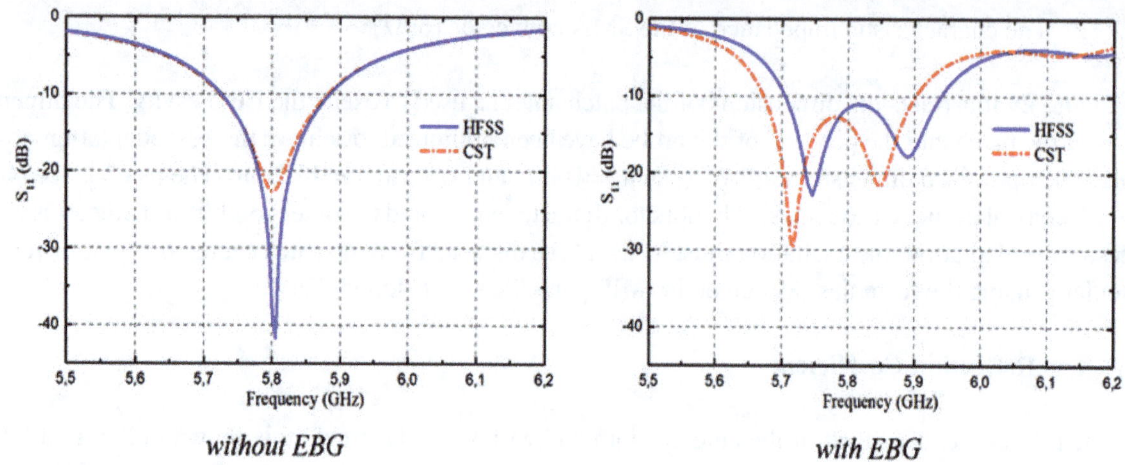

6. **The Directivity**

The directivity shown in Figure 31 is very high in all the bandwidth. It takes a value of 20 dB for the resonant frequency f_r chosen beforehand in the specifications.

7. **Radiation Pattern**

Figure 31. The directivity of the EBG antenna

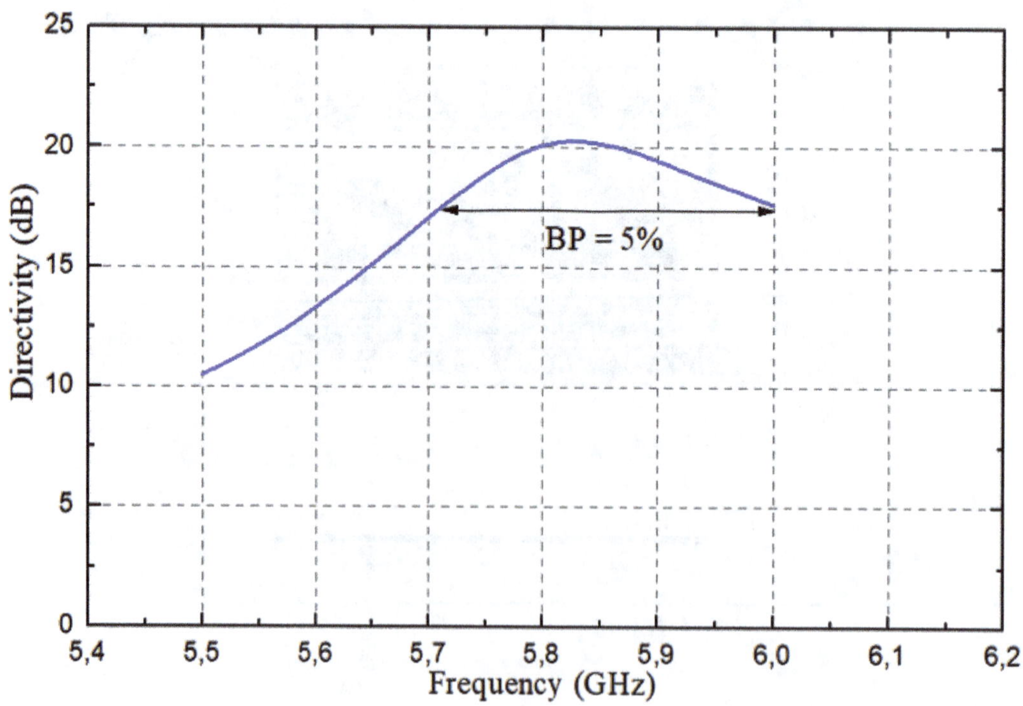

Figure 32. 3D radiation diagram

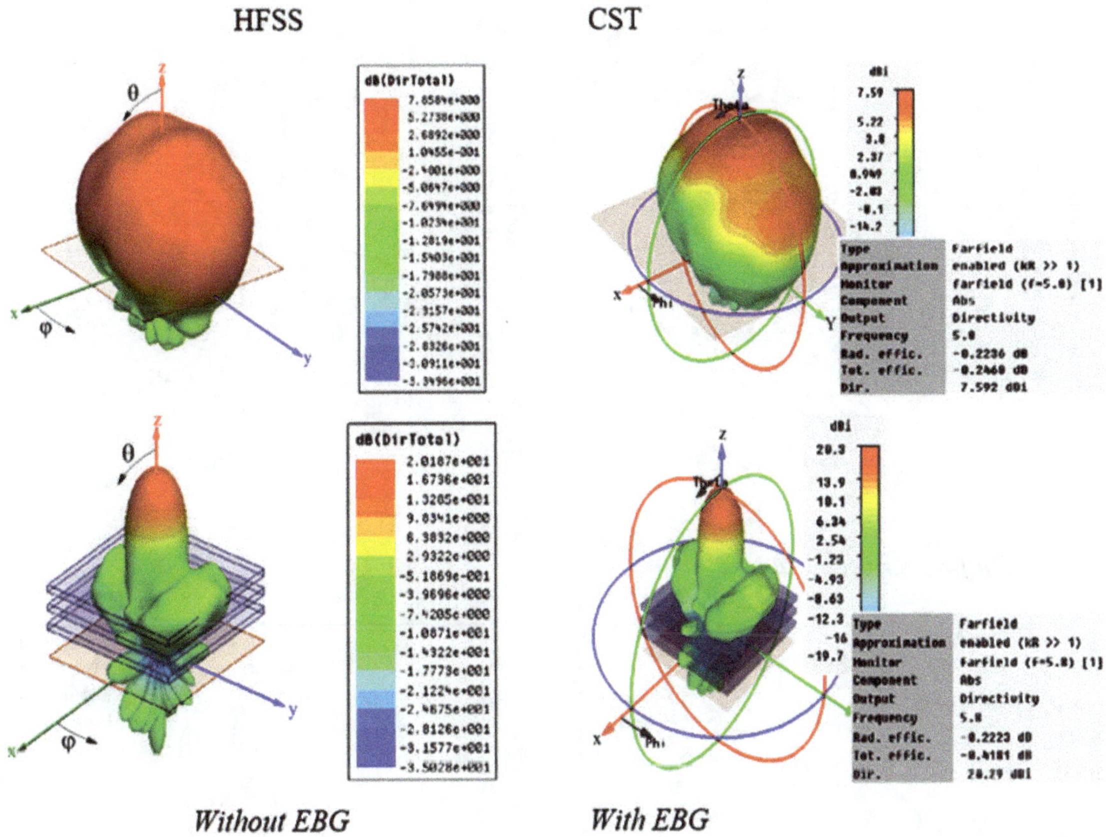

The $3D$ radiation pattern at 5.8GHz is shown in Figure 32. This figure shows the overall behavior for the antenna radiation. To present more useful and necessary information, the radiation pattern is presented in polar coordinates (see figure 33) or in 2D (see Figure 34).

From Figure 33, it can be seen that the maximum directivity is $20dB$. the secondary lobes are at $-20dB$ from the main lobe for the $H-plane$ and $-15dB$ for the $E-plane$.

Half-power beamwidth corresponds to the width of the main lobe, between the attenuation angles at $-3dB$ is narrower equal $16°$ (see Figure 34), which shows that periodic EBG structures make it possible to perform spatial filtering.

Figure 35 illustrates the mapping of the electric field of the antenna, with and without EBG, for different observation points. This figure shows that it is possible to favor a direction of radiation in space when the antenna is combined with EBG structure.

Figure 33. Radiation pattern of the EBG antenna at 5.8GHz

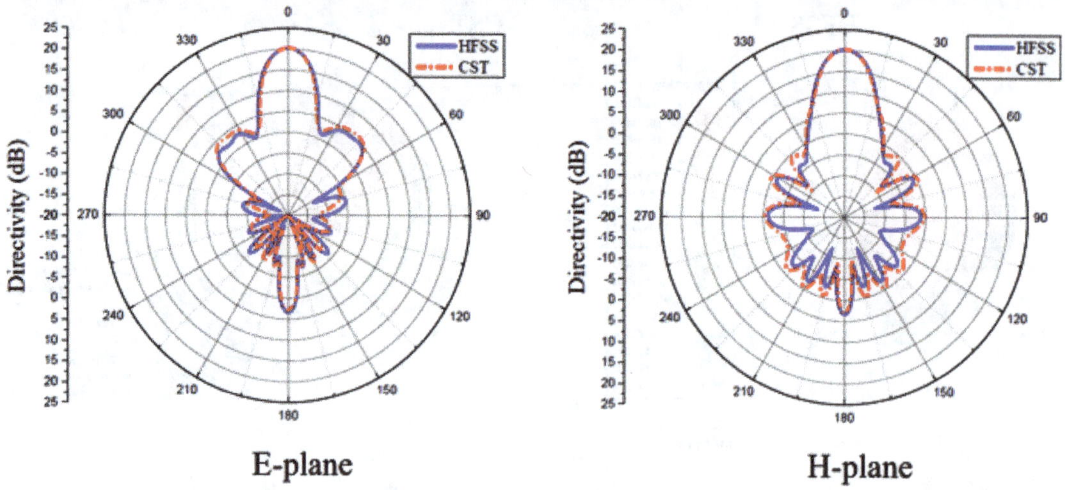

Figure 34. Half-power beamwidth

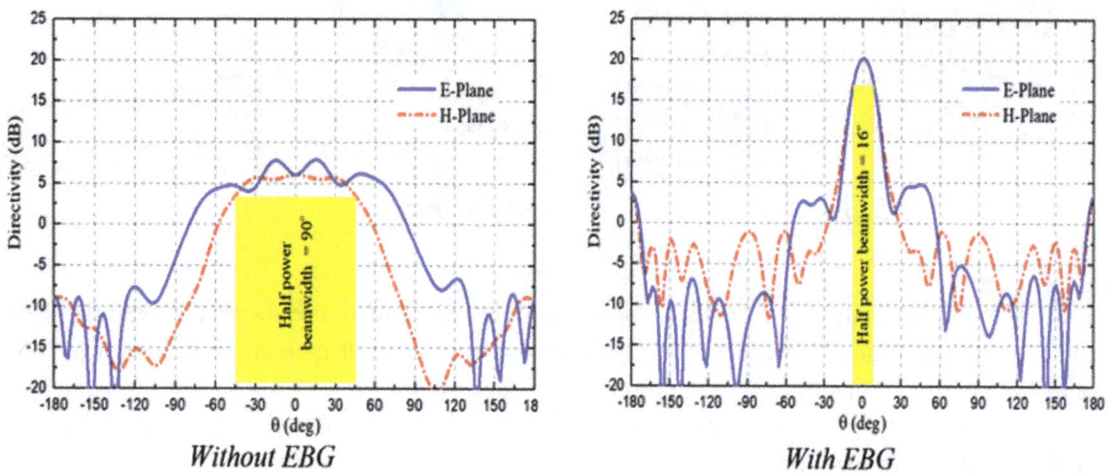

CONCLUSION

In this chapter, we have developed the design technique of the directive EBG antenna based on the use of one-dimensional electromagnetic band-gap structures. This technique has notably contributed to improving the analyze and to understand the behavior of this structure. This principle of operation can also be extended to the design of EBG antennas in two or three dimensions, and for the design of antennas with periodic metal structures called EBG-M antennas.

The EBG structure with defect, presents the symmetry plane. A metallic plane can replace this plane of symmetry, because the mapping of the electric field shows a cancellation of the tangential component of the E-field at this plane. Thus, by applying the electrical images theory, the behavior of the half-structure above the ground plane is equivalent to that of the complete EBG structure with defect. At the level of the ground plane, we can have a source of excitation that can be one or more patches. The planar

Figure 35. Mapping of the electric field of the antenna

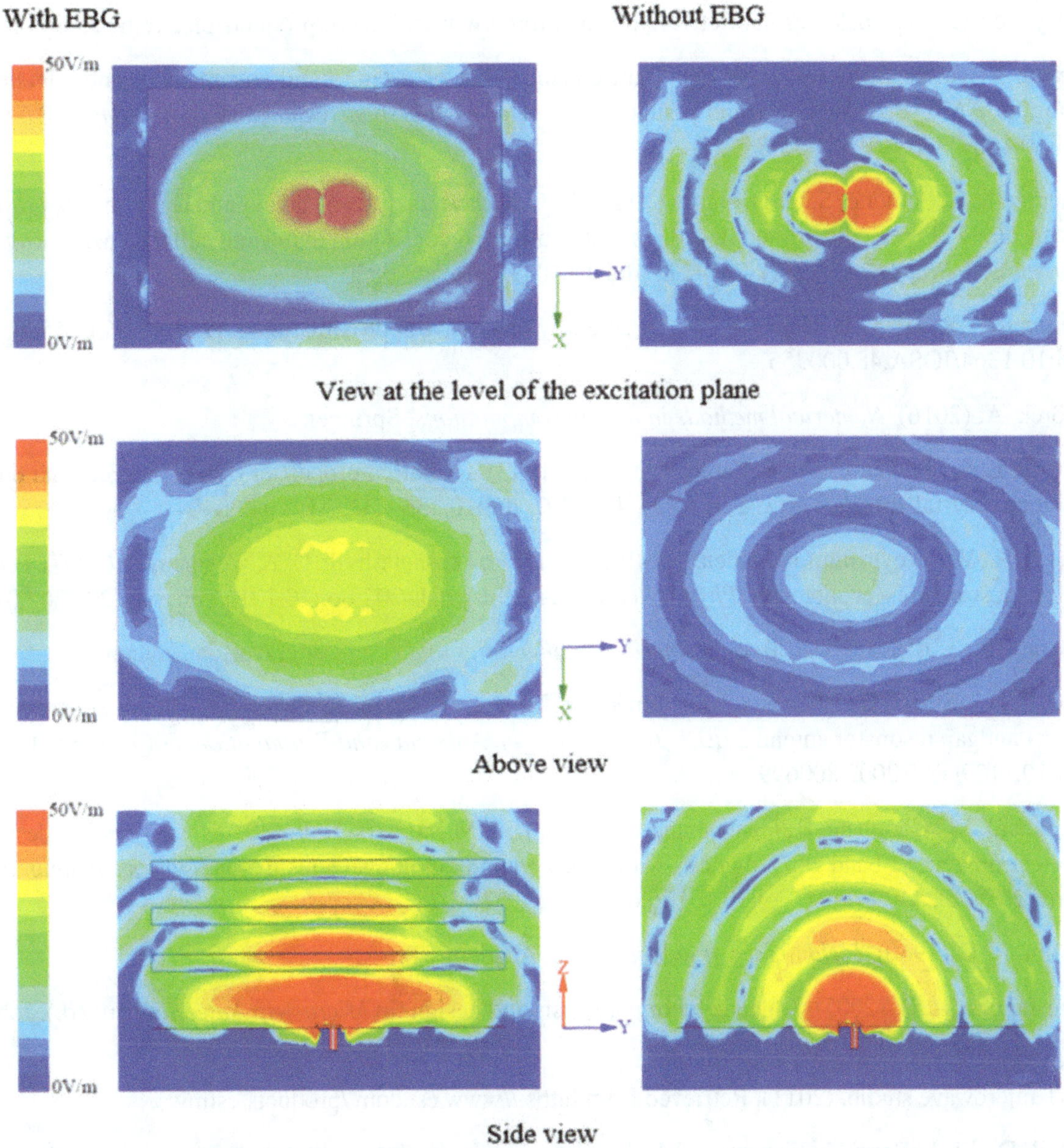

EBG antenna presented as a resonator composed of two mirrors; the first mirror generally constituted by a stack of dielectric layers is partially reflective, and has a reflection phase φ_{sup} whose profile is linearly decreasing and whose value is close to π. The second mirror being a metallic ground plane, totally reflective and has a reflection phase φ_{inf} equal to π. Increasing bandwidth is an important goal for the EBG antenna. An excitation source exhibits the broadband behavior connected with the EBG structure, which makes it possible to obtain directional radiation for a wide frequency band. The multi-source excitation can also reach the goal of increasing the bandwidth.

REFERENCES

Ansys Inc. (2011). *HFSS. v13.0.* Retrieved from http://www.ansys.com/products/electronics/ansys-hfss

Arand, B. A., & Bazrkar, A. (2015). Gain enhancement of a tetra-band square-loop patch antenna using an AMC–PEC substrate and a superstrate. *Wireless Personal Communications*, *84*(1), 87–97. doi:10.100711277-015-2595-8

Baccarelli, P., Paulotto, S., & Di Nallo, C. (2007). Full-wave analysis of bound and leaky modes propagating along 2D periodic printed structures with arbitrary metallisation in the unit cell. *IET Microwaves, Antennas & Propagation*, *1*(1), 217–225. doi:10.1049/iet-map:20050321

Baumeister, P. (1958). Design of multilayer filters by successive approximations. *JOSA*, *48*(12), 955–958. doi:10.1364/JOSA.48.000955

Björck, Å. (2016). *Numerical methods in matrix computations*. Springer.

Brown, E. R., Parker, C. D., & Yablonovitch, E. (1993). Radiation properties of a planar antenna on a photonic-crystal substrate. *JOSA B*, *10*(2), 404–407. doi:10.1364/JOSAB.10.000404

Busch, K., Von Freymann, G., Linden, S., Mingaleev, S. F., Tkeshelashvili, L., & Wegener, M. (2007). Periodic nanostructures for photonics. *Physics Reports*, *444*(3-6), 101–202. doi:10.1016/j.physrep.2007.02.011

Cheney, E., & Kincaid, D. (2012). *Numerical mathematics and computing*. Nelson Education.

Cheype, C., Serier, C., Thèvenot, M., Monédière, T., Reineix, A., & Jecko, B. (2002). An electromagnetic bandgap resonator antenna. *IEEE Transactions on Antennas and Propagation*, *50*(9), 1285–1290. doi:10.1109/TAP.2002.800699

Choubani, M., Choubani, F., Gharsallah, A., David, J., & Mastorakis, N. E. (2010). Analysis and design of electromagnetic band gap structures with stratified and inhomogeneous media. *Optics Communications*, *283*(22), 4499–4504. doi:10.1016/j.optcom.2010.04.078

Collin, R. E. (1960). *Field theory of guided waves*. Academic Press.

Constantine, A. B. (2005). Antenna theory: analysis and design. In *Microstrip Antennas* (3rd ed.). John Wiley & Sons.

CST microwave studio. (2011). Retrieved from https://www.cst.com /products/cstmws

Fattal, D., Li, J., Peng, Z., Fiorentino, M., & Beausoleil, R. G. (2010). Flat dielectric grating reflectors with focusing abilities. *Nature Photonics*, *4*(7), 466–470. doi:10.1038/nphoton.2010.116

Ghen, Z., Ney, M. M., & Hoefer, W. J. (1993). Absorbing and connecting boundary conditions for the TLM method. *IEEE Transactions on Microwave Theory and Techniques*, *41*(11), 2016–2024. doi:10.1109/22.273429

Hajj, M., Rodes, E., Serhal, D., Chantalat, R., Thevenot, M., Monediere, T., & Jecko, B. (2007, September). Metallic EBG sectoral antenna for base stations. In *Applied Electromagnetics and Communications, 2007. ICECom 2007. 19th International Conference on* (pp. 1-4). IEEE. 10.1109/ICECOM.2007.4544413

Hashmi, R. M., Zeb, B. A., & Esselle, K. P. (2014). Wideband high-gain EBG resonator antennas with small footprints and all-dielectric superstructures. *IEEE Transactions on Antennas and Propagation, 62*(6), 2970–2977. doi:10.1109/TAP.2014.2314534

Joannopoulos, J. D., Johnson, S. G., Winn, J. N., & Meade, R. D. (2011). *Photonic crystals: molding the flow of light*. Princeton University Press.

John, S. (1987). Strong localization of photons in certain disordered dielectric superlattices. *Physical Review Letters, 58*(23), 2486–2489. doi:10.1103/PhysRevLett.58.2486 PMID:10034761

Johnson, S. G., & Joannopoulos, J. D. (2000). Three-dimensionally periodic dielectric layered structure with omnidirectional photonic band gap. *Applied Physics Letters, 77*(22), 3490–3492. doi:10.1063/1.1328369

Kaabal, A., Ahyoud, S., Asselman, A., & Faize, A. (2013). Design Of High Gain Ultra Wide-Band Antenna For Wireless Communication Using EBG Structures. *European Scientific Journal, 9*(30).

Kushwaha, N., & Kumar, R. (2014). Study of different shape electromagnetic band gap (EBG) structures for single and dual band applications. *Journal of Microwaves, Optoelectronics and Electromagnetic Applications, 13*(1), 16–30. doi:10.1590/S2179-10742014000100002

Leger, L. (2004). *Nouveaux développements autour des potentialités de l'antenne BIE planaire* (Doctoral dissertation). Limoges.

Leger, L., Monediere, T., & Jecko, B. (2005). Enhancement of gain and radiation bandwidth for a planar 1-D EBG antenna. *IEEE Microwave and Wireless Components Letters, 15*(9), 573–575. doi:10.1109/LMWC.2005.855373

Leger, L., Serier, C., Chantalat, R., Thevenot, M., Monedière, T., & Jecko, B. (2004, March). 1D dielectric electromagnetic band gap (EBG) resonator antenna design. *Annales des Télécommunications, 59*(3-4), 242–260.

Li, Y. (2009, December). Investigation of minimum cavity height of small EBG-resonator antennas for maximum directivity. In *Microwave Conference, 2009. APMC 2009. Asia Pacific* (pp. 2687-2690). IEEE. 10.1109/APMC.2009.5384121

Li, Z. Y., & Lin, L. L. (2003). Photonic band structures solved by a plane-wave-based transfer-matrix method. *Physical Review. E, 67*(4), 046607. doi:10.1103/PhysRevE.67.046607 PMID:12786509

Lo, J. (2012). *Étude de la reconfigurabilité d'une structure à bande interdite électromagnétique (BIE) métallique par plasmas de décharge* (Doctoral dissertation). Université Paul Sabatier-Toulouse III.

Milligan, T. A. (2005). *Modern antenna design*. John Wiley & Sons. doi:10.1002/0471720615

Mishra, S., & Satpathy, S. (2003). One-dimensional photonic crystal: The Kronig-Penney model. Physical Review B, 68(045121).

Mizeikis, V., Seet, K. K., Juodkazis, S., & Misawa, H. (2004). Three-dimensional woodpile photonic crystal templates for the infrared spectral range. *Optics Letters, 29*(17), 2061–2063. doi:10.1364/OL.29.002061 PMID:15455780

Moghaddam, B. T., Entezar, S. R., & Adl, H. P. (2013). Nonlinear properties of a graded-index photonic heterostructure. *Pramana*, *80*(5), 887–894. doi:10.100712043-013-0523-6

Morente, J. A., Porti, J. A., & Khalladi, M. (1992). Absorbing boundary conditions for the TLM method. *IEEE Transactions on Microwave Theory and Techniques*, *40*(11), 2095–2099. doi:10.1109/22.168768

Rayleigh, L. (1888). XXVI. On the remarkable phenomenon of crystalline reflexion described by Prof. Stokes. *Philosophical Magazine and Journal of Science*, *26*(160), 256–265. doi:10.1080/14786448808628259

Russell, P. S. J., Birks, T. A., & Lloyd-Lucas, F. D. (1995). Photonic Bloch waves and photonic band gaps. In *Confined electrons and photons* (pp. 585–633). Boston, MA: Springer. doi:10.1007/978-1-4615-1963-8_19

Rylander, T., Ingelström, P., & Bondeson, A. (2012). *Computational electromagnetics*. Springer Science & Business Media.

Sakoda, K. (2004). *Optical properties of photonic crystals* (Vol. 80). Springer Science & Business Media.

Salah Toubeh, M. (2011). *Etude d'antennes BIE planaires de hauteur très inférieure à la longueur d'onde dite: The ULP EBG Antennas* (Doctoral dissertation). Limoges.

Sievenpiper, D. F. (2008). Artificial impedance surfaces for antennas. *Modern Antenna Handbook*, 737-777.

Tam, W. Y. (2007). Woodpile and diamond structures by optical interference holography. *Journal of Optics. A, Pure and Applied Optics*, *9*(11), 1076–1081. doi:10.1088/1464-4258/9/11/018

Thevenot, M., Cheype, C., Reineix, A., & Jecko, B. (1999). Directive photonic-bandgap antennas. *IEEE Transactions on Microwave Theory and Techniques*, *47*(11), 2115–2122. doi:10.1109/22.798007

Winn, J. N., Fink, Y., Fan, S., & Joannopoulos, J. D. (1998). Omnidirectional reflection from a one-dimensional photonic crystal. *Optics Letters*, *23*(20), 1573–1575. doi:10.1364/OL.23.001573 PMID:18091848

Xu, H., Zhao, Z., Lv, Y., Du, C., & Luo, X. (2008). Metamaterial superstrate and electromagnetic band-gap substrate for high directive antenna. *International Journal of Infrared and Millimeter Waves*, *29*(5), 493–498. doi:10.100710762-008-9344-y

Yablonovitch, E. (1993). Photonic band-gap structures. *JOSA B*, *10*(2), 283–295. doi:10.1364/JOSAB.10.000283

Yang, H. Y., Alexopoulos, N. G., & Yablonovitch, E. (1997). Photonic band-gap materials for high-gain printed circuit antennas. *IEEE Transactions on Antennas and Propagation*, *45*(1), 185–187. doi:10.1109/8.554261

Yeo, J., & Kim, D. (2009). Novel design of a high-gain and wideband Fabry-Perot cavity antenna using a tapered AMC substrate. *Journal of Infrared, Millimeter and Terahertz Waves*, *30*(3), 217–224. doi:10.100710762-008-9451-9

Zeb, B. A., Ge, Y., & Esselle, K. P. (2012, December). A single-layer thin partially reflecting surface for tri-band directivity enhancement. In *Microwave Conference Proceedings (APMC), 2012 Asia-Pacific* (pp. 559-561). IEEE. 10.1109/APMC.2012.6421663

Zeidler, E. (1995). Hilbert Spaces and Generalized Fourier Series. In *Applied Functional Analysis* (pp. 195–227). New York, NY: Springer. doi:10.1007/978-1-4612-0815-0_3

ENDNOTES

1 The linear medium is a medium whose response is proportional to the disturbance.

2 The isotope medium is a medium whose properties are identical regardless of the direction of observation.

Chapter 6
Miniaturization and Reduction of Mutual Coupling Between Antennas Arrays Using DGS and Planar EBG Structures

Ahmed Ghaloua
University of Hassan 1st, Morocco

Jamal Zbitou
University of Hassan 1st, Morocco

Larbi El Abdellaoui
University of Hassan 1st, Morocco

Mohamed Latrach
ESEO, France

ABSTRACT

As the size of the antenna often has a significant influence on overall dimensions of the wireless system, its reduction in size becomes a significant challenge. The objective of this chapter is to present new contributions made for reducing the size of the antenna array while maintaining excellent performance. An overview of the antenna array is introduced. Then, two designed and fabricated antenna arrays with compact size and good performances are exposed. The first microstrip patch antenna array is miniaturized using a novel shape of defected ground structure (DGS) etched in the ground plane of each radiating element of the antenna array. While the second one is two antenna arrays which are separated by two magnetic walls of a planar compact electromagnetic band gap (EBG) structure, with the aim to miniature and to reduce the mutual coupling between them, keeping both the antenna arrays separation smaller than 0.6λ5.8GHz. A full-wave electromagnetic analysis had achieved to evaluate the electrical performances of the proposed structures by using HFSS and CST-MWS.

DOI: 10.4018/978-1-5225-7539-9.ch006

INTRODUCTION

We define a communicating system by a multitude of physically dispersed objects, each having a data processing unit and access to a communication network. Several inventors have set numerous standards distinguished by the frequency of emission used, the speed and range of transmission of each technology. Wireless networks make it possible to connect different equipment, which is far from each other by many kilometers. In recent years, telecommunications systems have undergone rapid development, marked in particular by the creation of new standards such as GSM (Global System for Mobile communications), GPRS (General Packet Radio Service), EDGE (Enhanced Data Rates for GSM Evolution), UMTS (Universal Mobile Telecommunication System), LTE (Long Term Evolution), WiMAX (Worldwide Interoperability for Microwave Access)...etc. At the same time, new applications (GPS, Bluetooth, WiFi, 3G, 4G, etc.) have implemented on mobile terminals operating in specific frequency bands. In wireless transmission, the antenna is the transitional structure between free-space and a guiding device. A guided environment is a structure, where an electromagnetic wave is confined and propagates along the path defined by the environment. These media are usually in the form of a waveguide (Zarifi, Farahbakhsh, & Zaman, 2017), coaxial cable (Xing, Liu, Guo, Wei, Zhao, & Ma, 2017) or microstrip line (Basilio, Khayat, Williams, & Long, 2001). The antenna is a component used to transmit or receive the electromagnetic waves. It ensures the transition between the guided wave and the wave transmitted in the free space. We have a transmitting antenna when the antenna allows electromagnetic energy to radiate from a guided environment to free space. In the reverse case, the antenna is called a receiver. However, the antennas respect the principle of reciprocity. It is possible to deduct the characteristics of a receiver antenna by characterized its transmission performance and vice versa. The characterization of an antenna is made using several parameters, which controlled as a function of frequency. Among these parameters, there are the reflection coefficient, the radiation pattern and the polarization. Microstrip patch antennas play an integral part in many applications. Are widely preferred for wireless communication systems and navigation systems. Notably, these last two applications require antennas with low fabrication cost, thin profile and can handle the linear and circular polarization (Xue, Yang, & Ma, 2015). Unlike the conventional microwave antennas, the printed antennas are extremely compatible with the wireless devices such as cellular phones. Another major advantage associated with a planar antenna is that it could be easily integrated with the microwave integrated circuits (MICs). For various applications, it is necessary to design antennas which have a broad bandwidth and who radiate into total space not only into half space, with very high directive characteristics -very high gain- to meet the demands of long-distance communication. For several years, a lot of research had undertaken to overcome the limitations associated with these kinds of antennas. Actually, the gain of conventional microstrip patch antenna is limited a low level because of their limited size at resonance frequency. Some of the popular techniques proposed by researchers to increase the gain are: increase the electrical size of the antenna (Constantine & others, 2005), use a thin, low loss and low permittivity substrate (Alboni & Cerretelli, 2001). Another method proposes the use of artificial magnetic conductors in the design of low-profile high-Gain printed Antennas with high-permittivity dielectric superstrate, had mentioned in (Foroozesh & Shafai, 2009). While others (Zhang & Zhu, 2016) loading two sets of shorting pins along the two orthogonal diagonals of a square patch resonator. So far, the most common method of improving the antenna radiation characteristics, especially the gain, is to form an assembly of radiating elements -in one, two or three -dimensional- in an electrical and geometrical configuration. This new antenna formed had referred to as an array.

Figure 1. The topology of linear antenna array (left) and planar antenna array (right)

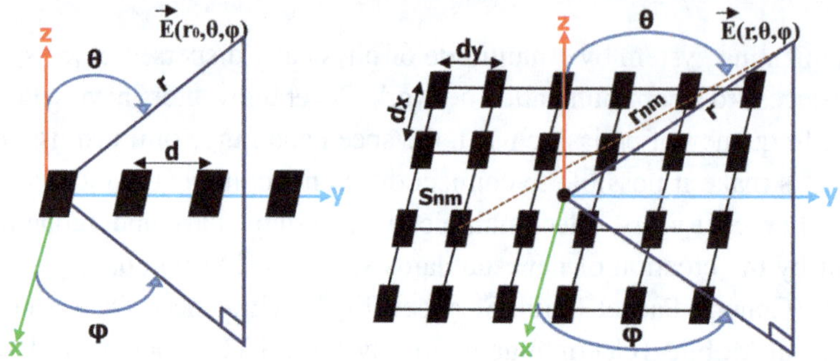

As we have already mentioned, the array in an association of several antennas compensates for the limitations of a single antenna. There are a several methods of assembly the microstrip antenna array. Among them, the linear and planar arrangements. The linear antenna array had been considered as the simplest antenna array. It is a linear assembly of N radiating elements arranged along a single axis (Fleuriault, 1996). While the planar antenna array configuration is two-dimensional generalization of a linear array. This type of array had usually formed by the use a matrix of N×M elementary antennas. The elements of the antenna array can be fed by a single line called the series-feed network (Yousefbeigi, Enayati, Shahabadi, & Busuioc, 2008) or by multiple lines called corporate-feed network (Combes, 1997). For the series feed microstrip antenna array, the antennas are linked in a chain using the high impedance transmission line and feeding the power at the first element (Constantine, 2005), while in the case of the corporate feeding, the distribution of information is done through by a succession of power dividers (Ding, Fang, Chen, & Wang, 2013). Consequently, the optimization can then be easily managed by realizing a balanced power splitter at different stages. Several constraints can influence the design of antenna array such as the mutual coupling, matching of different feed lines and the distance between the radiating elements. So, in the design process, it necessary to take into consideration all these constraints.

The use of antenna array in different applications to improve its transmission performances is a less expensive choice and easy to achieve, but perhaps impossible to use as a solution because of its geometric dimensions which are wider compared to its reserved location in the system box. Design an antenna array occupies a very small surface it is not an easy task because several factors come into consideration; for instance, it is necessary to take into thoughtfulness a major factor, which is the spacing between array elements that act directly on the radiation pattern and the gain. Noting that, when one increases the number of radiating element; we increase the gain, except that without taking into account the losses. If the

Figure 2. Feeding of microstrip antenna array: Series-feed network (left) and Corporate-feed network (right)

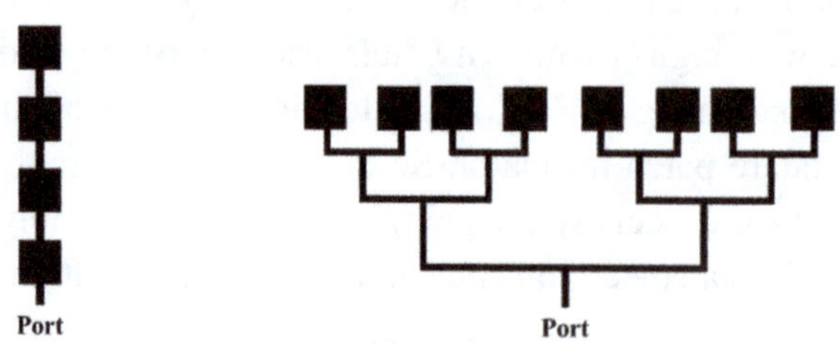

elements are too close to each other, a coupling phenomenon reduces the value of the gain, contrariwise when they are too far away sidelobes appear therefore reduce the directivity. After all this, the difficulty of achieving the goal was very clear; design of an antenna array always respecting the rules of design to have a good performance as well as keep the total focus on the objective, which is obtained an array very miniaturize compared to the others structures already exists.

Over the decades, several techniques have been exploited to miniaturize the antenna array, from structural modifications and lumped component loading, to the use of metamaterials. For instance a miniaturized dual-band antenna array using left-handed metamaterial unit cell located on the reverse of the antenna substrate combines the SRR, SERR, and rectangular ELC resonator for WiMAX, LTE, and WLAN applications had presented in (Alqadami et al., 2016). In (Qi, Liu, Yin, Zhao, & Kulesza, 2016) an asymmetrical coplanar strip (ACPS) wall is used as another decoupling method for reducing the space and to suppress the mutual coupling between two closely spaced 5.8GHz microstrip patch antennas, with an extremely close antenna separation of $0.28\lambda_0$ (center-to-center distance). In a recent paper by (Alsath, Kanagasabai, & Balasubramanian, 2013), a new configuration using slotted meander lines resonator (SMLR) has been proposed to block the surface current of the two patch antennas coupled along H-plane, separated by an edge-to-edge distance of 7mm ($\lambda_0/9$) and operating at the frequency of 4.8GHz. Always dealing with the problem that is related to the volume of the microstrip patch antenna array, another approach based on a parasitic isolator (PCI) was presented in (Cheng, Ding, Shao, & Wang, 2017). The purpose of this paper is to suppress the mutual coupling (MC) between two patches, which are very close to each other, center-to-center distance of 20.5mm (about $0.39\lambda_0$).

There has been a significant research effort to miniaturize antenna array using the electromagnetic band gap structure (EBG) material due to its unique properties (Sievenpiper, Zhang, Broas, Alexopolous, & Yablonovitch, 1999). Due to its uniqueness, various kinds of EBG material has presented and used by many researchers in a wide variety of applications, mainly in the microwave circuit (Baik, Han, Jeong, Jeong, & Kim, 2008) and in antennas technology (Ebadi & Semnani, 2014). Many configurations of electromagnetic band gap structure (EBG) has developed to reduce the mutual coupling in small-size antenna array. One is mushroom-like EBG surface (Ghaloua, Zbitou, Latrach, Abdellaoui, Tajmouati, & Abounaima, 2017) and the other is the uni-planar EBG (UC-EBG) surface (Farahani, Veysi, Kamyab, & Tadjalli, 2010). The authors (Ahmed Ghaloua et al., 2018) propose a new configuration of mushroom-like EBG structure exploiting the surface-wave suppression characteristics in order to reduce the mutual coupling between two antennas arrays and to obtain a very compact design. Without the EBG structure, there is a more or less strong mutual coupling of -33.24dB. By using, the novel configuration of mushroom-like electromagnetic band-gap (2D-EBG) structure an isolation of 26.12dB has recorded at 5.8GHz. Although much effort has been devoted to realize the small antenna array. Actually, the defected ground structure (DGS) are getting essential in different applications. He was used to design the compact antenna, to improve the radiation properties of microstrip antenna array and to reduce the mutual coupling between microstrip array elements. In order to miniaturize the number of patch layers and upgradable radiation directivity for outdoor and mobile systems, M. A. Aris et al (Aris, Ali, Rahman, & Pasya, 2016) propose (4x4) frequency reconfigurable antenna array using defected ground structures (DGS). This technique had applied to antenna array design to reduce mutual coupling between closely packed PIFAs (Zhu, Xu, & Xu, 2009).The proposed structure is a dumb-bell-like etched in the middle of the ground plane, exactly between the patches for suppressing the surface waves, as well as the mutual coupling reduction at the resonant frequency was 17.43dB.

The objective of this chapter is to design two miniaturized microstrip patch antennas arrays with good performances. The following work has organized as follow. The first part has devoted to design the circular patch antenna and two structures of the antenna array for a given choice of material. A second part will present a new defected ground structure used to miniaturize the printed antenna array for S-band at 2.45GHz. Subsequently, two columns of the proposed planar EBG structure had implemented between two antennas arrays to suppress the surface waves, in other words, this novel configuration of antennas arrays permits to have a miniature antennas arrays with a good mutual coupling coefficient. After the design and the simulation of the both proposed structures, and to verify the simulation results using electromagnetic solvers CST-MW studio and ANSYS HFSS, the proposed circuits are manufactured and tested.

DESIGN PROCEDURES

In this section, we will first introduce the basic microstrip patch antenna, based on circular shape microstrip antenna. In which a whole detail about the method used to design it. From this state of the art non-exhaustive of the design of the circular antenna, we will define the type and characteristics of the substrate that was used throughout this contribution as well as the dimensions of the elementary antenna used in the future antenna array. Linear and planar array arrangements with four elements had proposed to the design. The first antenna array design has fed with a microstrip feed and the second by using the coaxial probe feed. In addition, we will detail, in this step, the design approach that we followed to design two different configurations of the antenna array. Furthermore, because the matching circuit is required between the feed line and the antenna, the patch elements had matched by using a power divider based on quarter-wavelength impedance transformers.

Circular Patch Antenna Design

The metallic patch can take any forms, only the rectangular and circular configuration are the most commonly used configuration, because of ease of analysis and fabrication, and their attractive radiation characteristics. Changing the value of these two parameters length and width of the rectangular patch will change the mode of operation of the rectangular patch; actually, we have two degrees of freedom to control (length and width). Unlike the rectangular patch, the circular patch is only one degree of freedom to control and this is the radius of the patch. To calculate the radius of the patch, we have taken into account fringing in the circular patch. Fringing is the effect, which makes the patch look electrically larger than geometrical patch and it has taken into account by using an effective radius ae, to replace the actual radius a (Constantine, 2005). The radius (a) and effective radius (ae) of the elementary circular patch antenna can be found by using the given formulas (1) to (3). Microstrip circular antenna, shown in Figure 3 consists of a metallic patch, placed above a ground plane with a dielectric substrate sandwich between them. The procedure of the design assumes that the specified information includes the dielectric constant of the substrate (εr), the resonant frequency (fr), the height of the substrate h, and the thickness t of the metallic patch must defined at first.

The effective radius:

$$a_e = a\left\{1+\frac{2h}{\pi\varepsilon_r a}\left[\ln(\frac{\pi a}{2h})+1.7726\right]\right\}^{1/2} \quad (1)$$

The physical radius a of the patch:

$$a = \frac{F}{\left\{1+\frac{2h}{\pi\varepsilon_r F}\left[\ln(\frac{\pi F}{2h})+1.7726\right]\right\}^{1/2}} \quad (2)$$

where F is the fringing field:

$$F = \frac{8.791\times10^9}{f_r\sqrt{\varepsilon_r}} \quad (3)$$

The circular patch or disk had designed using the 3D EM simulation of high frequency components- CST Microwave Studio- has shown in Figure 3. The FR-4 board had used as a substrate material, which has dielectric constant of 4.4, thickness of 1.6mm and loss tangent 0.025 as well as the thickness t of the metallic patch is 0.035mm. In order to reach a good matching input impedance at 5.8GHz with a good return loss of the circular antenna, the dimensions of the antenna and the feed line had optimized minutely. This antenna is excited using transmission line of particular length and width, as such manner that this transmission line perfectly matches the impedance of the circular patch. The Physical area and the dimensions of the proposed printed circular antenna are Phyar=πx6.662mm^2, Ls=27.93mm, ws=35.74mm, Rc=13.32mm, Lf=5.8mm and wf=0.5mm. From the Figure 3, the resonance frequency of the proposed circular antenna had observed at around 5.8GHz with a good return loss of -30.77dB as well as the width of the bandwidth is 203MHz, which is from 5.9014GHz to 5.6984GHz.

Figure 3. The proposed microstrip circular antenna: Schematic representation (left) and Simulated return loss at 5.8GHz (right)

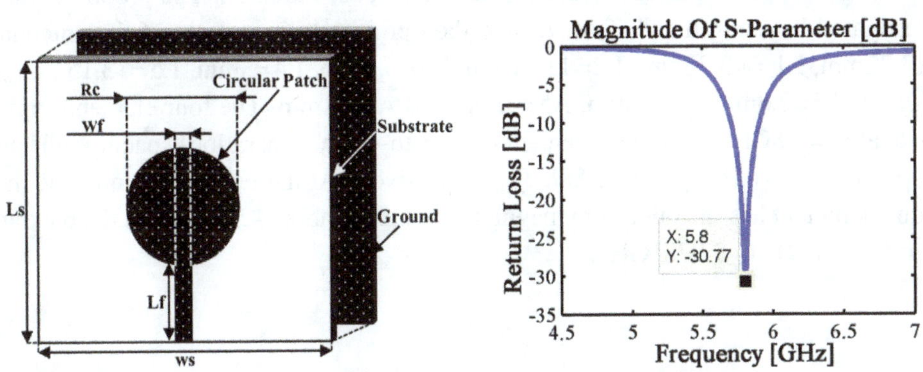

The Proposed Printed Antenna Array

Linear and planar array configuration had both fed by corporate feed arrangement using microstrip line feed and coaxial probe feed. This topology consists of transmission lines, bends, mitered bends, T-junction power divider or T-Junction with V groove and quarter wave transformers. The corporate-feed network gives the possibility to vary the space between the elements in order to obtain the optimal results. The benefits of this topology are not limited to this only but include the design simplicity, the broader bandwidth and easy to be integrated with other devices as the amplifier. The need for more space for the feed network is an inconvenience in large arrays, because the length of the feed lines becomes prohibitively long, which results in high insertion loss.

- **Linear Antenna Array Fed by Microstrip Line Feed:** The configuration of the linear patch antenna array of four elements with symmetrical corporate feeding networks and arrangements had shown in Figure 4. The antenna array elements are separated by an edge-to-edge distance of 20.29mm (d2=0.392$\lambda_{5.8GHz}$, where $\lambda_{5.8GHz}$ is the free space wavelength at 5.8GHz) and center-to-center separation of 33.61mm (d1=0.65$\lambda_{5.8GHz}$). The ground plane lies at the bottom of the substrate of a size of 53.43mmx135.96mm. The quarter wavelength matching and inset feed had used as the feeder for each patch to obtain the maximum matching impedance and provides a good antenna. The antenna array consists of four circular patches printed on the top of an FR-4 substrate already define. The physical dimension of the radiating array is Rp=13.32mm, Lf =5.8mm, Lfay=11.2mm, L1=27.22mm, L2=10mm, L3=8.5mm, L4=9.2mm, L5=11.61mm, L6=3.0mm, L7=6.5mm, L8=7mm, Wf=0.5mm, A1=3mm, A2 =1.6mm, A3=2.3mm, A4=3mm, A5=3mm, A6 =0.6mm, A7=3mm, and A8=3mm. Figure 5, shows the simulated reflection coefficient S11 of the proposed antenna array structure. It is observed that the antenna array resonates at 5.8GHz with the return loss better than -10dB. The frequency bandwidth (S11 ≤ -10dB) is 574.6MHz (5.3656GHz–5.9402GHz).
- **Planar Antenna Array Fed by Coaxial Probe Feed:** The 2 by 2 microstrip antenna array has designed by using the FR-4 board had shown in Figure 6. This configuration of the printed antenna array has selected for design, because this structure suffers from surface waves, which entails a mutual coupling effect between the elements of the antenna array. A 2x1 power divider/transmission lines consists of a corporate feed network fed from the center (using coaxial feed technique) had designed jointly, with attaching the elements of the antenna array at the end of the output lines. Thereafter, the impedance matching section composed by a quarter-wave transformer has been used to couple the power to each element, for better radiation. The proposed antenna array is investigated by using the CST-MWS and the optimized dimensions of the antenna array are Rp=13.32mm, L1 =43.38mm, L2=11.61mm, L3=6.5mm, L4=3mm, L5=13.15mm, A1=3mm, A2= 2mm, A3=2.6mm, A4=3mm, A5=3mm, and A6 =1mm. The four element array is seen to resonate at fr =5.8GHz with inter element (center-to-center separation) spacing in H and E plane corresponding to 0.65$\lambda_{5.8GHz}$ and 0.88$\lambda_{5.8GHz}$ respectively. At the resonant frequency, the input reflection coefficient has the minimum magnitude, which is about -25.8dB and the bandwidth range is from 5.6998GHz to 5.9781GHz.

Figure 4. Geometry of the proposed antenna array 4x1

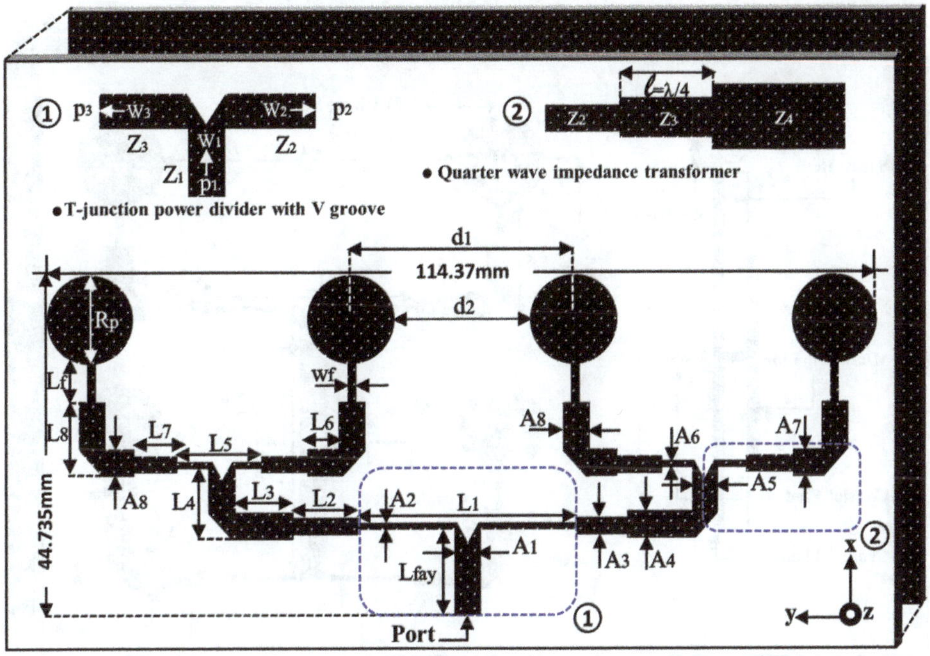

Figure 5. Simulated return loss |S11| of the proposed microstrip antenna array 4x1 resonating at 5.8GHz

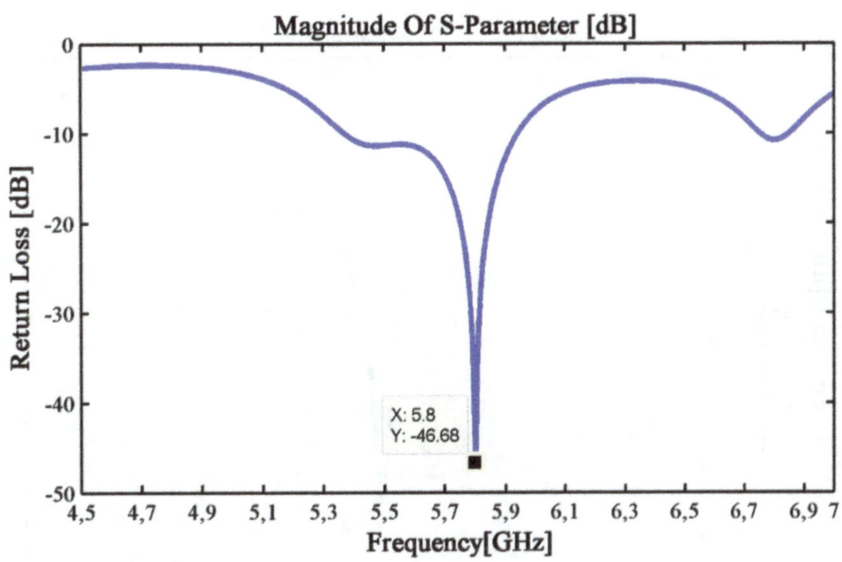

MINIATURIZED ANTENNA ARRAY BASED ON DEFECTED GROUND STRUCTURE

A new defected ground structure had employed to design a miniature antenna array resonating at 2.45GHz. For this purpose, a new shape of DGS had employed to shift the resonance frequency of the

Figure 6. Geometry of the designed antenna array with four elements 2x2

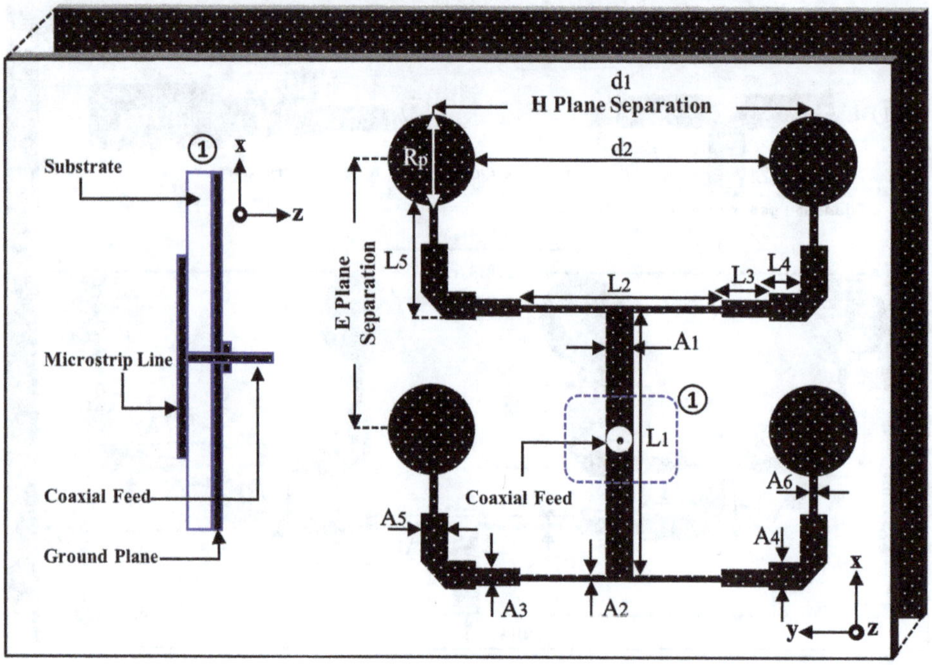

Figure 7. Simulated reflection coefficient |S11| of the proposed antenna array with four elements 2x2

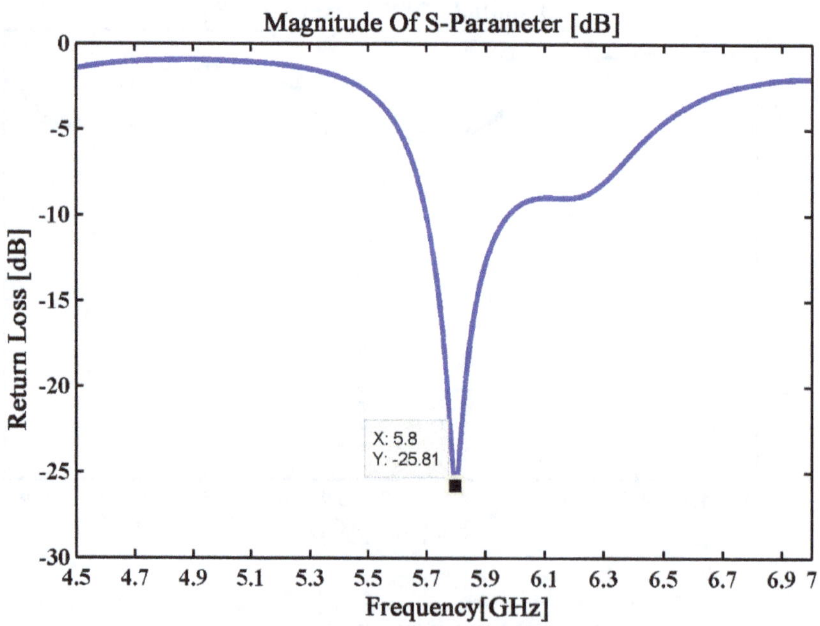

initial microstrip antenna array validated before at 5.8GHz to 2.45GHz by disturbing the antenna array current distribution. The use of the defection in the ground plane of each radiating element of the antenna array, permit to reduce the antenna array size up to 82.46% as compared with the antenna array without DGS resonating at 2.45GHz.

Defected Ground Structure

The notion of Defected Ground Structures (DGS) evolved in recent years primarily from the studies of Photonic Band Gap (PBG) structures in electromagnetics (Yang, Ma, Qian, & Itoh, 1999). Furthermore, DGS is an etched periodic or non-periodic cascaded configuration defect in the ground of a planar transmission line such as microstrip, coplanar and conductor backed coplanar waveguide, which disturbs the shield current distribution in the ground plane cause of the defect in the ground. DGS can be realized by etching of a simple shape in the ground plane as well as the shape of the defect may be changed from the simple shape to the complicated shape for the better performance (Tahar, Barakat, Saad, Yoshitomi, & Pokharel, 2017; Kun Wei, Li, Wang, Xing, & Xu, 2016).

This etched defect in the ground plane disturbs the shielding current distribution, which alters and increases the effective inductance and capacitance of the microstrip line (Guha & Antar, 2011). Depending on the shape and dimensions of the defect, the shielded current distribution in the ground plane disturbed and the band gap property relies on many design parameters such as lattice shape, number of the lattice, and lattice spacing.

Dumbbell DGS is the first that was proposed. It is composed of two 'a x b' rectangular defected areas, and a narrow connecting slot width 'g' etched in the metallic ground plane (Ahn et al., 2001). Until this moment, many variations of the shape of the DGS elements had proposed. They range from a simple rectangular slot to complicated configurations. In the literature different configurations of the DGS geometries exist, in addition to the dumbbell (Salehi & Tavakoli, 2006), there is also the circular head dumbbell, open loop dumbbell, spiral shaped, double-H shaped, concentric ring and square heads connected with U slots (Guha & Antar, 2011). On the other hand, various novel DGSs have been proposed and many applications have been explored extensively in microwave circuits, for instance, microstrip antenna, filter- lowpass filter, bandpass filter-, Wilkinson power dividers and microstrip diplexer design. Introduce the defected ground structure (DGS) in different applications is always to achieve a goal, either for harmonic control in active microstrip antennas (Chang & Lee, 2002) or suppress higher order harmonics for a microstrip antenna (Liu, Li, Sun, & Mao, 2005). Add to that, defected ground structure had used to minimize XP radiation for a microstrip patch (Ghosh, Chakraborty, Chattopadhyay, Nandi, & Basu, 2016), to reduce the mutual coupling between microstrip arrays elements (Wei, Li, Wang, Xing, & Xu, 2016) and to improve or modify the radiation properties of microstrip patch antenna (Biswas, Guha, & Kumar, 2013). Thereafter employing a defected ground plane beneath the microstrip patch antenna, help to increase the current path within the patch area. This helps in lowering the resonant frequency of the microstrip patch antenna and, therefore, leads to size-reduction.

The novel configuration of defected ground structure (DGS) is symmetric and consists of two incomplete ring torus shape connected by a narrow slot in backside metallic ground plane. After many optimizations by using CST microwave studio electromagnetic solver, the dimensions of the defected ground structure had shown in Figure 8: Rc1=3.8mm, Rc2=6mm, Ldg1=13.95mm, Ldg2=19.49mm, Ldg3=4.9mm, Wdg1=0.8mm and Wdg2=9.26mm.

Previous research has demonstrated that three types of equivalent circuits can mold the DGS: LC and RLC equivalent circuits, π shaped equivalent circuit and quasi-static equivalent circuit (Khandelwal, Kanaujia, & Kumar, 2017). The parallel LC circuit can represent the equivalent circuit of the proposed DGS section and one-pole Butterworth prototype of the LPF had presented in Figure 9. The LC parameters had determined by the shape and size of the defect geometry. The slotted area of the DGS is inversely proportional to the effective capacitance; indeed, a decrement in the DGS area reduces the effective capacitance. In contrast to this, the slotted area of the DGS is proportional to the effective inductance; an increment in slotted DGS area gives rise to the effective inductance. Two incomplete ring torus shape defected areas and one connecting slot correspond to the equivalently added inductance (L) and capacitance (C), respectively. The equivalent L and C lumped circuit elements are calculated by Equations (4) to (6) because two reactance values of Figure 9(a) and (b) must be equal at ωc,3dB as follows (Guha & Antar, 2011):

The reactance of Butterworth low pass filter can be obtained as:

$$X_{LC} = \frac{1}{\omega_0 C}\left(\frac{\omega_0}{\omega} - \frac{\omega}{\omega_0}\right) \tag{4}$$

The capacitance C, inductance L of the circuit are calculated as:

Figure 8. Geometry of the proposed defected ground structure (DGS)

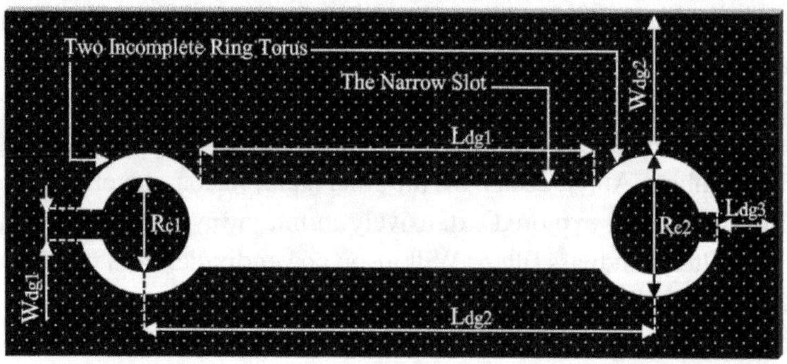

Figure 9. (a) Equivalent circuit of DGS, (b) Butterworth prototype of one-pole LPF circuit

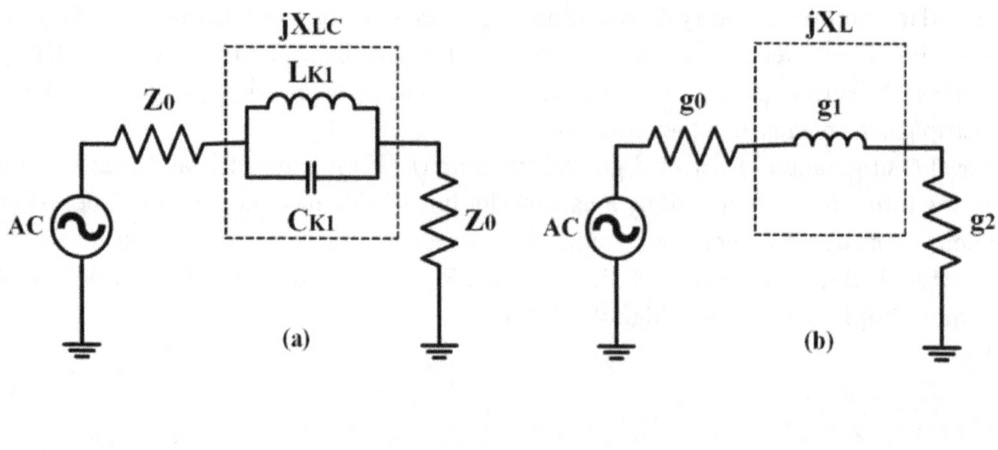

$$C = \frac{\omega_c}{Z_0 g_1}\left(\frac{1}{\omega_0^2 - \omega_c^2}\right) \tag{5}$$

$$L = \frac{1}{4\pi^2 f_0^2 C} \tag{6}$$

where ω_0, z_0, and g_1 denote the angular resonant frequency, input and output port impedances and the prototype element as described in (Constantine, 2005; Lim, Kim, Ahn, Jeong, & Nam, 2005) respectively.

Now, as shown in Figure 10, we introduce the novel geometry of DGS in order to shift the resonance frequency of the microstrip antenna array previously presented in Figure 4. In absence of DGS, we have seen in Figure 5 that the resonance frequency of our microstrip patch antenna array is at 5.8GHz. By the use, the defection in the ground plane of each radiating element of the antenna array and by varying the dimensions of the defected ground structure, the resonant frequency can be shifted without changing the length and width of each patch of the antenna array. The simulation result obtained for the return loss had shown in Figure 11. It had observed that the resonant frequency had shifted from 5.8GHz to 2.45GHz. The bandwidth of the antenna array without the DGS is 574.6MHz, while the antenna array bandwidth with the DGS is narrowed to 129.3MHz (from 2.3818GHz to 2.5111GHz), which is narrow than that of the antenna without DGS. Moreover, the simulated gain at 5.8GHz is 8.068dB, while the simulated gain of the antenna array with DGS at 2.45GHz is 6.681dB. The miniaturization of the antenna array in the design process diminishes the gain of the antenna array. The surface density plots of the designed antenna array with the defected ground structure at the resonant frequency are demonstrated in Figure 15. It had observed that the current had distributed uniformly along the surfaces of the microstrip patch antenna array. The maximum current density had observed focused along the DGS on the ground plane, as well as the power divider and at the ends of radiating elements of the antenna array. For the behavior of the of the antenna array in term of radiation, we can analyze the different radiation pattern of the array in the E and H plane as depicted in Figure 16. As seen, we have a bidirectional antenna array beam.

After the comparison of simulation results on CST Microwave Studio and HFSS, the antenna array structure is prototyped on the FR-4 substrate with a thickness of 1.58mm, a relative electric constant of 4.4, a loss tangent of 0.025 and a conductor thickness of 35um, using LPKF machine as presented in Figure 12. The entire area of the fabricated antenna array is (53.43x135.96mm^2). Measurement results was gotten with the network analyzer N5247A, US50470114 from the Agilent Technologies. Figure 13 displays a photograph of the measured return loss. It should be noticed that there is a deviation between the center frequencies (f_{Sim}=2.45GHz, f_{Mes}=2.32GHz) of the simulated and measured results (about 129MHz) which can be explained by the fabrication error or the connector use which is not considered during simulation or/and the change of the materiel properties. From the Figure 13, the impedance bandwidth (according to S11<10dB criterion) is about 6.81%. It can be observed that there is a difference of 28.7MHz between the simulated and measured bandwidth. It should be noted that further miniaturization can be achieved with other approaches which can be expensive or very difficult to achieve. The proposed structure is easy to fabricate and its design procedure is simple and relatively robust to distortion of design parameters.

Figure 10. Geometry of the proposed antenna array with defected ground structure (DGS)

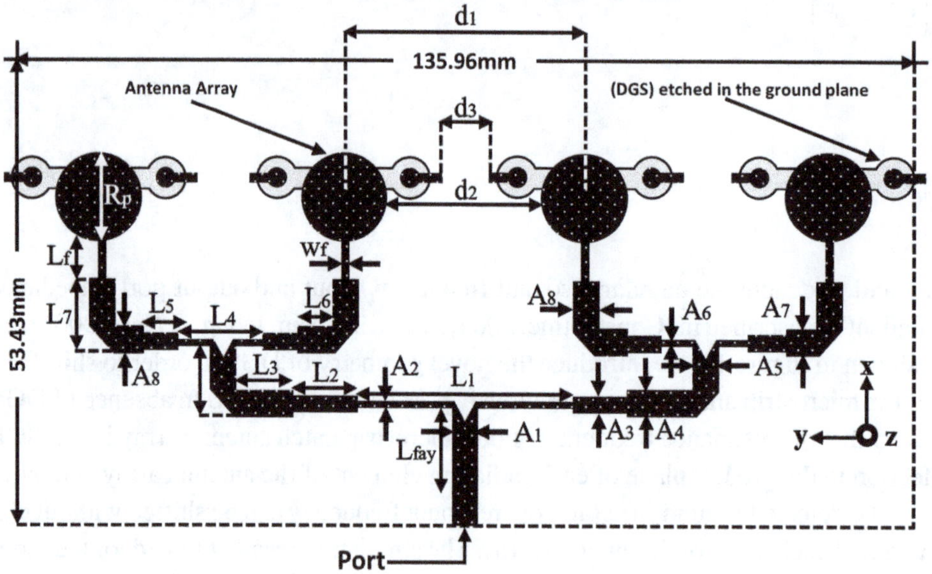

Figure 11. CST and HFSS return loss comparison of the proposed antenna array with DGS

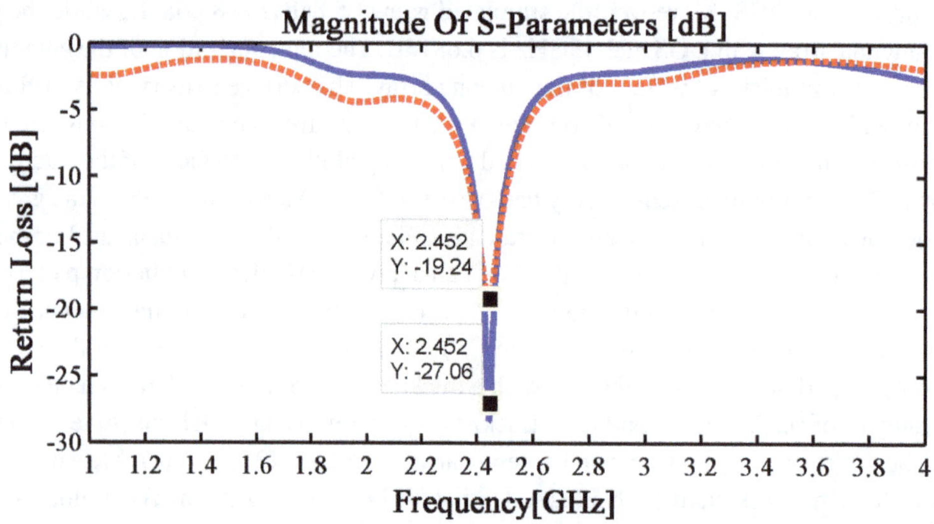

MINIATURIZATION AND REDUCTION OF MUTUAL COUPLING COEFFICIENT

Two columns of the planar compact electromagnetic bandgap (EBG) structure had designed to block the surface current at the resonant frequency of the two microstrip patch antennas arrays coupled along H-plane and operating at the frequency of 5.8GHz. The interelement isolation before and after the implementation of the planar compact electromagnetic bandgap (EBG) structure has been investigated. The design achieves a high isolation of about 49.09dB between the two antennas arrays separated by an edge-to-edge distance of $0.29\lambda_{5.8GHz}$ at 5.8GHz.

Figure 12. A photograph of the fabricated antenna array with DGS (left) and measured return loss (right)

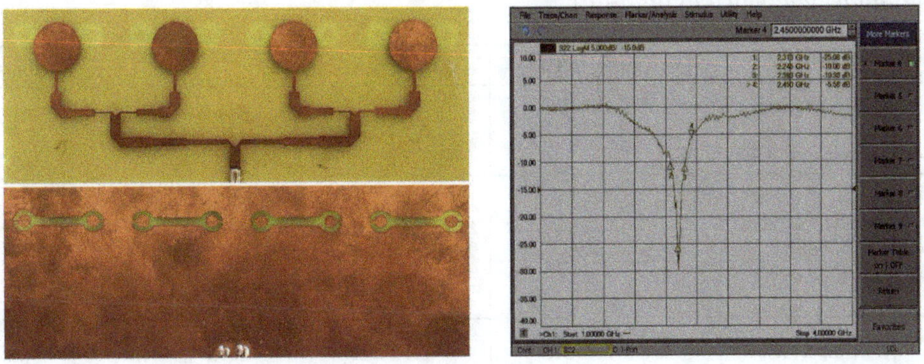

Figure 13. Simulated and measured results of the reflection coefficient for the proposed antenna array

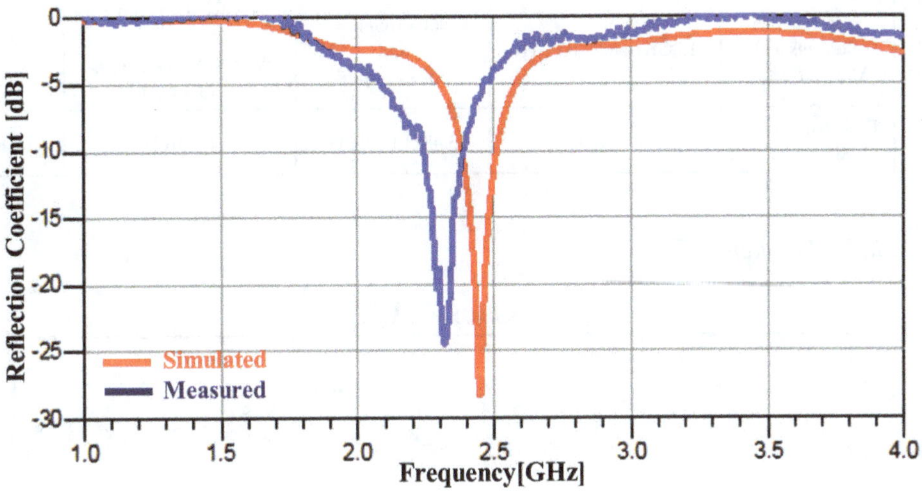

Mutual Coupling in Arrays

When the antennas are placed close to each other, and all / or some of them are transmitting or receiving the signals, in this case, a part of the energy is primarily intended for one ends up at the other. This interchange of energy is known as "mutual coupling," and depends primarily on the radiation characteristics or orientation or the distance between radiating elements (Craeye & González-Ovejero, 2011). Another factor that is among the origins of mutual coupling without setting up an array of radiating elements is the technology used. Recall that in this chapter, the radiating sources studied are microstrip antennas. Microstrip technology is in charge of for propagating different types of waves, like guided waves, radiated waves and surface waves. Furthermore, these three waves types are responsible for increasing mutual coupling. Excite the antenna produces a radiation emitted by the different metal parts of the antenna, at this moment part of the electric field will be trapped in the substrate between two conductive planes, namely between the antenna and the ground plane. This field belongs to the guided waves. As a result, when there is a very small distance between the plates of the radiating elements, the

Table 1. The state of the art comparison between the proposed miniature antenna array using the DGS and other solutions published in literature

	Technique	Frequency[GHz]	Dimension	Gain	Miniaturiza-tion
(Azimi Fashi, Kamyab, & Barati, 2010)	Metamaterial Zeroth-Order Resonator [ZOR]	4.25GHz	≈57x46mm²	6.6dB	50%
(Pavithra, Mohanageetha, Mary Anita, & Subramaniam, 2014)	Single-Layer Microstrip E-Shaped	2.45GHz	180×210mm²	6.7dBi	-
(Hao & Li, 2017)	ME-dipole (PCB)	36.8 to 53.3GHz	40×40mm²	11.6 ± 0.6dBi	-
(Foged et al., 2015)	Artificial Magnetic Materials	162MHz	500×500mm²	8.2/7.7/7.9dBi	50%
(Li, Li, Wang, Zhou, & Cheung, 2018)	BND antenna elements and AMC reflectors	1.8GHz and 2.05 GHz	560×165mm² and 558×164mm²	DCS band: [12.1-12.6 dBi] WCDMA band: [12.4-13.1 dBi]	46.3% and 19.9%
This Work	Defected Ground Structure	2.45GHz	53.43x135.96mm²	6.681dB	82.46%

Figure 14. Gain Vs Frequency

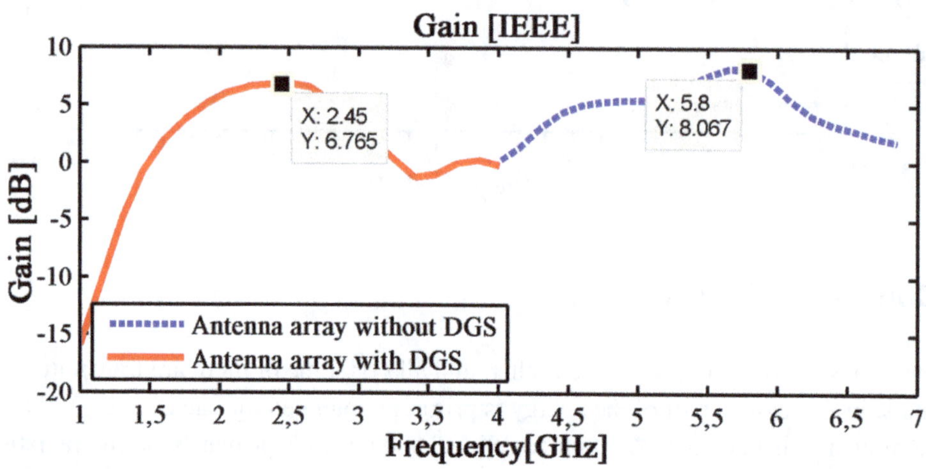

guided waves will flow between the adjacent antennas causing an increase in the mutual coupling. In another side, the propagation in the free space of the radiated waves can come into contact with other radiation from neighboring sources, thereby creating an induced current and disrupt the operation of the array. As a result, an aerial coupling is created. The surface waves pass through the substrate; exactly propagate in the dielectric with an angle of incidence, more precisely between the ground plane and the dielectric–air interface due to internal reflection between the two terminals of the substrate. At the ends of the microstrip structure, the surface waves are diffracted, reflected and dispersed in the free space,

Figure 15. Current distributions of the antenna array with DGS at 2.45GHz: Front view (left) and Back view (right)

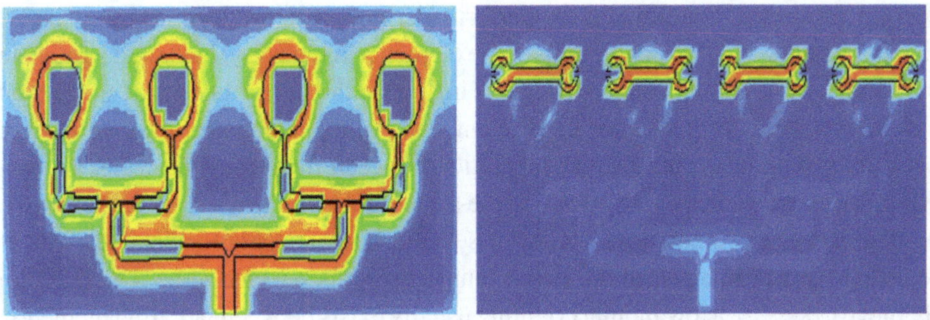

Figure 16. Simulated radiation patterns of the proposed antenna array in E-plane and H-plane: Without DGS (left) and With DGS (right)

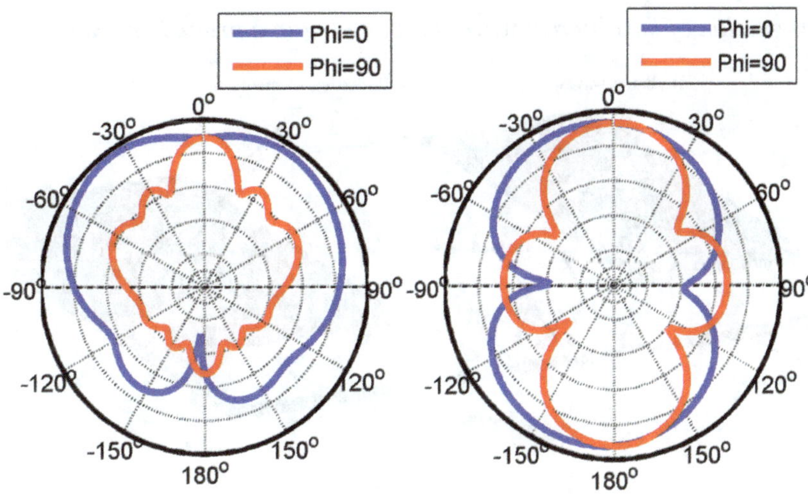

Figure 17. Measured radiation pattern of the realized antenna array at 2.45GHz in E-plane: E incident and perpendicular to the ground (left) and E incident (provided by the reference horn antenna) is parallel to the ground (right)

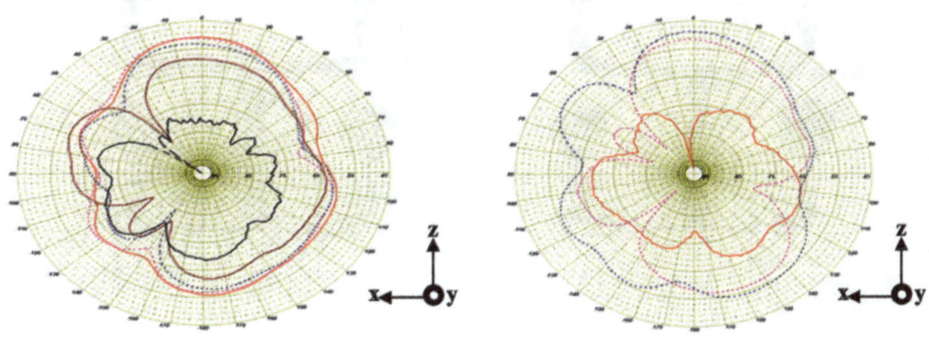

which results in significant gain reduction as well as a radiation pattern deform due to an increase in back radiation.

The effects of the mutual coupling on the performances of an array, depends upon the antenna type and its design parameters, relative positioning of the elements in the array, feed of the array elements and scan volume of the array. These design parameters influence the performances of the antenna array by varying its element impedance, reflection coefficient, and overall antenna pattern. For minimum mutual coupling between two antennas arrays, the separation between them must be wide, and whenever we reduce the distance between them, we notice that the mutual coupling tends to increase significantly. The mutual coupling between two antennas arrays for seven different separations had shown in Figure 20. At first, the initial separation between the two antennas arrays is selected to be d3=51.72mm ($1\lambda_{5.8GHz}$) and d3=36.2mm ($0.7\lambda_{5.8GHz}$), a low mutual coupling usually exists between radiating elements. Secondly, for d3=25.86mm ($0.5\lambda_{5.8GHz}$) and d3=20.69mm ($0.4\lambda_{5.8GHz}$) the both of antennas arrays are close to each other. Consequently the mutual coupling will be greater.

Figure 18. Representation of the different waves propagating in a printed antenna

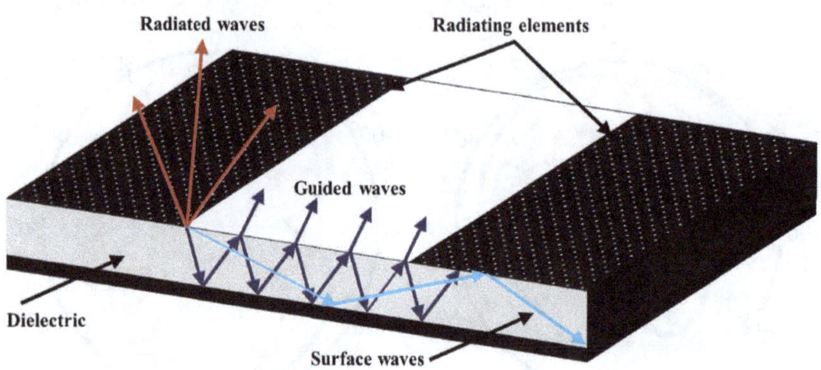

Figure 19. Top-view of the both antennas arrays configurations

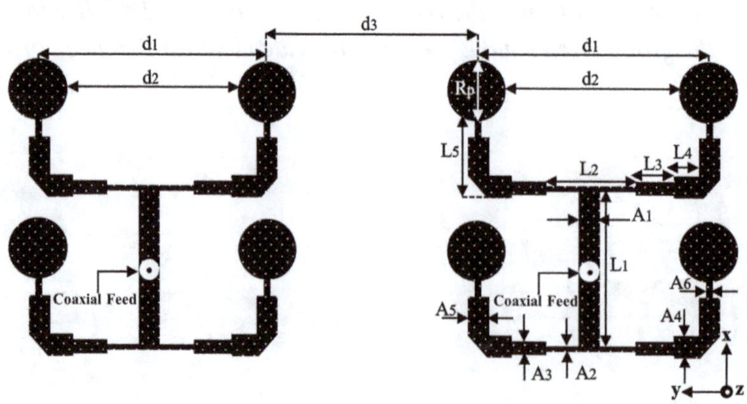

Figure 20. Coupling coefficient for different separations distance (d3) between the both antennas arrays

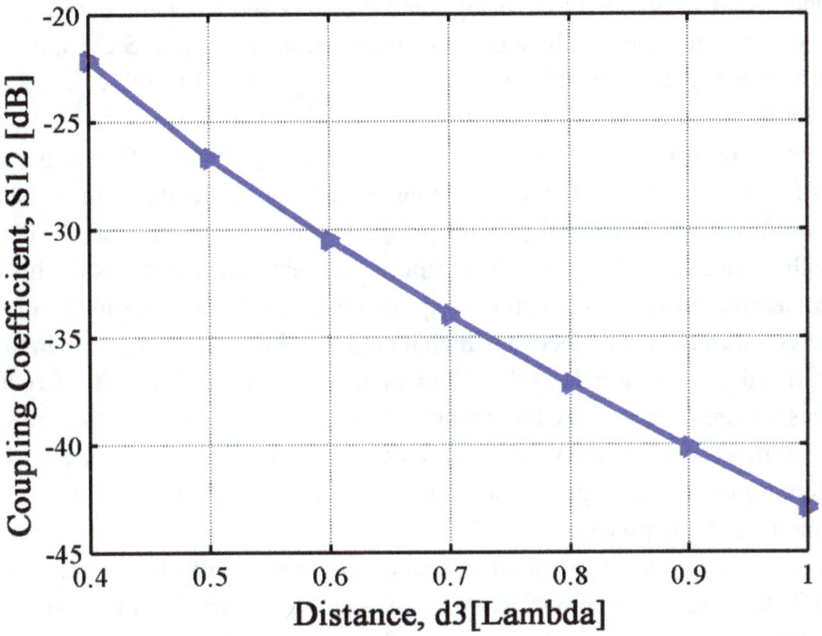

Electromagnetic Bandgap (EBG) Structure

EBG structures operating in the high frequency and the microwave frequency bands and their applications in antennas have become a new research direction in the antenna community, due to the growing interest in improving antenna's performance such as increasing the antenna gain and improve the radiation pattern of the antennas or reducing back radiation. The EBG structures also have the ability to suppress surface wave propagation in a frequency band on the ground plane of the antennas and thus minimize the mutual coupling in array configurations. EBG structures are based on the photonic bandgap phenomena in optics, that are typically realized by periodical structures (Abidin, 2012). These categories of the structures also known as photonic crystal (PC) or Photonic Band Gap (PBG) structures are broadly classified as metamaterials, which are realized by the periodic arrangement of dielectric materials and metallic conductors. Among the EBG configurations, there is multi-layer metallic tripod array and a woodpile structure consisting of square dielectric bars which are classified within 3-D EBG structures (Lee, Lu, Hao, Yang, Ubic, Evans, & Parini, 2007). Although the mushroom-like EBG structure that includes vias to the ground plane -not completely planar- and completely planar EBGs without vias or pins are categorized into two dimensional planar surfaces (Beiranvand, Afsahy, & Sharbati, 2017; Ghaloua, Zbitou, Latrach, Abdellaoui, Errkik, & Mandry, 2017). There is also a third configuration of this technology, which is the one-dimensional EBG transmission line (Radisic, Qian, Coccioli, & Itoh, 1998).

Planar Compact Electromagnetic Bandgap (EBG) Structure

The different geometries of EBG structures who have already proposed have the same common point: the forbidden frequency gap or stop band, which the frequency region where the incident waves cannot propagate through the structures, that giving the capability to block surface waves excitation in the operational frequency range of the phased arrays, and therefore reduce mutual coupling in the antenna arrays. The famous structure that is published by Dan Sievenpiper et al. in (Sievenpiper, Zhang, Broas, Alexopolous, & Yablonovitch, 1999) is the high impedance grounded metallic plates electromagnetic surface, known by the mushroom-like EBG structures. This structure is considered as a new type of metallic electromagnetic structure and is characterized by having a high surface impedance (HIS). The unit cell of mushroom-like EBG structure is comprised of square, rectangular or hexagonal shaped metal patches raised above a flat metal ground plane. These metal patches are connected to a PEC ground plane with vertical metallic vias. Furthermore, the fabrication process of this design is more complicated, because the mushroom structures require electrical connections to the PEC ground plane with the metallic vias through the dielectric, which it could lead to a slight increase in the discrepancies between the simulated and measured results. Compared to mushroom-like EBG structure, the planar EBG structure has a more compact size and completely planar, it does not require the vertical metallic via connections to the ground plane, which is an advantage from the point of view of manufacturing using printed-circuit board technology. These properties make it easy to be integrated with other radio frequency and microwave components.

When the periodicity (w+g) of a structure is small compared with the operating wavelength, the operation of EBG structure can be modeled by an LC equivalent circuit (Shahparnia, 2005). The capacitor C results from the gap between the patches and the inductor L results from the current along the adjacent patches. The parameters of this model are determined by the following formulas (7) and (8) (Tretyakov, 2003). In order to have a more compact structure at low frequencies, the equivalent capacitance or inductance should be increased. On the other hand, this study suggests a novel planar compact EBG structure. The schematic of the unit cell of suggested EBG structure is shown in Figure 21. In this proposed planar two-dimensional EBG unit cell, an I shaped slot X-oriented is added to the rectangular part in order to increase the shunt capacitance and series inductance. This EBG structure has a dimension of $6.2x5.5mm^2$, is considered as a periodic distribution of metallic elements which are printed on a dielectric substrate similar to the antenna array dielectric substrate. The dimensions of EBG unit cell are tabulated in Table 2.

Figure 21. Geometry of the suggested planar EBG unit cell

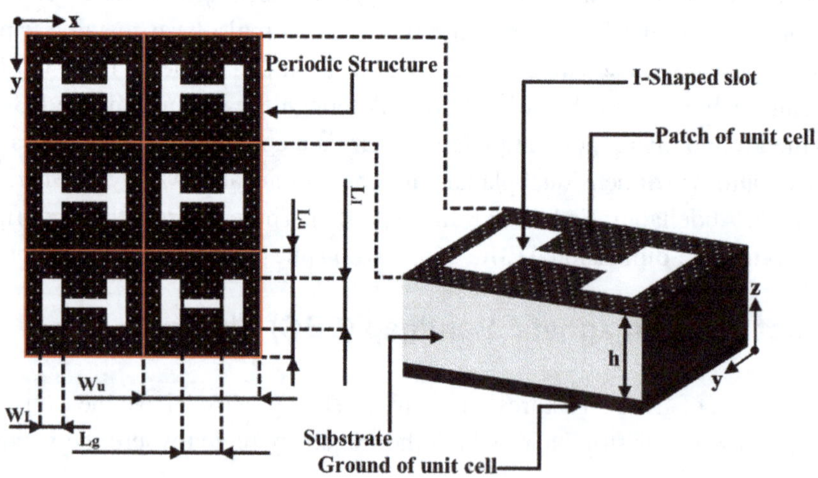

$$C = \frac{w\varepsilon_0(1+\varepsilon_r)}{\lambda}\cosh^{-1}\left(\frac{w+g}{g}\right) \tag{7}$$

$$L = \mu_r\mu_0 h \tag{8}$$

where μ_0 and ε_0 are the permeability and permittivity of free space, respectively. The surface impedance equals to the impedance of a parallel resonant circuit and the central frequency of the band gap had calculated as below:

$$Z_s = \left(\frac{j\omega L}{1-\omega^2 LC}\right) \tag{9}$$

$$\omega_0 = \frac{1}{\sqrt{LC}} \tag{10}$$

Microstrip Antenna Array With the Planar Compact EBG Structure

For two microstrip antennas arrays working in the same band (center frequency 5.8GHz), the coupling of two side-by-side elements is a function of the relative alignment. As shown in Figure 22, two antennas arrays are positioned collinearly along the x-axis, which is referred to the E-plane, and the elements are linearly polarised along the y-axis. The edge-to-edge distance between the patches of antennas arrays is 14.99mm (merely $0.29\lambda_{5.8GHz}$, where $\lambda_{5.8GHz}$ is the free-space wavelength at 5.8GHz) and the center-to-center distance between the circular patches is about $0.549\lambda_{5.8GHz}$. In addition to that, the gap between the edge of the antenna and planar compact EBG is dae=2.365mm. Moreover, the overall size of 14x2 EBG unit cells is 86.8x11mm^2. The two antennas arrays are fed with 50-Ω coaxial probes, and share a common ground plane of size 128.74x97.8mm^2.

To investigate the performance of the proposed planar compact electromagnetic bandgap (EBG) structure, a simulation study is carried out using CST Microwave Studio. The simulated scattering parameters of the proposed antennas arrays without and with one column or two columns of the planar EBG are depicted in Figure 23 and Figure 24, respectively. The |S12| comparison of microstrip antennas arrays with two columns of EBG to without EBG shows that mutual coupling is reduced from −28.79dB

Table 2. Dimensions of EBG unit cell and all dimensions are in mm

Wu	Lu	WI	LI	Lg
6.2	5.5	0.9	2.0	2.0

Figure 22. Integration of the suggested planar EBG structure between two antennas arrays

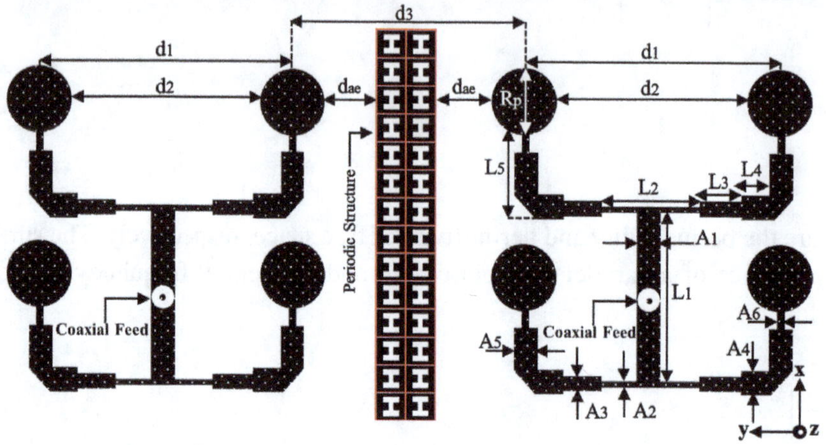

to −49.09dB, more than 20dB mutual coupling reduction between antennas arrays is achieved, which proves that the surface waves are suppressed. Furthermore, |S12| is lower than −37dB in the working band (S11 below −10dB). As Figure 24 indicates, employing one column of EBG cells does not provide considerable mutual coupling reduction. The proposed structure greatly improves the isolation without causing any significant degradation of the gain, on the contrary the antennas arrays achieves the high gain of 6.22dB with the excellent directivity of 9.584dBi. The current distributions of the antennas arrays with and without the planar EBG structure are shown in Figure 28. The results clearly show the isolating effect of the planar EBG structure; without it, a strong surface currents are induced on array 2, whereas when it is present, only weak surface currents are induced on array 2. This novel configuration of arrays provides smaller antennas arrays and noticeably affects the size of a system that has multiple antennas arrays with many elements. The dimensions of the two antennas arrays are reduced by ~24.3% by utilizing a planar EBG structure between arrays in the y-axis.

In order to validate the mutual coupling reduction performance of the proposed approach, a prototype of the highly miniaturized antennas arrays is fabricated on the FR-4 substrate and measured using the Network Analyzer N5247A, US50470114, from the Agilent Technologies. The practical scenario of excitation was made by using two SMA connectors, which are soldered at the center of the microstrip fed line (using coaxial feed technique) of each antenna array. The measurement is carried out by feeding only one element and by terminating the other element in a matched load (50 Ω). A photograph of the manufactured antennas arrays with the planar EBG structure is shown in Figure 25. Also, a comparison of simulated and measured reflection and transmission coefficient of the proposed circuit is depicted in the Figure 26. The significant difference observed between the simulated and measured scattering parameters is probably due to the fabrication tolerance or the uncertainty of the dielectric constant provided by the manufacturer or/and connector weldings. Inserting the proposed planar EBG between the two antennas arrays efficiently improves the isolation level. An average of 13dB mutual coupling reduction is obtained over the suggested bandwidth compared to the reference array. The measured results demonstrate the utility of the planar EBG structure, and this approach is potentially useful for a variety of array applications.

Figure 23. Simulated return loss |S11| of three studied cases of antennas arrays

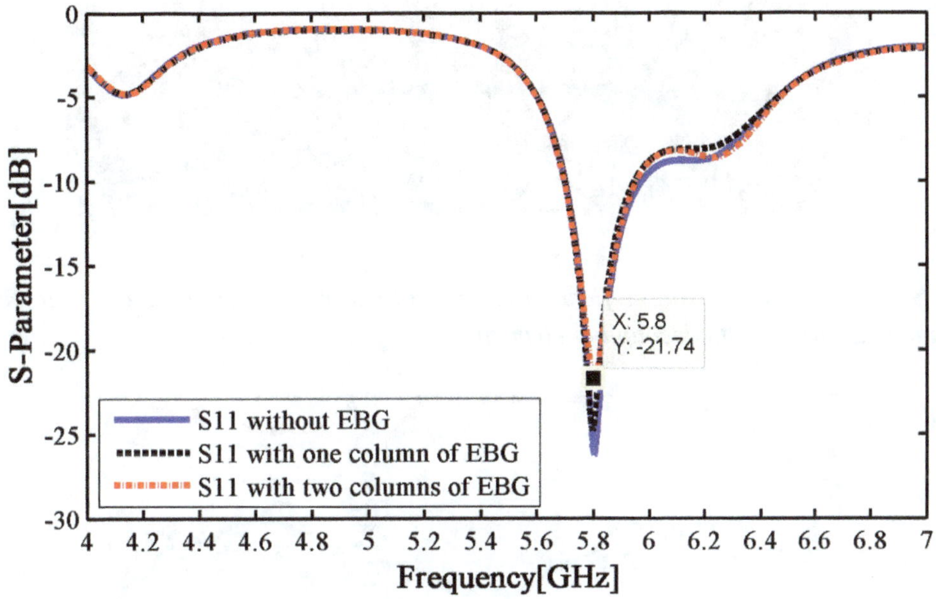

Figure 24. Simulated mutual coupling coefficient |S12| of three studied cases of antennas arrays

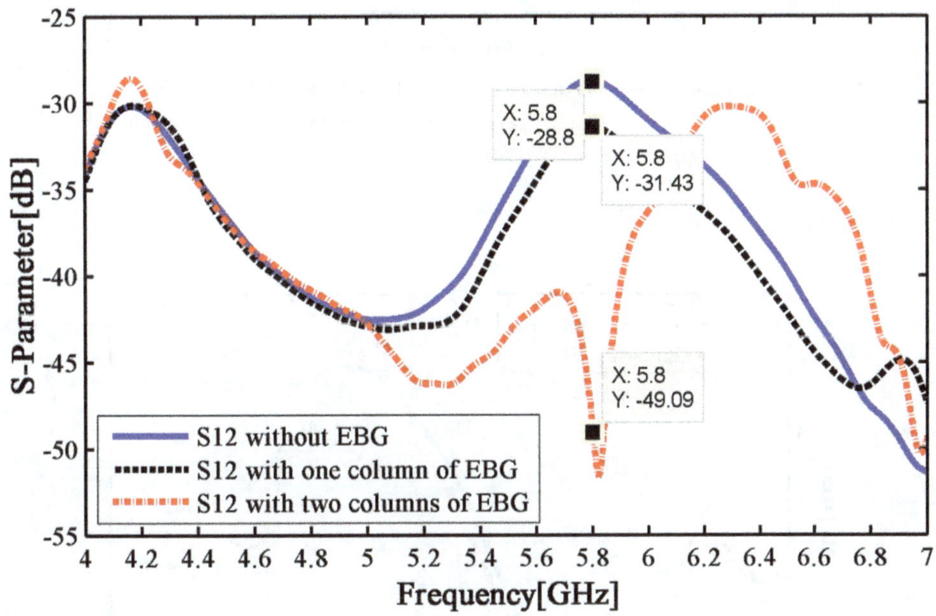

FUTURE RESEARCH DIRECTIONS

Throughout the development of this chapter, many possible lines have been opened to extend the study of different solutions applied for miniaturizing the antenna array. Despite some techniques have been tackled, there is still a lot of work to be done. We suggest here some attractive lines to study in the future:

Figure 25. A Photograph of the fabricated prototype

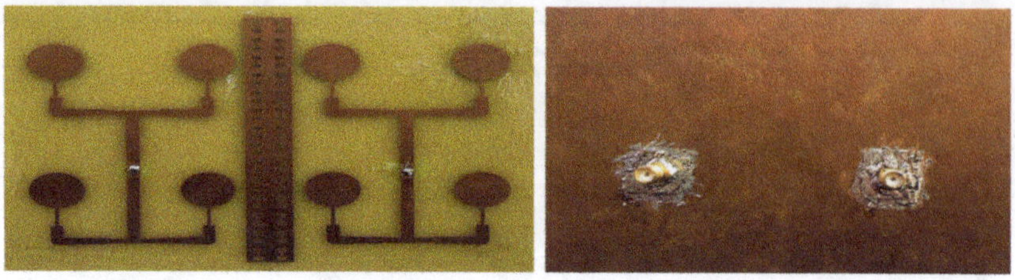

Figure 26. Simulated and measured reflection coefficient and mutual coupling characteristics of the proposed antennas arrays with decoupling elements

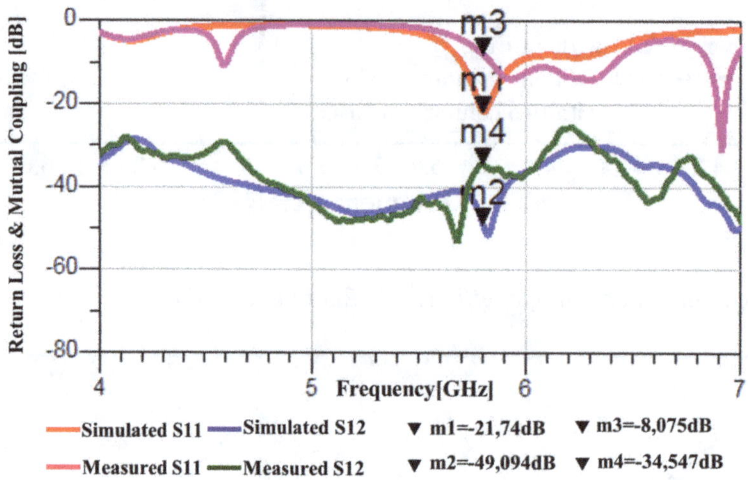

Figure 27. Gain Vs Frequency

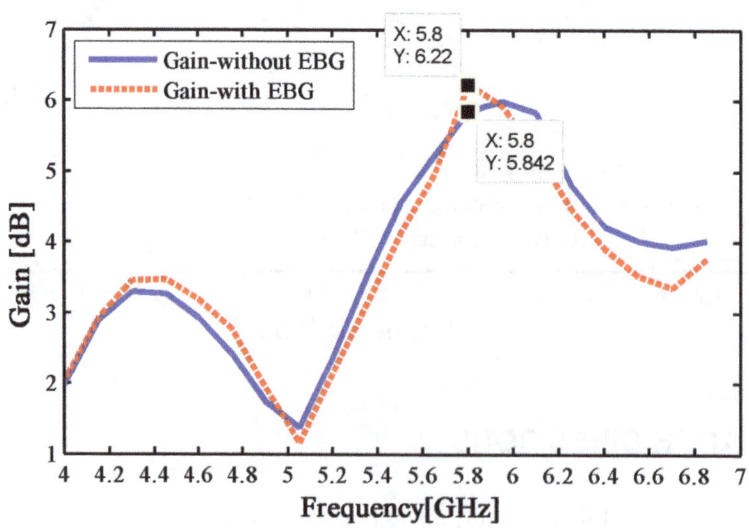

Figure 28. Current distributions of the antennas arrays: Without EBG (left) and With EBG (right)

Figure 29. Simulated radiation patterns of the antennas arrays in E-plane and H-plane: Without planar EBG (left) and With planar EBG (right)

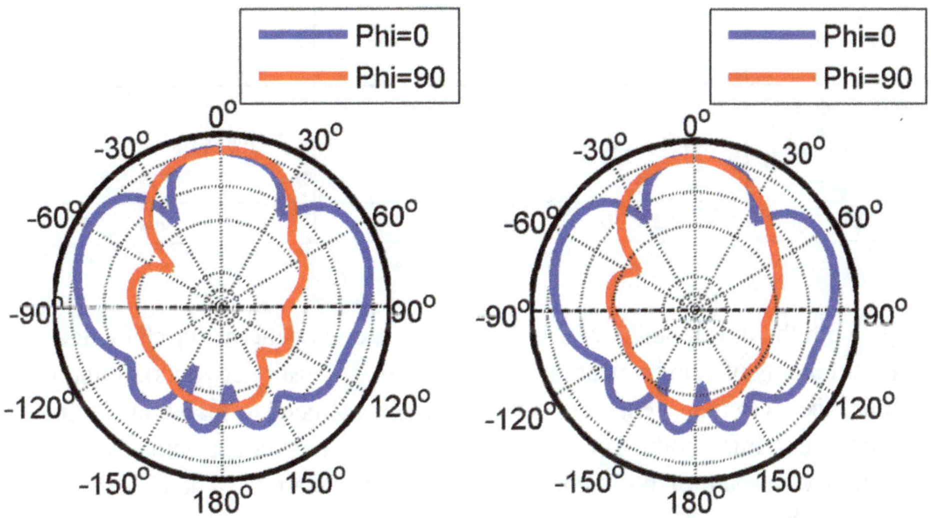

- In the last few years, there has been a growing interest in developing multifunction devices (reconfigurable antenna, tunable filter…etc.). Reconfigurable antenna array can improve the performance of future wireless networks at much lower cost than using complex phased networks. Reconfiguring an antenna has achieved through deliberately changing its frequency, polarization, or radiation characteristics. This change has achieved by many techniques, which distribute the antenna currents and thus change the electromagnetic fields of the effective antenna opening. To implement reconfigurable antennas, different approaches has used such as the antenna based on radio-frequency microelectromechanical systems (RF-MEMS) (won Jung, Lee, Li, & Flaviis, 2006), Pin diodes (Peroulis, Sarabandi, & Katehi, 2005), photoconductive elements (Tawk, Albrecht, Hemmady, Balakrishnan, & Christodoulou, 2010) and varactors (Jin & Mortazawi, 2003). Reconfigurable antennas have recently received much attention and growing applications in wireless communications, regarding to their reconfiguration flexibility in terms of frequency, polarization and radiation pattern. The two novel structures described in this work have the potential to be reconfigurable by inserting different electronic components or by using smart materials.

Table 3. The state of the art comparison between the proposed mutual coupling reduction technique and other solutions reported in literature

	Decoupling Structures	Resonant Frequency	Center-to-Center Spacing	Overall Structure Size	Improvement in S21 [dB]
(Habashi, Nourinia, & Ghobadi, 2011)	FSRRs	5.2GHz	$0.270\lambda0$	$68.0\times50.0mm^2$	30 to 41
(X. M. Yang, Liu, Zhou, & Cui, 2012)	WM	3.5GHz	$0.362\lambda0$	$76.4\times91.0mm^2$	6 to 20
(Alsath et al., 2013)	SML	4.8GHz	$0.380\lambda0$	$54.0\times45.0mm^2$	6 to 16
(Park & Son, 2016)	H-SCW	5.2GHz	$0.340\lambda0$	$54.0\times31.4mm^2$	51
(Hwangbo, Yang, & Yoon, 2017)	MCML	4.94GHz-4.99GHz	$0.250\lambda0$	$70.0\times50.0mm^2$	11 to 34.3
(Dhevi, Vishvaksenan, & Rajakani, 2018)	Asym-LR	3.4GHz & 4.2GHz	$0.420\lambda0$	$100\times60.0mm^2$	16.3 & 18.6
This Work	Planar EBGs	5.8GHz	$0.549\lambda_{5.8GHz}$	$128.74x97.8mm^2$	20.29

- Nowadays, with the development of wireless devices, one of the major challenges is to improve the energy autonomy of communicating devices (sensors, sensor networks). Sometimes, in certain applications, access to these sensors is difficult despite the need to maintain, for instance the sensors implanted in human body. The conventional feeding techniques using the wires or batteries remain restrictive, even difficult to consider for some applications. The transmission of information to a sensor has accomplished through electromagnetic waves transmitted by the antennas. Therefore, it is possible to use this radiated energy for supplying remote the low-power devices. The system that performs this operation has called MPT (microwave power transmission) (Wan & Huang, 2018). MPT can deliver energy for mobile platforms and to some areas where it is difficult or impossible for people to reach. For wireless power transmission (WPT) applications, a single rectenna element is not sufficient in supplying energy for reliable device operation (Taybi, Ghaloua, Tajmouati, Mandry, El Abdellaoui, & Latrach, 2017). To increase the received power level at the input of the rectifier, the antenna array proposed in this chapter can be used with an RF-DC rectifier to increase the received power level.

CONCLUSION

From the research that has been performed, it is possible to design the antenna array with a compact size and good operating performances. It was the main purpose of the first part to draw the attention to the importance of design the antenna array. In the second part, several antenna array miniaturization techniques have been proposed. Afterwards, two antennas arrays have designed and optimized in order to achieve compact dimensions and excellent performances. The third section of this chapter, introduce a novel design of miniaturized antenna array using defected ground structure for the ISM band applications. To get a small-size antenna array, resonating at a low-frequency band; we started in the first place to design an antenna array, which has small dimensions resonates at C-band 5.8GHz. Afterward, the defected ground structure has been etched in the ground plane of each radiating element of the antenna array to shift the resonance frequency to S-band at 2.45GHz. In this manner, we have been able to reduce the antenna array size up to 82.46% as compared with a conventional microstrip antenna array resonating at the same frequency (2.45GHz). In the next section, a new isolation structure for reducing mutual coupling between closely spaced antennas arrays has proposed. In addition, the antennas arrays performances before and after the implementation of the planar EBG structure has investigated at 5.8GHz. The goal of this present study is to achieve the highest radiation performances for microstrip antennas arrays, with minimal inter-element spacing and low mutual coupling. Finally, the proposed structures can be employed in various applications such as electromagnetic transponders, receivers, and transmitters for wireless systems and medical devices.

REFERENCES

Abidin, Z. Z. (2012). *Design, modelling and implementation of antennas using electromagnetic bandgap material and defected ground planes. Surface Meshing Analysis and Genetic Algorithm Optimisation on EBG and Defected Ground Structures for Reducing the Mutual Coupling between.* University of Bradford.

Ahn, D., Park, J. S., Kim, C. S., Kim, J., Qian, Y., & Itoh, T. (2001). A design of the low-pass filter using the novel microstrip defected ground structure. *IEEE Transactions on Microwave Theory and Techniques*, *49*(1), 86–93. doi:10.1109/22.899965

Alboni, E., & Cerretelli, M. (2001). *Microstrip Patch Antenna for GPS application.* Retrieved from http://www.elettromagnetismo.it/atti_rinem/2002S09A01.pdf

Alqadami, A. S. M., Jamlos, M. F., Soh, P. J., Rahim, S. K. A., Vandenbosch, G. A. E., & Narbudowicz, A. (2016). Miniaturized dual-band antenna array with double-negative (DNG) metamaterial for wireless applications. *Applied Physics. A, Materials Science & Processing*, *123*(1), 22. doi:10.100700339-016-0678-3

Alsath, M. G. N., Kanagasabai, M., & Balasubramanian, B. (2013). Implementation of Slotted Meander-Line Resonators for Isolation Enhancement in Microstrip Patch Antenna Arrays. *IEEE Antennas and Wireless Propagation Letters*, *12*, 15–18. doi:10.1109/LAWP.2012.2237156

Aris, M. A., Ali, M. T., Rahman, N. H. A., & Pasya, I. (2016). Development of antenna array using Defected Ground Structure. In *2016 10th European Conference on Antennas and Propagation (EuCAP)* (pp. 1–5). Academic Press. 10.1109/EuCAP.2016.7481577

Azimi Fashi, A., Kamyab, M., & Barati, M. (2010). A microstrip small-sized array antenna based on the meta-material zeroth-order resonator. *Progress in Electromagnetics Research*, *14*, 89–101. doi:10.2528/PIERC10032302

Baik, J. W., Han, S. M., Jeong, C., Jeong, J., & Kim, Y. S. (2008). Compact Ultra-Wideband Bandpass Filter With EBG Structure. *IEEE Microwave and Wireless Components Letters, 18*(10), 671–673. doi:10.1109/LMWC.2008.2003456

Basilio, L. I., Khayat, M. A., Williams, J. T., & Long, S. A. (2001). The dependence of the input impedance on feed position of probe and microstrip line-fed patch antennas. *IEEE Transactions on Antennas and Propagation, 49*(1), 45–47. doi:10.1109/8.910528

Beiranvand, E., Afsahy, M., & Sharbati, V. (2017). Reduction of the mutual coupling in patch antenna arrays based on EBG by using a planar frequency-selective surface structure. *International Journal of Microwave and Wireless Technologies, 9*(2), 349–355. doi:10.1017/S1759078715001440

Biswas, S., Guha, D., & Kumar, C. (2013). Control of Higher Harmonics and Their Radiations in Microstrip Antennas Using Compact Defected Ground Structures. *IEEE Transactions on Antennas and Propagation, 61*(6), 3349–3353. doi:10.1109/TAP.2013.2250240

Chang, I. C. I., & Lee, B. L. B. (2002). Design of defected ground structures for harmonic control of active microstrip antenna. *IEEE Antennas and Propagation Society International Symposium (IEEE Cat. No.02CH37313), 2*, 852–855. 10.1109/APS.2002.1016779

Cheng, Y. F., Ding, X., Shao, W., & Wang, B. Z. (2017). Reduction of Mutual Coupling Between Patch Antennas Using a Polarization-Conversion Isolator. *IEEE Antennas and Wireless Propagation Letters, 16*, 1257–1260. doi:10.1109/LAWP.2016.2631621

Combes, P. F. (1997). *Micro-ondes: Circuits passifs, propagation, antennes: cours et exercices*. Dunod.

Constantine, A. B., & ... (2005). *Antenna theory: analysis and design. In Microstrip Antennas* (3rd ed.). John wiley & sons.

Craeye, C., & González-Ovejero, D. (2011). A review on array mutual coupling analysis. *Radio Science, 46*(2), 1–25. doi:10.1029/2010RS004518

Dhevi, B. L., Vishvaksenan, K. S., & Rajakani, K. (2018). Isolation Enhancement in Dual-Band Microstrip Antenna Array Using Asymmetric Loop Resonator. *IEEE Antennas and Wireless Propagation Letters, 17*(2), 238–241. doi:10.1109/LAWP.2017.2781907

Ding, K., Fang, X., Chen, A., & Wang, Y. (2013). A Novel Parallel-Series Feeding Network Based on Three-Way Power Divider for Microstrip Antenna Array. *IEEE Antennas and Wireless Propagation Letters, 12*, 757–760. doi:10.1109/LAWP.2013.2270938

Ebadi, S., & Semnani, A. (2014). Mutual coupling reduction in waveguide-slot-array antennas using electromagnetic bandgap (EBG) structures. *IEEE Antennas & Propagation Magazine, 56*(3), 68–79. doi:10.1109/MAP.2014.6867683

Farahani, H. S., Veysi, M., Kamyab, M., & Tadjalli, A. (2010). Mutual coupling reduction in patch antenna arrays using a UC-EBG superstrate. *IEEE Antennas and Wireless Propagation Letters, 9*, 57–59. doi:10.1109/LAWP.2010.2042565

Fleuriault, J.-M. (1996). *Synthèse du diagramme de rayonnement d'un réseau de sources*. These l'Universitede Rennes I.

Foged, L. J., Giacomini, A., Saccardi, F., Tancioni, L. M., Schirosi, V., Di Cintio, A., ... Baracco, J.-M. (2015). Miniaturized Array Antenna Using Artificial Magnetic Materials for Satellite-Based AIS System. *IEEE Transactions on Antennas and Propagation, 63*(4), 1276–1287. doi:10.1109/TAP.2015.2398462

Foroozesh, A., & Shafai, L. (2009). Effects of Artificial Magnetic Conductors in the Design of Low-Profile High-Gain Planar Antennas With High-Permittivity Dielectric Superstrate. *IEEE Antennas and Wireless Propagation Letters, 8*, 10–13. doi:10.1109/LAWP.2008.916684

Ghaloua, A., Zbitou, J., El Abdellaoui, L., Latrach, M., Tajmouati, A., & Errkik, A. (2018). Reduction of Mutual Coupling between Closely Spaced Microstrip Antennas Arrays using Electromagnetic Band-gap (2D-EBG) Structures. *Telkomnika (Telecommunication Computing Electronics and Control), 16*(1).

Ghaloua, A., Zbitou, J., Latrach, M., El Abdellaoui, L., Errkik, A., & Mandry, R. (2017). Mutual coupling reduction and miniaturization arrays antennas using new structure of EBG. In *2017 International Conference on Wireless Technologies, Embedded and Intelligent Systems (WITS)* (pp. 1–5). Academic Press. 10.1109/WITS.2017.7934593

Ghaloua, A., Zbitou, J., Latrach, M., El Abdellaoui, L., Tajmouati, A., & Abounaima, M. C. (2017). Miniaturization and reduction of mutual coupling for four arrays antennas using new structure of EBG. In *2017 International Conference on Wireless Technologies, Embedded and Intelligent Systems (WITS)* (pp. 1–5). Academic Press. 10.1109/WITS.2017.7934595

Ghosh, A., Chakraborty, S., Chattopadhyay, S., Nandi, A., & Basu, B. (2016). Rectangular microstrip antenna with dumbbell shaped defected ground structure for improved cross polarised radiation in wide elevation angle and its theoretical analysis. *IET Microwaves, Antennas & Propagation, 10*(1), 68–78. doi:10.1049/iet-map.2015.0179

Guha, D., & Antar, Y. M. M. (2011). *Microstrip and printed antennas: new trends, techniques and applications*. John Wiley & Sons.

Habashi, A., Nourinia, J., & Ghobadi, C. (2011). Mutual Coupling Reduction Between Very Closely Spaced Patch Antennas Using Low-Profile Folded Split-Ring Resonators (FSRRs). *IEEE Antennas and Wireless Propagation Letters, 10*, 862–865. doi:10.1109/LAWP.2011.2165931

Hao, Z. C., & Li, B. W. (2017). Developing Wideband Planar Millimeter-Wave Array Antenna Using Compact Magneto-Electric Dipoles. *IEEE Antennas and Wireless Propagation Letters, 16*, 2102–2105. doi:10.1109/LAWP.2017.2697903

Hwangbo, S., Yang, H. Y., & Yoon, Y. K. (2017). Mutual Coupling Reduction Using Micromachined Complementary Meander-Line Slots for a Patch Array Antenna. *IEEE Antennas and Wireless Propagation Letters, 16*, 1667–1670. doi:10.1109/LAWP.2017.2663114

Jin, Z., & Mortazawi, A. (2003). An L-band tunable microstrip antenna using multiple varactors. In *IEEE Antennas and Propagation Society International Symposium. Digest. Held in conjunction with: USNC/CNC/URSI North American Radio Sci. Meeting (Cat. No.03CH37450)* (Vol. 4, pp. 524–527). IEEE. 10.1109/APS.2003.1220325

Khandelwal, M. K., Kanaujia, B. K., & Kumar, S. (2017). Defected ground structure: Fundamentals, analysis, and applications in modern wireless trends. *International Journal of Antennas and Propagation*.

Lee, Y., Lu, X., Hao, Y., Yang, S., Ubic, R., Evans, J. R. G., & Parini, C. G. (2007). Directive millimetre-wave antenna based on freeformed woodpile EBG structure. *Electronics Letters*, *43*(4), 195–196. doi:10.1049/el:20073798

Li, M., Li, Q. L., Wang, B., Zhou, C. F., & Cheung, S. W. (2018). Miniaturized Dual-Band Base Station Array-Antenna Using Band Notch Dipole Antenna Elements and AMC Reflectors. *IEEE Transactions on Antennas and Propagation*, *1*. doi:10.1109/TAP.2018.2819890

Lim, J.-S., Kim, C.-S., Ahn, D., Jeong, Y.-C., & Nam, S. (2005). Design of low-pass filters using defected ground structure. *IEEE Transactions on Microwave Theory and Techniques*, *53*(8), 2539–2545. doi:10.1109/TMTT.2005.852765

Liu, H., Li, Z., Sun, X., & Mao, J. (2005). Harmonic suppression with photonic bandgap and defected ground structure for a microstrip patch antenna. *IEEE Microwave and Wireless Components Letters*, *15*(2), 55–56. doi:10.1109/LMWC.2004.842809

Park, C. H., & Son, H. W. (2016). Mutual coupling reduction between closely spaced microstrip antennas by means of H-shaped conducting wall. *Electronics Letters*, *52*(13), 1093–1094. doi:10.1049/el.2016.1339

Pavithra, R., Mohanageetha, D., Mary Anita, E. A., & Subramaniam, M. (2014). A New Compact Microstrip Integrated E-Array Patch Antenna with High Gain and High Aperture Efficiency. *Wireless Personal Communications*, *78*(2), 1011–1020. doi:10.100711277-014-1798-8

Peroulis, D., Sarabandi, K., & Katehi, L. P. B. (2005). Design of reconfigurable slot antennas. *IEEE Transactions on Antennas and Propagation*, *53*(2), 645–654. doi:10.1109/TAP.2004.841339

Qi, H., Liu, L., Yin, X., Zhao, H., & Kulesza, W. J. (2016). Mutual Coupling Suppression Between Two Closely Spaced Microstrip Antennas With an Asymmetrical Coplanar Strip Wall. *IEEE Antennas and Wireless Propagation Letters*, *15*, 191–194. doi:10.1109/LAWP.2015.2437995

Radisic, V., Qian, Y., Coccioli, R., & Itoh, T. (1998). Novel 2-D photonic bandgap structure for microstrip lines. *IEEE Microwave and Guided Wave Letters*, *8*(2), 69–71. doi:10.1109/75.658644

Salehi, M., & Tavakoli, A. (2006). A novel low mutual coupling microstrip antenna array design using defected ground structure. *AEÜ. International Journal of Electronics and Communications*, *60*(10), 718–723. doi:10.1016/j.aeue.2005.12.009

Shahparnia, S. (2005). *Electromagnetic bandgap structures for broadband switching noise mitigation in high-speed packages*. Academic Press.

Sievenpiper, D., Zhang, L., Broas, R. F. J., Alexopolous, N. G., & Yablonovitch, E. (1999). High-impedance electromagnetic surfaces with a forbidden frequency band. *IEEE Transactions on Microwave Theory and Techniques*, *47*(11), 2059–2074. doi:10.1109/22.798001

Tahar, F., Barakat, A., Saad, R., Yoshitomi, K., & Pokharel, R. K. (2017). Dual-Band Defected Ground Structures Wireless Power Transfer System With Independent External and Inter-Resonator Coupling. *IEEE Transactions on Circuits and Wystems. II, Express Briefs*, *64*(12), 1372–1376. doi:10.1109/TCSII.2017.2740401

Tawk, Y., Albrecht, A. R., Hemmady, S., Balakrishnan, G., & Christodoulou, C. G. (2010). Optically Pumped Frequency Reconfigurable Antenna Design. *IEEE Antennas and Wireless Propagation Letters*, *9*, 280–283. doi:10.1109/LAWP.2010.2047373

Taybi, A., Ghaloua, A., Tajmouati, A., Mandry, R., El Abdellaoui, L., & Latrach, M. (2017). A New Design of Rectenna Array with High Performances. In *Proceedings of the 2Nd International Conference on Computing and Wireless Communication Systems* (p. 1:1--1:7). New York, NY: ACM. 10.1145/3167486.3167487

Tretyakov, S. (2003). *Analytical modeling in applied electromagnetics*. Artech House.

Wan, S., & Huang, K. (2018). Methods for Improving the Transmission-Conversion Efficiency From Transmitting Antenna to Rectenna Array in Microwave Power Transmission. *IEEE Antennas and Wireless Propagation Letters*, *17*(4), 538–542. doi:10.1109/LAWP.2018.2801320

Wei, K., Li, J., Wang, L., Xing, Z., & Xu, R. (2016). S-shaped periodic defected ground structures to reduce microstrip antenna array mutual coupling. *Electronics Letters*, *52*(15), 1288–1290. doi:10.1049/el.2016.0667

Wei, K., Li, J. Y., Wang, L., Xing, Z. J., & Xu, R. (2016). Mutual Coupling Reduction by Novel Fractal Defected Ground Structure Bandgap Filter. *IEEE Transactions on Antennas and Propagation*, *64*(10), 4328–4335. doi:10.1109/TAP.2016.2591058

Won Jung, C., Lee, M., Li, G. P., & Flaviis, F. De. (2006). Reconfigurable scan-beam single-arm spiral antenna integrated with RF-MEMS switches. *IEEE Transactions on Antennas and Propagation*, *54*(2), 455–463. doi:10.1109/TAP.2005.863407

Xing, K., Liu, B., Guo, Z., Wei, X., Zhao, R., & Ma, Y. (2017). Backlobe and Sidelobe Suppression of a Q-Band Patch Antenna Array by Using Substrate Integrated Coaxial Line Feeding Technique. *IEEE Antennas and Wireless Propagation Letters*, *16*, 3043–3046. doi:10.1109/LAWP.2017.2759909

Xue, H. G., Yang, X. X., & Ma, Z. (2015). A Novel Microstrip-CPW Fed Planar Slot Antenna With Broadband and Circular Polarization. *IEEE Antennas and Wireless Propagation Letters*, *14*, 1392–1395. doi:10.1109/LAWP.2015.2407904

Yang, F.-R., Ma, K.-P., Qian, Y., & Itoh, T. (1999). A uniplanar compact photonic-bandgap (UC-PBG) structure and its applications for microwave circuit. *IEEE Transactions on Microwave Theory and Techniques*, *47*(8), 1509–1514. doi:10.1109/22.780402

Yang, X. M., Liu, X. G., Zhou, X. Y., & Cui, T. J.. (2012). Reduction of Mutual Coupling Between Closely Packed Patch Antennas Using Waveguided Metamaterials. *IEEE Antennas and Wireless Propagation Letters*, *11*, 389–391. doi:10.1109/LAWP.2012.2193111

Yousefbeigi, M., Enayati, A., Shahabadi, M., & Busuioc, D. (2008). Parallel-series feed network with improved G/T performance for high-gain microstrip antenna arrays. *Electronics Letters*, *44*(3), 180–182. doi:10.1049/el:20083604

Zarifi, D., Farahbakhsh, A., & Zaman, A. U. (2017). A Gap Waveguide-Fed Wideband Patch Antenna Array for 60-GHz Applications. *IEEE Transactions on Antennas and Propagation*, *65*(9), 4875–4879. doi:10.1109/TAP.2017.2722866

Zhang, X., & Zhu, L. (2016). High-Gain Circularly Polarized Microstrip Patch Antenna With Loading of Shorting Pins. *IEEE Transactions on Antennas and Propagation*, *64*(6), 2172–2178. doi:10.1109/TAP.2016.2552544

Zhu, F., Xu, J., & Xu, Q. (2009). Reduction of mutual coupling between closely-packed antenna elements using defected ground structure. In *2009 3rd IEEE International Symposium on Microwave, Antenna, Propagation and EMC Technologies for Wireless Communications* (pp. 1–4). IEEE. 10.1109/MAPE.2009.5355659

KEY TERMS AND DEFINITIONS

AMC Reflector (Artificial Magnetic Conductors Reflector): Consists of the AMC surface, the metallic ground plane, and the air substrate between them. Each unit cell of the AMC surface consists of a square patch and a pair of crossed-stubs inside the slot on the patch.

Asym-LR (Asymmetric Loop Resonator): Consists of an asymmetric loop resonator with a coupled microstrip line, deployed in microstrip antenna arrays to suppress the surface waves. The characterization of the asymmetric loop resonator is performed using the popular Erentok's method.

FSRRs (Folded Split-Ring Resonators): Artificial material, which has a negative permeability, composed of single or double rows, etched into the ground plane for more reduction of the mutual coupling and spacing between antennas.

H-SCW (H-Shaped Conducting Wall): Consists of an H-shaped copper strip printed on a dielectric substrate, and a row of periodic vias through which the strip is connected to a ground plane. This structure suppresses the mutual coupling and improves the isolation between closely spaced MSAs.

MCML (Micromachined Complementary Meander Line Slots): A pair of micro-fabricated point symmetric meander line slots (ML) inserted on the ground plane between narrowly spaced two patches to suppress mutual coupling.

Chapter 7
Design of New Reconfigurable and Miniature Microstrip Planar Antennas

Ridouane Er-Rebyiy
University of Hassan 1st, Morocco

Jamal Zbitou
University of Hassan 1st, Morocco

Abdelali Tajmouati
University of Hassan 1st, Morocco

Mohamed Latrach
ESEO, France

ABSTRACT

Recently, wireless communication systems have developed rapidly and have become more mobile and small. This necessarily requires the adequacy of its design. The materials used should be as much as possible small and at a lower cost. So low cost with reduced volume and low weight are some of the major challenges that must be faced by the designer of modern telecommunications equipment. The microstrip planar antenna is an attraction and the most crucial component used in mobile systems, and its miniaturization is one of the research challenges. This chapter focuses on the concept of miniature reconfigurable antennas by presenting and discussing the state of art with contributions in designing reconfigurable miniature printed antennas.

INTRODUCTION

The need of both mobility and communication leads to the integration of miniature wireless communication devices. This type of wireless components reduces the complexity and performance compared to the cable system. Antennas are necessary and critical components of the communication chain. It can be said that different types of antennas have proliferated during the past 50 years in both wireless

DOI: 10.4018/978-1-5225-7539-9.ch007

communication and radar systems. These varieties include dipoles/monopoles, loop antennas, reflector antennas, microstrip antennas, log periodic antennas, helical antennas, etc (Abla, 2010). The need of miniaturizing microstrip patch antennas to allow their integration with small objects is increasing and its design is one of the most important current challenges for designers.

The miniaturization of antennas can be done by reducing the size and keeping the fidelity of performance characteristic. Among the different existing techniques, we can find methods based on adding inductance, altering the geometry of the structure, adding additional components, or modifying the material characteristics (Roy, 2014).

However, the miniaturization of antennas is generally due to a degradation of its efficiency and its bandwidth. In order to overcome this bandwidth problem and to achieve the multi-frequency behavior (multiband antenna), the frequency agility technique or reconfigurability may be useful and suitable in wireless communication system. Reconfigurability, when used in the context of antennas, is the capacity to change an individual radiator's fundamental operating characteristics (Mosallaei, 2002; Bernhard, 2003).

Reconfigurable antennas can be classified into different categories: the frequency reconfigurable antenna is a structure that is able to change its operating frequency, radiation pattern reconfigurable antenna is a structure that is able to tune its radiation pattern and polarization reconfigurable antenna that is a structure that can change its polarization (Huff, 2006).

Frequency agility is obtained by modifying the resonant frequency of the resonator. For this purpose, the antenna is loaded with active elements whose resistance can be controlled electronically by using (varicap diode), or by components acting as a switch (PIN, FET or MEMS …) (Brown, (2001)).

This chapter will be divided into two parts. The first one will be about the most common used techniques of antenna miniaturization and a comprehensive study of the frequency agility techniques. The second part will focus on some contributions and current results of new miniature antennas by using Defected Ground Structure (DGS) technique, and other contributions of reconfigurable antennas based on PIN diodes.

MICROSTRIP PATCH ANTENNAS

Introduction

In its most basic form, a microstrip patch antenna consists of a radiator mounted on an FR4 substrate associated to a ground plane as shown in Figure 1 (Carver, 1981).

The different parameters of the antenna are as follows: L is the length of the patch, w is the width; h is the thickness of substrate and t is the metallization thickness of the patch.

The patch is generally square, rectangular, circular, triangular, and elliptical or other shapes as illustrated in Figure 2. Square, rectangular, and circular are the most celebrated because of ease of analysis, fabrication, and their attractive radiation characteristics, especially low cross-polarization radiation (Bahl, 1980; Wheeler, 1959; Abla, 2010).

Feeding Methods

Microstrip planar antennas can be fed by four fundamental methods. These methods can be distributing into two categories. For the first type, the power is fed directly to the microstrip patch by using a con-

Figure 1. Structure of a Microstrip Patch Antenna

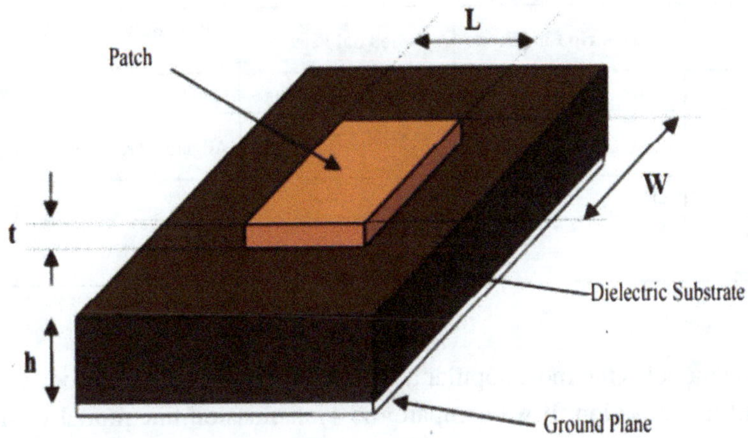

Figure 2. Representative Shapes of microstrip patch elements. (Bahl, 1980)

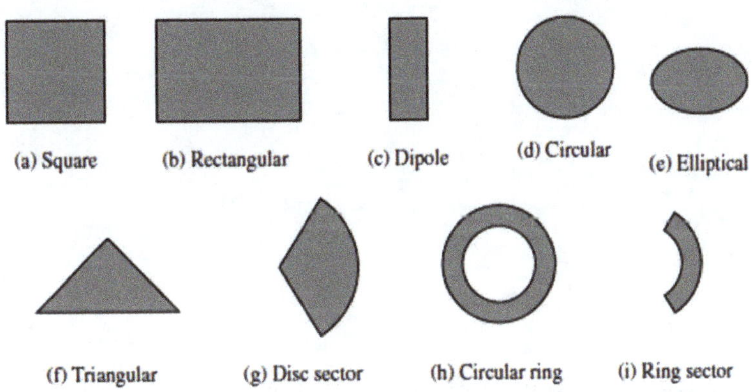

necting element such as a microstrip line or coaxial probe. But for the second type, we use a coupled line to feed the radiator (Pozar, 1992).

Table 1 present a summary which shows the characteristics of each method (Pozar & Daniel, 1992).

MICROSTRIP PATCH ANTENNA PARAMETERS

Method of Analysis

There are three methods to analyze the data of microstrip planar patch antenna.

- Transmission Line Model
- Cavity Model
- Full wave Model

Table 1. Below summarizes the characteristics of the different feed techniques

Characteristics	Microstrip Line Feed	Coaxial Feed	Proximity Coupled feed	Aperture Coupled Feed
Reliability	Better	Poor due to soldering	Good	Good
Ease of fabrication	Easy	Soldering and drilling needed	Alignment required	Alignment required
Impedance Matching	Easy	Easy	Easy	Easy
Bandwidth (achieved with impedance matching)	2-5%	2-5%	13%	2-5%

Transmission line model is the most popular ad the easiest to use. It gives good physical acumen but it is fewer in term of modelisation. If we compare the Transmission line model with cavity model and full wave model, it is more exact but more complexes (Godara, 1997).

The dimensions (width (W) and the length (L)) of the rectangular patch antenna can be designed by using the following equations formulas given in (Godara, 1997).

$$W = \frac{c}{2f}\left(\frac{\varepsilon_r + 1}{2}\right)^{-\frac{1}{2}} \tag{1}$$

$$L = \frac{c}{2f\sqrt{\varepsilon_e}} - 2\Delta L \tag{2}$$

Where: c is the free space velocity of light, f is the resonant frequency, εr is the relative permittivity of substrate, and εe is the effective dielectric constant calculated by:

$$\varepsilon_e = \frac{\varepsilon_r + 1}{2} + \frac{\varepsilon_r - 1}{2}(1 + \frac{12h}{W})^{-\frac{1}{2}} \tag{3}$$

The length extension is given by:

$$\Delta L = h.(0.412)\frac{(\varepsilon_e + 0.3)(\frac{W}{h} + 0.264)}{(\varepsilon_e - 0.258)(\frac{W}{h} + 0.8)} \tag{4}$$

Input Impedance

The Input impedance is established as the "the impedance presented by an antenna at its terminals, or other words is the ratio of the voltage to current at a pair of boundary. Another definition of the input impedance is the ratio of the appropriate components of the electric to magnetic fields at a point. Zin = (Rin + jXin)

Rin is the real part, representing the power dissipated though heat or through radiation losses.

Xin is the imaginary part, representing the reactance of the antenna and the power stored in the near field of the antenna.

Return Loss

An important parameter to measure when testing a microstrip patch antenna is the Return loss. It is related to impedance matching and the maximum transfer of power.

The return loss (RL) is defined by the ratio of the incident power of the antenna P_{ref} to the power reflected back from the antenna of the source P_{in} (Sechrist, 1969).

RL defined as a negative of the magnitude of the reflection coefficient in dB.

The mathematical expression is:

$$RL = 10\log_{10}\frac{P_{in}}{P_{ref}} dB \tag{5}$$

Frequency Bandwidth

Frequency bandwidth is the spectrum of frequencies into which the performance of the antenna, with respect to some characteristic, conforms to an itemized standard. The bandwidth can be considered to be the range of frequencies, on either hand of the focus frequency, where the antenna characteristics are during an acceptable value of those at the center frequency. In microstrip technology, the antenna is required to provide a return loss less than -10dB over its frequency bandwidth. The frequency bandwidth of a microstrip patch antenna can be expressed as either absolute bandwidth (ABW) or fractional bandwidth (FBW).

The ABW is defined as the difference of the two value f_H and F_L with f_H and f_L are respectively the upper edge and the lower edge of the antenna bandwidth. But the FBW is designated as the percentage of the frequency difference over the center frequency, as given in Equation 6 and 7, respectively.

$$ABW = f_H / f_L \tag{6}$$

$$FBW = [f_H - f_L / f_H + f_L]100 \tag{7}$$

The Radiation Pattern

According to references, the radiation pattern is defined as "a mathematical function or a graphical representation of the radiation properties of the antenna with respect to a private type of coordinates". In terms of efficiency; radiation pattern can be defined as the power radiated from the microstrip element divided by the power received by the input to the element (Sechrist, 1969).

We generally consider spherical coordinates as the ideal antenna is supposed to radiate in a spherically symmetrical pattern. Practically in all cases the radiation pattern is determined in the far field.

General expression of the radiation efficiency can be found in most books of antennas including references of this research (Chu, 1948).

$$e = \frac{p_{rad}}{p_{rec}} \tag{8}$$

Where:P_{rad} Power radiated by the antenna.P_{rec} Power accepted by the antenna.Efficiency can also be expressed in terms of the quality factor Q as follows (Harrington, 1960; Collin, 1964):

$$e = \frac{Q_t}{Q_{rad}} \tag{9}$$

Q_t Total quality factorQ_{rad} Quality factor due to radiation (space wave) losses.

Antenna Gain and Directivity

The term of the gain is defined as a ratio of the intensity of an antenna radiation in the direction of strongest, to that of a reference antenna, when both the antennas are fed by the same input power. If the antenna reference is an isotropic, the term of gain is frequently expressed in units of dBi.

The expression for the maximum gain of an antenna is as follows:

$$G = \eta \text{ x } D \tag{10}$$

η – The efficiency of the antennaD – Directivity

Directivity is the ability of an antenna to focus energy in a particular direction. According to IEEE, the definition of the directivity is the ratio of the radiation intensity in a given direction from the antenna to the radiation intensity averaged over all directions" (McLean, 1996; McLean, 1994; Wheeler, 1947).

The directivity of the antenna depends on the form of the radiation pattern. The measurement is done taking a reference of isotropic point source from the response. The quantitative measure of this response is known as the directive gain for the antenna on a given direction.

The Polarization

The term of the Polarization of an antenna in a given direction is defined as "the polarization of the wave transmitted or radiated by the antenna" (Mendez, 2006).

The Polarization of a planar antenna is a property of its electromagnetic wave; it describes the magnitude and direction of the electric field vector as a function of time, with the orientation of the electric field for a given position in space (Dey, 1997).

The Polarization of the planar antenna can be classified as circular, linear, and elliptical. The first type is the Circular polarization when the antenna radiates power in all planes in the direction of propagation (vertical, horizontal, and between them). The second type is the linear polarization when the antenna radiates power in the plane of propagation, only in one plane (Zhang, 1998).

A receiving antenna has to be in the same polarization as the transmitting antenna otherwise it will not resonate.

Advantages and Disadvantage

Microstrip planar antennas are growing more popular for the use in wireless applications. Thus, they are extremely compatible for embedded antennas in manipulate wireless devices, such as: cellular phones, Wireless routers etc. These are equally useful for spacecraft, aircraft, and satellite applications. Some of their advantages are given below:

- Low volume and Light weight
- Linear as well as circular polarization
- Ease of fabrication and low fabrication cost
- Mechanically robust specially on inflexible surface
- Can be facilely integrated with microwave integrated circuit (MICs)

Microstrip antenna has some major disadvantage like:

- Narrow Bandwidth
- Low efficiency
- Low power
- Surface wave excitation

Applications of Microstrip Patch Antenna

The Microstrip planar antennas are very famous for their performance, sturdy design, fabrication and important usage. The applications are in several areas such as: the satellites, medical applications and of course the military systems -just like in aircrafts missiles, the rockets, etc. The uses of the planar antennas are spreading broadly in all the fields and now they are booming in the commercial look due to their low cost of fabrication. Some of the applications for the Microstrip planar Antenna are as follows (Noghanian, 2005):

- Command and control systems
- Remote sensing and instrumentation
- Antennas array
- Mobile radio
- Biomedical radiators and intruder alarms
- Satellite communication, direct broadcast services (DBS)

Today, some researchers in medicinal field found that in the treatment of malignant tumors, the use of microwave energy is said to be the best effective way of inducing hyperthermia. The design of the particular antenna, which is to be used for this purpose, should possess light weight, easy in manipulation and to be robust. Only the planar radiator satisfies these requirements. The initial designs for the radiator used for this object was based on the design of a planar printed dipoles and annular rings, which were designed on S-band. And other design was based on the circular patch disk at L-band. This simple operation is used to measure the temperature inside the human body.

Main Techniques of Antenna Miniaturization

In this part we will present and discuss the main techniques reported in the literature for antenna miniaturization. These miniaturization techniques include reshaping the antenna geometry, reactive loading, shorting and folding, material loading, top loading, the use of metamaterials and modification of ground plane.

This miniaturization technique is currently the most famous among others. We have listed in this paragraph different antenna structures models based on the modification of geometry.

Adding Truncation: Slots, Notches and Cuts

One of the techniques used to lengthen the electric path of currents is the introduction of slots in the radiating structure. The principle used in this technique consists in forcing the currents to bypass the slots, and also to induce capacitive and inductive effects modifying the input impedance of the antenna. Figure 3 corresponds to the structure studied by Nguyen et al. (Nguyen, 2005).

By changing the characteristics of the slot, the resonance frequency is changed. The length of the slot increases and the resonance frequency decreases. The same results are found for the slot width. There are also folded slots in U, H and butterfly but also curvilinear form.

The resonance frequency is changed, by changing the characteristics of the slot, for this goal the authors in" Stacked H-shaped microstrip patch antenna", (Oraizi, 2001) propose a method for miniaturizing the microstrip antenna and generate multi band characteristics by cutting helical slots in the patch. A final prototype of a miniature proposed antenna is shown in Figure 4.

Fractal

One of the interesting techniques to provide miniature antenna is the use of fractal structures. In this type, the antenna can be miniaturized by making changes in the shape of its geometry and its feed line. The fractal geometry depict a recursively generated shape that has fractional dimensions, each fraction is a reduced copy of this shape. The mathematician French Mandelbrot is the first scientist who used the

Figure 3. Notch modifications in H-shape antenna to reduce its resonant frequency

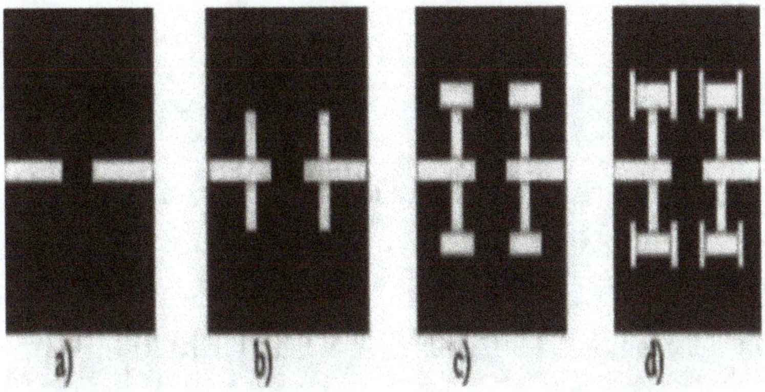

Figure 4. The realized prototypes of the proposed antenna in Oraizi, 2001

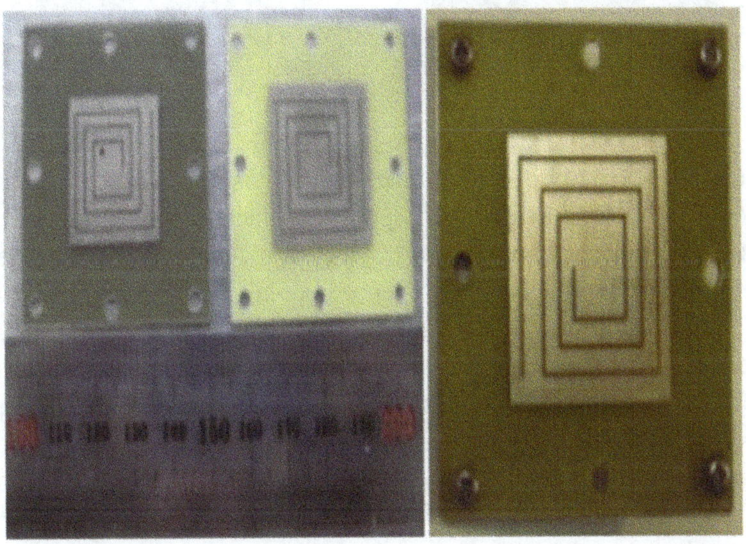

'fractal' term derived from the Latin word 'fractus' meaning 'broken' in his book "les objets fractals" (1975). However, fractal geometry was previously known in mathematics, but without an application domain. Sierpinski triangle, Koch curve or Minkowski curve was Shapes existed since 1916.

Abdelati Reha, in his paper named "Fractal Antennas: A Novel Miniaturization Technique for wireless Networks," (Reha, 2014), provided some generalities of fractal geometries and he described some linear fractal geometries such as Sierpinski, Koch, Minkowski, Circular, after that he discussed the applications of these geometries in the some contributions design of the miniaturized antennas.

Another paper (Ihamji, 2017) presents a miniature and multiband rectangular patch antenna based on the Koch fractal structure for Radio Frequency Identification (RFID) reader applications. The antenna is suitable to operate at 915 MHz, 2.45 GHz, and 5.8 GHz frequencies. The size reduction of the proposed microstrip antenna is achieved by etching the patch edges according to Koch curve. The form Koch fractal is shown in Figure 5.

Figure 5. The antenna prototype: (a) Top Face; (b) Back Face

In the paper (Basil, 2013), Durer pentagon fractal concept is introduced as a shaped patch, in order to obtain miniature and multiband antenna. The proposed fractal antenna operates in multiple bands at 1.8 GHz and 2.45GHz. In terms of patch size, 25.56% of reduction size is achieved by the proposed fractal antenna in comparison to the first patch antenna operating in the initial resonant frequency band.

The addition of Koch and Sierpinski fractal techniques gives rise to better impedance bandwidth and return loss characteristics. The proposed antenna show resonant frequency at 5.48 GHz with a return loss of -31.9dB to satisfy the band specifications for LAN applications and WiMAX application at 5GHz. The multi-fractal concept can be used in designing of the miniaturized antennas as the application of the fractal iteration is establish to reduce the frequency. Further iterations can be found to obtain antennas with smaller size and special characteristics.

A compact antenna based on the Hilbert curve fractal is presented in (Murad, 2006). The basic antenna is designed at 2.4 GHz, for the frequencies used in RFID applications. The paper show that the antenna size can be reduced up to ten times in the third iteration and the resonant frequency has shifted down by approximately 0.09 GHz from the first iteration. Figure 6, presents the Hilbert curve fractal at different iteration levels.

Use of Genetic Algorithm

Optimization is a problem that is very present in many areas, for example in electronics to miniaturize the size of circuits, in the field of finance to develop models of economic analysis. The Genetic Algorithm (GA) is a powerful optimization technique that has proven useful in optimizing microwave components. Genetic algorithms were originally developed by John Holland and were popularized through Goldberg's book (Holland, 1992; Goldberg, 1989). The genetic algorithm is a stochastic algorithm; his principle is based on the mechanisms of genetics (Mendel's laws) and the fundamental principle of natural selection

Figure 6. Hilbert curves at different iteration levels

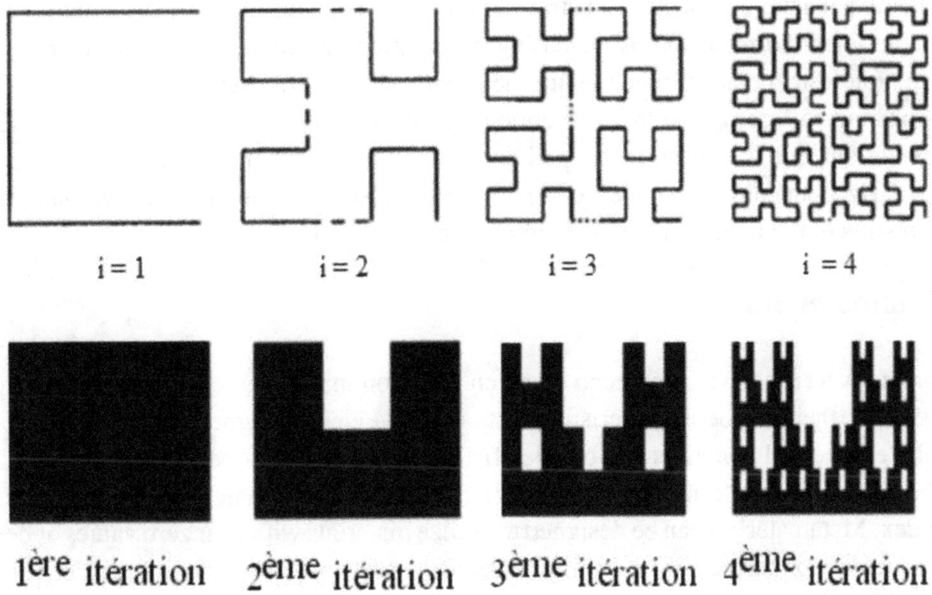

elaborated by Charles Darwin (Darwin, 1859). Stochastic laws of this algorithm can handle electromagnetic optimization problems that are not easily handled by other traditional methods.

In the paper (Herscovici, 2001) a Genetic Algorithms optimization procedure was used in the design of a microstrip patch antenna. The miniaturization of a rectangular piece has been designed to resonate at a frequency lower than the initial resonant frequency with the conservation of the same physical volume. The patch antenna is fed by a perpendicular coaxial probe and the reduction in the resonant frequency achieved was about 37%.

A genetic algorithm optimization is applied to optimize PIFA antenna parameters for wireless application in (Wakrim, 2014). The paper presented a short overview of the GA optimization procedure with the design of two examples of optimal antennas. The simulated results show the optimized antennas realize a minimization of reduction in size of the antenna reach 89.93% and 88.55% with the reflection coefficient maintaining good value of adaptation.

Material Loading

The use of a substrate with a high relative permittivity (εr) is one of the simplest ways to miniaturize a microstrip patch antenna. Its principle serves to easily generate the tuning capacity in the structure. Various investigations have studied different materials as well as configurations to effectively use this approach in order to always miniaturize a Microstrip patch antenna (MPA). The recent availability of low-loss ceramic materials, high-contrast furnishes us with possibilities for significant antenna miniaturization. The paper (Kula, 2006), explores the use of low-temperature co-fired-ceramic (LTCC) substrates of εr = 100, in producing a miniaturized patch-antenna design. A reduction of the area of the patch was achieved as compared with that of a conventional patch printed on an FR-4 substrate. By using these proposed approach, reduction size factor of more than 8 was achieved, with a return-loss greater than 9%. Moreover, respectable gain was maintained, given the achieved miniaturization.

The paper (Farzami, 2012) discussed theoretically and experimentally a miniature rectangular planar antenna by using a magneto-dielectric substrate. A compact magneto-dielectric substrate is designed in order reduce the antenna dimensions by increasing the constitutive parameters of the substrate. Moreover, the size of the microstrip planar antenna with the proposed magneto-dielectric substrate at 2.4 GHz is reduced up to about 65% compared to a conventional dielectric antenna. The bandwidth of the proposed miniature planar antenna is almost unchanged due to the increase of the magnetic permeability at the designed 2.4 GHz frequency. An achieved version of the proposed antenna is tested and measured in finally. The results of the measurement and simulation are in good agreement.

Use of Metamaterials

Nowadays, metamaterials have been received much attention in wireless communications and antenna applications, due to their unique characteristic and its attractive electromagnetic properties. Metamaterials are artificially engineered materials exhibit specific electromagnetic properties not commonly make in nature like backward propagation, reverse Doppler Effect and reverse Vavilov –Cerenkov Effect, negative refractive index. Metamaterials can be designed to realize materials with near zero values of permittivity; negative permittivity or permeability; or simultaneous negative permittivity and permeability (McVay & Hoorfar, 2007). The Russian Victor Veselago was the first physicist theorist who introduced this new invention of metamaterials in 1967 (Veselago, 1968), and based on this theory, Pendry proposed a new technique and rediscovered same characteristic for its practical implementation. Victor and Pendry suggested an artificial thin wire medium to give negative permittivity in 1996 and split ring resonators (SRR) to give negative permittivity in 1999 (Pendry, 1996; Pendry, Holden, Robbins, & Stewart, 1999), but the first practical demonstration was made in 2000 by Smith (Smith, 2000). In the design of microstrip patch antenna, there are several metamaterials types such as a split ring resonators (SRRs) and its dual counterpart (CSRRs) that can be useful. (Cui & Smith, 2010)

The paper (Amir & Saghanezhad, 2015) presents a new dual band miniature planar antenna loaded with 4-cell metamaterials. The planar structure consists of coplanar waveguide at input, microstrip line, patch, metamaterials loaded substrate under patch and back conductor. To simplify the construction procedure, the substrate under the patch is loaded by shapes metamaterials in U and inverted U made in two layers.

These antennas work at two frequency bands, which at the first band radiates as a monopole antenna with miniaturization of about 35%, and at the second band as a wideband patch antenna with about 77% miniaturization compared to the conventional patch antenna.

(Caloz & Itoh, 2005) proposed a novel design of a circular MPA based on metamaterials (MTM) substrate. In this paper, the FR-4 substrate was changed by single split ring unit cell of metamaterials. At the resonance frequency, the new metamaterials confirmed to act with the negative permittivity equal to -3.09. The new MTM substrate was able to improve the return loss from -22 dB to -24 dB and also was able to reduce size for more than 50% of the circuit.

Defected Ground Structure

Defected Ground Structure (DGS) is one of the methods, which are used to miniaturize the size of microstrip antennas, DGS consist of etching a variety of a simple shape in the ground plane, as well as complex shapes for the better performances such as split-ring resonators (SRRs) and dumbell-shaped.

We presented in Figure7, some example of DGS specific shapes etched on the ground plane of microstrip circuits (Aris & Ramli, 2015).

The author in (E l Ftouh, 2014) presents a final design of a miniature microstrip patch antenna resonating at 3GHz. For this goal, defected ground structure (DGS) has been employed as a method to shift the resonance frequency from 5.7GHz to 3GHz. The proposed shaped DGS is incorporated in the ground plane under the patch antenna and a miniaturization up to 50%, with any change in the dimensions of the initial microstrip antenna resonating at 5.7 GHz. The measurements results were in good agreement with simulation results.

The objective of the paper presented in (Guha, 1999) is aimed to address the topic in a comprehensive way, and to present a major constraint of antenna miniaturization designs using Defected Ground Structures, and to review the state-of-the-art advances in the area of DGS. Debatosh Guha, in this article describes the Defected Ground Structure (DGS) approach for designing low profile antennas such as microstrip and dielectric resonator antennas.

Overview of Reconfigurable Planar Antennas

After the presentation of state-of-the-art of miniature antenna structures that used in frequency bands fixed by the final dimensions of antenna. The second part will address an overview of the most famous used techniques for planar antennas to become active or reconfigurable, which means that with the same antenna structure while by integrating electronic components that can be active and that are capable of making the antenna reconfigurable and adapted to several application bands

Figure 7. Example of defected ground structures (DGS): (a) Concentric ring shaped; (b) Dumbbell-shaped; (c) Arrow head dumbbell; (d) Spiral-shaped; (e) U-shaped; (f) Circular head; g) Split-ring resonators; (c) H-shaped dumbbell; (i) Cross-shaped, and; (j) Meander line
(Aris & Ramli, 2015)

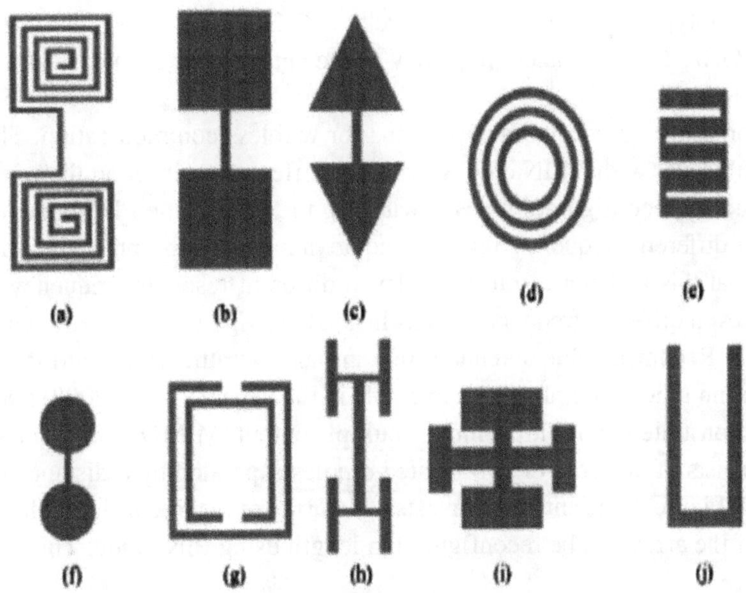

Reconfigurable antenna, it is an antenna which can change its characteristics after its manufacture, by the introduction of an extern command, this definition specific to this manuscript and without reference. According to the reference (Bernhard, 2003), reconfigurability, when used in the context of antennas, is the capacity to change an individual radiator's fundamental operating characteristics through electrical, mechanical, or other means. The first patent on reconfigurable antennas appeared in 1983 by Schaubert (Schaubert, 1983). DARPA (the Defense Advanced Research Projects Agency) in 1999 triggered a program between 12 universities, research institutes, and companies in the United States under the name Reconfigurable Aperture Program (RECAP), in order to investigate and to develop reconfigurable antennas and their potential applications (Smith, 1999).

There are actually many ways to classify reconfigurable antennas; it can be done, for example which is able to change its operating or notch frequency, this type is called frequency reconfigurable antenna. The second type is the radiation pattern reconfigurable antenna, for this category, the antenna radiation pattern changes in terms of direction, or gain. Another case of reconfigurable antenna is the antenna can change, for example, from vertical to left-hand circular polarization, this type is called polarization reconfigurable antenna.

In this chapter we adopt a classification according to the frequency, or so-called frequency reconfigurable antenna and in the following part we will expose and discuss diverse techniques reported as o stat of art in the literature for frequency reconfigurable antenna.

Reconfigurable Antennas Based on PIN Diodes

Frequency reconfigurable antenna based on PIN diodes is discussed in the paper (El Maleky, 2017). The paper proposes a miniature microstrip planar antenna with a simple geometry and simple concept of frequency agility. The T-shaped antenna consists of the insertion of 6 PIN diodes, in order to redistribute the surface currents. The reconfiguration is achieved through the change of the state of the PIN diodes ON-OFF. A miniature antenna have a dimensions 20 mm*20 mm*1.6 mm printed on FR4 substrate with 4.4 permittivity and with 0.04 loss tangent, the planar antenna can be adopted in many communication devices and can operates in S-Band with an acceptable band and gain. The simulated and de measured prototype frequency is respectively 3.12 GHz, 3.2 GHz, 3.31 GHz, 3.44 GHz, 3.56 GHz, 3.76 GHz, 4 GHz. The resonance frequency of the antenna, for each case and the bandwidth are respectively presented in Table 2.

(Chitra, 2014) presents a reconfigurable antenna for wireless communication. The proposed structure consists of double slot with 6 PIN diodes placed at different position on the ground plane in order to achieve the frequency reconfigurability. By switching the state of the PIN diode, the antenna can be operating at eleven different frequency ranges. The advantages of the proposed structure antenna in (Chitra, 2014), is that this antenna resonant at eleven different resonant frequency. Single antenna is used to radiate at eleven different frequency bands Instead of using two or three antenna, and the size of the antenna is small. Reconfigurable antenna arrays are also an attractive solution for MIMO systems to maintain good communication links, as an example in the paper (Piazza, 2008) a new reconfigurable antenna array is demonstrated for multiple inputs multiple output (MIMO) communication systems. The antenna system consists of an array of two printed dipoles separated by a distance of a quarter wavelengths, as shown in Figure 8, the antenna consists of an array of two reconfigurable microstrip dipoles; the two elements of the array can be reconfigured in length using PIN diode. This paper demonstrated

Table 2. Resonant frequency and band for different case

State of PIN Diode	Resonance Frequency (GHz)	Bandwidth (GHz)
Case 1	3.11	2.95 to 3.95
Case 2	3.24	3.03 to 3.43
Case 3	3.30	3.19 to 3.84
Case 4	3.43	3.22 to 3.66
Case 5	3.56 and 5.89	3.31 to 3.82 and 5.73 to 6.22
Case 6	3.77 and 5.97	3.45 to 4.08 and 5.67 to 6.22

Figure 8. Prototype of fabricated reconfigurable antenna array (Piazza, 2008)

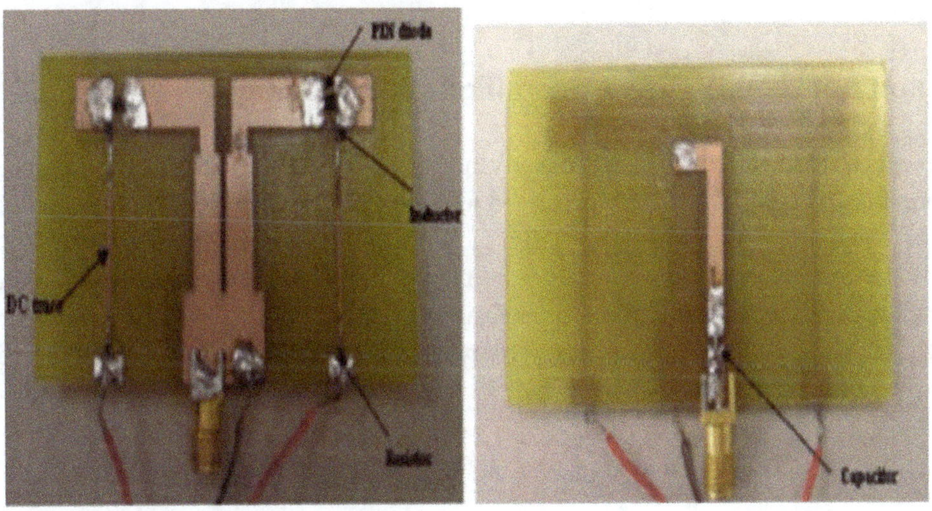

the advantage of switching between different antenna configurations based on the spatial characteristics of the MIMO channel.

Reconfigurable Antennas Based on Varactor Diodes

(Hum & Xiong, 2010) Presents the design of microstrip patch antenna based on varactor diodes in order to obtain a frequency-agile. In this paper the patch antenna was fabricated on a Rogers RO3203 substrate with $\varepsilon r = 3.02$. Frequency agility can be achieved by tuning the series and shunt groups of capacitors, 0.4 nH is the value of the package inductance of varactor diode was included in the simulations. So in this paper the frequency changes with the changes of capacitance value. For simplicity, in this simulation a series RLC circuit used to implement the varactor diodes in the model.

A new design of Single- and dual-polarized antennas with wideband tuning based on varactor diodes have been demonstrated (White & Rebeiz, 2009). The antenna tunes from 0.95 to 1.8 GHz with a return loss less than -10 dB, and polarization of the antenna tune from 0.93 to 1.6 GHz. The different value of the varactor diodes capacitance varies from 0.45 to 2.5 pF. The antennas are printed on 70 * 70* 0.787 mm^3substrates with a dielectric constant of 2.2. The antenna can be made frequency agile, or can operate at two independent frequencies on two orthogonal polarizations. The corresponding structure for this antenna is shown in Figure 9.

Reconfigurable Antennas Based on Smart Materials

Another concept of antennas reconfigurable is also made through a change in the substrate characteristics. These antennas use special materials such as ferrites (Hu & Ismail, 2007); Liu and Langley, (2008) or liquid crystals (Pozar, 1998; Dixit & Pourush, 2009). At the design level of these antennas, the change in the substrate characteristics is achieved by a change in the relative electric permittivity or magnetic permeability. Actually, by altering the orientation of the molecules, a liquid crystal is a material whose dielectric constant can be changed under different voltage levels. As in the use of a ferrite material, the relative permittivity/permeability can be changed by a static applied electric/magnetic field (Costantine, 2009).

Comparison Between Different Reconfigurable Techniques

Table 3 presents a summary which shows the advantages and disadvantages of each method.

In the previous paragraphs, we have presented major techniques that have been proposed in the literature for antenna miniaturization and reconfigurable antenna. But many articles in the literature have not clarified the methodology used to design the physical antenna geometries and also they did not present the characteristic principles of the antenna after miniaturization. In order to intervene with new approach, the authors in the in the following parts present their contributions either in the miniaturization of antennas or in reconfigurable antennas.

Figure 9. (a) Single-polarized slot-ring antenna, (b) Dual-polarized slot-ring antenna (White & Rebeiz, 2009)

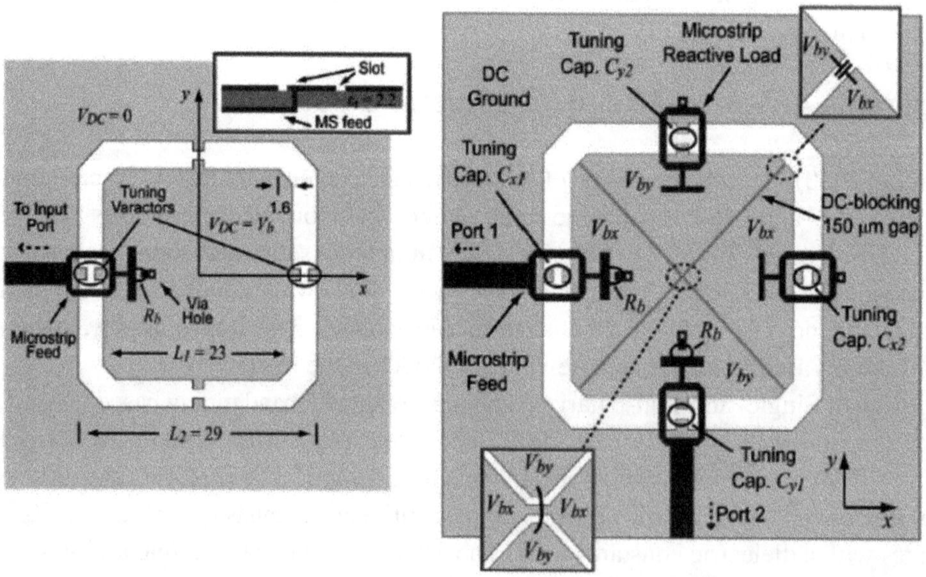

Table 3. Comparaison between different reconfigurable techniques

Method	Advantages	Disadvantage
PIN Diode	• Good isolation • Low-loss property • Easy integration into antenna structures • Low bias voltage	• Sometimes complex electronic control.
Varactor Diode	• Good isolation • Low-loss property • Easy integration into antenna structures	• Losses introduced by the electronic component. • Polarization voltage up to 30V for Varactor diodes.
Smart Material	• Small footprint	• Difficulty making the substrate agile. • Complex devices for feeding. • Very high cost.

Contribution to the Miniaturization of Microstrip Patches Antennas

In order to contribute to the miniaturization of antennas, the paper published in (Er-Rebyiy & TELKOMNIKA, 2017), uses defected ground structures (DGS) in order to miniaturize a microstrip patch antenna.

The DGS structure is incorporated, in the ground plane in the aim to shift the resonance frequency from 5.8 GHz to 2.5 GHz, with a miniaturization up to 83%. This antenna is mounted on an FR-4 substrate having a dielectric constant 4.4, a loss tangent tan (ϕ) = 0.025, thickness of 1.6 mm with the area of 34 X 34 mm^2. The antenna is designed, optimized, and miniaturized by using the CST MW-studio electromagnetic solver. The proposed antenna structure with DGS is fabricated by using LPKF machine and tested by using Vector Network Analyzer (VNA) from Rohde and Schwarz. After the achievement and the test of the final miniature patch antenna we have obtained good agreement between simulation and measurement results. The final circuit is suitable for ISM (Industrial, Scientific and Medical) applications at 2.5 GHz with S11 ≤ (-10) dB. More details of the proposed antenna design procedures, simulated and measured results are presented and discussed below.

For a better performance, the aim was to design an antenna that functions in a high frequency band at 5.8 GHz, in order to have miniature dimensions of the planar antenna. The design of the first antenna resonating at 5.8 GHz is shown in Figure 10(a) (Er-rebyiy & TELKOMNIKA, 2017). The different parameters of the designed antenna are as follows: WP=16mm, LP=10.89mm, L1=7mm, L2=7mm Wf=3mm and metallization thickness of t=0.035mm. The dimensions of the ground plane are WG=34mm and LG=34 mm.

Figure (10(b)), shown the position of DGS in the ground plane. The located place of DGS plays an important role in the miniaturization of the microstrip antenna resonating at 5.8 GHz. The fabricated antenna is shown in Figure 11.

Simulation and measurement results of S11 parameter are in good agreement as shown in Figure 12. The result presented in this Figure shows that the proposed antenna can be suitable for ISM applications. Figure13 (a) and (b) shows the measured radiation patterns in E- and H-plane respectively at 2.45 GHz.

Consequently, in this study we have validated a printed antenna which have miniature dimensions thus by using DGS technique. The final circuit is suitable for ISM applications.

To clarify the importance of the DGS method in the miniaturization of the antennas, another study has been conducted in (Er-Rebyiy & IJMOT, 2017). In this paper, a novel etched shape is proposed as a

Figure 10. (a) Geometry of the proposed antenna, and; (b) The position of the proposed DGS inserted with the microstrip patch antenna
(Er-rebyiy & TELKOMNIKA, 2017)

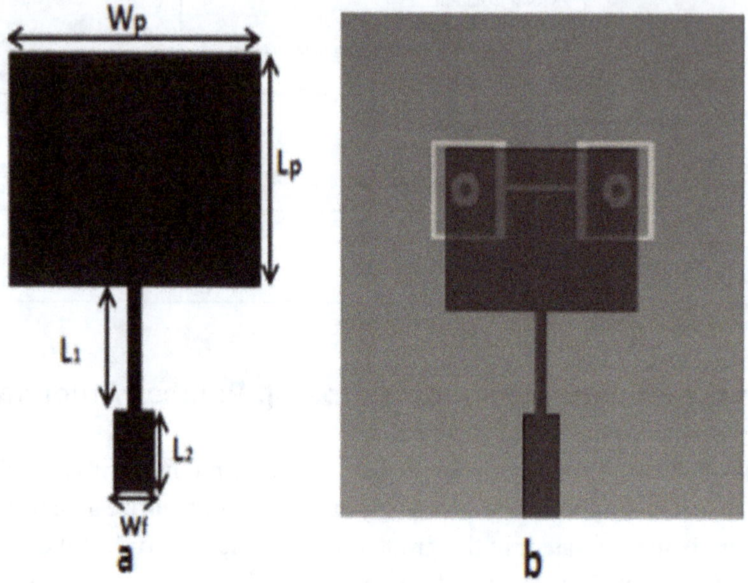

Figure 11. (a) A photograph of the fabricated antenna, and; (b) The position of the proposed DGS inserted with the microstrip patch antenna
(Er-Rebyiy & TELKOMNIKA, 2017)

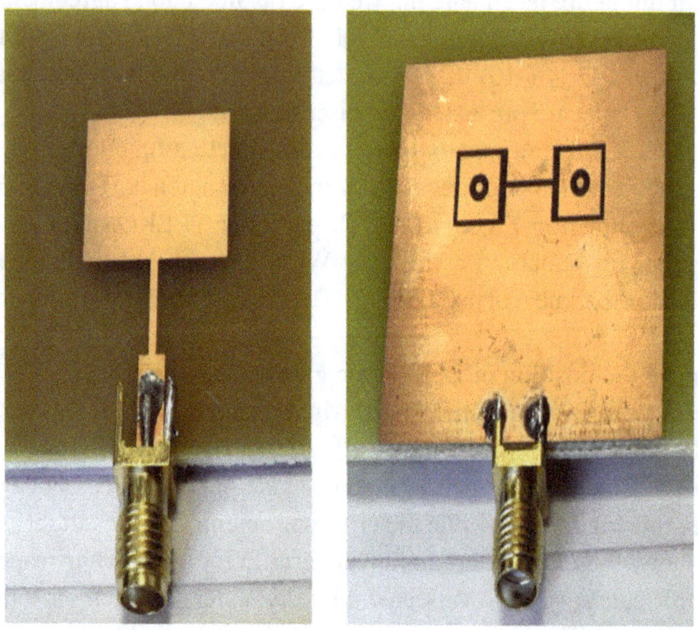

Figure 12. Comparison between simulation and measurement of the S11 parameter

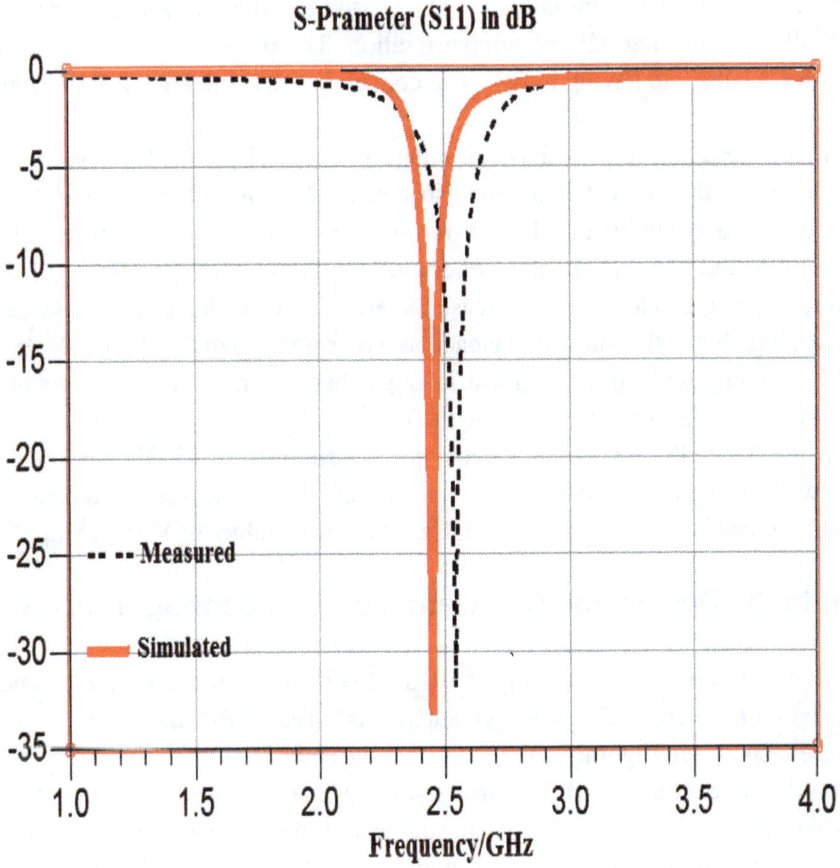

Figure 13. (a) Measured Radiation pattern in E-plane at 2.45 GHz, and; (b) Measured Radiation pattern in H-plane at 2.45 GHz

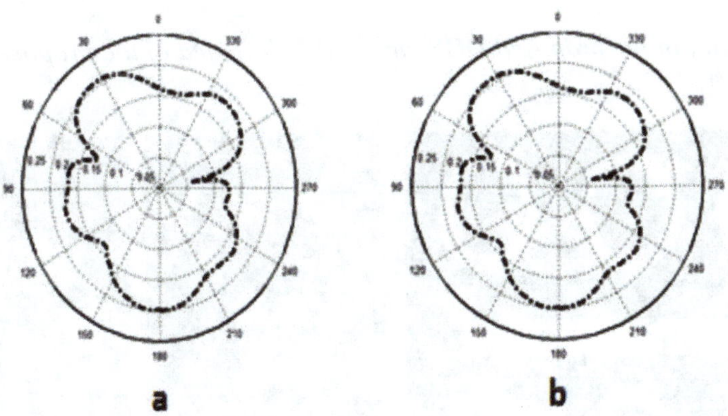

defected ground structure (DGS) always for the purpose of miniaturization of the antenna. The proposed microstrip patch antenna resonates initially at 10 GHz, and the final designed antenna is validated at 3.5 GHz, which allows achieving 92% of miniaturization. The proposed antenna with DGS has been fabricated on an FR-4 substrate having a dielectric constant of 4.4, thickness of 1.6 mm and its size is 27 X 30 mm^2.

The good agreement between the results of simulation obtained by using CST-MW and the measured results, permit validating the microstrip antenna structure. The novel etched shape disturbs the shield current distribution in the ground plane. This disturbance will change characteristics of a transmission line such as inductance and line capacitance. Otherwise, we can say that the defect etched in the ground plane of the microstrip patch antenna can increase the effective capacitance and inductance.

In Figure14 (a), it is shown that the shape etched as a defected ground structure (DGS) in (Er-Rebyiy & IJMOT, 2017), is composed from 6 concentric rings shaped with a rectangular slot. Figure14 (b), shows the prototype of the fabricated antenna with DGS.

After introducing the DGS, the resonant frequency is shifted from 10 GHz to 3.5 GHz. As we can see in Figure 15 we have a good agreement between simulation and measurement results. These results permit to validate the fabricated antenna at 3.5 GHz which is suitable for WiMAX applications.

Contribution to the Design and Realization of the Reconfigurable Antennas

The first contribution is studied in (Er-Rebyiy & IREE, 2017). It presents a simple geometry of a planar miniature reconfigurable antenna. The proposed antenna is based on PIN diodes in order to make the U-shaped antenna frequency reconfigurable. We found that the insertion of switches between the two forms, allows increasing the concentrations of the current surface. The location of the switches is determined following a study of the distribution of the surface current of the suggested antenna. The metallic tapes replace the use of the PIN diodes and the final circuit is mounted on an FR-4 substrate with a thickness of 1.58 mm, dielectric constant of 4.4, tangent loss of 0.025, fed by a microstrip line and its size is 56 X 40 mm2. The originality of this study is to develop a new reconfigurable structure which is multiband, and miniature. This concept gives an optimal solution in terms of hardware complexity which can re-

Figure 14. (a) Geometry of the proposed DGS, and; (b) Fabricated of the proposed antenna
(Er-rebyiy & IJMOT, 2017)

Figure 15. Comparison between simulated and measured reflection coefficient V_s frequency

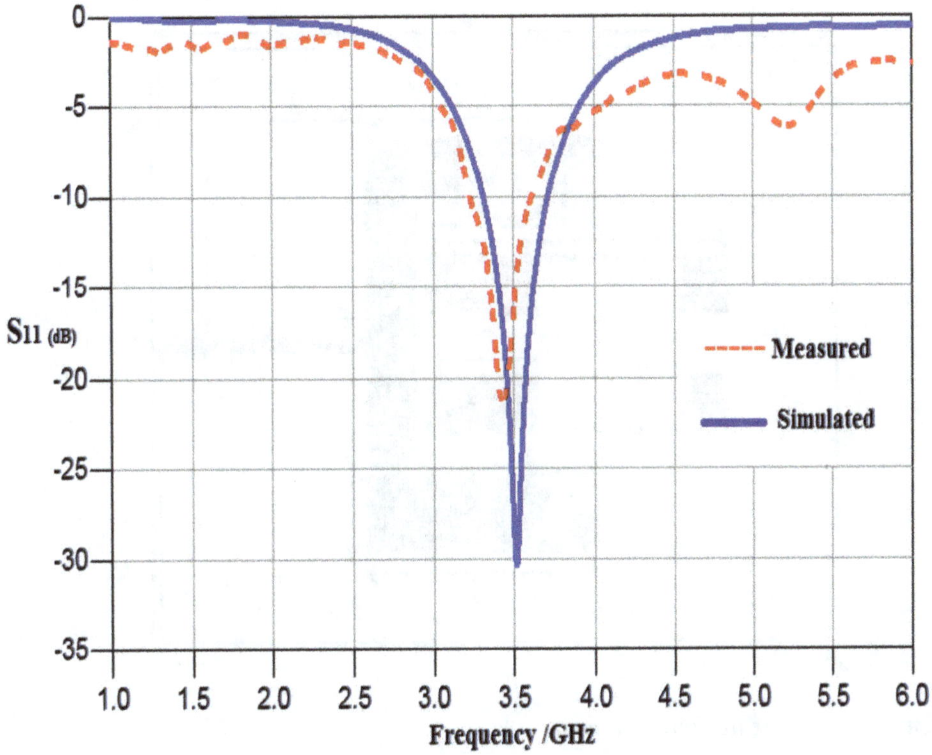

duce the number of components and cost compared to old antenna technology. The Figure 16 describes the geometry of our printed antenna, it's a printed antenna on a substrate FR-4, it has a length of L-sub = 40 mm and a width of W-sub = 56 mm. The antenna has a complete ground plane with an area of 56x40 mm2. The antenna fed by a 50 Ω microstrip fed line. After many series of optimization by using the electromagnetic solver CST- MW, Table 4 illustrates the final dimensions of the proposed antenna.

The idea aim consists to increase the movement of the current surface by integrating two PIN diodes (the diodes here are replaced by a simple metallic tape). Firstly, we simulate the antenna without a switch and then we select two optimal positions for reconfiguration. ON state of the PIN diode is replaced by the presence of the small piece of metal with the area W* L = 0.3 * 0.9 mm2 as depicted in Figure17. The methodology is well detailed in (Wang, Xiao & Wang, 2007).

By changing the states of the microstrip line (a simple metallic tape) integrated into the position ("On" or "off"), the antenna is capable of redirecting the current flow and switching between 4 frequencies bands (4 different configurations). Figures 18 (a, b), shows the simulated and measured reflection coefficients of the reconfigurable antenna at the selected two state (ON-OFF and ON-ON). A good agreement can be observed between the simulation and the measurements. As we can see in the figure 18, the frequency agility is achieved. By a single antenna it is possible to increase the number of resonance frequencies. The proposed antenna resonates initially without PIN Diodes at 5.8 GHz, after the insertion of the PIN Diodes, the frequency of resonances decreases towards lower values at 2.6 GHz, 2.9 GHz, 3.6 GHz and 3.9 GHz, depending on the states of the switch; it shows the role of the PIN diodes to obtain frequency agility of microstrip patch antenna. In Figure 18, we have a clear displacement in

Figure 16. Geometry of the proposed antenna

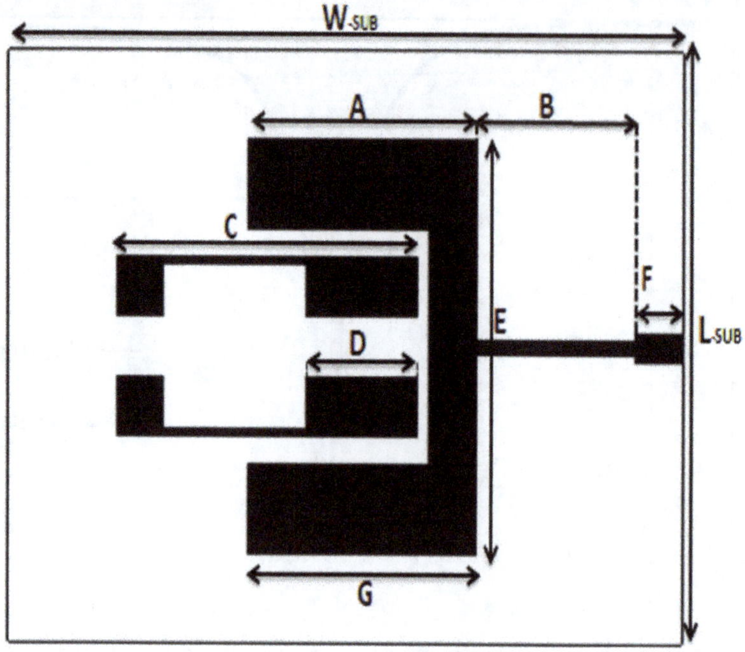

Table 4. Dimensions of the designed antenna

Parameters	Values (mm)
A	19
B	13
C	25.1
D	9.1
E	28
F	4
G	19

the peak of the frequency resonance of the proposed antenna by passing from one value to another; this displacement increases the bandwidth of the reconfigurable antenna.

To understand the influence of the PIN diode on the antenna, figure 19 shows the distribution of the current surface in the two states: (a) OFF-OFF and; (b) ON-ON. As we see the insertion of the switch increases the current surface generating different frequency responses which mean a change of the resonant frequencies.

In this paper, a U-shaped frequency reconfigurable antenna has been designed for wireless applications. By changing the state of the PIN diode, the antenna has the capability to be reconfigurable.

Figure 17. Fabricated of the proposed antenna, (a) on-on state; (b) off-off states, and; (c) off-on state

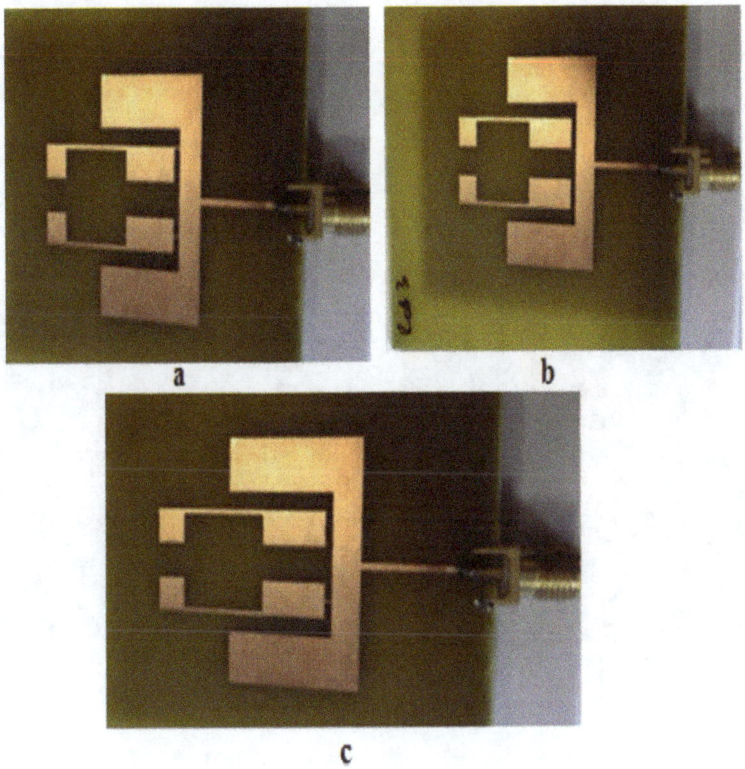

Figure 18. Comparison between simulation and measurement results of the reflection coefficient S11 versus frequency: (a) in on-on state, and; (b) in off-on state

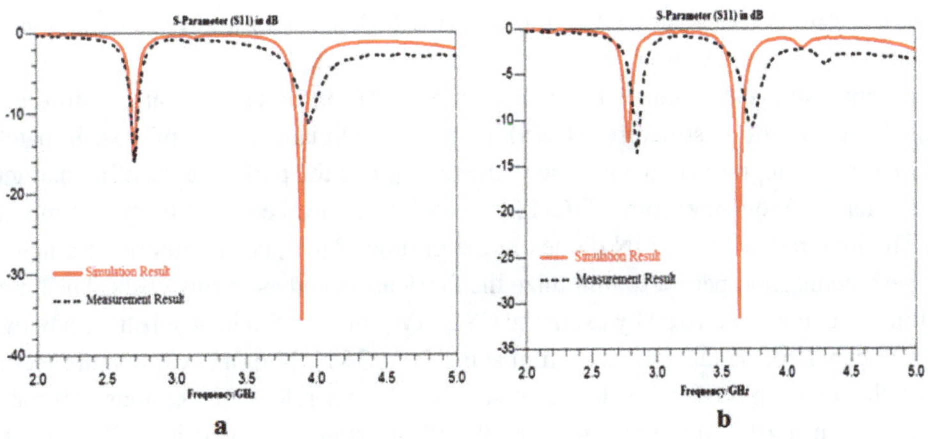

Figure 19. Surface current of the proposed antenna at: (a) 5.8 GHz (OFF-OFF) and; (b) at 3.9 GHz (ON-ON)

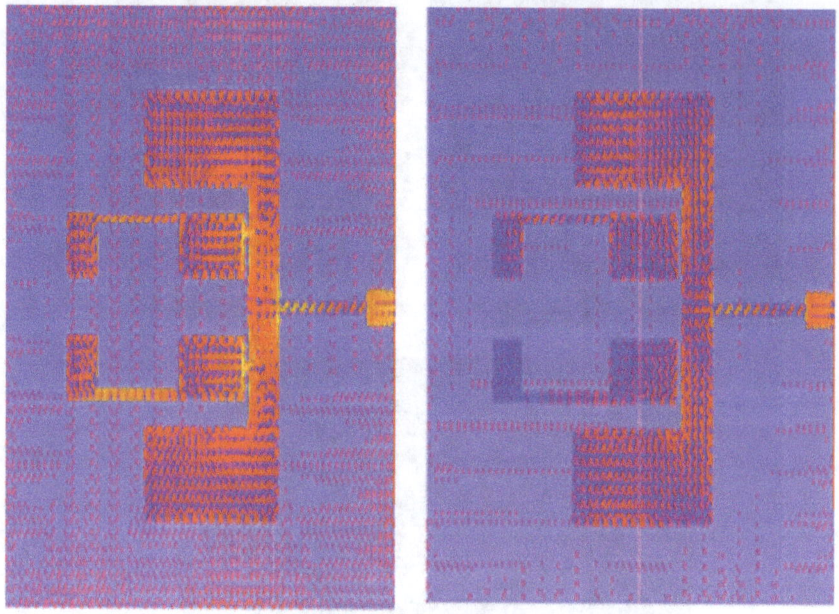

New Design of Miniature Reconfigurable Antenna: New Approach

In this chapter, the performances obtained comply with the fundamental limits of the antennas, particularly with a decrease in the bandwidth of the antenna related to miniaturization. Therefore, we are studying in this part a new structure miniaturized by the DGS method and in order to overcome the problem of the reduced bandwidth, we have used the insertion of the PIN Diodes.

Firstly, the microstrip patch antenna resonates at 5.8 GHz in order to obtain small dimensions of the antenna. Defected ground structures (DGS) are used to miniaturize the microstrip patch antenna, The DGS structure is integrated in the ground plane to improve the performance of planar antenna, and shifted the resonance frequency from 5.8 GHz to 2.45 GHz, with better performance miniaturization up to 83%. The integration of the PIN diodes in the ground plane permits to obtain a new frequency reconfigurable antenna, and permit to minimize the fabrication and assembly costs. The antenna is designed, optimized, and miniaturized by using the CST MW-studio, with FR4 substrate having dielectric constant 4.4, thickness of metallization 1.6 mm and size of 34 X 34 mm^2. A miniature reconfigurable microstrip patch antenna has been developed and analyzed for wireless applications. After many series of optimization by using the different methods of optimization integrated in CST-MW, as shows in Figure 20, the different parameters of the antenna are as follows: Wp=16mm, Lp=10.89mm, L1=7mm, L2=7mm Wf=3mm,W1=0.85mm and metallization thickness of t=0.035mm. The dimensions of the ground plane are WG=34mm and LG=34 mm.

Figure 21, shows the simulated return loss obtained for this antenna. As we can see, we have an antenna which functions at 5.8GHz with a reflection coefficient under -30dB.

Figure 20. Geometry of the proposed antenna

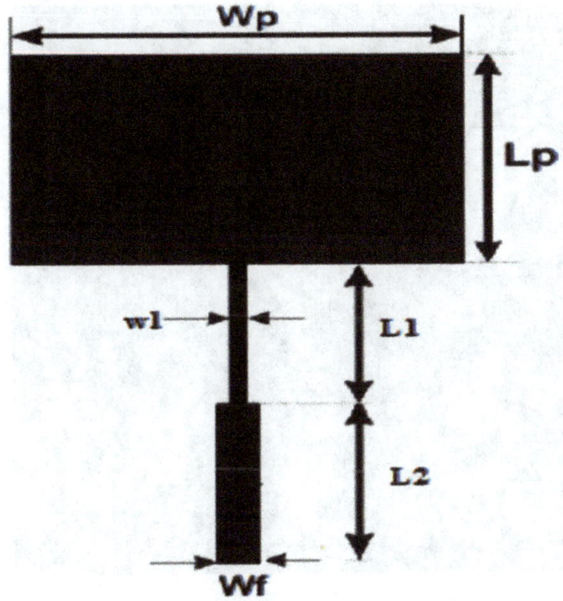

Figure 21. Simulation of return loss for the antenna at 5.8 GHz using CST-MW

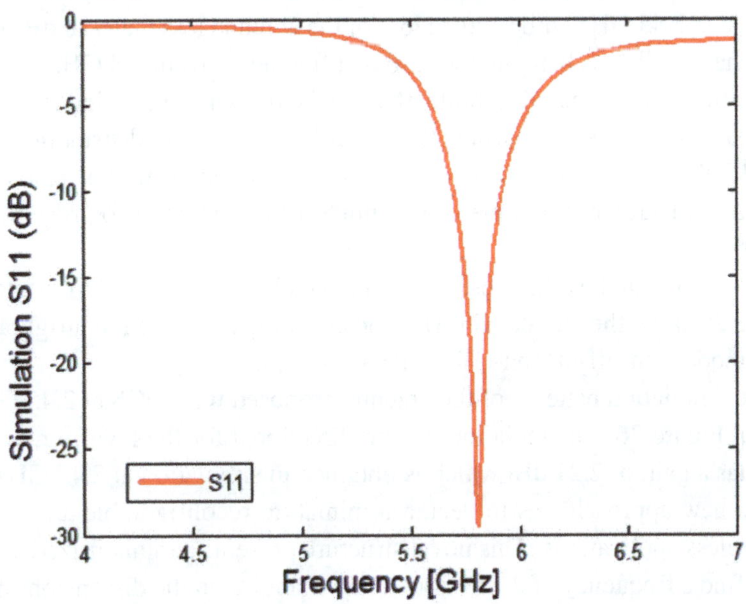

Figure 22, shows the located place of DGS in the metallic ground plane, and illustrates the details of the DGS parameters permitting to shift the resonance frequency of the microstrip antenna previously presented in Figure 20.

The length and the width of ground plane are L_G= 34mm, W_G=34mm, and different parameter of the shaps are as follows: a=4 mm, b=6 mm, c=1.8 mm, d=2 mm.

Figure 22. Geometry of different parameters of the proposed DGS

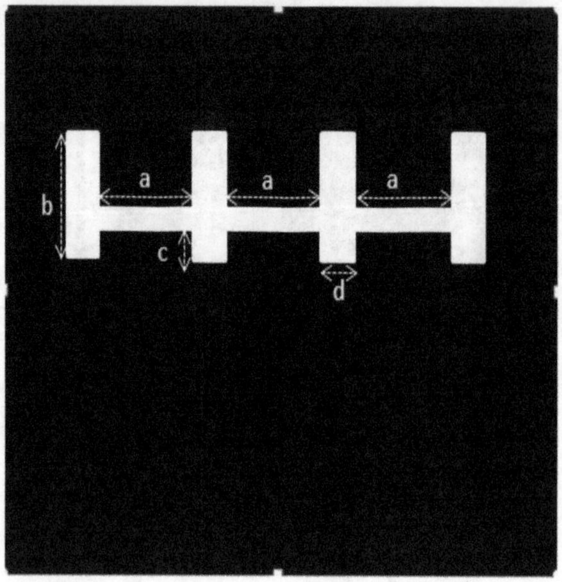

Based on the antenna simulation designed in the CST-MW solver, the optimized simulation results are shown in Figure 23. From the simulation, the resonance frequency of the antenna after integration of DGS was obtained at 2.45 GHz with a return loss of the antenna less than -20 dB. We can see that the effect of the DGS shaped allows changing the resonant frequency from 5.8 GHz to 2.45 GHz.

In order to make the antenna reconfigurable, the idea is to insert 3 PIN diodes into the ground plane as mentioned in figure 24. This study permits the control of the current distribution path.

We have placed 3 PIN Diodes to obtain six additional frequencies. In the same time we have modeled the PIN diodes by metal tapes of dimension 1mm × 1.5mm in order to prove the concept .(Wang, Xiao & Wang, 2007).

The "ON" state is illustrated by the presence of the metal line, and the "Off" state is replaced by an open circuit. Figure 25 show the simulated reflection coefficients of the reconfigurable antenna at the selected operating modes: on-off-off, on-off-on and on-on-on.

The 2 D simulated radiation patterns of the antenna proposed with DGS at 2.45 GHz in the H and E planes are shown in Figure 26. The radiation is omnidirectional for the low frequency 2.45 GHz. The proposed antenna has a gain of 2.21 dBi which is obtained in simulation at 2.45 GHz using CST-MW.

The goal of this new approach was to design a miniature reconfigurable antenna based on active components for wireless applications. This novel structure presents a miniaturized antenna by the DGS method in order to find a frequency of 2.45 without any changing in the dimensions of the first antenna validated at 5.8 GHz, and also the use of the active components PIN diodes permits to obtain a reconfigurable antenna.

Figure 23. Simulation of return loss for the antenna at 2.45 GHz using CST-MW

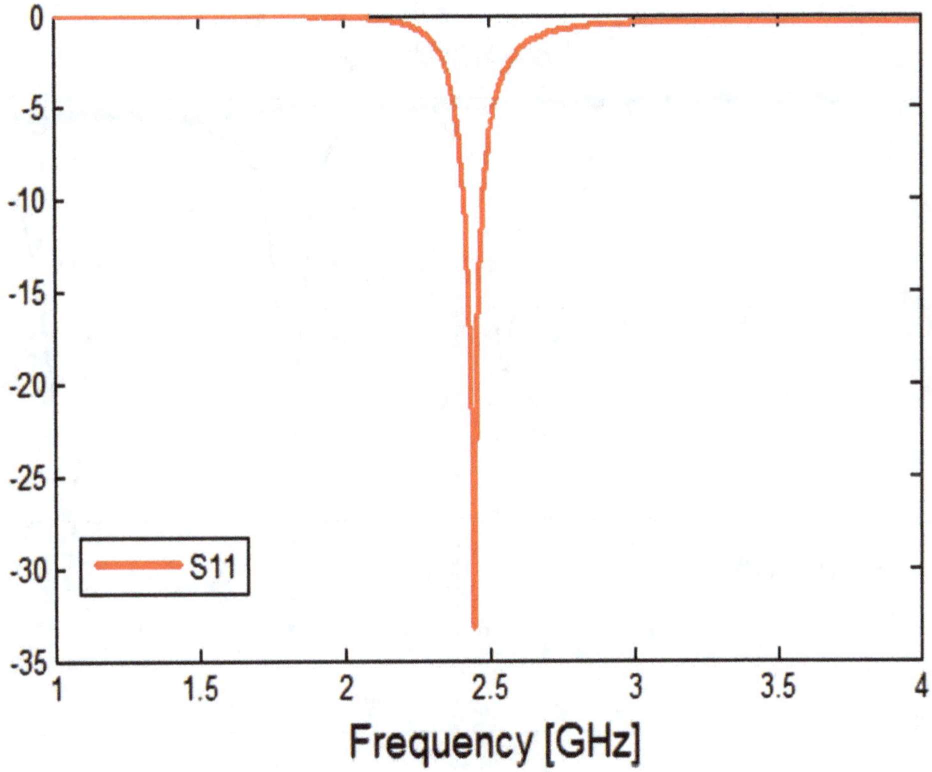

Figure 24. Located place of the different PIN diodes

Figure 25. Simulation of return loss for the antenna at the selected operating modes: on-off-off, on-off-on and on-on-on

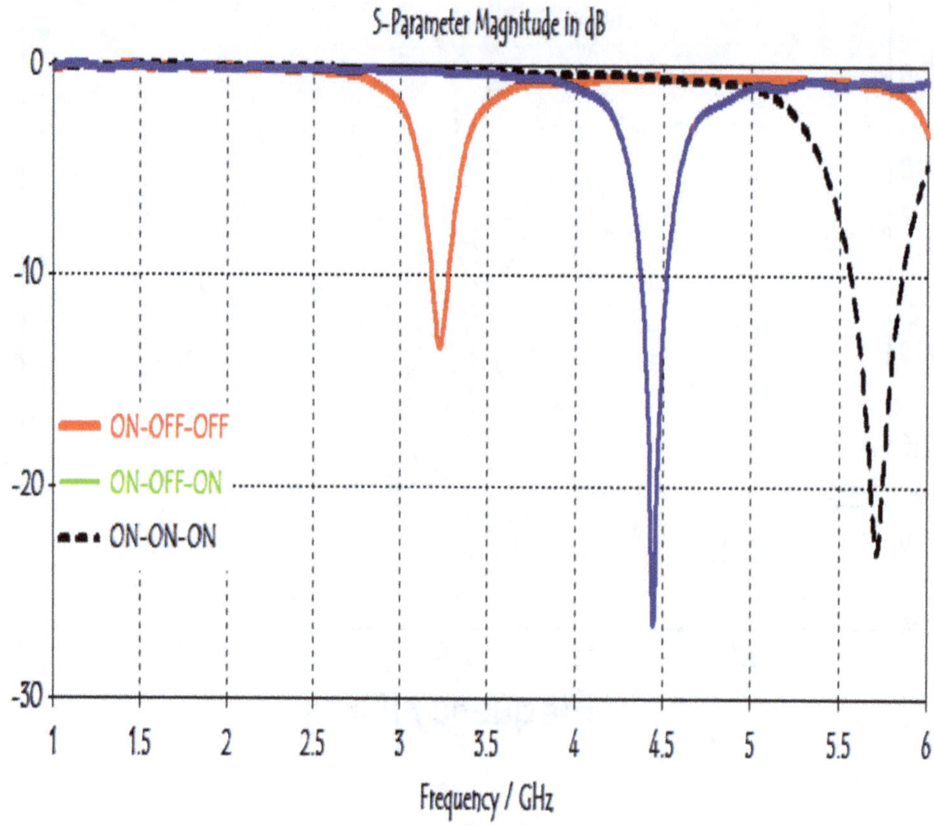

Figure 26. (a) Simulated Radiation pattern in E-plane at 2.45 GHz, and; (b) Simulated Radiation pattern in H-plane at 2.45 GHz

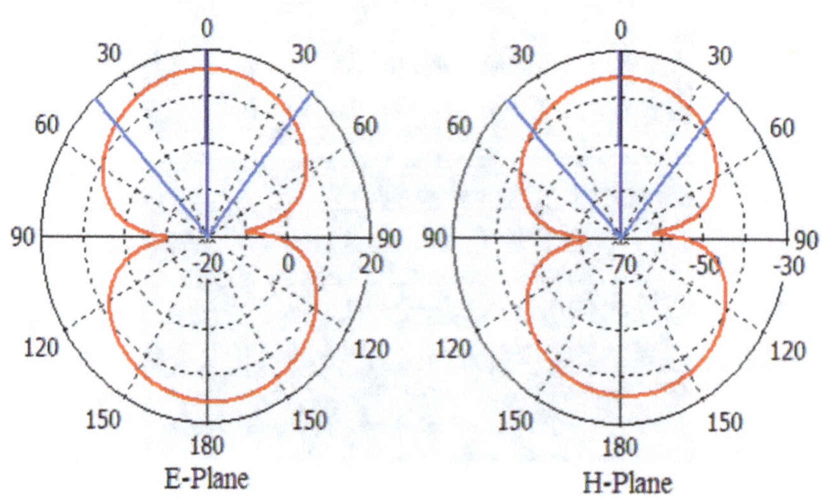

Comparison Between the Obtained Results With Published Literature

In the previous sections, we have presented the results of simulation and realizations of antennas structures proposed in this chapter. The chart below present comparisons between the antennas studied and the other antennas structures validated in bibliography in the term of performance.

CONCLUSION

In the first part of this chapter, we have presented an extensive study on planar antennas with the current state of the art for the miniaturization techniques of these antenna types. These varieties of techniques which are used in the literature, including material loading, reactive loading, reshaping the antenna geometry, use of metamaterials and the modification of ground plane (DGS). The first part also includes an overview with a lot of literature references about reconfigurable antenna precisely the antennas agile in frequency. This study also allowed us to count the considered different techniques and technologies used

Table 5. Comparisons between the proposed antennas with published literature

Antenna Studied	Initially Resonance Frequency	Finally Resonance Frequency	Miniaturization	The Method Used
Reference (Herscovici, 2001)	2.96 GHz	1.8 GHz	37%	Genetic Algorithm
Reference (Farzami, 2012)	2.48 GHz	2.38 GHz	-	magneto-dielectric substrate
Reference (El ftouh H, 2014)	5.7 GHz	3 GHz	50%	DGS
Our work (Er-Rebyiy R & TELKOMNIKA, 2017)	5.8 GHz	2.5 GHz	83%	DGS
Our work (Er-rebyiy R & IJMOT, 2017)	10 GHz	3.5 GHz	92%	DGS

Table 6. Comparisons between the proposed antennas with published literature

Antenna Studied	Frequency Agility	Number of Switches	Size
Reference (El Maleky O, 2017)	3.12 GHz to 4 GHz	6 PIN diodes	20 *20 mm^2
Reference (Hum, 2010)	2 GHz to 4 GHz	6 varactor diodes	63*63 mm^2
Reference (Hu, 2007)	9 GHz to 10 GHz	Smart materials	200*200 mm^2
Our work (Er-Rebyiy & IREE, 2017)	2.6 GHz to 5.8 GHz	2 PIN diodes	40*56 mm^2
Our work New approach	3.2 GHz to 5.8 GHz	3 PIN diodes	34*34 mm^2

to make an antenna agile in frequency. This led us to propose in the second part original contributions in order to validate these techniques in the improvement of the performances of the printed antennas. The first two structures proposed consist of miniaturized planar antennas by the DGS method. In the other two structures, on the other hand, we proposed antennas reconfigurable into frequencies by the integration of the PIN diodes. The proposed structures were fabricated and measured in order to validate this approach.

This work has allowed us to create prototypes with low costs and low RF losses. Through the use of FR-4 type substrates and the use of high performance PIN diodes; these devices can withstand high RF power without distorting the modulated RF signal. These prototypes are simple to manufacture and use, which can be easily industrialisable.

REFERENCES

Abla, B. (2010). *Contribution à l'étude des antennes microstrip Supraconducteur de forme triangulaire.* Mémoire de Magistère en Micro-ondes. Université Mentouri.

Amir, S., Saghanezhad, H., & Atlasbaf, Z. (2015). Miniaturized Dual-Band CPW-Fed Antennas Loaded With U-Shaped Metamaterials. *IEEE Antennas and Wireless Propagation Letters, 14*, 658–661. doi:10.1109/LAWP.2014.2376554

Aris, M., & Ali, M. T. (2015). *Frequency Reconfigurable Aperture-Coupled Microstrip Patch Antenna Using Defected Ground Structure.* 2015 IEEE International RF and Microwave Conference (RFM 2015), Kuching, Malaysia.

Bahl, I. J., & Bhartia, P. E. (1980). *Microstrip antenna*. Dedham, MA: Artech House.

Basil, Jeemon, Shambavi, & Alex. (2013). A Multi-fractal Antenna for WLAN and WiMAX application. *Proceedings of IEEE Conference on Information and Communication Technologies (ICT 2013).*

Bayatmaku, N., Lotfi, P., Azarmanesh, M., & Soltani, S. (2011). Design of simple multi-band patch antenna for mobile communication applications using new E-shape fractal. *IEEE Antennas and Wireless Propagation Letters*, *99*, 1.

Bernhard, J. T. (2003). Reconfigurable antennas and apertures: State-of-the-art and future outlook. *Proceedings of SPIE Conference on Smart Electronics, MEMS, BioMEMS, and Nanotechnology, 5055*, 1–9.

Brown, E. R. (2001). on the gain of a reconfigurable-aperture antenna. *IEEE Transactions on Antennas and Propagation*, *49*(10), 1357–1362. doi:10.1109/8.954923

Caloz and Itoh. (2005). *Electromagnetic Metamaterials: 'Transmission Line Theory and Microwave Applications*. Piscataway, NJ: Wiley- IEEE.

Carver, K. R., & Mink, J. W. (1981). Microstrip Antenna Technology. *IEEE Transactions on Antennas and Propagation*, *AP-29*(1), 2–24. doi:10.1109/TAP.1981.1142523

Chitra, J., & Nagarajan, V. (2014). Frequency Reconfigurable Antenna Using PIN Diodes. *IEEE Transactions on Antennas and Propagation*, *53*(2).

Chu, L. (1948). Physical Limitations of Omni-Directional Antennas. *Journal of Applied Physics, 19*(12), 1163–1175. doi:10.1063/1.1715038

Collin & Rothchild. (1964). Evaluation of Antenna Q. *IEEE Trans. AP, AP-12*, 23–27.

Costantine, J. (2009). *Design, optimization and analysis of reconfigurable antennas* (Ph.D. dissertation). Electr Comput, Eng. Dept., Univ. New Mexico (UNM), Albuquerque, NM.

Cui, T. J., Smith, D. R., & Liu, R. (2010). *Metamaterials Theory Design and Applications*. Springer.

Darwin, C. (1859). On the origin of species. John Murray.

Dey, S., & Mittra, R. (1997). Compact microstrip patch antenna. *Microwave and Optical Technology Letters, 13*(1), 12–14. doi:10.1002/(SICI)1098-2760(199609)13:1<12::AID-MOP4>3.0.CO;2-Q

Dixit, L., & Pourush, P. K. (2009), Radiation characteristics of switchable ferrite microstrip array antenna. *Inst. Electr Eng Proc V Microw Antennas Propag., 147*(2), 151–155.

El Maleky, O., Ben Abdelouahab, F., Essaaidi, M., & Ajana, L. (2017). *Reconfigurable T-Shaped Antenna for S-Band Applications*. IEEE Xplore.

Elftouh, H., Touhami, N., Aghoutane, M., El Amrani, S., Tazon, A., & Boussouis, M. (2014). Miniaturized Microstrip Patch Antenna with Defected Ground Structure. *Progress In Electromagnetics Research C, 55*, 25–33.

Er-rebyiy, R., Zbitou, J., Latrach, M., Errkik, A., Tajmouati, A., & Abdellaoui, L. E. L. (2017a). New Design of a Miniature Microstrip Patch Antenna Integrating Defected Ground Structure. *International Journal of Microwave And Optical Technology, 12*(5), 354–359.

Er-rebyiy, R., Zbitou, J., Latrach, M., Errkik, A., Tajmouati, A., & Abdellaoui, L. E. L. (2017b). A Novel Structure of a Reconfigurable Printed Antenna for Wireless Applications. International Review of Electrical Engineering, 12(5), 387-392.

Er-rebyiy, R., Zbitou, J., Latrach, M., Tajmouati, A., Errkik, A., & Abdellaoui, L. E. L. (2017, September). New Miniature Planar Microstrip Antenna Using DGS for ISM applications. *TELKOMNIKA, 15*(3), 1149–1154. doi:10.12928/telkomnika.v15i3.6864

Farzami, F., Forooraghi, K., & Noroozíarab, M. (2012). Miniaturization of a microstrip antenna using a compact and thin magneto-dielectric substrate'. *IEEE Antennas and Wireless Propagation Letters, 10*, 1540–1542. doi:10.1109/LAWP.2011.2181968

Godara. (1997). Performance improvement feasibility and System considerations. *IEEE, 85*(7).

Goldberg, D. E. (1989). *Genetic Algorithms in Search, Optimization & Machine Learning*. Addison Wesley.

Guha, Biswas, & Kumar. (1999). *Printed Antenna Designs Using Defected Ground Structures: A Review of Fundamentals and State-of-the-Art Developments*. Forum for Electromagnetic Research Methods and Application Technologies (FERMAT).

Harrington, R. (1960). Effects of Antenna Size on Gain, Bandwidth, and Efficiency. *J Nat Bur. Stand., 64-D*, 1–12.

Herscovici, N., Osorio, M., & Peixeiro, C. (2001). Minimization of a rectangular patche using genetic algorithms. *Antennas and Propagation Society International Symposium IEEE, 4*, 34 – 37.

Holland, J. H. (1992). Genetic algorithms. *Scientific American, 267*(1), 66–72. doi:10.1038cientificam erican0792-66

Hu, W., Ismail, M. Y., Cahill, R., Encinar, J. A., Fusco, V., Gamble, H. S., ... Rea, S. P. (2007). Liquid-crystal-based reflect array antenna with electronically switchable monopulse patterns. *Electronics Letters, 43*(14), 744. doi:10.1049/el:20071098

Huff, G. H., & Bernhard, J. T. (2006). Integration of packaged RF-MEMS switches with radiation pattern reconfigurable square spiral microstrip antennas. *IEEE Transactions on Antennas and Propagation, 54*(2), 464–469. doi:10.1109/TAP.2005.863409

Hum, S. V., & Xiong, H. Y. (2010). *Analysis and design of a differentially-fed frequency agile microstrip patch antenna. IEEE Transactions on Antennas and Propagation.*

Ihamji, M., Abdelmounim, E., Latrach, M., & Bennis, B. (2017, August). Design of Compact Tri-Band Fractal Antenna for RFID Readers. *Iranian Journal of Electrical and Computer Engineering, 7*(4), 2036–2044.

Kula, J., Psychoudakis, D., Liao, W., Chen, C., Volakis, J., & Halloran, J. (2006). Patch antenna miniaturization using recently available ceramic substrates. *IEEE Antennas & Propagation Magazine, 48*(6), 13–20. doi:10.1109/MAP.2006.323335

Kula, J., Psychoudakis, D., Liao, W. J., Chen, C. C., Volakis, J., & Halloran, J. (2006). Patch antenna miniaturization using recently available ceramic substrates. *IEEE Antennas & Propagation Magazine, 48*(6), 13–20. doi:10.1109/MAP.2006.323335

Liu, L., & Langley, J. R. (2008). Liquid crystal tunable microstrip patch antenna. *Electronics Letters, 44*(20), 1179–1180. doi:10.1049/el:20081995

McLean, J. S. (1994). The radiative properties of electrically-small antennas. IEEE Int Sympos on Electromagnetic Compatibility, 320–324.

McLean, J. S. (1996). A re-examination of the fundamental limits on the radiation Q of electrically small antennas. *IEEE Trans. AP, 44*(5), 672–676. doi:10.1109/8.496253

McVay, J., & Hoorfar, A. (2007). Miniaturization of top-loaded monopole antennas using Peano-curves. *IEEE Radio and Wireless Symp*, 253-256. 10.1109/RWS.2007.351816

Mendez, T., Jardon, J., Aguilar, H., & Iturbide-Sanchez, F. (2006). Application of the defected microstrip structure as a tuning technique for rectangular printed antennas. *Microwave and Optical Technology Letters, 48*, 16–19.

Mosallaei, M., & Sarabandi, K. (2002). Periodic meta-material structures in electromagnetics: Concept, analysis, and applications. *IEEE AP-S International Symposium.* 10.1109/APS.2002.1016103

Murad, N., Esa, Y. M., & Ali, S. (2006). Hilbert Curve Fractal Antenna for RFID Application. *Proceedings of IEEE International RF and Microwave Conference,* 182-186. 10.1109/RFM.2006.331065

Nguyen, H. T., Noghanian, H., & Shafai, L. (2005). Microstrip patch antenna miniaturization by slots loading. *IEEE Antennas and Propagation Society International Symposium*, 215-218. 10.1109/APS.2005.1551525

Oraizi, H., & Hamidkhani, M. (2001). Design of miniaturized multi band microstrip antennas with helical slot patterns on patches. *Microwave Conference (EuMC), 41st European.*

Pendry, J. B., Holden, A. J., Robbins, D. J., & Stewart, W. D. (1999). Magnetism from conductors and enhanced nonlinear phenomena. *IEEE Transactions on Microwave Theory and Techniques*, *47*(11), 2075–2084. doi:10.1109/22.798002

Pendry, J. B., Holden, A. J., Stewart, W. J., & Youngs, I. (1996). Extremely Low Frequency Plasmons in Metallic Mesostructures. *Physical Review Letters*, *76*(25), 4773–4776. doi:10.1103/PhysRevLett.76.4773 PMID:10061377

Piazza, D., Kirsch, N. J., Forenza, A., Heath, R. W., & Dandekar, K. R. (2008). Design and evaluation of a reconfigurable antenna array for MIMO systems. *IEEE Transactions on Antennas and Propagation*, *56*(3), 869–881. doi:10.1109/TAP.2008.916908

Pozar & Daniel. (1992). *The Analysis and Design of Microstrip Antennas and Arrays.* Academic Press.

Pozar & Sanchez. (1998). Magnetic tuning of a microstrip antenna on a ferrite substrate. *Electron Lett*, *24*(12), 729–731.

Pozar, D. M. (1992). Microstrip antennas. *IEEE*, *80*(1), 79-91.

Reha, A., El Amri, A., Benhmammouch, O., & Oulad Said, A. (2014). Fractal Antennas: A Novel Miniaturization Technique for wireless Networks. *Transactions on Networks and Communications*, *2*(5), 165–193. doi:10.14738/tnc.25.566

Roy, Choudhary, Anand, Sarkar, & Bhunia. (2014). A novel approach on miniaturization of microstrip patch antenna with loaded strips, electronics. Communication and Instrumentation (ICECI), International Conference, 1-4, 16-17.

Schaubert. (1983). *Frequency-agile polarization diversity microstrip antennas and frequency scanned arrays*. U.S. Patent 4 367 474.

Smith, D. R., Padilla, W. J., Vier, D. C., Nemat-Nasser, N. C., & Schultz, S. (2000). Composite Medium with Simultaneously Negative Permeability and Permittivity. *Physical Review Letters*, *84*(18), 4184–4187. doi:10.1103/PhysRevLett.84.4184 PMID:10990641

Smith, J. K. (1999). Reconfigurable aperture antenna (RECAP). *DARPA*. Retrieved from www.darpa.mil

Veselago, V. G. (1968). The electrodynamics of substances with simultaneously negative values of ε and μ. *Physics Uspekhi*, *10*(4), 509–514. doi:10.1070/PU1968v010n04ABEH003699

Wakrim, L., Ibnyaich, S., & Hassani, M. M. (2014). Optimization by genetic algorithm of a PIFA antenna parameters for Wifi application. *IEEE Multimedia Computing and Systems International Conference*, 1501 – 1505. 10.1109/ICMCS.2014.6911223

Wang, B. Z., Xiao, S., & Wang, J. (2007). Reconfigurable patch-antenna design for wide band wireless communication systems. *IET Microwaves, Antennas & Propagation*, *1*(2), 414–419. doi:10.1049/iet-map:20050349

Wheeler, H. (1947). Fundamental Limitations of Small Antennas. *Proc. IRE*, 1479–1484.

Wheeler, H. A. (1959). The radian sphere around a small antenna. Proceedings IRE, 1325-1331.

White, C. R., & Rebeiz, G. R. (2009). Single and dual-polarized tunable slot-ring antennas. *IEEE Transactions on Antennas and Propagation*, *57*(1), 19–26. doi:10.1109/TAP.2008.2009664

Zhang, X.-X., & Yang, F. (1998). Study of a slit cut on a microstrip antenna and its applications'. *Microwave and Optical Technology Letters*, *18*(4), 297–300. doi:10.1002/(SICI)1098-2760(199807)18:4<297::AID-MOP14>3.0.CO;2-1

KEY TERMS AND DEFINITIONS

ADS: Advanced design software.
AEP: Antenne électriquement petite.
CAO: Conception assistée par ordinateur.
CST: Computer simulation technology.
DC: Direct current.
DCS: Digital communication system.
DGS: Defected ground structure.
E: Champ électrique.
EM: Electromagnetic.
FDTD: Finite difference in time domain.
FET: Field effect transistor.
FIT: Finite integration technique.
GPS: Global positioning system.
GSM: Global system of mobile communication.
H: Champ magnétique.
IEEE: Institute of Electrical and Electronics Engineers.
ISM.: Industrial, scientific, medical.
J: Densité du courant électrique, [A/m2].
LTE: Long-term evolution.
M: Densité d'un flux de magnétique, [V/m2].
MEMS: Micro-electro-mechanical systems.
MIC: Microwave-integrated circuit.
MIMO: Multiple-input multiple-output.
MPA: Microstrip patch antenna.
MTM: Métamatériaux.
PIN: Positive intrinsic negative.
RC: Radio cognitive.

RF: Radio frequency.
RFID: Radio frequency identification.
RMPA: Rectangular microstrip patch antenna.
SDR: Software-defined radio.
SISO: Single-input single-output.
SRR: Complementary split ring resonators.
TE: Transverse electric.
TEM: Transverse electromagnetic.
TM: Transverse magnetic.
TR: Tuning range.
TS: Total spectrum.
VNA: Vector network analyzer.
Wi-Fi: Wireless fidelity.
WiMAX:: Worldwide interoperability for microwave access.
WLAN: Wireless local area network.
WPAN: Wireless personal area network.

Chapter 8
A Comparison Between the Microstrip and the Co-Planar Wave-Guide Antennas in Ultra-Wide-Band Applications by Using Fractal Geometry

Akram El Hamdouni
University of Hassan 1st, Morocco

Abdelali Tajmouati
University of Hassan 1st, Morocco

Hamid Bennis
University of Hassan 1st, Morocco

Mohamed Latrach
ESEO, France

ABSTRACT

This chapter describes a comparison study between the techniques of coplanar waveguide (CPW) and microstrip line applied to antenna in the ultra-wide band by analyzing the different parameters achieved into simulation and fabrication. Fractal geometry has been chosen to design the radiating patch of both types of antennas by including two electromagnetic solvers based on two different numerical methods: CST of microwave studio and ADS. The parameters S11, current density, gain, and the radiation pattern have been achieved into simulation and measurement in the frequency range 3.1 – 10.6GHz released by the Federal Communications Commission (FCC) as a commercial UWB. The photolithographic technique, the network analyzer, and the anechoic chamber have been involved to perform the fabrication and the measurement of the validated microstrip and CPW antennas.

DOI: 10.4018/978-1-5225-7539-9.ch008

INTRODUCTION

With the enormous progress of wireless communications, a growth demand of Ultra-Wideband (UWB) applications appeared, which push researchers to design the UWB systems due to a number of advantages that make them attractive such as low complexity, low cost and a good time domain resolution allowing for location and tracking applications. In the past twenty years the UWB was used only in Radars, sensing and military communications, therefore UWB and since February 2002 when the Federal Communications Commission (FCC) released the regulation (2002 a, b) the UWB become dramatically involved in the data communications. The famous passive devices that can be used in this kind of systems are the Microstrip antenna (MSA) and the Coplanar waveguide Feed line antenna (CPW). Those categories of antennas will be deeply detailed in this chapter in order to compare the advantages and the disadvantages between them and to identify the appropriate applications for each type of antenna.

Firstly the MSA in its simplest definition considered as a radiating conductor element on the top side of a dielectric and a ground plane conductor on the bottom side, the radiator can be a geometric shape, such as a square, a circular, a triangular, a semi-circular, a sectorial and an annular ring. The first concept of this kind of patch antenna appeared in 1953 by Deschamps, after that Munson and Howell designed the first practical MSA antenna in 1970. The MSA antenna has many advantages such as low profile, light weight, small size and ease of manufacturing, which make it suitable for personal and mobile communications, therefore the MSA suffer from the narrow bandwidths by nature, therefore many techniques are used in order to improve the bandwidth some of them is to increase the thickness or decrease the dielectric constant of the substrate, use a partial ground plane and modify the shape of the radiating element by using the fractal geometry for example, which is now more popular in the MSA.

Secondly, the CPW antenna defined as a planar transmission line located symmetrically or asymmetrically in between two grounds collocated in the same side of the dielectric material with the fed line and the radiator, which is totally opposite of the MSA antenna concept. The CPW technique has been appeared in 1969 by Cheng P. Wen in RCA Sarnoff Laboratories in New Jersey. The supports of CPW can be TE Mode in higher frequencies, therefore are shown slightly TEM Mode in lower frequencies, also the dielectric thickness has no impact factor on the characteristic impedance of the CPW antenna however the CPW antenna depends to the width and the length of the substrate element. Regarding the benefits, the CPW antenna presents a low dispersion, a facility of manufacturing due the exiting of all antenna elements in the same side, which help the CPW antenna to be integrate easily in the passive and active RF circuits. But the CPW is not suitable candidate in terms of circuit size compare to the MSA, so one of the methods to reduce the antenna size is to etch the patch by using the fractal technique.

The fractal geometry defined as rough or fragmented shape that can subdivided to copies of the original form with reduced sizes, this geometry has two principal properties which are self-similarity and space filling, the first one is used in MSA in order to increase the input bandwidth matching in some specific wavelengths therefore the other one is used in CPW antenna to get miniaturization.

This chapter, in the beginning, detailed the definition of the ultra-wide band (UWB), the microstrip transmission line antenna, the coplanar waveguide antenna and the fractal geometry, than depicted a study of an original CPW antenna validated into simulation and measurement by the way described the conception of a microstrip antenna by using the computed and measured parameters. At the end a brief comparison between the two kind of antennas will presented in summarized table based on the studied parameters for each antenna in order to identify the appropriate applications for the microstrip antenna and the CPW antenna.

UWB DEFINITION

In the past the UWB has been used in some limited applications such as the radars, the sensing and the military communications. However UWB and since February 2002 when the Federal Communications Commission (FCC) released the regulation (2002 a, b), the UWB become massively used in the data communications.

The UWB systems remains the most attractive for the customer communications due to a several advantages such as; low complexity, low cost, resistance to the multipath and jamming, noise-like signal and good time domain resolution allowing for location and tracking applications.

Many terms which are related to Ultra-wideband such as impulse short-pulse, super wideband, fast frequency chirp and mono - pulse.

The radio communication systems and radars using the impulse on transmission are based on short pulse which can be considered as ultra-wideband spectrum. This technique of communication can be applied as modulation method based on the pulses because as known the pulse position modulations (PPM) is the introduction of the data modulation.

Regarding the influence of the noise on the UWB signal, the operations of interception and detection can be performed hardly due to the power spectral density of the UWB signal which is low however the interference between the UWB signals and the narrow band radio systems is negligible. So it will be recommended to the national and international authorities of regulation to allow the uses the of UWB

The carrier-less baseband transmission is can considered as the behavior of the time-modulated (TM) impulse radio signal. The main specification, which can be considered as ultra-wideband method, differentiating the impulse radar and the impulse radio transmission from narrow-band applications, direct sequence (DS), spread spectrum (SS) and multi-carrier (MC) transmissions is the absence of carrier.

Other techniques able to generate the UWB signals are the fast slewing chirps and the exponentially damped sine waves.

The UWB become suitable for the customer of radio communications applications, due to several advantages such as; low profile, low cost, low complexity, noise-like signal, resistance to hard multipath and jamming and good time domain resolution allowing for location and tracking applications.

The low profile, low cost and low complexity offered by the UWB systems came basically from the nature of the UWB signal transmission, which based on baseband propagation. In the opposite of the radio applications, a very short time domain pulse has been generated by the UWB transmitter, this kind of pulse can be transmitted without any mixing stage with the radio frequency (RF) signal. As known the frequency modulation or the RF mixing stage is performed on a baseband signal which can be injected into a carrier frequency or can be translated to a frequency suitable for the far field propagation. By nature the wider band allows to UWB signal the possibility to propagate without the necessary to the amplification or the use the RF mixing with the carrier frequencies. Also in the reception the inversion operation of modulation is not applied which means an omission of the local oscillator and by the bay the delay complex and tracking phase loops are removed. So the UWB systems can be integrated in the circuits with low cost and low power.

The frequency allocation remains the most important issue in the UWB radio communication. In USA some companies tried to escape the restrictions established by the FCC to use the UWB in the radio systems. To perform this idea these companies created a group called the Ultra-Wideband Working Group (UWBWG) in order to push the FCC to remove the restrictions. The same story has been done in

Europe with the ETSI (European Telecommunications Standards Institute) or ITU (International Telecommunications Union) to fix the the restriction related the frequency allocation and the radio interference.

AMERICAN UWB REGULATION

The first report related to regulation of the UWB application it was on 2002 by the Federal Communications Commission FCC. An important effort has been provided to push the FCC to publish the UWB standard under the FCC Part 15 regulation limitations and the UWB applications to be free licensed, below some recommendations issued in the FCC Part 15:

- **47 C.F.R. 15.1:** No frequency license and no frequency coordination are recommended for the operation of radio frequency devices classes of UWB.
- **47 C.F.R. 15.5:** Ensure that the radio interference from the UWB systems to the other systems is probably low.
- **47 C.F.R. 15.209:** It permits to the intentional radiators to operate in the frequency range allowed the signal emissions and within the required limits.
- **47 C.F.R. 15.205:** It not permits to use the sensitive or safety-related frequency spectrums from UWB intentional radiator.

The notice of inquiry issued by the FCC in the 1998s, had as goal to depict the complain of the very low transmission power levels related to the existing UWB systems used in the civilian radio communications. The majority of the claims are related to the interferences appeared in the systems operating with the restricted frequency ranges such as TV broadcast bands and frequency bands reserved for radio astronomy and GPS. The Federal Aviation Administration (FAA) complained also the existing of interferences in the aeronautical safety systems and the direction of the UWB transmitters. The FCC Notice of Inquiry exist also on the Internet (Ultra-Wideband Working group, 1998, 1999, 2004). However a large possibility to utilize the UWB technology for innovative products has been proved by several organizations. At that time no definition has been proposed by the FCC for the UWB civilian applications.

The first definition of UWB signal has been proposed by The Defense Advanced Research Projects Agency (DARPA) by using the fractional bandwidth Bf of the UWB signal given by Taylor formula in the 1995s by $B_f=2(f_H - f_l)/(f_H + f_L)$, where respectively f_L is the lower and f_H is the higher 3 dB point in a spectrum. So a signal is identified as UWB when B_f is greater than 0.25.

CURRENT UWB DEFINITION

The FCC has been released in February 2002 the commercial version of the UWB technology by identifying the radiation limitations for UWB signals. The final report published during April 2002 by the FCC as Federal Communications Commission, 2002 a, b which contains the four different categories proposed for UWB applications and the radiations masks for the four categories.

The current definition proposed by FCC has been also based on the above Tylor formula. The signal is seen as UWB when Bf is greater than 0.20 decreased by 0.05 compare to the definition given by The

Defense Advanced Research Projects Agency (DARPA). A second definition of UWB has been issued by FCC based on the bandwidth of the signal in order to be equal or more than 500MHz.

EUROPEAN UWB REGULATION

The European UWB standard has been based on the technical report which describes the impact of the UWB signal on the up running systems. So the European authority was prudent during the release the UWB regulation compare to the American authority by obliging the new UWB applications to without any influence on the existing systems. Even with that the vision of FCC in the regulation of the UWB impact the European organizations, so the responsible of the short-range standardization is the European Conference of Postal and Telecommunications (CEPT) and the working group is Ultra- Wideband Working Group, 1999 under the name of CEPT/ ERC/ REC 70- 03. The ETSI/CEPT in generally use the same way followed by the FCC in the UWB regulation but keeping in the same time the European specifications.

The Figure 1 presents the limits of the spectral mask proposed by the European standard compared to the masks proposed by the FCC. The upper graphs present the masks of the data communication systems for indoor and outdoor applications, while the lower graphs present the masks of radar and sensing published by the FCC standardization.

The FCC part 15 recommendations have been respected in term of the maximum average power spectral density published under the Federal Communications Commission,2004. The European standardization follows the recommendations of working groups for UWB called ERM/TG31A in order to cover UWB generic and ERM/TG31B in order to cover the automotive applications operating in the higher frequencies.

UWB ANTENNAS

The antenna behaves as filter for the transmitted signal operating in the UWB by allowing to the radiated components of the signal to pass. Also the antenna acts as differentiator of the transmitter and receiver. This functionality has an impact on the duration of the pulse in transmit and receive. So the time of resolution can be decreased by the extending of the duration. The antenna influences deeply the UWB signal compares to the narrow band signal, due the huge frequency spectrum of the UWB signal.

The Conventionally, the antenna operating in the UWB when the frequency band demand is within 6:1 or more according to Taylor in the 1995s, which can be explained by the highest frequency that should be equal or more than six times the lower frequency in the spectrum (Low, Cheong & Law, 2005.).

The ringing is one the problem that can be appeared during the excitation of the antenna in the very short domain pulse. In the Figure 2 is presented the response of an antenna, by using the Bessel formula to express the ringing effect (Rahim, Aziz, & Abdullah, 2005; Levy, Kumar & Dinh, 2013).

The resistive antenna with low Q-values remain the best solution to avoid the ringing effect, the unwanted signal component are generated by the resistive loading. To increase the antenna bandwidth it will be necessary to have a low Q-value which means that the bandwidth is inversely proportional to Q-value. The Q- value is given by the expression: $Q = f_0/(f_H-f_L)$ where f_0, f_H, f_L are respectively the centre, the upper and lower frequencies at 3 dB.

Figure 1. UWB radiation mask defined by FCC and the existing CEPT proposal

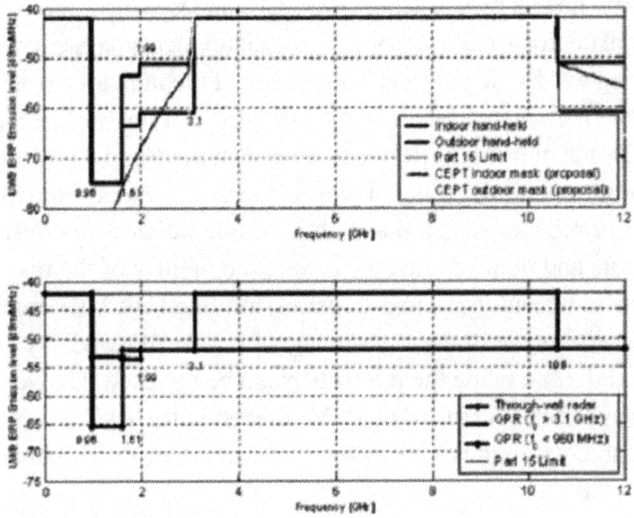

MICROSTRIP ANTENNA (MSA)

In the 19th century the transmission line has been obtained by using the coaxial cable or what we called the parallel two wires. This concept it was behind the realization of almost radio frequency (RF) and microwave devices, however this construction remains mechanical and request too much effort. In the 20th century the printed circuit board has been involved massively in the development of the transmission line by the way the production of the radio frequency and microwave components become easily than before. The technology planar applied to design of the printed circuits similar to the coaxial cable was

Figure 2. The response of an antenna to impulse excitation showing the ringing effect

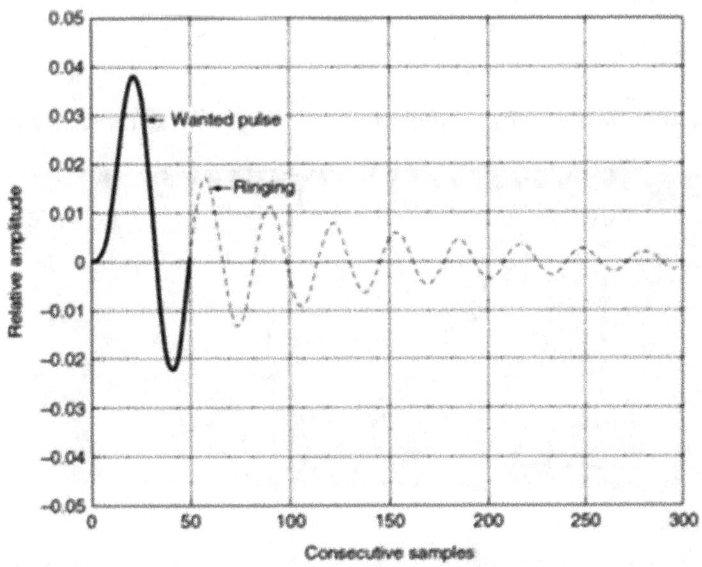

called stripline with a ground plane image in order to realize a virtual conductor, which is called later the microstrip transmission line (James & Hall, 1989; Gronau & Wolff, 1986).

In the 1950s the original microstrip geometry able to radiate electromagnetic waves has been developed while in the 1952s by Grieg and Englemann the connection of the radiator to the microstrip transmission has been established. The first antenna realized by using the microstrip line with a matched radiating patch has been done by Deschamps in the 1953s as shown in the Figure 3(Kumar & Ray, 2003).

Early the radiators connected to the transmission lines have been developed just in the laboratories without any commercial printed antennas, but with controlled dielectric constants, so the conception of the microstrip resonators and their efficiency become the inquiry of the majority of designers in the propagation field. The stripline and the planar transmission line have been more interesting due to the ease analysis of the transverse mode of the electromagnetic wave allowed. By analogy the stripline can be matched with the coaxial cable while the microstrip can be matched the two wires transmission line. The antecedents merits have been confirmed by R. M. Barrett in the 1955s from the stripline, microstrip system and the coaxial cable, two wires transmission line system (Mahatthanajatuphat, Saleekaw, Akkaraekthalin & Krairiksh, 2009).

From the 1950s, the microstrip remains not popular due the above reasons, but the development has been continued by Wheeler and Purcel in the 1960s in terms of analysis and design methods of the microstrip transmission line.

Previously, it was known that only 50% og the energy circulated in the microstrip resonator can be radiated until the 1969s when Denlinger has been designed the rectangular and circular microstrip radiators with highest efficiency in term of radiation by describing the mechanism based on the distribution of power density in the discontinuities at each end of a truncated microstrip transmission line. The multiple of a half wavelength can be applied to separate the two discontinuities which help to depict the complete microstrip radiator. It will be necessary to mentioned that the percentage of the power radiated, compare to the total input power, with the thickness of the dielectric material are proportionally increased. So Denlinger has been proved that slightly 70% of the input power can be radiated into the space in case of rectangular microstrip resonator, while 75% of the input power can be radiated into the space in case of circular microstrip resonator. The resonant modes of the circular radiator have been detailed later by Watkins (Hildebrand & McNamara, 1995).

Figure 3. Original ntework designed by Deshamps in 1953 fed with microstrip transmission line

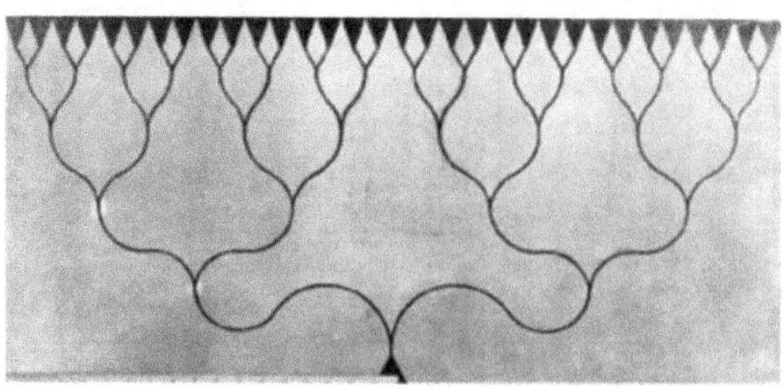

The final conception of the microstrip antenna has been validated in the 1970s when it became involved in the aerospace applications, like missiles and spacecraft, which push the researchers to deeply investigate this type of antenna. In the 1972s Howell achieve the basic rectangular radiating patch with the microstrip transmission line with non-negligible radiation loss known in the time being as the microstrip antenna. So it was unbelievable that this type of antennas can be used to radiate the waves in the space with approximatively 90% of efficiency. The only handicap of the microstrip antenna in that time it was the narrow band offered which limited the possibilities of applications. By 1981, the microstrip antenna became more and more studied and it development became the interest of the majority of antennas designers (Bancroft, 2009; Rama Rao, Alekhya & Sridevi, 2013; Denlinger, 1969).

In the time being the design of the microstip antenna has been dramatically developed by using several techniques to enhance its performances. The Figure 4 presents the conventional microstrip antenna geometry based on the conductor patch on the top of the dielectric substrate which can take several shapes, but generally rectangular or circular and the ground plane on the bottom of the dielectric element.

The resonant length of the rectangular microstrip antenna is conventionally calculated by using the one half wavelength and the substrate effective dielectric constant. Based on the last introduction, the analysis methods have as target to determine an approach to calculate the resonant resistance of a rectangular microstrip radiator. So the transmission line model has been applied as the first method able to calculate the approximate value of the resonant resistance, this model has been introduced by Munson in order to design a simplest microstrip antenna, which is considered as not sufficient model due to the presence of multiple resonant modes of the waves. In the 1970s a new model, based on the resonant cavity of the rectangular microstrip antenna, has been involved as an exact method of analysis. In the 1980s the method of moment (MoM) has been introduced the first numerical method to perform a computed design of the microstrip antenna. So with the improvement of the performances of the computers by increasing the memory and establishing the speedy CPUs, other techniques of simulation have been introduced based on the finite difference time domain (FDTD) method and finite element method (FEM) (Guy, Vandenbosch & Vasylchenko, 2011; Mosig & Gardiol, 1983; Mosig & Gardiol, 1985; Mosig, 1988).

The microstrip antenna remains the suitable type of printed antennas due to several advantages such as low cost of manufacturing, ease implementation in the curved surface of vehicle and product, high resistance to shock and vibration, large designed model for linear or circular polarization, many options of gain in range between 2.5 to 10dBi, ease integration with circuits realized by the technique of microstrip without any extra steps, low profile and small thickness antenna substrate. However the microstrip antenna suffers from some disadvantages such as the narrow bandwidth from 5% to 10% with

Figure 4. The geometry of the microstrip antenna

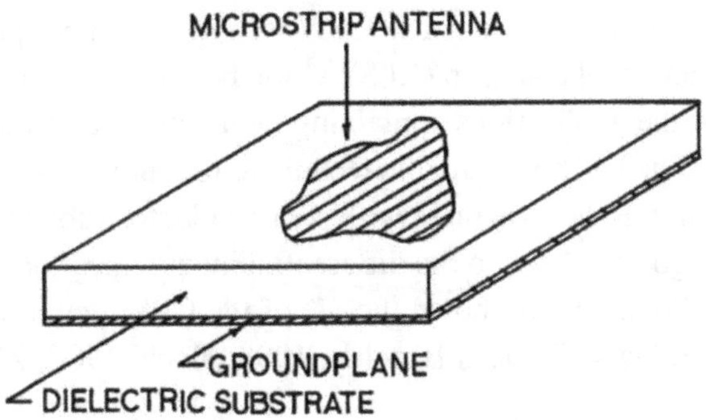

voltage standing wave ratio (VSWR) variated between 2 and 1, poor antenna efficiency due the large losse in the dielectric and the conductor and sensitivity to weather criteria like the temperature and the humidity (Howell, 1972; Katehi & Alexopoulos, 1984).

The microstrip antennas have been introduced in a large number of commercial applications such as the Global Positioning System (GPS), Zigbee, Bluetooth, WiMax, WiFi applications, 802.11a,b,g, the most useful microstrip antenna is the one with the rectangular radiating patch. The massive demand of the antennas it was in the GPS application spatially the one used in the vehicles and marines tracking. The rectangular microstrip antennas are the most involved which have been changed in order to operate at 1.575 GHz at right-hand circular polarization (RHCP). Some commercial patch antennas used the ceramic as substrate with high dielectric constant of $\varepsilon r = 6, 20, 36$ to increase the miniaturization of the microstrip antenna in order to be conformal with the proposed applications. The integration of the patches with circuits can be done with low noise amplifiers. The rectangular micrsotrip antenna has been applied in the Bluetooth devices with right-hand circular polarization (RHCP) at the frequency 2.4 GHz.

The notched annular microstrip antenna at the frequency 2.338 GHz with the printed monopole permit to have the left-hand circular polarization (LHCP). This combination provides a strict radiation pattern able to be used in the Satellite Digital Audio Radio Services (SDARS) applied as alternative to AM and FM FM commercial broadcasts in automobiles (Lo, Solomon, & Richards, 1979).

The WLAN known as wireless local area networks produce a short-range and high speed data in order to connect devices, like laptop computer, and the access points in indoor wireless area of 100 to 300 feet and 2000 feet in outdoor. The data circulated in the WLAN use the IEEE Standards 802.11 a, b, g. the most popular WLANs use the unlicensed 2.4 GHz band (802.11b and 802.11g). The 802.11a standard uses the 5 GHz unlicensed frequency band. So a microstrip diplexer has integrated with the Multiband printed antennas that are integrated into ceiling tiles in order to ensure the combine of the signal from Global System for Mobile communication (GSM) cell phones (860 MHz band), personal communications services (PCS) cell phones (1.92 GHz band), and 802.11a WLAN service (2.4 GHz band) provided by two integrated microstrip dipoles.

COPLANAR WAVEGUIDE (CPW) ANTENNA

The term of coplanar waveguide (CPW) classify a group of the planar microwave transmission devices characterized by various arrays of conductors designed to be in the same side of the dielectric material. So the CPW and the microstrip structures are produced in opposite way, because the micorstrip required a singular ground plane on the bottom side of the dielectric material, while in CPW the ground planes principally have chosen adjacent to the active conductor, due to this configuration the signal circulate laterally in the overall surface of the antenna. The CPW technique can be splited to two different structures which are the slotline and coplanar strip (CPS); while both of the two structures are balanced line and have been integrated in the applications containing elements able to radiate or to transit the electromagnetic waves. The slotline has been suitable technique for antenna and ferrite circuits, while The CPS has been involved mainly in the structures with semiconductor substrates, for the transmission of digital signal chosen with high speed due to the differential interconnection or differential line offrered by the CPS technique. The Figure 5 has shown the 3D of the CPW devices (Joseph, Paul, Mridula & Mohanan, 2013; Katsibas, Balanis, Tirkas a Birtcher, 1998; Koch, 1906; Zbitou, Latrach, & Toutain, 2006; Yoo & Chang, 1992).

Figure 5. Structure of: (a) slotline, and; (b) differential line or CPS

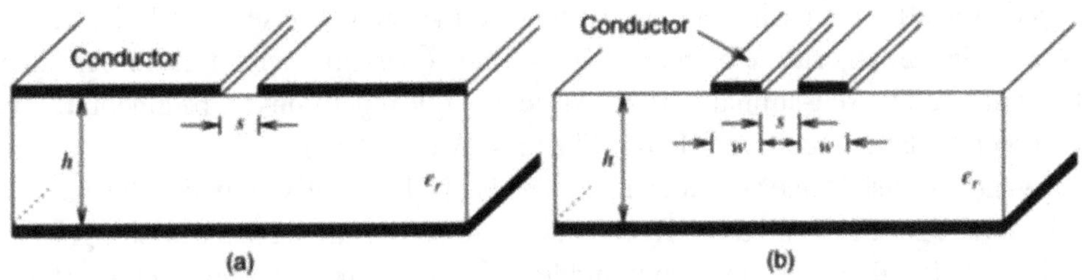

The CPW has been introduced in the 1969 by C.P. Wen, however no relationship between the abbreviation of coplanar waveguide which is CPW and the name the inventor C.P. Wen. The original CPW structure designed with symmetrical strip width and assumed to equal the longitudinal gap separation has been presented in the Figure 6(a).

Theoretically the ground plane of the CPW, in the same side of the conductor, is infinity considered. The finite-ground CPW (FGCPW) line is one of the variant of the CPW structure as shown in the Figure 6(b), which permits to have a lower coupling though the ground planes which adjacent to the transmission line.

C.P Wen has been developed the formulations for CPW based on an infinite substrate in term of dimension which created a problem in the fabrication and integration of the structure in the circuits. The CPW technique presents large advantages such as lower cost of manufacturing, ease ground planes design in the same side of the transmission line, lower dispersion in small shapes, minimal losses in the radiation, higher directivities for couplers and low dependence to the thickness of the substrate in the fabrication with the photolithographic method (Lee, Yook, & Park, 2002; Li-na, Shu, Run-nan, Guan-long & Wen-bin, 2011; McSpadden, Yoo & Chang, 1992).

FRACTAL GEOMETRY

The fractal geometry defined as rough or fragmented shape that can be subdivided to copies of the original form with reduced sizes, therefore the geometry in different scales looks in identic view. B.Mandelbrot was introduced the first mathematical shape called 'Fractal Curve' in order to define a family of geometrical which remains without description in the standard Euclidean geometry. The self-similarity

Figure 6. Structure of a CPW: (a) conventional, and; (b) FGCPW

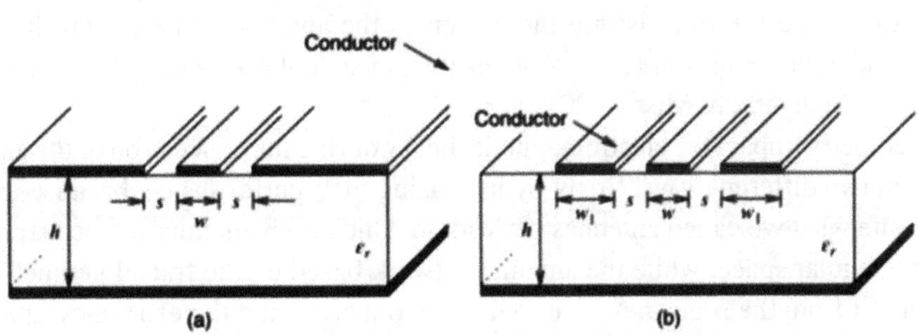

behavior can be considered as the main property of the fractal curve which give to the object the ability to be unchanged even after increasing or contracting its dimension. The self-similarity of an object can be computed by using an algorithm, by repeating the same operation many times to get small or large scales until obtain a perfect self-similarity of the structure, this repetition can be algebraic, symbolic, or geometric (Werner & Gangul, 2003; Li, Jiang, Cheng & Wang, 2013).

Proposed by the Swedish Helge von Koch in the 1904s, the Koch curve remains the classical example of the fractal curve. The idea behind the Koch curve is to start with segment or straight line which is called the 'initiator' after that is create from it middle third an equilateral triangle called the 'generator', this process can repeated for all straight segments infinitely as shown in the Figure 7(a). Due to the first iteration the length of the fractured line is given by $(4/3)^2$. By iterating n times the generator of Koch we have as result the Koch snowflake presented in the Figure 7(b), with a perimeter given by the expression $(4/3)^{n}$ (Mandelbrot, 1975; Manimegalai, Raju & Abhaikumar, 2009) .The perimeter can be more longer by applying the Koch fractal generator in some geometry without limitation such the shapes in the Figure 7(c) and (d).

The length expression L(r) of the smooth curve can be written by using the N which is the number of the straight line segments of and the length r need to step along the curve from one end to the other end, so the formula of the length is L(r) = N.r, L(r) converge to a finite limit when r approaches to zero, which equal approximatively to the length of the curves. However for the fractal geometry when r converges to zero the expression N.r diverges infinitely, due to that the curves shown more complicated. A critical exponent called the Hausdorff dimension has been introduced to stabilize the divergence of the product N.r by using the formula $N.r^{DH}$ the expression of D_H is given for nth generation of the Koch curve or the snowflake construction by:

$$\mathbf{D}_{\mathrm{H}} = \lim_{r \to 0} \frac{\log(N)}{\log(1/r)}$$

If $\mathbf{r} = \frac{\mathrm{r0}}{3^{\mathrm{n}}}$ then the number N of pieces will be proportional to 4^{n} and D_H will be given by

$$\mathbf{D}_{\mathrm{H}} = \frac{\log(4)}{\log(3)} = 1.26\ldots$$

$D_H = 1$ for the smooth curve, $D_H = 2$ for the smooth surface, and $1 < D_H < 2$ for the Koch and other fractal geometry. The dimension remains the important behavior of the fractal geometries and represents the main cause to involve the fractals into the design of the antennas. The fractal dimension impact significantly the radiation pattern in terms of intensity and spatial (Dhrgham, Naji, Jaber; Aziz, Raad, & Fyath, n.d.; Jena, Mangaraj & Mishra, 2013).

The fractal geometry impact the electromagnetic theory during the conception of the antennas and the antennas arrays in two different ways, firstly by increasing their performance, because in the array, the antennas composite what we called antennas network in which each antenna can be arranged randomly or according to a regular space, while the antenna network based on the fractal geometry to juxtapose the antennas benefit from the robustness due to the random array and the efficiency due to the regular array. Secondly, due to a non-uniform of the radiation pattern of the fractal antennas array. The sum of the radiated power density of each element in the antennas network equal to a total at chosen point with a vectored amplitude and phase therefore in antennas array performed by using the fractal it will enough to study the properties of only one fractal antenna (Haider, Al Sabbagh, Farhad, Mahmood, Edwards & Brister, 2012; Homayoon & Shahram, 2012; Lizzi & Oliveri, 2011; HUSSEINI, RAMADAN, EL HAJJ & KABALAN, 2009).

The energy radiated in the far field is given by:

$$\mathbf{R}\left(\mathbf{x},\mathbf{y},\mathbf{z}\right) \propto \int \mathbf{E}^2\mathbf{dt}$$

For single dipole, the array factor can be normalized by maximum and R can be written according to the below formula:

$$\mathbf{R}_0 \simeq \frac{\beta^2 \boldsymbol{I}_0 \boldsymbol{A}}{4\left(1-\beta \boldsymbol{a}\right)^2 \boldsymbol{h}}$$

where

$$\beta = \frac{\mathbf{v}}{\mathbf{c}}$$

where v is the celerity of the current thought the wire of the dipole, h is the height of the detector and Δr equal to the difference of distance between the bottom and the top point of the dipole.

GEOMETRY AND DESIGN OF THE CPW ANTENNA

The radiating patch of the proposed CPW antenna has been chosen circular with a fractal geometry performed by merging the cross shapes with circulars iterated four times and in different sizes with D1, D2 and D3 are respectively, from the grand to the small, the diameters of the circular shapes inside the radiator as shown in the Figure 8 (a), (b), (c) and (d).

The transmission line has been designed in the same side of the substrate with two partial ground plane in order to respect the CPW technique and to be matched at 50Ω.

The dielectric substrate is an Epoxy FR4 with the below features:

- **Substrate Thickness:** h=1.58mm.
- **Relative Dielectric Permittivity:** Ɛr=4.4.
- **Dielectric Loss:** Tan (δ) =0.025.
- **Metallic Thickness:** t=35μm.

Figure 7. (a)Koch fractal; (b)Koch snowflake, and; (c) Infinite Koch fractal

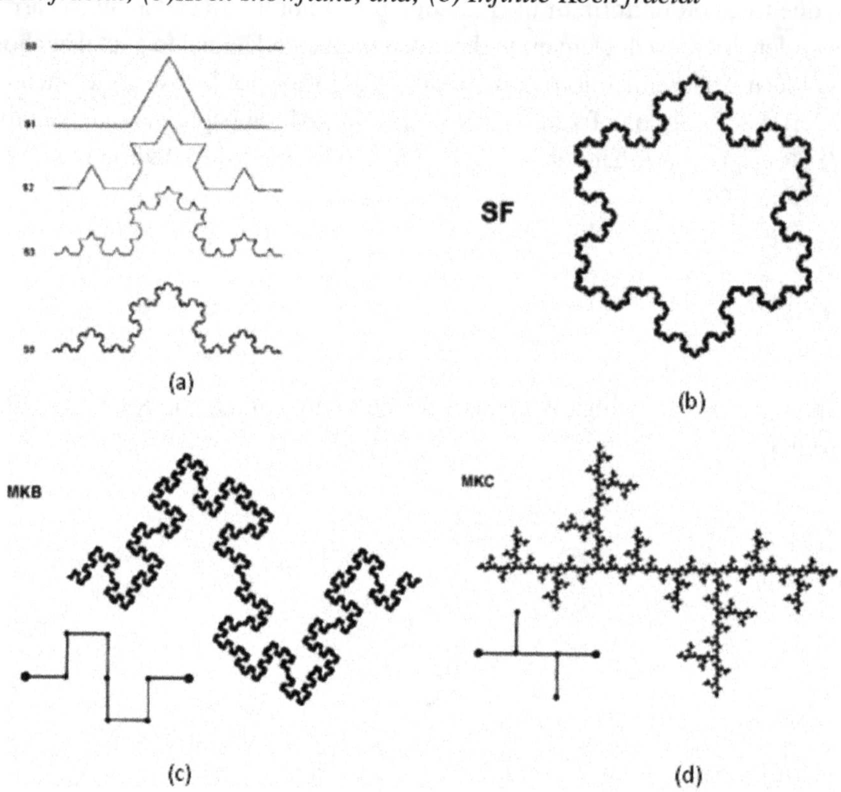

The final CPW antenna is shown in the Figure 9 with different parameters of the radiating patch, the feed line, the ground planes and the substrate.

The Table 1 contains the values different parameters of the designed antenna which are optimized by using the electromagnetic solver.

The simulation of the CPW structure has been performed by using mainly the electromagnetic simulator CST of Microwave Studio based on The finite-difference time-domain (FDTD) method included in it Transient solver, which can be applied to simulate the arbitrary material and shape of antennas. To confirm the design of the CPW antenna studied in this chapter a second simulator has been involved

Figure 8. The geometry of the CPW antenna (a) first iteration; (b) second iteration; (c) third iteration and; (d) fourth iteration

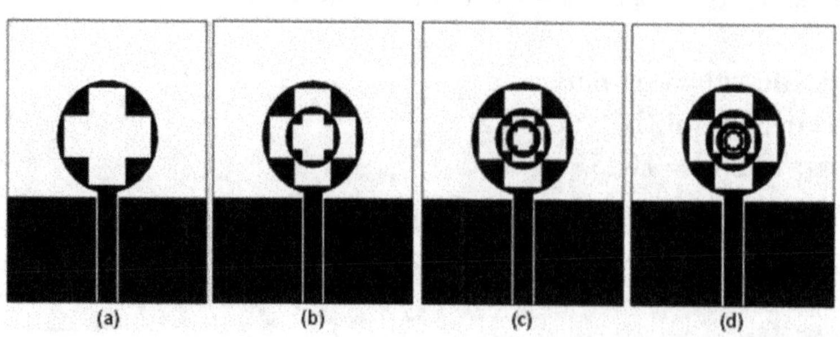

Figure 9. The Final geometry of the CPW antenna

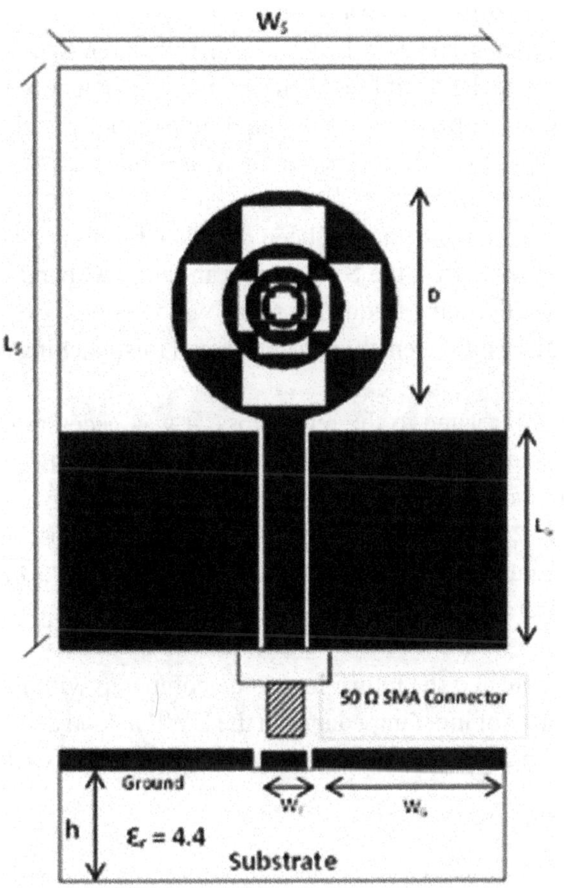

Table 1. The optimized parameters of the CPW antenna

Parameter	Value (mm)
L_S	43
W_S	34
L_G	16
W_G	15
W_F	3
D	16.4
D_1	7.2
D_2	3.81
D_3	2.23

based on the method of the moment (MoM) which is ADS of Agilent by defining the grid density of the structure and considering an infinite dielectric layer.

The return loss has been computed firstly by CST in different iterations as shown in the Figure 10 in order to determine the impact of the fractal geometry on the coefficient of reflection.

According to the graphs an improvement of the input impedance matching appeared from iteration to the other to have finally in the fourth iteration a CPW antenna with S_{11} less than -10db in between the frequency range 3.1 – 10.6 GHz.

The return loss of the proposed antenna has been calculated in the two different mesh densities 10 and 70 lines per wavelength to confirm the S_{11} behavior in the FCC band. The Figure 11 presents the two graphs plotted together in the same frequency range with a good agreement. The S_{11} simulated by using 70 lines per wavelength as mesh density present a small mismatching in terms of input impedance in the frequency 9.5 GHz.

To complete the simulation related to the return loss, it was necessary to check the impact of the dielectric thickness on the input impedance matching at 50Ω in the UWB. In order to be in concordance with the reality two different substrate thickness of the Epoxy FR4 have been chosen which are 1.58 mm and 0.8mm, the Figure 12 presents the two graphs of the return loss in the 2 – 12GHz band with the values of the antenna simulated with thickness of 0.8mm above -10dB in the frequency range 4.5 - 6.5GHz, therefore the substrate epoxy FR4 with dielectric thickness equal to 1.58 mm will be chosen during the manufacturing.

ADS has been used to simulated the return loss to be compared with the one computed by CST as shown in the Figure 13. In terms of input impedance in the UWB a good agreement exist between the two graphs of S_{11}, with small mismatching in the frequency 7GHz and 9GHz of the graph simulated by ADS.

Figure 10. The S_{11} versus the frequency of the fourth iterations of the designed CPW fractal antenna

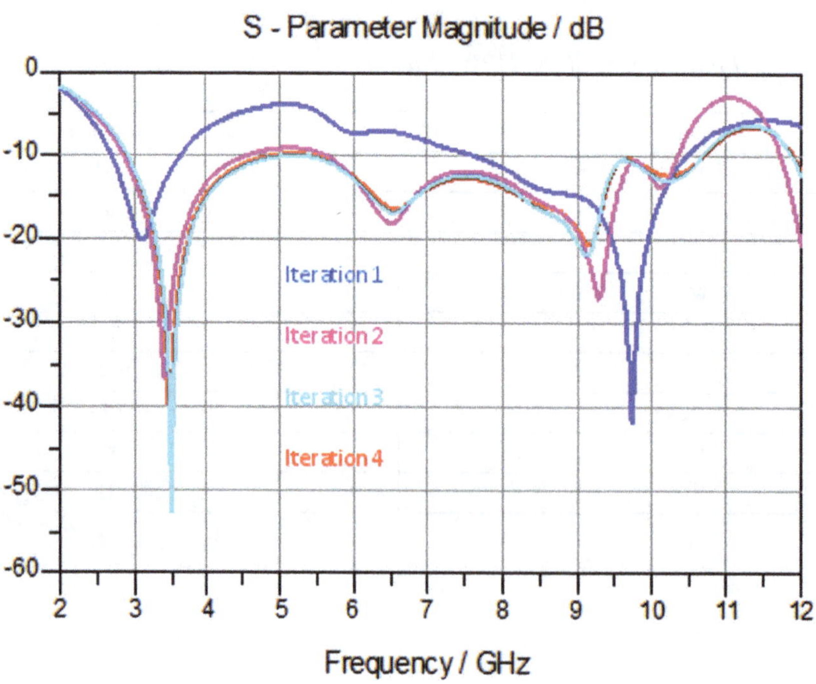

Figure 11. The S_{11} versus the frequency in mesh density 10 and 70 lines per wavelength

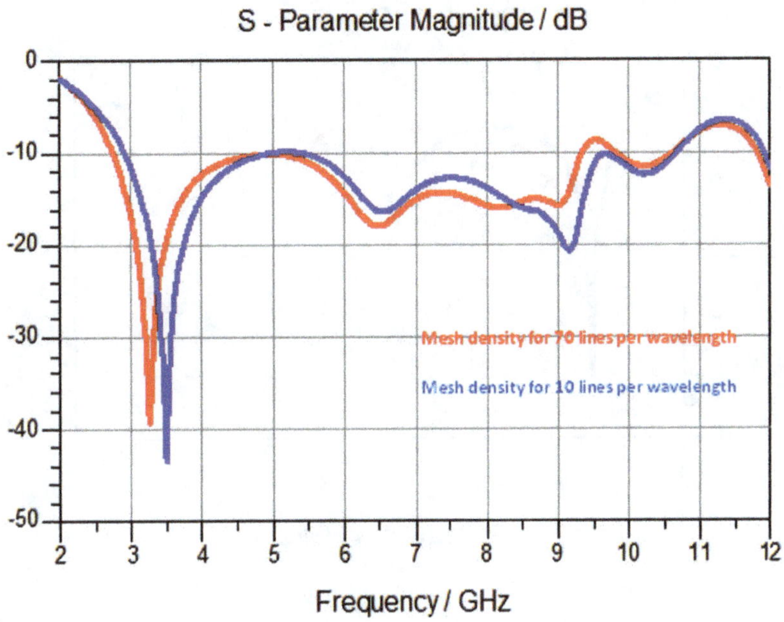

Figure 12. The S_{11} versus the frequency simulated with two dielectric constants 0.8mm and 1.58 mm

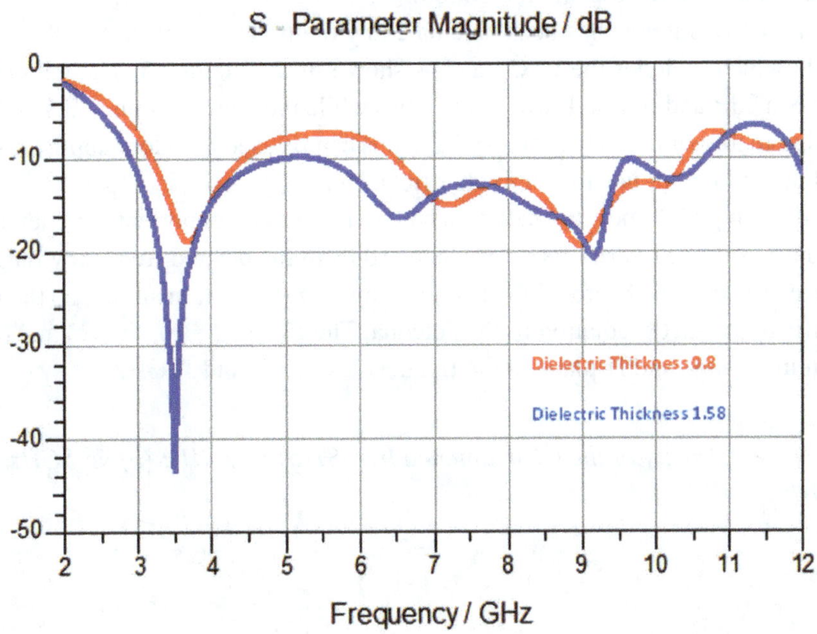

The current density of the CPW antenna has been achieved into simulation by CST in four different frequencies covering the FCC band which are 3, 5, 7 and 10 GHz as presented in the Figure 14(a), (b), (c) and (d).

Figure 13. The S_{11} versus the frequency simulated by CST and ADS

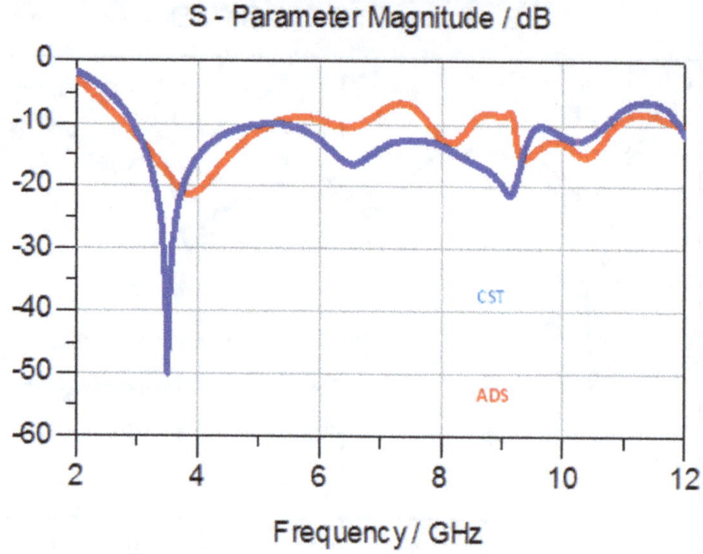

The current distribution appeared mainly in between the peripheries of the transmission line and the two partial ground planes in all frequencies, however a low density of the current circulate on the radiating patch especially in the top and the bottom.

The gain of the CPW antenna has been calculated by the transient solver of CST in the frequency range 2 – 12GHz which includes the FCC band as shown in the Figure 15. The gain of the proposed structure is between 2dB and 3dB in lower frequencies while the gain is around 4dB and 5dB in higher frequencies, facing to this situation the achieved antenna can be used in indoor applications in the lower frequencies and in outdoor applications in higher frequencies.

To complete the computed study, the radiation pattern remains an important parameter in the analysis of antenna, because even the return loss is less than -10dB in the desired frequency range, the antenna can behave as a resistance and absorb all the energy coming from the transmitter, so the radiation is the only way to confirm the waves radiation of the antenna. The Figure 16 (a), (b), (c) and (d) presents the radiation pattern in E-plane and H-plane in the frequencies 3, 5, 7 and 10GHz.

Figure 14. The current density of the CPW antenna by CST(a) @ 3GHz; (b) @ 5GHz; (c) @ 7 GHz, and; (d) @ 10GHz

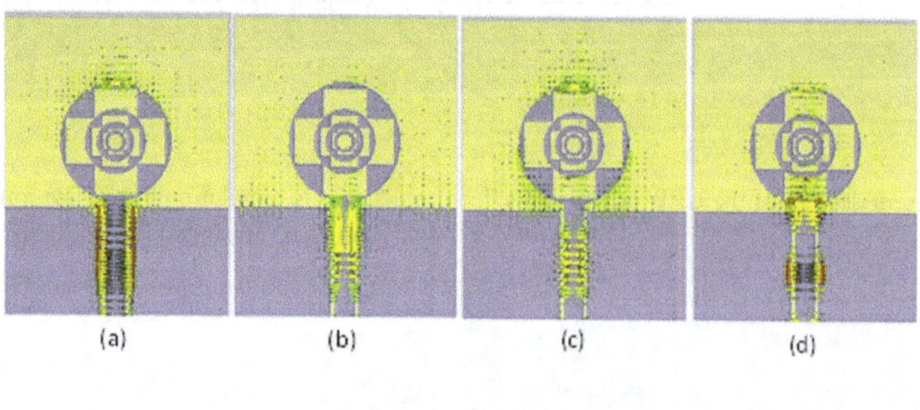

Figure 15. The gain vs frequency of the CPW antenna simulated by CST

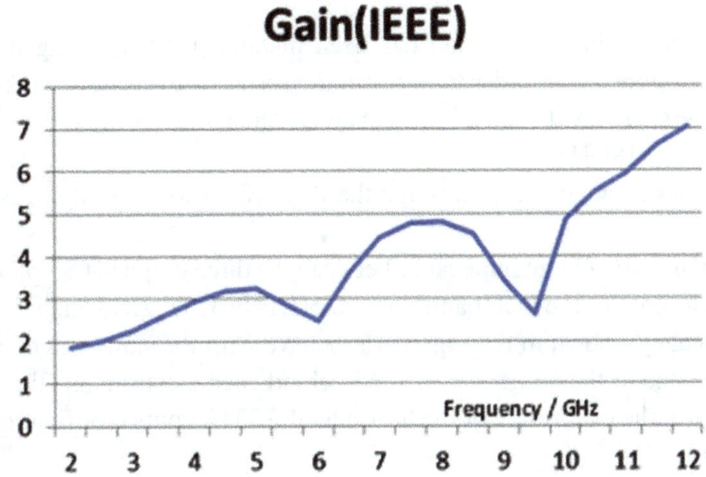

Figure 16. The Radiation Pattern in E-Plane and H-Plane by CST; (a)@3GHz; (b)@5GHz; (c)@7GHz, and; (d)@10GHz

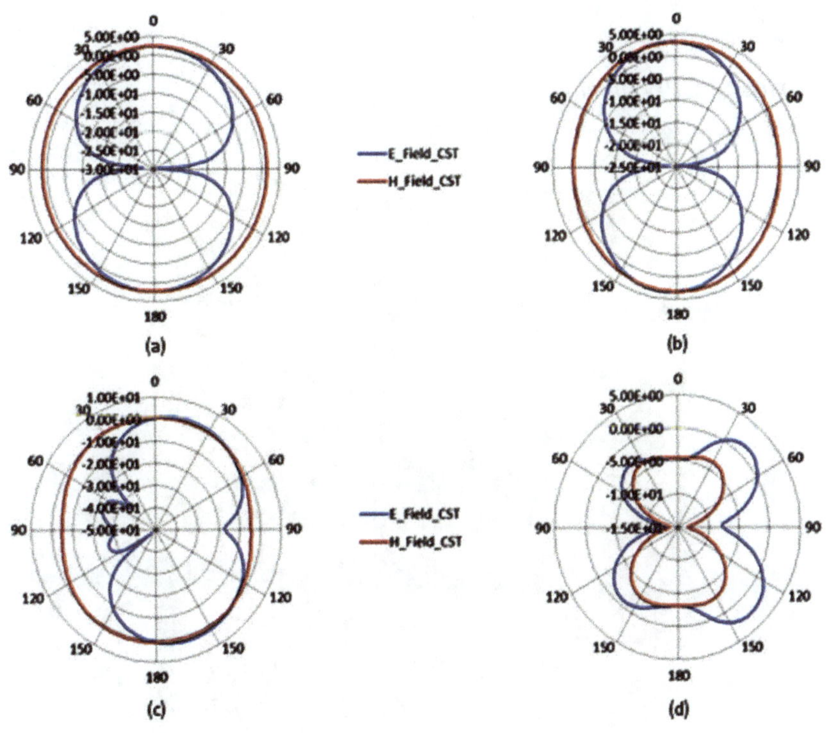

The radiation pattern in E-plane is bidirectional in all frequencies with some variation in the graph of the radiation in 7 and 10GHz, while the antenna keeps the unidirectional behavior in H-plane with a changing of the radiation pattern from unidirectional to bidirectional in the frequency 10GHz.

EXPERIMENTAL RESULTS AND DISCUSSION OF THE CPW ANTENNA

The achieved CPW antenna into simulation has been manufactured by using the photolithographic process, shown in the Figure 17, in order to perform the series of measurements to be compared with the simulated parameters already depicted in this chapter. The female SMA connector connected to the structure can operate until 18GHz.

The S_{11} parameter has been generated through the PNA-X Network Analyzer N5247A as shown in the Figure 19(a).

A good agreement in term of input impedance between the three graphs of S_{11} plotted in the frequency range 2 - 12GHz which contains the FCC band 3.1 – 10.6GHz by CST, ADS and into measurement. Also the three S11 vs frequency, shown in the Figure 18, behave with the same way in all frequencies with some small difference due to the numerical method behind each electromagnetic solver or difficulties of the measurement such the calibrating of the tools and the SMA connector fixation to the fed line and the two ground planes.

The measurement of the radiation pattern has been used by establishing a microwave link of 3m inside the anechoic chamber to avoid the reflections of the electromagnetic waves, as presented in the Figure 19 (b)

Figure 17. The fabricated CPW antenna

Figure 18. The S11 versus the frequency simulated by CST- ADS and Measured

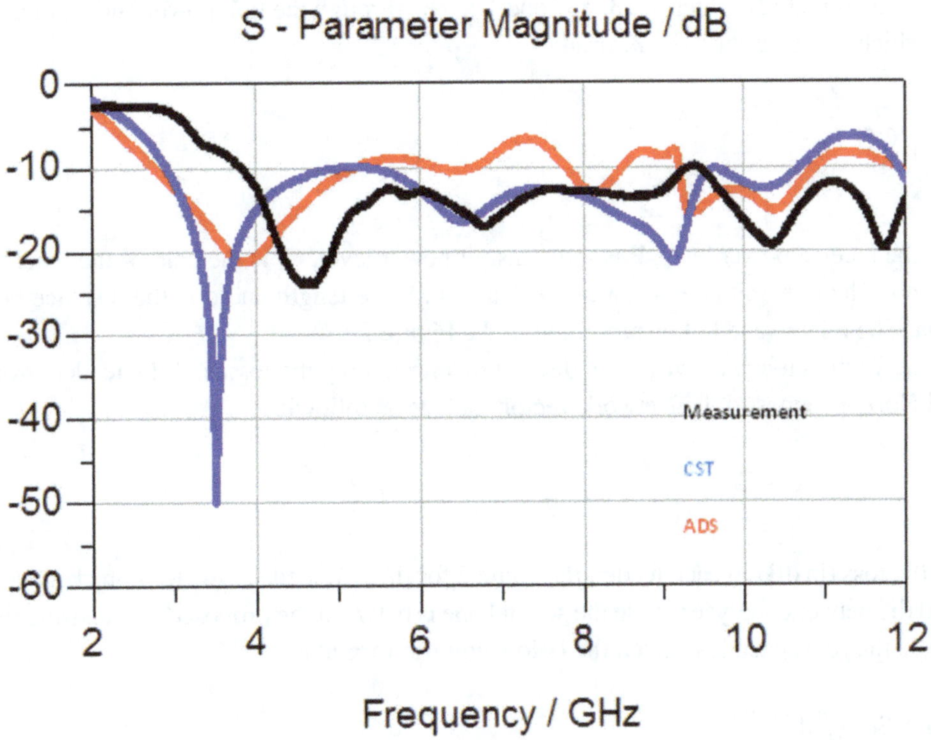

Figure 19(a). The Measurement of S11 by PNA-X Network Analyzer N5247A of the proposed Antenna, and; (b) The measurement of the CPW antenna in the anechoic chamber

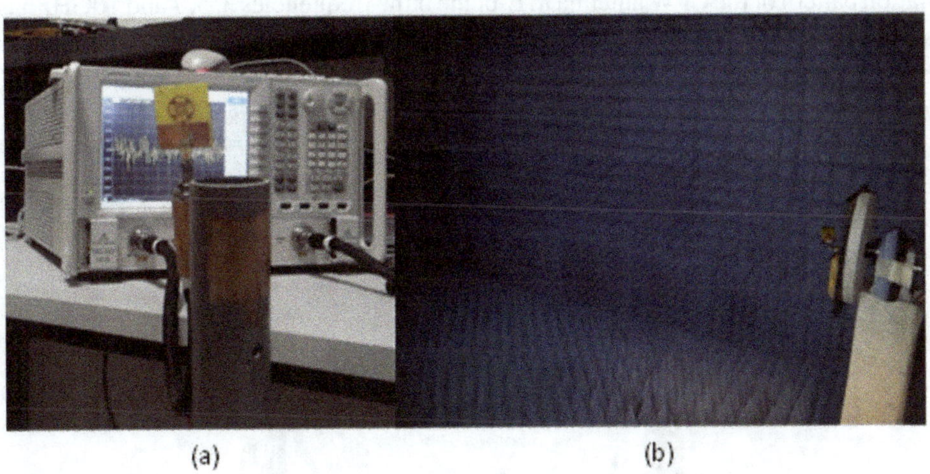

Before plotting the gain radiation pattern it was necessary to calculate the gain vs the angle in each frequency during this study which are 3, 5, 7 and 10GHz, through the link power budget of a microwave link of 3m which by using the Friis Formula:

$$P_r = \left(\frac{\lambda}{4\pi d}\right)^2 G G_r P$$

where P_r is the receive power level, P is the transmit power level, G_r is the gain of the receive antenna, G is the gain of the transmit antenna, λ is the Transmit wave length and d is the distance between the two antenna. The microwave link is presented in the Figure 20.

The above formula it can be written in decibel by introducing the losses of the feeders from the fabricated and Horn antennas to the Network vector analyze as following:

$$P_r = P - 2L + G - A + G_r$$

where L is the loss (in dB) in each feeder, the value 2 for the two feeders, the first one between the horn antenna and the network analyzer while the second one is between the proposed antenna and the network analyzer. A is the free space loss given the below empiric formula:

$$A = 92.45 + 20\log\left(d.f\right)$$

where d is the distance between the two antennas in (km) while f is the operating frequency in (GHz), the above formula it can be used if $f < 15$GHz, which is the situation in our case.

The radiation pattern of the CPW antenna in E-plane if the frequencies 3, 5, 7 and 10GHz has presented in the Figure 21 (a), (b), (c) and (d) respectively. The bi-directionality behaviour has been respected into measurement and simulation buy CST slightly in all frequencies with some variations in the measured diagrams which are results of different causes during the laboratory experiences.

Figure 20. The Microwave link of 3m for Radiation Pattern Measured

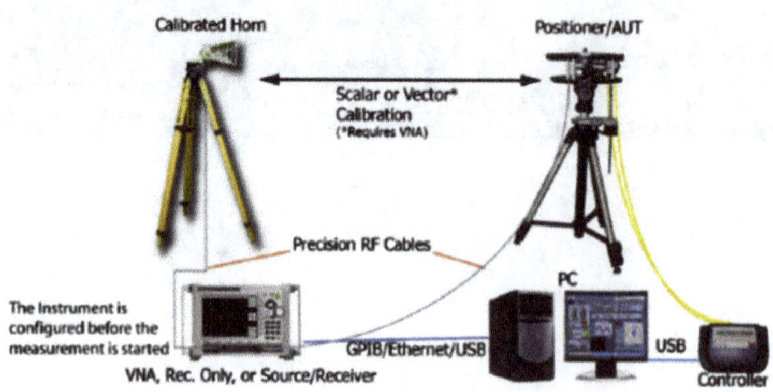

Figure 21. The Radiation Pattern in E-plane by CST and into Measurement; (a)@3GHz, (b)@5GHz, (c)@7GHz and (d)@10GHz

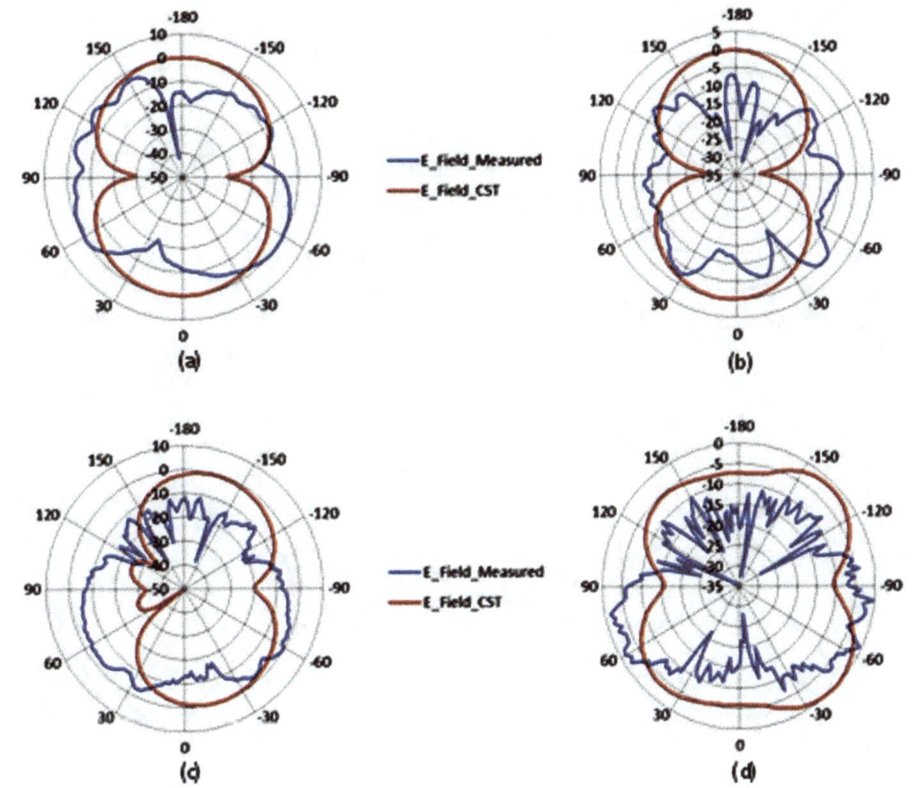

The radiation pattern in H- plane of the studied antenna presented in the Figure 22 (a), (b), (c) and (d) has been performed for the same four frequencies which are 3, 5, 7 and 10GHz. The measured and simulated diagrams are unidirectional in term of radiations.

GEOMETRY AND DESIGN OF THE MSA

The proposed antenna, designed with a microstrip line technique, has been presented in the Figure 23 by using same dielectric substrate used for the CPW antenna and with same features.

The transmission line of the MSA chosen with two parts, the first part is a matched section of 3 mm, however the second is a tapered section of 1.4 mm connected to the radiator in order to match the patch to the feed line over an ultra-wide frequency range, this types of feed lines connected directly to the patch give the opportunity to establish a smooth path of current.

The radiating patch has been of MSA is an hexagon with star shapes inscribed with different sizes to invent the fractal geometry and benefit from the self-similarity and the space filling offered.

The ground plane has partially designed on the bottom of the substrate in order to create a virtual image the conductor on the top which let the MSA behaving as coaxial cable.

Figure 22. The Radiation Pattern in H-plane by CST and into Measurement; (a)@3GHz, (b)@5GHz, (c)@7GHz and (d)@10GHz

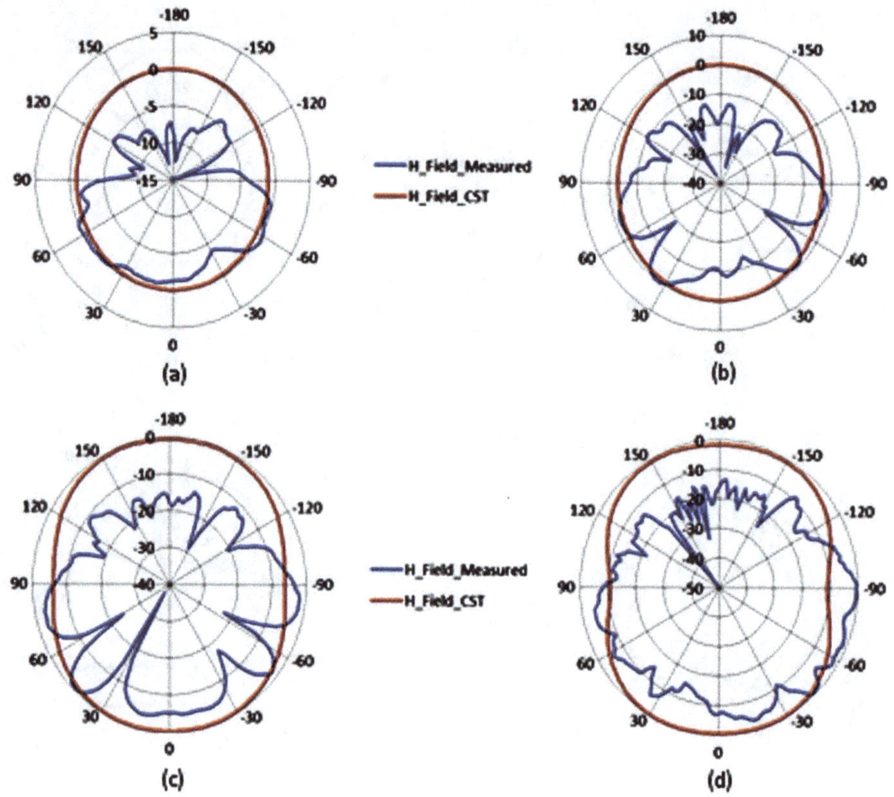

Figure 23. The Geometry of the MSA

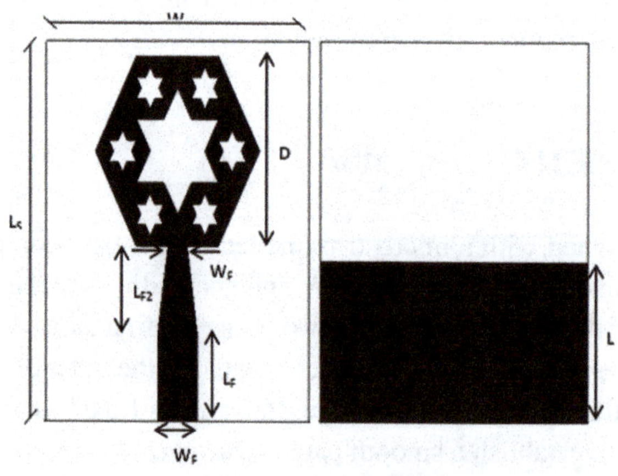

Table 2. The optimized parameters of the MSA

Parameter	Value (mm)
L_S	25
W_S	20
L_G	10.5
W_{F1}	3
W_{F2}	1.4
L_{F1}	5.5
L_{F2}	6
D	12.5

The optimized dimensions of the MSA are shown in Table.2, the optimization performed through the methods of calculation implemented in the transient solver of CST.

The Figure 28 presents the S_{11} computed by using the two electromagnetic tools CST and ADS, the return loss simulated by CST is less than -10dB in the frequency range 3.4 – 11GHz, However the one calculated by ADS is less than -10dB in the frequency range 3.3 – 9.6GHz and from 10.4GHz to 12GHz. As conclusion the proposed MSA shows a good matching of input impedance in the whole FCC frequency band.

The current density of the MSA has been simulated by using CST in the four frequencies 3, 5, 7 and 10GHz to check the current distribution behavior in the FCC band as shown in the Figure 24 (a), (b), (c) and (d).

The current density is mainly present in the periphery of the radiating patch in the frequency 3GHz with low density in the border of the microstrip line. While in the frequencies 5 and 7GHz the current distribution appeared with high density in the periphery of the transmission line and in the lower part of the radiator. In the frequency 10GHz the current density is only around the border of the feed line.

The gain of the MSA presented in the Figure 25 variated from 1.3dB to 3.1dB in the FCC band which means that the proposed MSA can be used in indoor applications.

Figure 24. The current density of the CPW antenna by CST(a) @ 3GHz; (b) @ 5GHz; (c) @ 7 GHz, and; (d) @ 10GHz

(a) (b) (c) (d)

The radiation pattern has been computed by using the CST in the E-plane and H-plane in the same frequencies mentioned before to coverage the FCC band. The diagrams has been shown in the Figure 26 (a), (b), (c) and (d), in E- plane the bidirectional behavior has been confirmed in the UWB while in H-plane the antenna present an unidirectional radiation in the frequencies 3, 5, 7 and 10GHz .

Figure 25. The gain vs frequency of the MSA

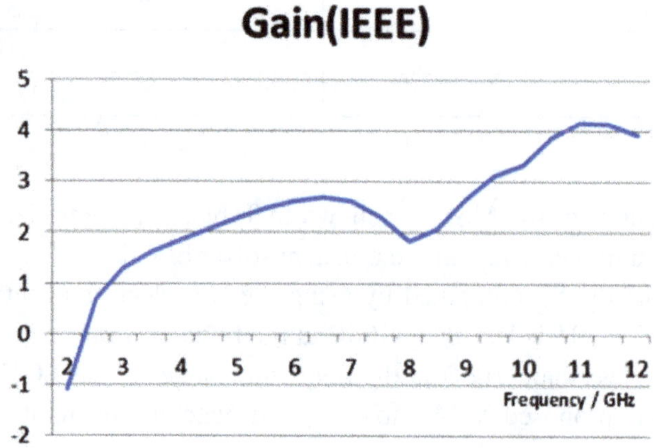

Figure 26. The Radiation Pattern in E-Plane and H-Plane by CST; (a)@3GHz; (b)@5GHz; (c)@7GHz, and; (d)@10GHz

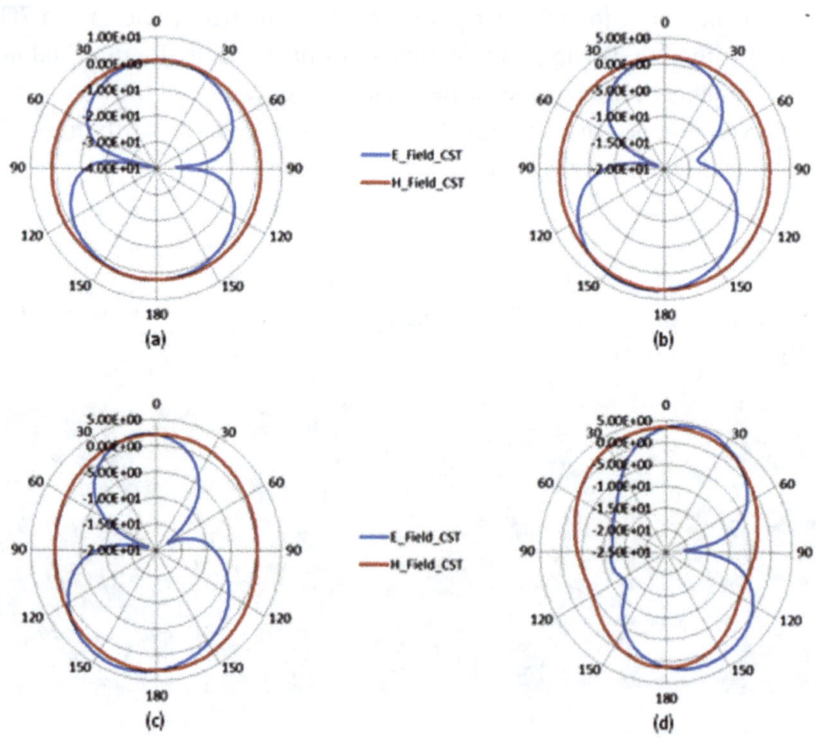

EXPERIMENTAL RESULTS AND DISCUSSION OF THE MSA

With the same method as mentioned before for the CPW antenna which is the photolithographic the fabrication of the MSA has been performed, Figure 27, in order to establish the measurement of the parameters S_{11} and the radiation pattern f.

The three graphs of S_{11} achieved into simulation by using CST and ADS and into measurement and plotted in the figure 28 are matched in the frequency range 3.1 – 10.6GHz in term of input impedance at 50Ω due to the values of return loss shown below the -10dB in the UWB, which means that the MSA will radiate the maximum energy coming from the transmitter with a minimum reflection.

The MSA radiation pattern in E-plane has presented in the Figure 29 (a), (b), (c) and (d) respectively for the frequencies 3, 5, 7 and 10GHz. The MSA behave as bidirectional antenna in the diagram generated into measurement which is matched with the simulation performed by CST in all chosen frequencies with some variations in the measurement due some difficulties during the experience.

The radiation pattern in H- plane of the studied antenna presented in the Figure 30 (a), (b), (c) and (d) has been performed for the same four frequencies which are 3, 5, 7 and 10GHz. The measured and simulated diagrams are unidirectional in term of radiations.

COMPARISON BETWEEN THE MSA AND CPW ANTENNA

The comparison between the MSA and CPW will based on the parameters already analyzed for both antennas which are the antenna dimensions, the gain, the current density, the efficiency and the coefficient of reflection. The comparison based on those parameters can be the important criteria to decide the adequate application for each antenna.

Figure 27. The fabricated MSA

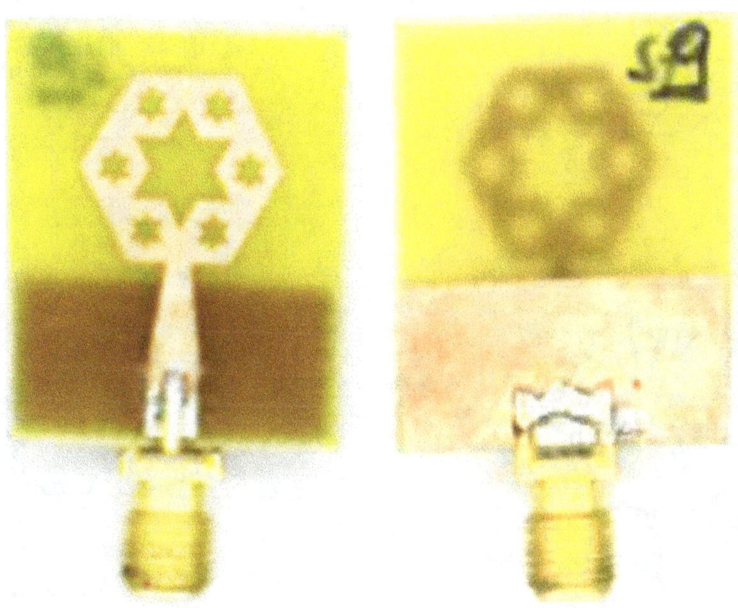

Figure 28. S_{11} vs frequency of the MSA by CST, ADS and Measurement

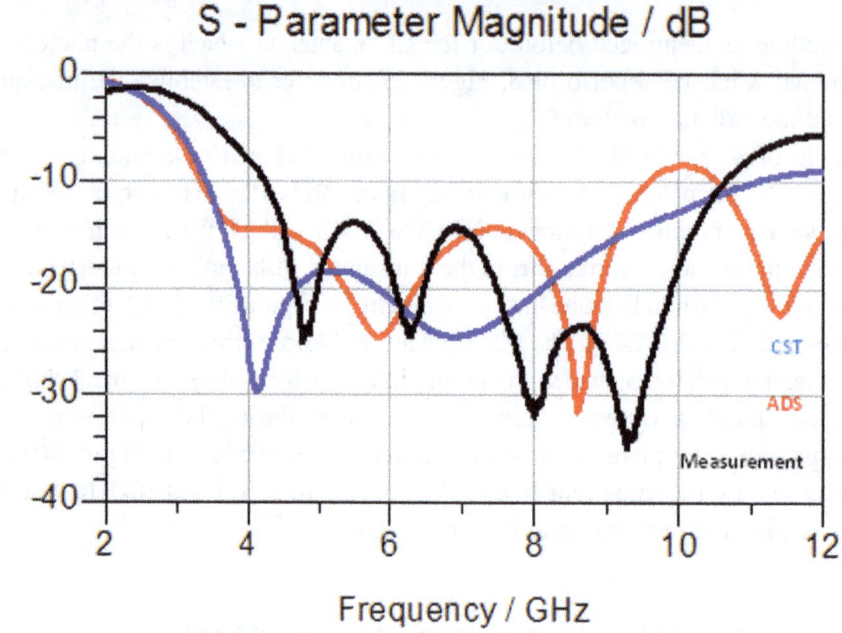

Figure 29. The Radiation Pattern in E-plane by CST and into Measurement; (a)@3GHz, (b)@5GHz, (c)@7GHz and (d)@10GHz

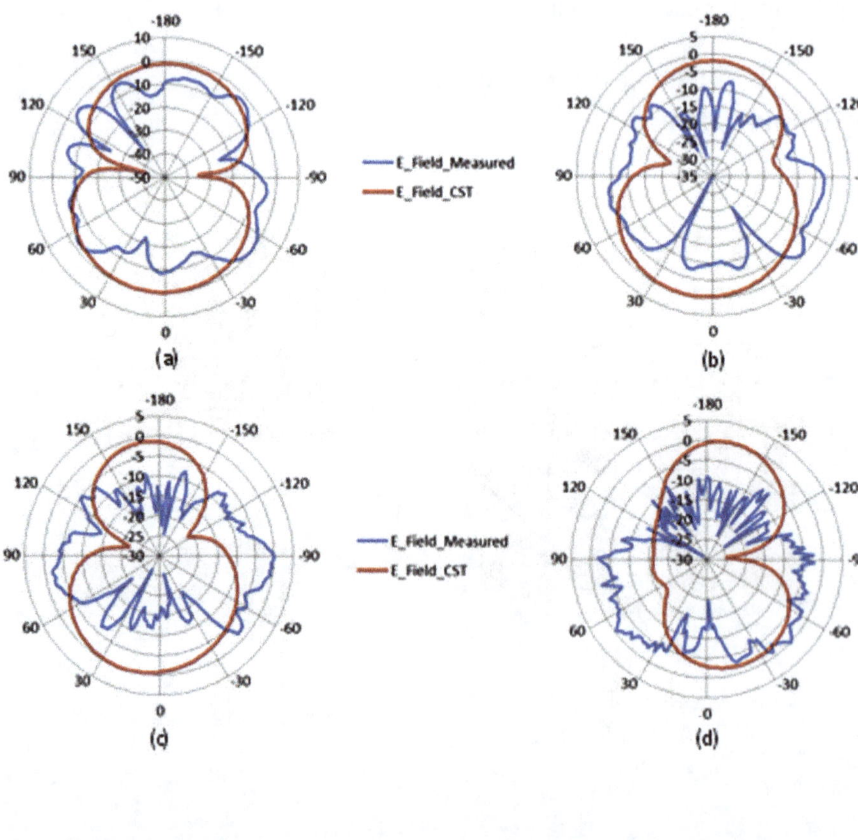

Figure 30. The Radiation Pattern in E-plane by CST and into Measurement; (a)@3GHz, (b)@5GHz, (c)@7GHz and (d)@10GHz

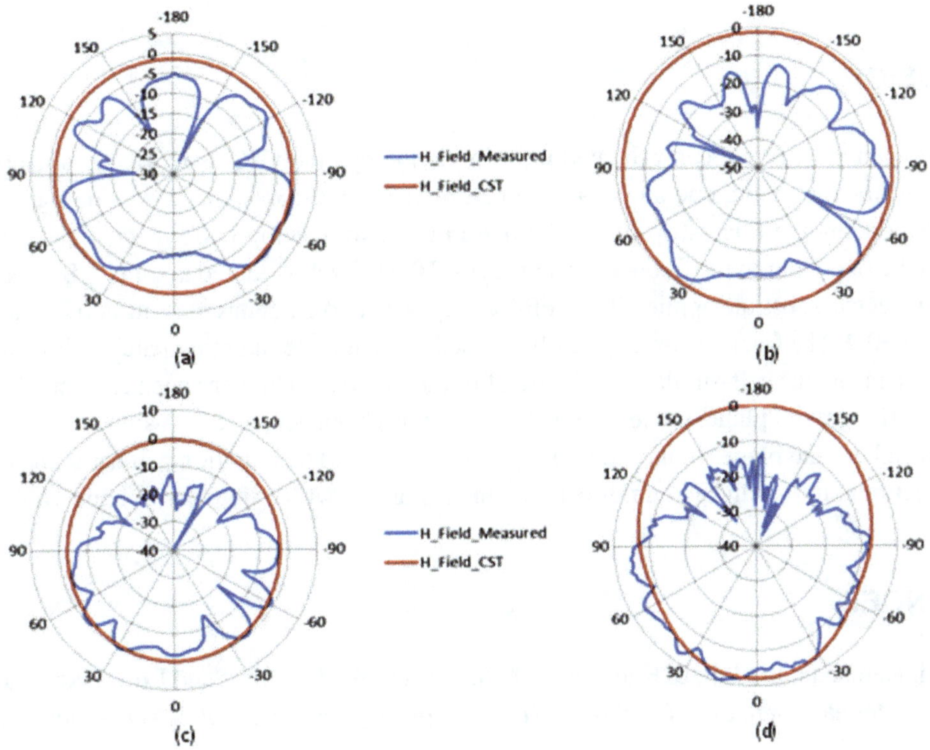

Table 3. The comparison between MSA and CPW antenna

	CPW Antenna	Microstrip Antenna
Antenna dimensions	• 43 x 34 mm^2 • No ground plane on the bottom of the substrate	• 25 x 20 mm^2 • Partial ground plane on the bottom of the substrate
Return Loss (S_{11})	• less than -10dB in the FCC band • Important matching in the input impedance at 50Ω	• less than -10dB in the FCC band • Very important matching in the input impedance at 50Ω
Current Density	In the periphery of the transmission line and the partial ground plane in all frequencies of the UWB.	• In the periphery of the radiating patch in the lower frequencies • In the periphery of the transmission line in the higher frequencies
Gain	• Between 1.3 - 2.5dB in Lower frequencies. • Between 2.5 - 3.1dB in Higher frequencies.	• Between 2 - 3.5dB in Lower frequencies • Between 3.5 - 5dB in Higher frequencies
Radiation Pattern	• The radiation pattern is bidirectional in E-plane. • The radiation pattern is unidirectional in H-plane.	• The radiation pattern is bidirectional in E-plane. • The radiation pattern is unidirectional in H-plane.

The Table 3 summarizes the parameters listed above for the two types of antennas studied in this chapter.

CONCLUSION

Two different antennas have been validated into simulation by involving two different solvers CST and ADS and into measurement, one antenna designed by using the coplanar waveguide method and the other by introducing the microstrip line technique in order to develop two low cost and miniaturized structures to be operating in the frequency range 3.1 – 10.6 GHz considered as an UWB released by the FCC. The dielectric substrate applied is the FR4-epoxy. The two antennas present good matching input impedance at 50Ω. The fractal geometry has been used to design the radiating patch of both antennas in order to benefit from the self-similarity behavior. For the microstrip line has been chosen with a tapered part and a partial ground plane, while the CPW antenna has been achieved by using a transmission line with two partial ground planes in the same side of the substrate. The radiation pattern of the both antennas is bidirectional in E-plane and unidirectional in H-plane in whole impedance bandwidth.

REFERENCES

Al Husseini, Ramadan, El Hajj, & Kabalan. (2009). Design of a Compact and Low-Cost Fractal-Based UWB PCB Antenna. *Sciences of Electronic, Technologies Information and Telecommunications.*

Bancroft, R. (2009). *Microstrip and printed antenna design.* SciTech Publishing.

Denlinger, E. J. (1969, April). Radiation from microstrip radiators. *IEEE Transactions on Microwave Theory and Techniques, 17*(4), 235–236. doi:10.1109/TMTT.1969.1126937

Gronau, G., & Wolff, I. (1986). Aperture-Coupling of a Rectangular Microstrip Resonator. *Electronics Letters, 22*(10), 554–556. doi:10.1049/el:19860378

Haider, Sabbagh, Mahmood, Edwards, & Brister. (2012). A UWB Fractal Antenna for Body Area Network Applications. Loughborough Antennas and Propagation Conference, Loughborough, UK.

Hildebrand, L. T., & McNamara, D. A. (1995). A guide to implementational aspects of the spatial-domain integral equation analysis of microstrip antennas. *Applied Computational Electromagnetics Journal, 10*(1), 40–51.

Homayoon, O., & Shahram, H. (2012). Miniaturization of Microstrip Antennas by the Novel Application of the Giuseppe Peano Fractal Geometrices. *IEEE Transactions on Antennas and Propagation, 60*(8), 3559–3567. doi:10.1109/TAP.2012.2201070

Howell, J. Q. (1972). Microstrip antennas. *IEEE International Symposium on Antennas and Propagation*, 177–180.

James, J. R., & Hall, P. S. (1989). *Handbook of Microstrip Antennas* (Vols. 1–2). Academic Press.

Jena, M. R., Mangaraj, B. B., & Mishra, D. (2013). Bandwidth and Gain Enhancement of Multiband Fractal Antenna Based on The Sierpinski CARPET GEOMETRY. *ICTACT Journal on Communication Technology*, *4*(1), 669–674. doi:10.21917/ijct.2013.0095

Joseph, S., Paul, B., Mridula, S., & Mohanan, P. (2013). A Novel Planar Fractal Antenna with CPW Feed for Multiband Applications. *Radioengeneering*, *22*(4), 1262–1266.

Katehi, P. B., & Alexopoulos, N. G. (1984). On the Modeling of Electromagnetically Coupled Microstrip Antennas-The Printed Strip Dipole. *IEEE Transactions on Antennas and Propagation*, *32*(11), 1179–1186. doi:10.1109/TAP.1984.1143235

Katsibas, K. D., Balanis, C. A., Tirkas, P. A., & Birtcher, C. R. (1998). Folded Loop Antenna for Mobile Handheld Units. *IEEE Transactions on Antennas and Propagation*, *46*(2), 260–266. doi:10.1109/8.660971

Koch, H. V. (1906). Une méthode géométrique élémentaire pour l'étude de certaines questions de la théorie des courbes planes. *Acta Mathematica*, *30*(1), 145–174. doi:10.1007/BF02418570

Kumar & Ray. (2003). *Broadband Microstrip Antennas*. London: Artech House.

Lee, Yook, & Park. (2002). Broadband planar antenna having round corner rectangular wide slot. *IEEE Antennas and Propagation Society International Symposium*.

Levy, Kumar, & Dinh. (2013). A novel fractal UWB antenna for earthquake and tsunami prediction application (LETPA). *IEEE Canadian Conference of Electrical And Computer Engineering (CCECE)*.

Li, J., Jiang, T., Cheng, C., & Wang, C. (2013). Hilbert Fractal Antenna for UHF Detection of Partial Discharges in Transformers. *IEEE Transactions on Dielectrics and Electrical Insulation*, *20*(6), 2017–2025. doi:10.1109/TDEI.2013.6678849

Li-na, W., Shu, L., Run-nan, C., Guan-long, H., & Wen-bin, Z. (2011). *Multiband Printed Monopole Antenna with Square-Nested Fractal*. Paper presented at 6th International ICST Conference on Communications and Networking (CHINACOM), China.

Lizzi, L., & Oliveri, G. (2011). *Hybrid Design of a Fractal-Shaped GSM/UMTS Antenna*. Technical Report, DISI-11-089.

Lo, Y. T., Solomon, D., & Richards, W. F. (1979). Theory and experiment on microstrip antennas. *IEEE Transactions on Antennas and Propagation*, *AP-27*(2), 137–149. doi:10.1109/TAP.1979.1142057

Low, Z. N., Cheong, J. H., & Law, C. L. (2005). Low-Cost PCB Antenna for UWB Applications. IEEE Antennas and Wireless Propagation Letters, 4.

Mahatthanajatuphat, C., Saleekaw, S., Akkaraekthalin, P., & Krairiksh, M. (2009). A rhombic patch monopole antenna with modified minkowski fractal geometry for UMTS,WLAN, and mobile WiMAX application. *Progress in Electromagnetics Research, 89*, 57–74. doi:10.2528/PIER08111907

Mandelbrot, B.B. (1975). *Les objets fractals: forme hasard et dimension.* Nouvelle bibliothèque scientifiques.

Manimegalai, B., Raju, S., & Abhaikumar, V. (2009). A multi fractal cantor antenna for multiband wireless applications. *IEEE Antennas and Wireless Propagation Letters, 8*, 359–362. doi:10.1109/LAWP.2008.2000828

McSpadden, J. O., Yoo, T., & Chang, K. (1992). Theoretical and experimental investigation of a rectenna element for microwave power transmission. *IEEE Transactions on Microwave Theory and Techniques, 40*(12), 2359–2366. doi:10.1109/22.179902

Mosig, J. R. (1988, February). Arbitrarily shaped microstrip structures and their analysis with a mixed potential integral equation. *IEEE Transactions on Microwave Theory and Techniques, 36*(2), 314–323. doi:10.1109/22.3520

Mosig, J. R., & Gardiol, F. E. (1983). Analytical and numerical techniques in the Green's function treatment of microstrip antennas and scatterers. *IEE Proceedings, 130*(2), 175–182. 10.1049/ip-h-1.1983.0029

Mosig, J. R., & Gardiol, F. E. (1985). General integral equation formulation for microstrip antennas and scatterers. *IEE Proceedings, 132*(7), 424–432. 10.1049/ip-h-2.1985.0076

Naji, Aziz, & Fyath. (n.d.). Design and Simulation of RFID Aperture Coupled Fractal Antennas. *International Journal of Radio Frequency Identification & Wireless Sensor Networks.*

Rahim, M. K. A., Abdul Aziz, M. Z. A., & Abdullah, N. (2005). Wideband Sierpinski Carpet Monopole Antenna. *Proceedings of Asia-Pacific Conference on Applied Electromagnetics*, 62-65.

Rama Rao, Alekhya, & Sridevi. (2013). Design of Circular with triangle Microstrip Fractal Antenna for Dual band and UWB applications. *International Journal of Scientific & Engineering Research, 4*(7).

Vandenbosch & Vasylchenko. (2011). *A Practical Guide to 3D Electromagnetic Software Tools*. InTech.

Werner, D. H., & Ganguly, S. (2003, February). An Overview of Fractal Antenna Engineering Research. *IEEE Antennas & Propagation Magazine, 45*(1), 38–57. doi:10.1109/MAP.2003.1189650

Yoo, T., & Chang, K. (1992). Theoretical and experimental development of 10 and 35 GHz rectennas. *IEEE Transactions on Microwave Theory and Techniques, 40*(6), 1259–1266. doi:10.1109/22.141359

Zbitou, J., Latrach, M., & Toutain, S. (2006). Hybrid Rectenna and Monolithic Integrated Zero-Bias Microwave Rectifier. *IEEE Transactions on Microwave Theory and Techniques, 54*(1), 147–152. doi:10.1109/TMTT.2005.860509

KEY TERMS AND DEFINITIONS

CPW Antenna: Planar transmission line located symmetrically in between two grounds collocated in the same side with the fed line and the radiator.

Fractal: A fragmented shape that can be subdivided to copies of the original form with reduced sizes.

Koch Curve: A fractal shape started with straight line that divides into three equal segments and where the middle segment is replace by two sides of an equilateral triangle of the same length as the segment being removed.

Microstrip Antenna: Radiating conductor element on the top side of a dielectric and ground plane conductor on the bottom side.

Sierpinski Carpet: A fractal shape base on a square subdivided recursively into smaller squares.

Sierpinski Triangle: A fractal shape based on an equilateral triangle subdivided recursively into.

Section 2

Antenna Design for RFID Applications

Chapter 9
Printed Antenna for UHF and SHF RFID Applications:
Reader and Tag Antenna's Design

Ahmed Elhamraoui
University of Hassan 1st, Morocco

Elhassane Abdelmounim
University of Hassan 1st, Morocco

Hamid Bennis
Moulay Ismail University, Morocco

Mohamed Latrach
ESEO, France

ABSTRACT

In this chapter, new microstrip antennas for RFID system are presented. The chapter is split into two sections. The first section deals with the design of two dual-band antennas for handheld RFID readers, and the second section is dedicated to antennas for RFID tag. In the first section, the authors describe theory and principle characteristics of microstrip antenna and the fundamentals of multiband behavior. The two proposed antennas are designed to operate in the ISM (industrial, scientific, and medical) band at 2.45 and 5.8 GHz bands. In the second section, the authors present the principles of matching techniques for tag antenna design. Then, they present two examples of tag antenna. The first one is a flexible antenna mounted on paper substrate, and the second one is designed for the identification of metallic object.

DOI: 10.4018/978-1-5225-7539-9.ch009

INTRODUCTION

The Radio Frequency identifications (RFID) is an automatic wireless technology for data gathering that has been deployed in a various ranges of applications including logistics, transport, animal tagging and asset tracking offering the capabilities of the identification in different weather conditions, the identification of many items simultaneously, identifying items within a vicinity between a few centimeters to several meters and enabling the off line-of-sight communication, hence RFID is becoming more attractive (Finkenzeller, 2010). The two basic components of an RFID system are the tag and the reader; the tag consists of an antenna and a microchip transponder. The three main classes of RFID tags are: passive semi passive and active (Brown, 2007). Once activated, the content stored in the internal memory of the tag is communicated to interrogator by backscattering modulation of the signal. The reader is connected to the host computer which is used to program the reader and store information received from the transponder. The reader is a radio transceiver connected to transmit and receive antennas (Paret, 2008).

The exchange of information takes place due to magnetic or electromagnetic coupling between reader and tag. The inductive coupling operating in the near field is based on the transfer of energy from one device to another through a shared magnetic field (transformer principle). The magnetic alternating field of the reader induces at resonance frequency a voltage in the transponder coil that is used to provide the power supply. While the far-field coupling is based on the electromagnetic wave propagation and the reflection of electromagnetic wave back in the direction from which it was received. This backscattering modulation technique permits to recover data contained in the tag to the reader receiver (Nieto, 2011).

The RFID applications occupy four frequency bands over the radio frequency spectrum. The Low-frequency (LF, 125; 134 kHz) and high-frequency (HF, 13.56 MHz). These applications are based on magnetic field coupling and most of them are related to "passive tag" system showing good performance when tags are close to the reader coil. RFID systems at Ultra-high frequency (UHF, 860; 960 MHz) and microwaves (SHF, 2.45 GHz and 5.8 GHz) involve electromagnetic coupling between antennas offering higher data rates and establishing a communication link at a longer distance (Foster & Bueberry, 1999).

Amid the components of the RFID system, antenna is considered as an essential element and has been paid a great attention that is why antenna design becomes very critical and very important in the performance of the whole system due to many reasons (Rao & Nikitin, 2005);

- Tag antenna permit to recover and transfer power to the chip,
- The detection range and accuracy are directly dependent on the performance of tag antenna.
- Reducing tag fabrication cost by using low cost substrates compatible with those manufacturing process are employed.

In case of UHF and SHF frequency bands, the reader and tag antenna design depends on specific requirements such as gain, overall dimension, effective aperture and reading range. Therefore the design is mostly based on the use of dipole, helix and recently compact printed microstrip antenna (Chandra, 2010; Maicas, 2011).

The growing demand for small and compact wireless devices has increased the need for small antennas that can be easily integrated while providing acceptable overall performance, especially for the applications involving RFID handheld reader offering many advantages in terms of compactness and maneuverability. Also to comply with the expectations of the users of RFID devices that require the multi-standard capability, high data performance, security protocols and compact profile, the need to reduce the size of the antenna is highly essential. But reducing the size of antenna limits its performances. In this context multi-band antennas with a large impedance bandwidth and an excellent radiation performance have been designed (Hebib, 2008). Because of their low cost, light weight, and good radiation characteristics, the printed monopole antennas have been widely used in wireless communication terminals such as RFID systems (Necibi & Hamzaoui, 2016; Zhang, 2009). Many techniques have been introduced to achieve the multi-band behavior for the compact printed antennas such as the fractal geometry, square ring slot and rectangular slot, the use of multiple branches, the slot loading technique by using circular ring slot, the use of an outer loop strip shorted which is connected on both sides of the ground with CPW-fed and the monopole with notch structure (Chen, 2004).

Depending on the items in which the tag will be attached, the antenna part should be able to perform well meaning that the tag should show a good readable range, be matched to the tag chip for maximum power transfer. A typical example is when the tag is attached to metallic objects such as containers, cars etc, this configuration can affect the performances of the RFID system and create destructive interference (Eunni, 2004). A solution to this problem is the use of the microstip patch antenna. When attaching this antenna to the surface of metal; this surface acts as an extension of the antenna's ground plane. On the other hand when tags are mounted on plastic items, clothes or human bodies they should be flexible and environment friendly, for these reasons the use of meander printed dipole on organic substrates such as paper is a good solution permitting fast printing processes, such as conductive paste inkjet-printing, that can be used instead of metal etching techniques (Yasar, 2013; Su & Lai, 2008).

The proposed chapter is divided on two sections. In the first section two validated examples of printed microstrip dual band antenna for handheld readers will be presented. The first one is a dual-band Coplanar Waveguide (CPW) monopole antenna operating at 2.45 GHz and 5.80 GHz. This antenna is printed on a FR-4 substrate with dielectric permittivity constant 4.4, thickness of 1.6mm and loss tangent of 0.025. with a total area of 30×35 mm^2 and consists of a vertical strip combined with two symmetrical horizontal stubs inserted on the radiating element which permits to generate the two desired frequencies. The second one is a dual band monopole antenna fed also by a coplanar waveguide (CPW) line and covering free ISM bands of 2.45 GHz and 5.8 GHz. The antenna uses a shorting stub between the monopole and the ground plane in order to produce the dual-band behavior. A good impedance bandwidth of 300 MHz is achieved in measurement (from 2.3 GHz to 2.6 GHz for the lower band), while the upper band covers 450 MHz (from 5.4 GHz to 5.85 GHz). The total volume of the proposed antenna is 33×30×1.6 mm^3. The two reader antennas proposed in this chapter present the advantages to be compact, low cost and offering a multiband operation behavior.

In the second section, authors present two validated examples of RFID tag antenna; the first structure presents a new design of a low cost passive UHF RFID tag antenna using paper substrate. The design is based on a two meandered rectangular dipoles inductively coupled, in order to reduce the size of the design. By choosing a proper dimension of the rectangular loop, a good and broadband impedance matching can be achieved. This antenna has a total area of 72×26 mm^2 and mounted on a paper substrate HP photo paper, with a constant dielectric permittivity, equal to 3.3, a thickness of 0.5 mm and a loss tangent of 0.04.The simulated bandwidth at the UHF RFID band is about 56 MHz (894–950 MHz). The second antenna was designed for metallic objects. The proposed antenna tag is based on the geometry of the nested-slot suspended-patch in order to ensure the complex impedance matching between antenna and chip thanks to the inductive reactance of a non-resonant slot. This antenna has a total area of 57×26 mm^2 and mounted on a FR4 substrate with a constant dielectric permittivity, equal to 4.4, a thickness of 0.5 mm and a loss tangent of 0.025. The simulated bandwidth at the UHF RFID band is about 28MHz

(900–928 MHz). The two tag antennas presented in this chapter present the advantages to exhibit a good impedance matching, no vias are used which makes them easy to fabricate, low cost and covering the entire UHF-RFID North and South of America band.

SECTION I: MICROSTRIP DUAL BAND ANTENNAS FOR RFID READER

Microstrip Patch Antenna

Originally proposed in the 1950's, microstrip antennas have since received a great attention from a number of researchers in the telecommunication field (Bahl & Bhartia, 1981). The inherent advantages like low profile, lightweight and ease of fabrication facilitate their integration in active devices. In its fundamental form, a microstrip antenna is made of a dielectric substrate inserted between two conducting surfaces: the antenna plane and the ground plane. The radiating patch has generally common shape such rectangular, circular, square, triangular, and elliptical. The Figure 1 shows a simplified structure of microstrip patch antenna.

Microstrip patch antennas radiation results from the fringing fields between the patch edge and the ground plane. Since the propagating EM fields lay, both in the substrate and in free space, a quasi-TEM mode is generated. The two methods commonly used to describe the functioning mechanism of microstrip antennas are the transmission line model and the cavity model. Microstrip antenna performances are affected by the patch geometry, substrate properties and feed techniques. A dielectric substrate with properties such as low loss, low dielectric constant and sufficiently thick can provide maximum radiation efficiency and bandwidth. However, the antenna dimensions are large when low dielectric substrates are used which increase the antenna quality factor (Q), resulting in narrow bandwidth (Balanis, 2003).

The microstrip antenna can be excited directly either by a coaxial probe or by a microstrip line. It can also be excited indirectly using electromagnetic coupling or aperture coupling, in which case electromagnetic field coupling is done to transfer power between the microstrip line and the radiating patch and a coplanar waveguide feed. Feeding technique influences the input impedance and characteristics of the antenna. The common feeding techniques are shown in Figure 2.

Microstrip patch antennas suffer from some drawbacks compared to conventional antennas. Some of them are: Narrow bandwidth, low efficiency, low gain and low power handling capacity.

Figure 1. Basic structure of Microstrip Patch Antenna

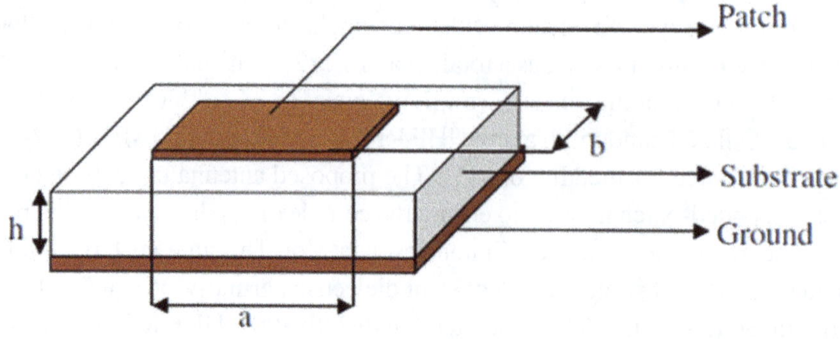

Figure 2. Feeding Methods in MPA

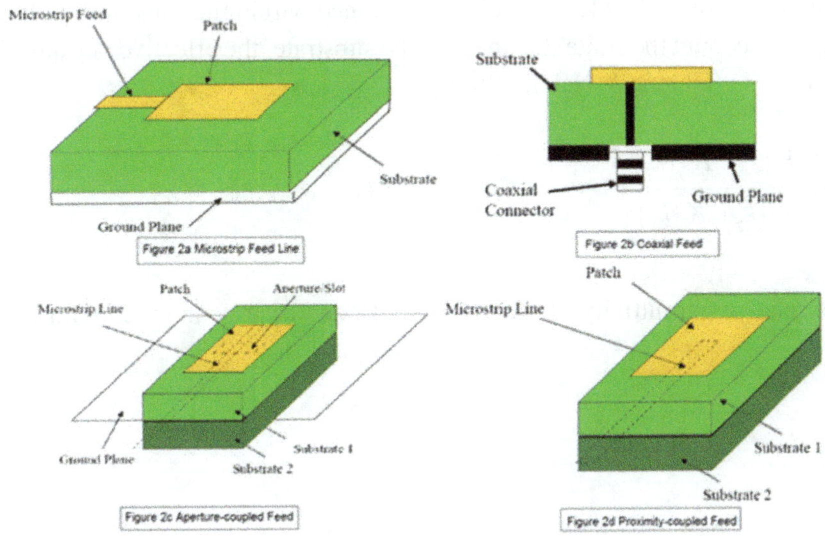

In this chapter the proposed dual band antennas are based on CPW feed line. As shown in the Figure 3, the conductors form a center strip separated by a narrow gap from two ground planes on either side (Simons, 2001). The dimensions of the center strip, the gap, the thickness and permittivity of the dielectric substrate determined the effective dielectric constant, characteristic impedance and the attenuation of the line. The gap in the coplanar waveguide is usually very small and supports electric fields primarily concentrated in the dielectric. With little fringing field in the air space, the coplanar waveguide exhibits low dispersion.

Where:

W=width of the center conductor.
H=height of the substrate.
S=Slot width made between center conductor and ground plane.

Figure 3. Cross section of a classic Coplanar Waveguide (CPW) structure

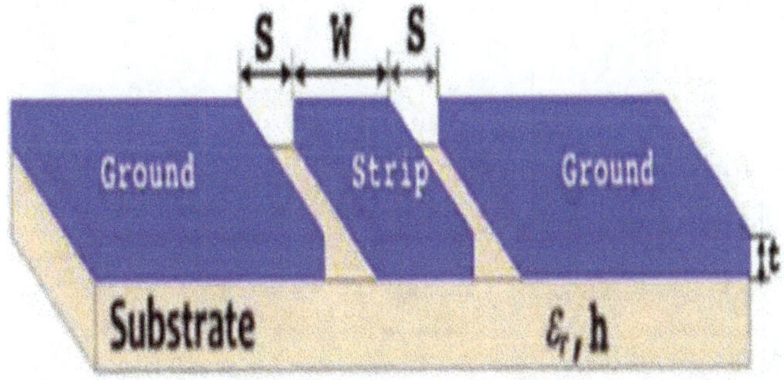

In Microstrip development a new concept of Effective Dielectric Constant ε_{eff} was introduced, which takes into account the most of the electric fields constrained within the substrate.

When taking into account the finite thickness of the substrate, the effective permittivity can be computed with Equation (1) (Ghione & Naldi, 1984).

$$\varepsilon_{re} = 1 + \frac{\varepsilon_r - 1}{2} \frac{K(k_2)}{K'(k_2)} \frac{K'(k_1)}{K(k_1)} \tag{1}$$

where Ɛre is the relative permittivity of the substrate, k_1 and k_2 are computed with Equations (2) and (3) respectively:

$$k_1 = \frac{W}{W + 2S} \tag{2}$$

$$k_2 = \frac{\sinh\left(\frac{\pi W}{4h}\right)}{\sinh\left(\frac{\pi(W + 2S)}{4h}\right)} \tag{3}$$

The strip width (Ws) and gap (S) of the Coplanar Waveguide (CPW) feed are derived using standard design equations for 50Ω input impedance.

The characteristic impedance of a conventional coplanar waveguide structure can be computed using Equation (4).

$$Zc = \frac{30 \times \pi}{\sqrt{\varepsilon re}} \cdot \frac{K'(k1)}{K(k1)} \tag{4}$$

where: $K(k_1)$ is the complete elliptic integral of the first kind and $K'(k_1)$ its complement.

The ratio $\frac{K'(k_1)}{K(k_1)} \frac{\mathrm{K}'(k_1)}{K(k_1)}$ can be easily computed using Equation (5).

$$\frac{\mathrm{K}'(k_1)}{K(k_1)} = \left\{ \frac{\pi}{\ln\left(2 \frac{1+\sqrt{k}}{1-\sqrt{k}}\right)} \text{for} 0 \le k \le \frac{1}{\sqrt{2}} \right\} \tag{5}$$

From the above equation, we notice that the characteristic impedance is controlled by the separation of metallic traces and not by substrate thickness. This can be advantageous, especially when making connections at high frequencies.

Multi-Band Behavior Techniques

The multiband behavior defines the capability of the antenna to operate in a finite bandwidth at different frequencies. This kind of configuration gives the best solution to the increased demand for microwave and wireless communication systems in various applications including the Wireless Local Area Networks (WLAN) (Molisch, 2011), the Worldwide Interoperability for Microwave Access (WiMAX) and Radiofrequency identification (RFID), Wi-Fi and Bluetooth in 2400 – 2483 MHz, permitting to reduce the number of embedded antennas by combining several applications on the same antenna.

In literature several methods are used to reach the multiband behavior. Among these techniques, we use:

- **Higher Order Resonances:** This method is based on the use of monopole antenna as their length is increased in λ/4 increments. For such configuration, the antenna resonates at f_0 while the electric field at the feed is at the minimum and the current density is at the maximum. For microstrip patch antenna, a simple rectangular patch can be regarded as a cavity with magnetic walls on the radiating edges.
- **Multi-Patch Stacked Antennas Approach:** The use of two or more patches permits to reach multi-band behavior. The dual-frequency behavior of these antennas is obtained by means of multiple radiating elements, each of them supporting strong currents and radiation at the resonance (Huang & Boyle, 2008).
- **Combination of Different Radiating Elements:** This may be obtained with a probe-fed configuration, which is displaced from the two principal axes of the patch. In this case two or more resonant structures can be closely located or even co-located with a single feed point in order to achieve multiband operation. This is illustrated in (El Hamraoui, 2016): the longer element is operational at the lower frequency f_1, while the smaller one is operational at the higher frequency f_2. It is also possible to use planar multi-resonator structures fed by microstrip line or CPW coplanar line.
- **PIFA (Planar Inverted F Antenna):** The decrease of the PIFA size is achieved by a short circuit placed along the line where the electric field between the radiating plate and the ground plane is zero for the fundamental mode of the printed antenna. In order to create the multi-band behavior, several techniques can be used:
    - Modification of the radiating element (in the form of triangles, meanders ...).
    - Adding short circuits.
        - Adding parasitic resonators.
    - Inserting slots in the radiating element.
- **Fractal Antennas:** Fractal geometry is an extension of the Euclidean geometry. Its introduction was an opportunity for the researchers to discover new configurations of antennas structures. Fractal geometry is based on fragmented geometric shape that can be subdivided in parts, each of which is (at least approximately) a reduced-size copy of the whole, it's principally characterized by two properties: self-similar and independence of scale used to obtain multi-band antennas. The classical examples of such repetitive construction are the Von Koch curve, Sierpinski carpet and

the Minkowski curve. The fractal solutions propose antennas that are both compact and multi-band. At different operating frequencies, usually log-periodically distributed, the radiation patterns of fractal antennas are often not very similar.

Design and Optimization of New Microstrip Dual-Band Antennas for Handheld RFID Reader

In this section, two low cost dual-band Coplanar Waveguide (CPW) monopole antennas structures have been developed and designed. The proposed antennas are based on FR4 substrate with 4.4 relative permittivity (Ɛr), 1.6 mm thickness (H) and 0.025 loss tangent (tan (δ)). This structure is validated in the ISM (Industrial Scientific and Medical) band at 2.45 GHz and 5.8 GHz. The main purpose of the two works is to develop an antenna that can be easily integrated in handheld RFID reader.

Antenna Design 1

This antenna consists of a vertical strip combined with two symmetrical horizontal stubs inserted on the radiating element which permit to generate the two desired frequencies.

The Figure 4 shows the geometry of the proposed dual band CPW-fed monopole antenna (left) and the CPW feeding line (right). This antenna has a total area of 30×35 mm² (Ws × Ls) and a metallization thickness t=0.035 mm. The antenna is fed by a 50 Ohm coplanar line with a fixed strip width W_f=3.8 mm and a gap S=0.5 mm. The strip width W_f and gap S of the Coplanar Waveguide (CPW) feed are derived using standard design equations for 50Ω input impedance, in order to match the characteristic impedance of transmission line. The characteristic impedance of a conventional coplanar waveguide structure can be computed using standard equations.

The structure optimization is conducted by using CST Microwave Studio which provides different techniques and optimization methods. After many series of optimization by using CST, the final dimensions of the proposed antenna are reported in Table 1.

The Figure 5(a) describes the design evolution of the proposed antenna; the design was achieved through three steps. Firstly, we start with a straight rectangular slot (Figure 5a.D-1). The second step consists of adding a central vertical stub in the rectangular slot which permits to reduce the dimension of

Figure 4. Geometry of the proposed antenna Top view of the designed antenna (left) and Side view antenna (right)

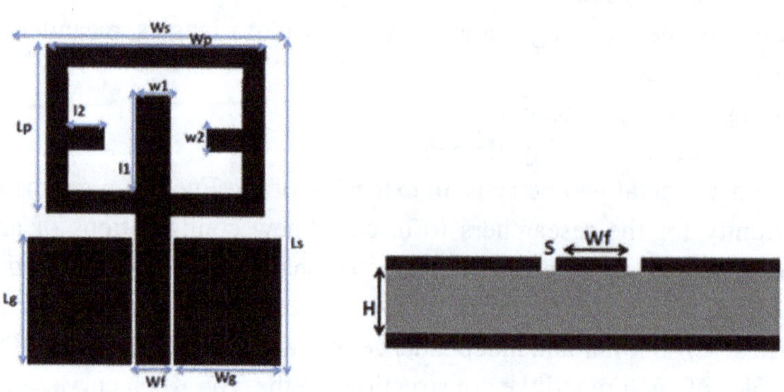

Table 1. The dimensions of the designed antenna (unit in:mm)

Parameter	Length
L_{sub}	35
W_{sub}	30
L_g	16
W_g	12.9
L_p	17
W_p	27.8
l_1	9
l_2	4
w_1	3.8
w_2	2

the antenna (Figure 5a.D-2). At the end, the final dual-band antenna is achieved by inserting two symmetrical stubs in the sides of the rectangular patch (Figure 5a.D-3). Thus, the matching input impedance of the final antenna structure is achieved respectively in frequency bands -2.45GHz and 5.8GHz with a return loss less than -10 dB. Figure 5(b) shows the simulated return losses for successive/successful cases of the conception of the final dual-band antenna.

Before the achievement of the antenna, a comparison of the return loss using second electromagnetic solver ADS software "Advanced Design System" is performed. This step permits to confirm the obtained result shown in Figure 5. Figure 6 shows the comparison of the simulated return loss versus frequency for the proposed antenna obtained using both software's.

We can observe a difference in return loss obtained with CST and ADS due to the technique of calculation used in each simulation software. CST is 3D EM simulator based upon Finite Integration Technique (FIT) while ADS is 2D EM simulator based upon the Method of Moment (MoM). *Figure 6*

The 2D radiation pattern is given by Figure 7 in the E-plane and the H-plane, which show a stable and bi-directional radiation pattern for the two resonant frequency bands. The radiation at 2.45 GHz

Figure 5. (a) The Design evolution of the proposed antenna; (b) and return loss vs frequency of the proposed antenna for different cases on CST

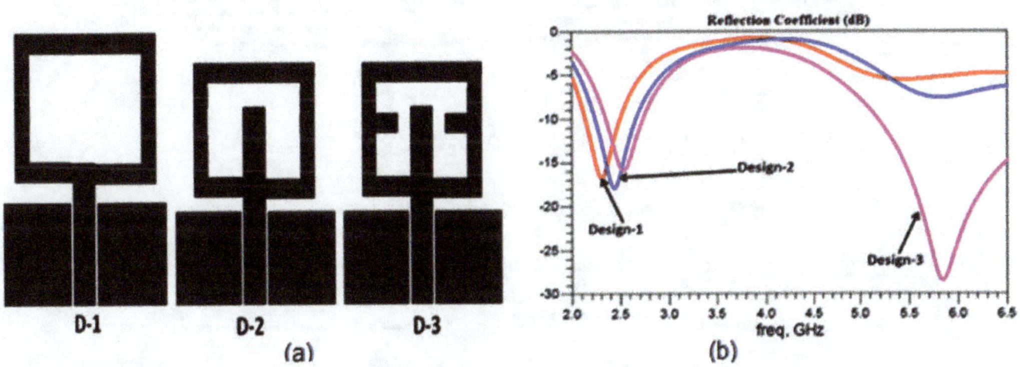

and 5.8 GHz is over z-axis. The antenna aperture angle at 2.45 GHz is approximately 80° and 140° at 5.8 GHz. The antenna gain is 1.9 dBi at 2.45 GHz and 2.7 dBi at 5.8 GHz.

The prototype of the investigated antenna fabricated by using the Chemical etching machine is given in Figure 8 (a). After the calibration, the return loss for the achieved antenna is tested. At the same time, both the simulations on ADS and CST with measurement results are compared as shown in the Figure 8 (b). Small disparities between the measured and simulated results are observed, due to cable effects, SMA connector and fabrication imperfection.

The reflection coefficient curve shows that the present antenna is fed at 2.45 GHz with a –10 dB return loss bandwidth of 200 MHz (2.3–2.5 GHz) and at 5.8 GHz with an impedance bandwidth of 800 MHz (5.2–6GHz).

The measured radiation pattern characteristics of the proposed antennas in E-plane and H-plane at 2.45GHz and 5.8GHz are presented in Figure 9. The measured results shows that the good omni-directional patterns in the H-plane and the nearly bidirectional patterns in the E-plane are obtained for the two working frequency bands.

Antenna Design 2

The second design is a compact dual-band monopole antenna fed by a CPW line, using a shorting stub between the monopole and the ground plane in order to produce the dual-band behavior. The designed and optimized antenna presents a good matching in the desired ISM bands. The first band [2.3–2.6] GHz with a bandwidth of 300 MHz covers the ISM 2.45 band, and the second band [5.4–5.85] GHz with an impedance bandwidth of 450 MHz covers the ISM 5.8 band. The radiation pattern presents acceptable characteristics for RFID applications.

Figure 7. Simulated 2D radiation pattern for the designed structure in E–plane at (a) 2.45 GHz and; (b) 5.8 GHz and in H-plane at; (c) 2.45 GHz and; (d) 5.8GHz

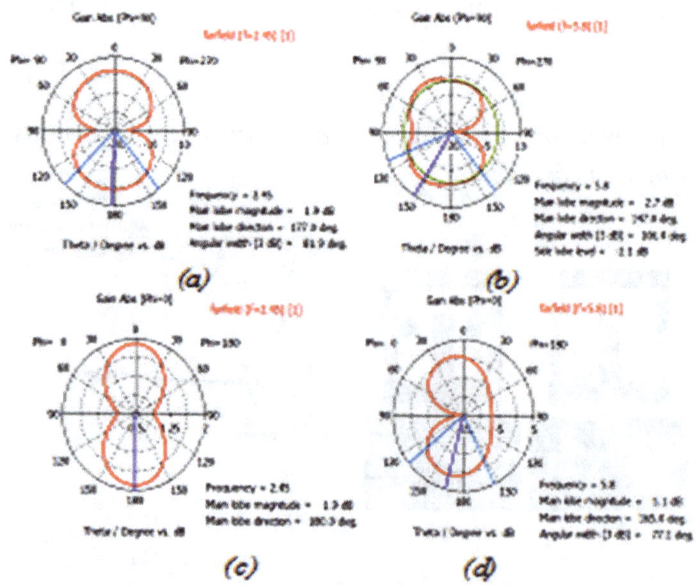

Figure 8. (a) Photograph of the fabricated structure; (b) and Comparison of simulated and measured return loss

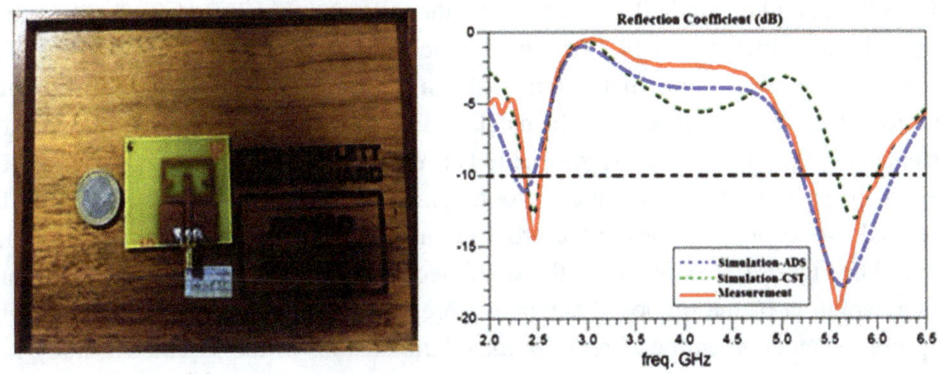

(a) (b)

Figure 9. Measured 2D radiation pattern for the designed structure in E–plane at (a) 2.45 GHz and; (b) 5.8 GHz and in H-plane at; (c) 2.45 GHz and; (d) 5.8GHz

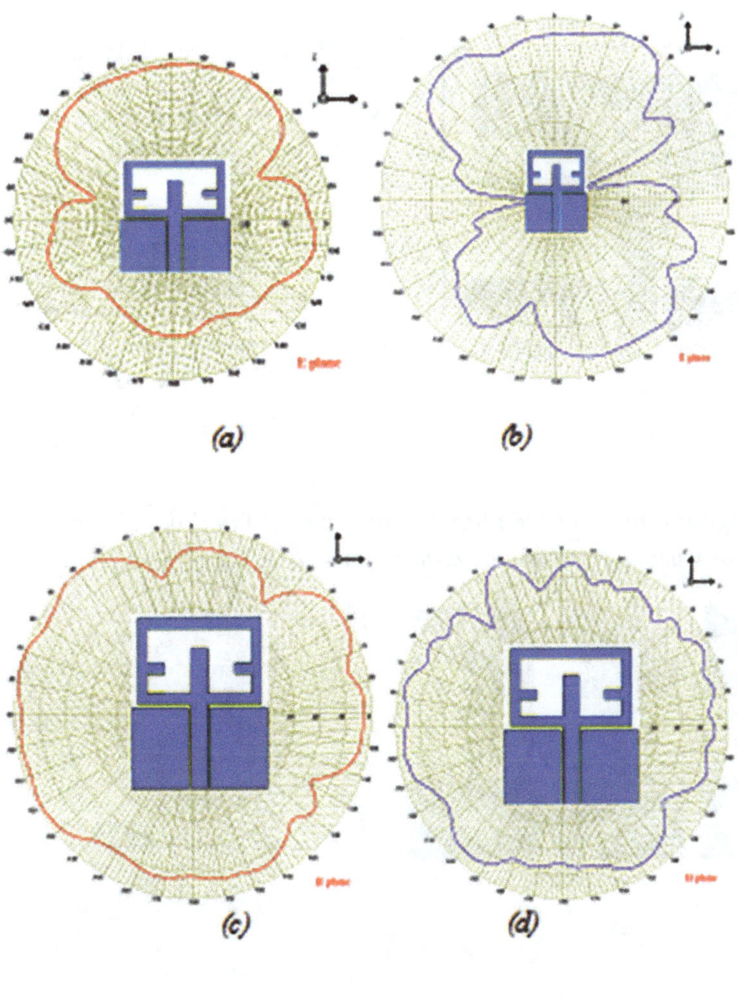

(a) (b)

(c) (d)

The configuration of the proposed planar dual-band antenna is illustrated in Figure 10 (a). This antenna has a total area of 33×30 mm^2 (Ws × Ls). The antenna is fed by a 50 Ohm coplanar line with a fixed strip G=3.8 mm and a gap S=0.5 mm between the strip and the ground. Two symmetrical finite ground planes with the same size of 14.1×9 mm^2 are etched on each side of the CPW line. The design of this antenna consists of a rectangular patch with some rectangular slots inserted in the upper side which permit to reduce the overall seize of the patch and better matching input impedance at the higher resonance frequency, also in order to exhibit the lower resonance mode we have introduced a shorting stub between the patch and the left side of the ground plane. By selecting a proper length of the ground plane, the desired two resonant modes of the proposed antenna can be excited with good input impedance matching. The Figure 10 (b) illustrates the simulated return loss resulting from simulation on CST.

The design evolution of the proposed antenna is presented in Figure 11 (a). The simulated return loss for the proposed antenna with different optimized dimensions of the parametric study is shown in Figure 11(b). Table 2 presents a synthesis of the return loss results for the two frequency bands. The final optimized dimensions of the antenna through EM simulations according to the Figure 10 (a) are listed in Table 3.

Figure 10. (a) Geometry of the CPW monopole antenna; (b) and Simulated return loss for the proposed antenna

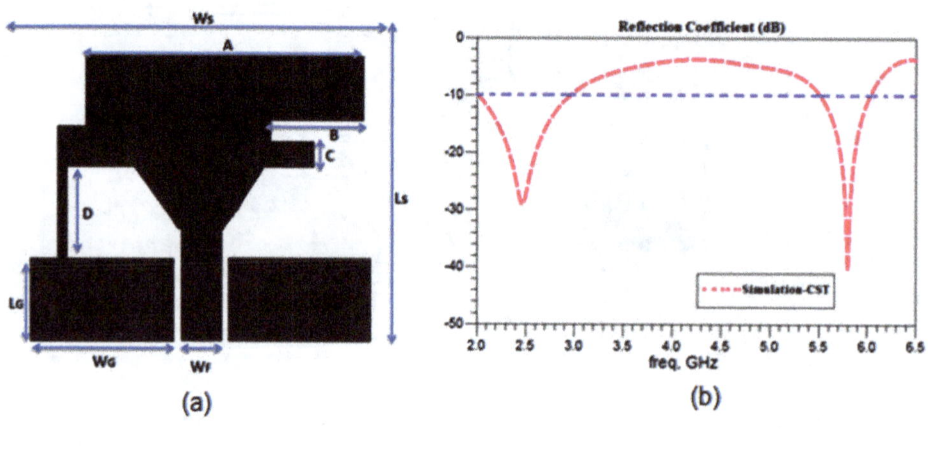

Figure 11. (a) Design evolution of the proposed antenna; (b) and the return loss vs frequency of the proposed antenna for various steps of the optimizations cases on CST antenna

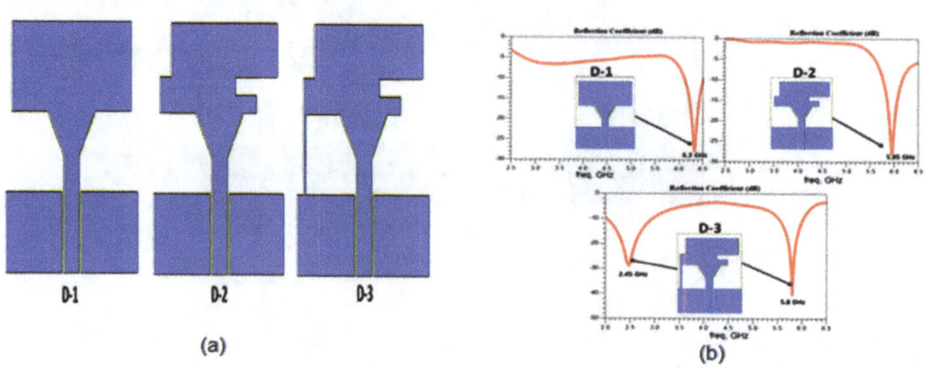

Table 2. Synthesis of simulation results using CST-MW

Operating Frequency Band		Resonant Frequency (GHz)	Return Loss (dB)	Bandwith (GHz)
Simulated results by CST-MW	First resonant	2.48	-30	2-2.8
	Second resonant	5.78	-41	5.4-6.02

Table 3. The dimensions of the CPW antenna (unit in:mm)

Parameter	Length
L_{sub}	30
W_{sub}	33
L_g	9
W_g	14.1
A	24
B	11
C	2
D	9
W_F	3.8

The 2D radiation pattern is given by Figure 12 in the E-plane and the H-plane. In order to complete the study of the structure a calculation of the gain variation over the operating frequency band is performed through the Far Field by using CST-MW at 2.45 GHz and 5.8 GHz as shown in Figure 13.

The gain of the proposed antenna is about 1.8 dBi and 4.6 dBi at the resonant frequencies of 2.45 GHz and 5.8 GHz respectively.

The prototype of the developed antenna is illustrated in Figure 14 (a). The return loss was measured by using Vectorial Network Analyzer (VNA) PNA-X from Agilent Technologies. A comparison between the simulated and measured results is shown in Figure 14 (b). The result presents good concordance between simulation and measurement.

The measured radiation pattern in E-plane and H-plane at 2.45 GHz and 5.8 GHz are presented in Figure 15. The measured results show that this antenna has a stable monopole-like radiation pattern with bidirectional radiations in the E-plane and a nearly omnidirectional pattern in the H-plane. The proposed antenna has an acceptable quasi omnidirectional and stable radiation pattern required to receive information signal.

In order to highlight the interesting performances of the designed antennas in term of size, resonance frequency and gain, a comparative study of the proposed antennas has been conducted with similar antennas presented in bibliography and the results are reported in the Table 4. It is clear from this table that the proposed works exhibit good features appropriate for dual-band RFID applications.

Figure 12. Simulated 2D radiation pattern for the designed structure in E–plane at: (a) 2.45 GHz and; (b) 5.8 GHz and in H-plane at; (c) 2.45 GHz and; (d) 5.8 GHz

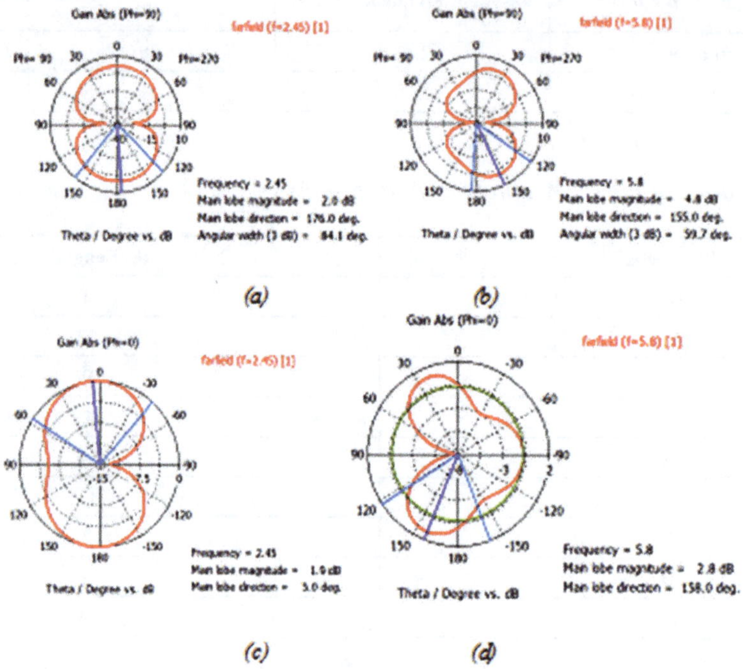

Figure 13. Simulated antenna gain vs frequency

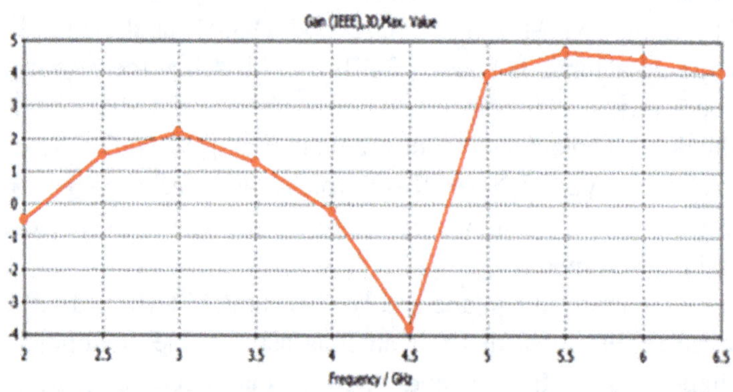

SECTION II: PRINTED ANTENNAS FOR RFID TAG

An RFID tag is formed by an integrated circuit connected to an antenna. In case of passive tags that are not fitted with the internal power supply, the power transfer between the tag antenna and the chip is an important factor in the design. The increasing of the available power at the tag will increase the read range (Ahson & Ilyas, 2008) which is a key factor in RFID system. In this context antenna has been paid great attention and the requirements for RFID tag antenna are the following (Dobkin, 2008):

Figure 14. (a) Prototype of the structure; (b) and Comparison of simulated and measured return loss

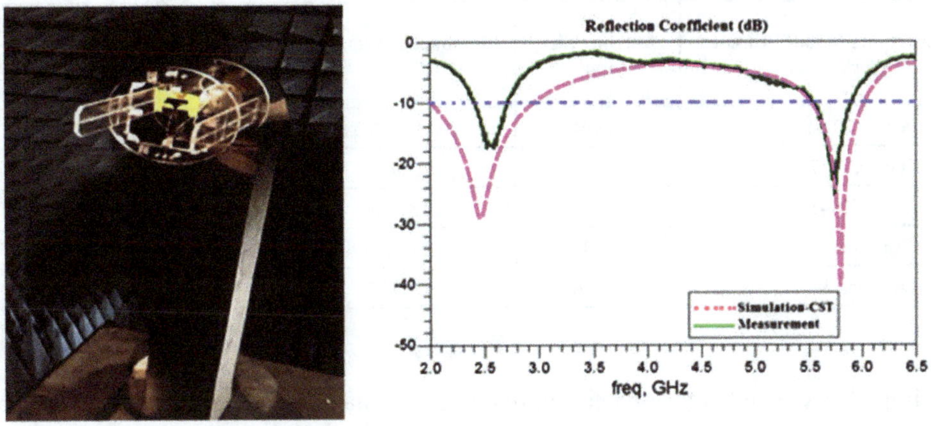

Figure 15. Measured 2D radiation pattern for the designed structure in E–plane at: (a) 2.45 GHz and; (b) 5.8 GHz and in H-plane at; (c) 2.45 GHz and; (d) 5.8 GHz

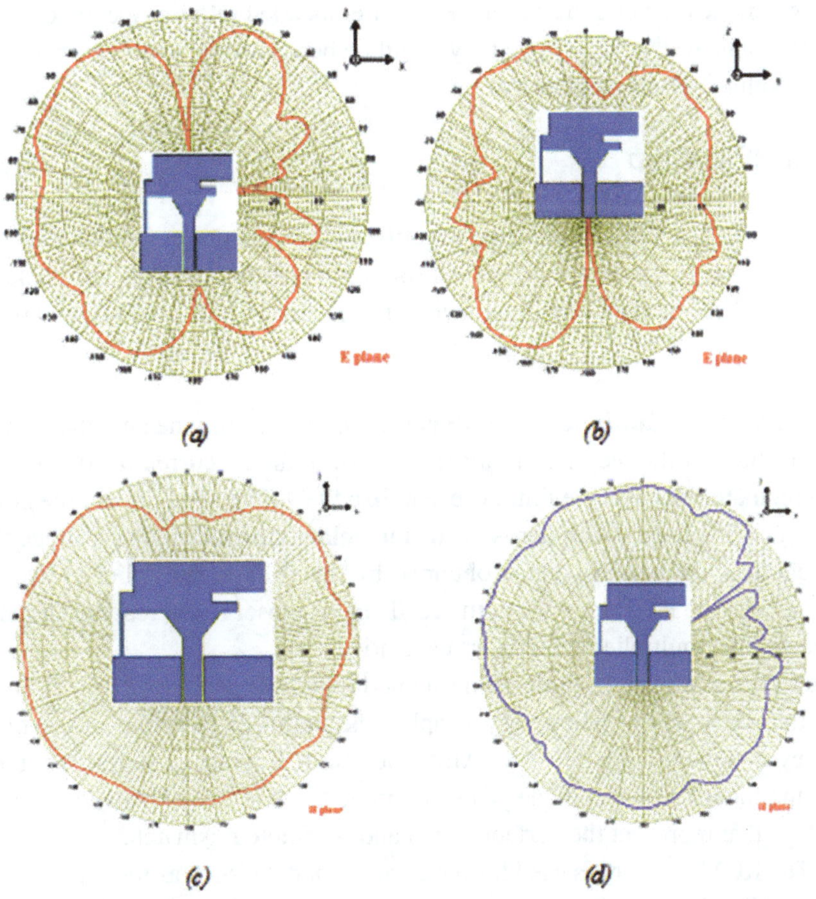

Table 4. The two designed antennas performances compared with other antennas in literature

Published Literature Versus Proposed Work	Antenna Size (mm^2)	Resonance Frequency (GHz)	Return Loss (dB)		Gain (dBi)	
			F1	F1	F2	F2
(Ihamji, 2017)	71×94	0.915/2.45/5.8	-11	-18.6	2.2	2.31
(Ennajih, 2017)	36×36	2.45/5.8	-14.33	-12.9	2.68	3
(Yang, 2014)	30×44	2.45/5.8	-18	-12	2.4	3.6
Antenna 1	35×30	2.45/5.8	-15	-17	1.9	2.7
Antenna 2	33×30	2.45/5.8	-13	-21	1.8	4.6

- Good impedance matching for receiving maximum signals from the reader to power up the chip.
- Insensitive to the attached object (tag placement) to keep the performance consistent, especially for metallic objects or near water.
- The antenna gain should be relatively high to achieve a long reading range.
- Required radiation patterns (omnidirectional, directional or hemispherical).
- Small enough and low profile to be attached to or embedded into the specified object.
- Robust in mechanical structure (since they could be bent in some applications).
- Low cost in both materials and fabrication.

Description of RFID Chip

A typical chip IC consists of four components: a rectifier, a modulator/demodulator, a control logic unit and a memory unit. Each component's power consumption or power transfer efficiency can influence the threshold power. These modules are illustrated in the Figure 16 and are discussed in the subsections below (Turcu, 2009).

- **Rectifier and Voltage Multiplier:** The RF power collected from the antenna is transmitted to the inside circuit through the rectifier circuit that converts the ac current to dc. The rectifier is followed by a regulator or a voltage limiter to stabilize the voltage and protect the chip from the risk of over-voltage. A voltage multiplier is needed to enlarge the induced ac voltage then the voltage regulator regulates the output voltage obtained by the voltage multiplier to avoid over-voltages that could damage the RFID chip when the RFID tag is closer to the reader. The impedance at the input of the voltage multiplier is therefore capacitive.
- **Modulator/ Demodulator Circuits:** The demodulation is performed by an envelope detector which has the advantage of being very simple. The uplink is therefore in the majority of cases performed by a modulation type ASK (Amplitude Shift Keying). The reverse link is done using backscattering modulation technique. The chip impedance is varied between two different states, Z_{chip1} and Z_{chip2} that represent the perfect match and complete mismatch.
- **Memory:** The ROM memory is used for the permanent data, such as the identification code and a RAM or EEPROM for managing the data in the case of tags in read / write mode.
- **Digital Control Unit:** This module allows the control and the execution of all the processes and commands of the integrated IC. This unit architecture is based on a microprocessor for analyzing

the received instructions, coding/decoding the information and responding by sending these data to the HF interface.

Fundamental Parameters of Tag Antennas

- **Link Budget in RFID Systems:** The analysis of wireless communication systems require the use of the Friis equation which express the received power P_r by an antenna in reception mode as a function of the power transmitted P_t by an antenna in transmission mode located in the far field. This configuration is illustrated in Figure 17 (Youbok, 2003).

The complete formula of Friis is given by Equation (6).

$$\frac{P_r}{P_t} = e_{\mathrm{cdt}} e_{\mathrm{cdr}} \left(1-\left|\Gamma_t\right|^2\right)\left(1-\left|\Gamma_t\right|^2\right)\left(\frac{\lambda}{4\pi R}\right)^2 D_t\left(\theta_t,\phi_t\right) D_r\left(\theta_r,\phi_r\right)\left|\widehat{p_t}\widehat{p_r}\right|^2 \quad (6)$$

Figure 16. RFID tag block diagram containing the RF-analog front end

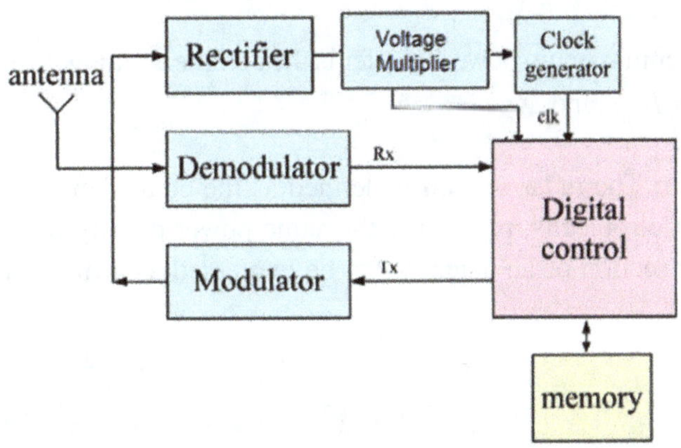

Figure 17. Illustration of a transmission antenna and reception antenna

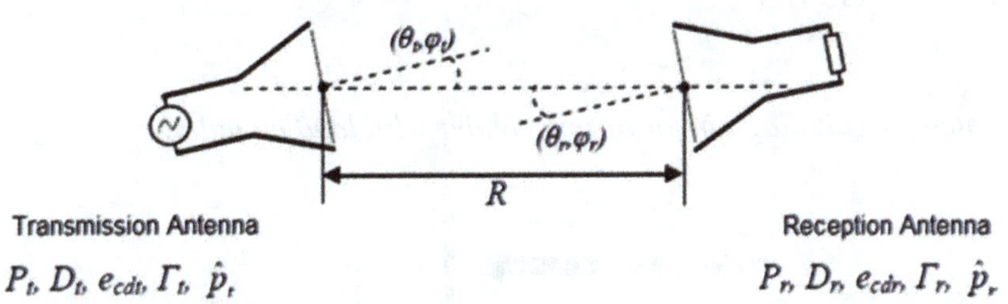

In this equation Γ_t, Γ_r are the reflection coefficients of the transmitting antenna and the receiving antenna respectively, D_t and D_r are the directivity of the transmitting and the receiving antenna respectively and e_{cdt}, e_{cdr} denote the antenna efficiency and $\left|\hat{p}_t \hat{p}_r\right|^2$ denote the polarization losses PLF (Polarization Loss Factor)

The term $(\frac{\lambda}{4\pi R})^2$ is called the loss factor of free space (or PL for Path Loss).

- **Power Transmission in a Tag:** The passive tag in receiving mode is represented by the Thevenin equivalent circuit. The RFID tag antenna is receiving signal from a reader's antenna and the signal is powering the chip in the tag. $Z_{ant} = R_{ant} + jX_{ant}$ is the input impedance of the antenna. The receiving antenna is connected to the load (chip) $Z_{chip} = R_{chip} + jX_{chip}$ by a transmission line of which the characteristic impedance is Z_0.

The power delivered to the chip is given by equation (7)

$$P_r^{chip} = \frac{\left|I\right|^2 R_{chip}}{2} = \frac{\left|V_{in}\right|^2 R_{chip}}{2\left[(R_{tant} + R_{chip})^2 + (X_{tant} + X_{chip})^2\right]} \tag{7}$$

The load impedance for maximum power transfer is the complex conjugate of the antenna impedance. Thus in this case $R_{chip} = R_{tant}$ .and $X_{chip} = -X_{tant}$.

- **The Radar Section:** The radar section is defined as the equivalent area intercepting a quantity of power radiated isotropically, producing the same power density reflected by the target at the receiver. The radar section of an antenna can be expressed as a function of the antenna gain as (Ghiotto, 2008):

$$\sigma = \left|1 - \Gamma^2\right|^2 \frac{\lambda^2}{4\pi} G^2 \tag{8}$$

- **The Load Modulation:** This technique is performed by modulating the impedance of the chip as described in the Figure 18.

Figure 18. Equivalent circuit diagram for a transponder with load modulator

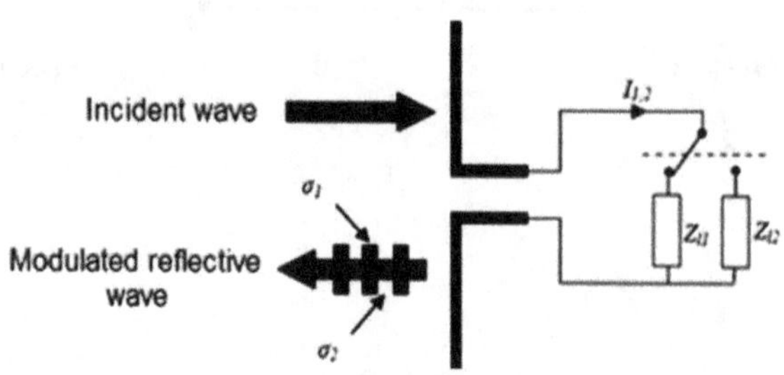

We consider a charge modulation between the impedances Z_1 and Z_2 corresponding to the two reflection coefficients Γ_1 and Γ_2. Since the reflected field is proportional to the current in the antenna, the differential reflected power equivalent to the radiated power by the antenna traversed by the difference of the complex currents I_1 and I_2 corresponding to the two impedance states Z_{l1} and Z_{l2}.

$$P_{\text{diff}} = \frac{1}{2}\left|I_1 - I_2\right|^2 R_a G \tag{9}$$

And:

$$I_{1,2} = \frac{V_0}{Z_a + Z_{l1,2}} = \frac{V_0}{2R_a}\left(1 - \Gamma_{1,2}\right) \tag{10}$$

Chip Impedance Properties

To achieve a maximum power transfer between the antenna and the chip, the impedance of the RFID IC must be conjugate of the antenna impedance. The passive RFID tag antenna impedance cannot be taken directly since RFID chip input impedance differs significantly from normal input impedance of 50 Ω. Each chip exhibits a specific complex input impedance depending on its internal architecture design, packaging and mounting procedure. Consequently, the modeling and measurement of the chip input impedance is crucial.

RFID chip input impedance is a complex value and presents a negative imaginary part. This impedance can be modeled either by a parallel or a series of equivalent circuit Rp, Cp, Rs, Cs, these parameters can be calculated according to Equations (11) (12) (13) (14) described in Figures 19 (a) and (b).

$$R_p = \frac{(\mathrm{Im}(Z_c)^2 + R_e\left(Z_c\right)^2)}{R_e\left(Z_c\right)} \tag{11}$$

$$C_p = \frac{\mathrm{Im}\left(Z_c\right)}{2\pi f(\mathrm{Im}(Z_c)^2 + R_e\left(Z_c\right)^2)} \tag{12}$$

$$R_s = R_e\left(Z_c\right) \tag{13}$$

$$C_s = \frac{1}{2\pi f.\mathrm{Im}\left(Z_c\right)} \tag{14}$$

Table 5 shows the modeling parameter of a standard RFID chip an "Impinj Monza 1a" (Impinj, 2006). passive tag microchip having input impedances of 36-j117Ω at 866MHz in Europe, 33-j112Ω at 915MHz in North and South of America and 30-j108Ω at 953MHz in was used in Japan. This microchip is compatible with the EPC Global Class-lGeneration-2 UHF RFID specification and the minimum threshold power of this RFID chip is -9 dBm (126 µW). The parameters of the equations are as follows:

Direct chip attachment requires high degree of precision and cannot be done manually. For this reason we use strap attachment where the RFID IC is connected to two mounting pads with a thin superstrate, the strap is then attached manually to the antenna by conductive epoxy. This technique requires an estimation of the strap impedance when attached with conductive epoxy.

Conjugate Impedance Matching

The high capacitive impedance of the chip requires a complex conjugate impedance matching at the antenna terminals in order to maximize the power transfer between the antenna and the chip. This becomes a real problem because the chip impedance changes at different power levels and frequencies (Pillai, 2006). As a solution, the IC impedance should be calculated at minimum threshold power and the antennas must be highly inductive for matching to the input impedance of the transponder IC. Due to the requirements such as easy manufacture, low cost and small size, adding the matching network such as inductance is infeasible. To solve this problem, the antenna should be able to match the chip directly by adjusting its structure. In this section authors present some of the common matching techniques used for this purpose.

- **Shunt Stub (T-) Matching:** A typical example consists on the use of structure combination of inductances mounted in parallel or in series. Such approach is shown in Figure 20. These inductances can be made by conductive lines connecting the antenna to the chip.

Figure 19. Equivalent circuit of the chip impedance: (a) Parallel circuit, (b) Series circuit

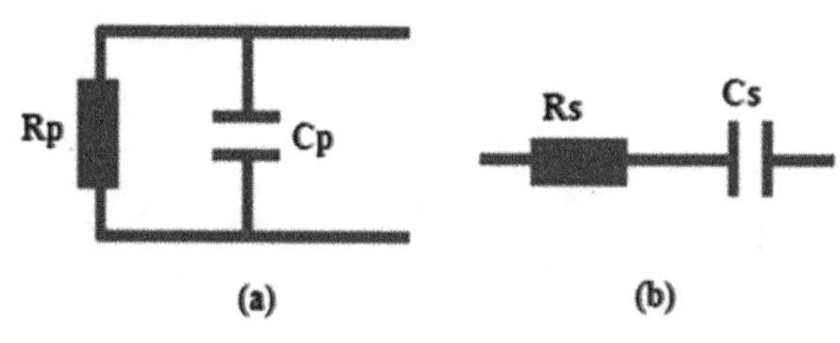

Table 5. The Model Parameter of the proposed Chip

Impinj Monza 1a	915 MHz
Z_C	33-j112Ω
R_S; C_S	33 Ω ; 1.553 pF
R_P; C_P	413.12 Ω ; 1.429 pF

Figure 20. Inductive stub matching of the antenna against the capacitive chip

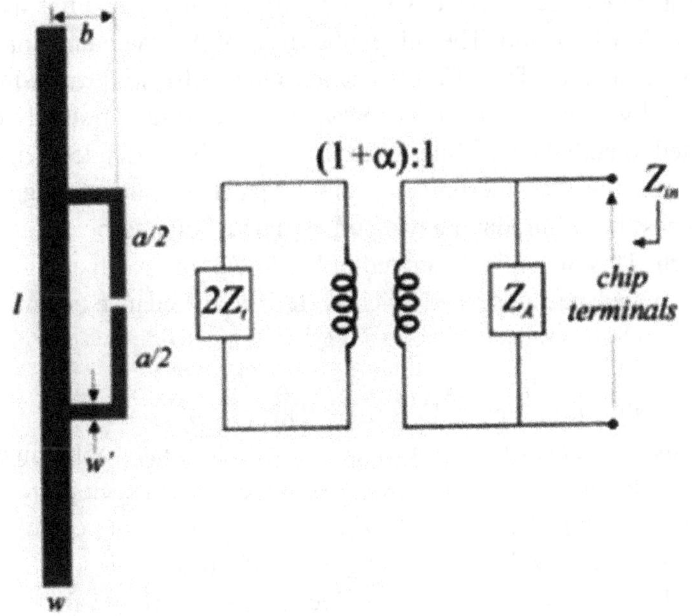

- **Inductive Loop Matching:** In this configuration we use a metal loop close to the antenna to exploit the mutual coupling property to adjust the impedance of the structure (Figure 21). The parameters of the impedance match are the loop and antenna dimensions and the spacing between the loop and the antenna. The input impedance of the antenna and the self-inductance of the loop are calculated from the analytical formulas (Son & Pyo, 2005).
- **Nested Slot Matching:** This approach is based on inserting slots inside the metal surface. Slots are equivalent to inductances when the operating frequency is high enough (UHF band) which permits to store and radiate electromagnetic energy. The slot shape is an important parameter since it's considered as a line-slot impedance transformer (Marrocco, 2008). The sizes of the slots can be comparable to the surface of the patch, so that the radiation characteristics are related to the two objects (patch, slots). In other words, the maximum gain of the antenna is fixed mainly by the patch length, while the impedance value is controlled by the dimensions of the slot. The equivalent inductance of the slots can be used to cancel the capacitance of the chip.

Figure 21. An inductive loop near the main radiator (left) and a circuit interpretation (right)

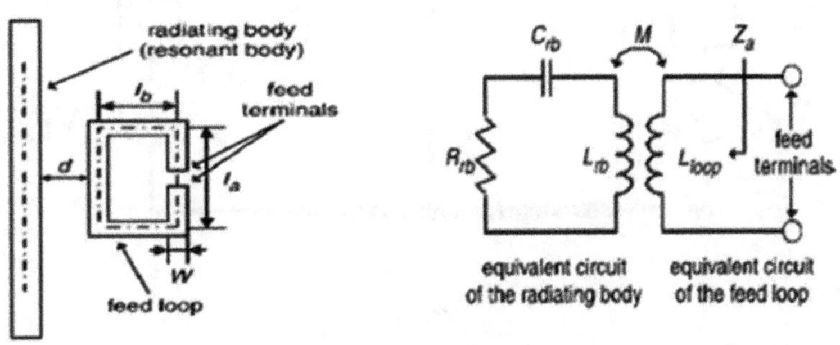

- **Single Shorted Matching Stub for Balanced Feed Microstrip Patch Antenna:** An RFID chip is considered as a symmetrical device with two ports; in this case a balanced feeding is used to connect a symmetric RFID chip. The differential feed of the patch antenna is achieved by using the two symmetric feed lines. The RFID IC is connected to the two feeds. Since both the feed lines need to be matched we must either use two matching networks or simply construct a matching network connected to both the feed lines. The shorting stub is extended symmetrically to connect with the second feed line. Considering a rectangular patch antenna of length L and width W with a dielectric thickness h, by introducing two ports P_1 and P_2 symmetrically at a distance l_2 from the center of the patch. This model is illustrated in Figure 22, the patch antenna is equivalent to two radiating slots of impedance Z_R spaced by characteristic impedance transmission line Z_A (Eunni, 2004).

The matching circuit shown in Figure 23 is used. This circuit consists of an impedance transmission line Z_F. For reasons of symmetry, the impedance seen at the center of the line Z_S is zero. The total impedance seen between P_1 and P_2 is equal to twice the impedance at each port.

The determination of the impedances Z_{in}, Z_F, Z_S, Z_a and Z_R permits to calculate the lengths l_1, l_2, l_3 and l_4 in order to carry out the impedance matching of the antenna.

In practice, this analytical method makes it possible to have a starting value for the different lengths of lines. The existing coupling between the lines requires the use of simulation tools to optimize the line lengths taking into account neglected effects in this model.

Miniaturization Techniques for Tag Antennas Design

Currently the miniaturization of RFID tag antennas has been paid a great interest for many reasons:

Figure 22. Circuit model of a rectangular microstrip patch with plane of symmetry with two symmetrical ports P1 and P2

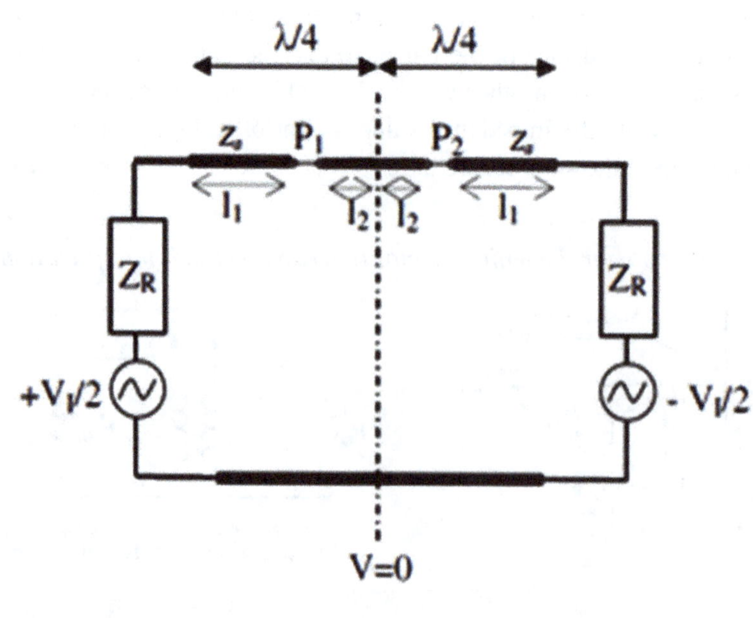

Figure 23. Circuit model of microstrip patch antenna with balanced feed matching network

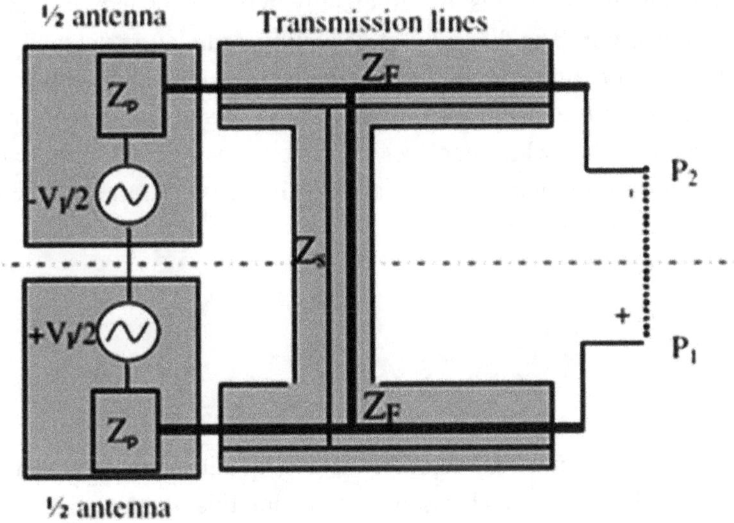

- The surface of the tag is mainly formed by the antenna; hence the reduction of tag size depends on the reduction of antenna size.
- Decreasing the cost of RFID tags by reducing the quantity of material involved in the fabrication process and by allowing higher production rates.
- Allowing the identification of small objects and enabling their individual identification.

Actually, many techniques of antenna size reduction have been introduced; some of them are used especially for RFID tag antennas design. In this section three size reduction strategies successfully used for RFID tags are reviewed.

- **Fractal Geometry:** By using pre-fractal structure the total dimension of the antenna can be reduced while keeping the same effective length. The matching of a fractal antenna to an RFID IC chip is obtained using the principle described in (Salama & Quboa, 2008). This principle is based on the use of a conductive strip placed in parallel with the pre-fractal structure which affects the antenna impedance by increasing the inductance. Thus by optimizing the distance between the element and the antenna, it is possible to obtain the impedance matching of the RFID chip.
- **Meandering:** By folding the straight main radiator along a meandered path, a wire configuration is produced with distributed capacitive and inductive reactance's that globally affect the antenna's input impedance. Resonances are achieved at much lower frequencies than in the case of a straight dipole of the same height at the expense of a narrow bandwidth and low efficiency. It has also been shown in (Marrocco, 2003; Tang & Li, 2015) that the resonant frequency is considerably affected by the lengths of the horizontal segments.
- **Insertion of Slots:** This technique consists of the addition of one or more slots on the radiator. This slot has the effect of extending the current lines resulting in a reduction of the resonant frequency. In the works of (Dehbashi, 2006), a comparison of the original antenna to the miniatur-

ized antenna by inserting a U-shaped slot shows that the reduction rate of about 58% and 40%. However, the miniaturized antenna is less efficient with reduced bandwidth and small gain.

- **Limits of Miniaturization:** The difficulty of designing a miniature antenna is related to making a compromise between obtaining a good efficiency and a wide bandwidth while remaining miniature. In the works of Chus and Wheeler's (Chu, 1948) an antenna is considered as very small if there exists a sphere of radius which encloses this antenna $a = \lambda / 2\pi$. In this case the fundamental limit as a function of the minimum quality factor Q_{min} based on its largest dimension a is expressed by equation (15) (McLean, 1996):

$$Q_{in} = (\frac{1}{\mathrm{ka}})^3 + \frac{1}{\mathrm{ka}} \; k = \frac{2\pi}{\lambda} \quad (15)$$

By decreasing the size of the antenna, its coefficient of quality increases, as the bandwidth of an antenna depends on its quality factor which results in reducing this parameter.

The equation (16) gives the limit of the maximum gain of an antenna according to its largest dimension. It is expressed by (Harrington, 1958):

$$G_{max} = (\mathrm{ka})^2 + 2\mathrm{ka}\left(\mathrm{dBi}\right) \quad (16)$$

Thus, we see that the miniaturization of antennas has limitations in terms of quality coefficient, bandwidth, gain and efficiency. It is about making compromises according to the expected performances.

Design and Optimization of New Printed Antennas for RFID Tag

The key challenge for RFID tag design is to maximize the reading distance under the various environment constraints (such as performing in complex environments containing water based or metallic items, and cost), such configuration depends mainly on the antenna performances along with the selected microchip (Ghiotto & Cantalice, 2008; Yang & Basat, 2006). In this section the authors propose two antenna designs, the first one is based on flexible organic substrates and the second design is suitable for RFID tags operating close to metallic objects.

Antenna Design 1

The proposed antenna is shown in Figure 24. It consists of a thin copper layer printed on a cheap HP Adv. Photo-paper substrate with a dielectric constant =3.3, dielectric loss tangent $\delta = 0.04$ and thickness H=0.5 mm. The overall size of the proposed tag antenna is relatively small (72×26 mm^2) compared with that of most commercial RFID tags. The dimensions of the antenna are listed in the Table 6.

The antenna is comprised of a double quarter wavelength monopoles ($\lambda/4 = 75$mm at 915 MHz) in which a meandering technique is applied to reduce the overall dimension at the working frequency and a rectangular loop for feeding the microchip to which the meandered structure is coupled inductively. By adjusting the separation L_5 and meandered lines lengths (L_1, L_2) of the two radiation elements, the

Figure 24. Geometry of proposed tag antenna

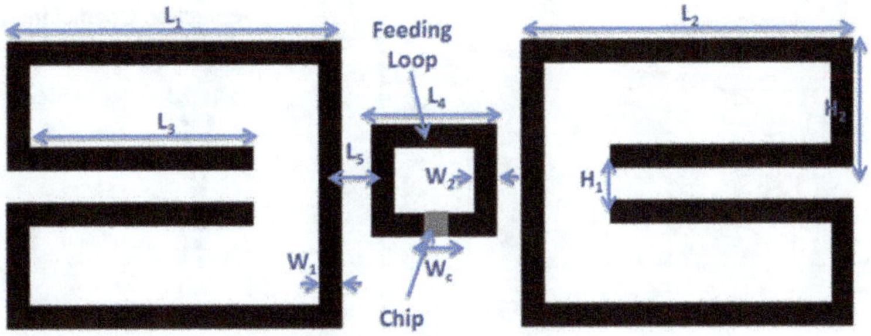

antenna impedance can be adjusted to match the specific chip impedance. For this design, an "Impinj Monza 1a" microchip is used.

Figure 25 shows the simulated reflection coefficient versus frequency of the proposed tag antenna obtained by using CST Microwave Studio (left) and the simulated reflection coefficient versus frequency of the proposed tag antenna obtained by using ADS software (right). We can observe a difference in return loss obtained with CST and ADS due to the technique of calculation implemented in these two softwares.

The simulated bandwidth obtained by CST microwave studio and ADS simulator is 56MHz (0.892-0.948 GHz) and 55 MHz (0.865-0.920 GHz), respectively, which covers the operating frequency of UHF RFID in North and South of America (902-928 MHz).

Figure 26 shows the simulated reflection coefficient S11 (dB) of the RFID tag antenna in free space for different values of the distance L_5 separating the feeding loop from the two dipole arms. As it can be seen from this study, the resonant frequency remained unchanged but the return loss varied between -13.4dB and -33dB.

The impedance responses of the RFID UHF tag antenna obtained by using CST Microwave Studio for the commercial RFID chip are shown in Figure 27. The impedance at the center frequency of 915MHz is 33.4+j111.3 Ω, which means that there is a good conjugate matching between the RFID chip and the antenna.

Table 6. The dimensions of the meander dipole (unit in:mm)

Parameter	Length
L_1	27.9
L_2	27.9
L_3	18.6
L_4	12.2
L_5	1.5
W_1	2
W_2	1.8
W_C	4
H_1	11
H_2	2

Figure 25. Simulated return loss S11 using CST (left) and simulated return loss using ADS (right)

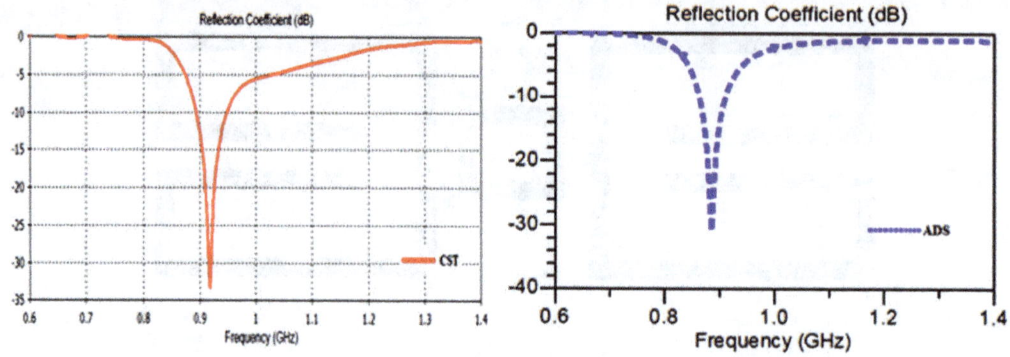

Figure 26. Simulated reflection coefficient versus frequency for different L_5 values

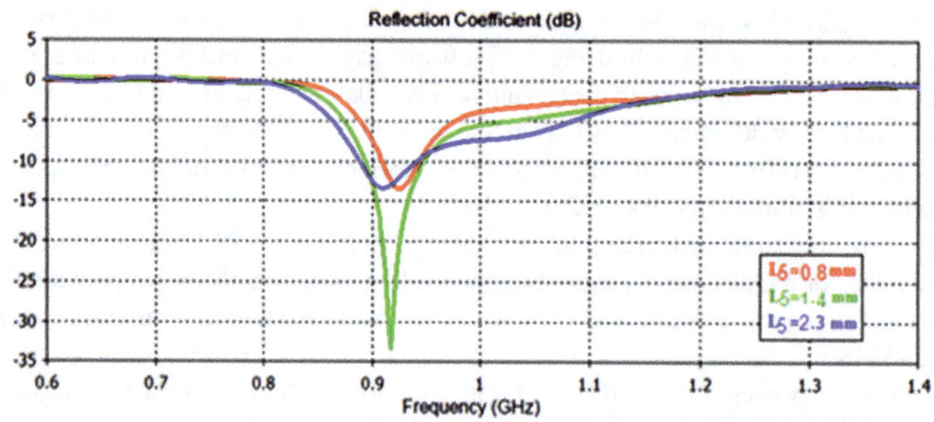

Figure 27. Simulated impedance of the proposed RFID tag antenna (left) Real part, (right) Imaginary part

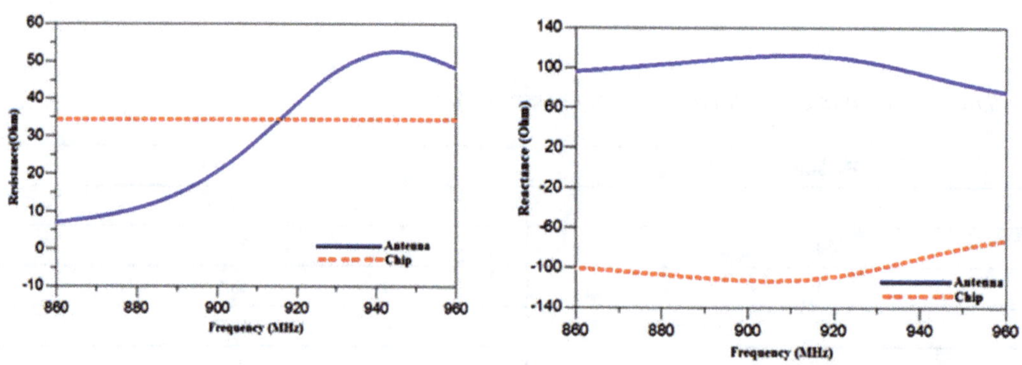

Next, we have simulated the reading range, which is the most important feature to get an understanding about the practical capability of the proposed RFID tag antenna. The read range can be obtained using Friis free space simplified formula:

$$r = \frac{\lambda}{4\pi}\sqrt{\frac{p_t G_t G_r \tau}{p_{th}}} \tag{17}$$

Where r is the theoretical maximum read range, λ is the wavelength, P_r is the radio power of the reader, G_r and G_t are the antenna gains of the reader and tags antennas, respectively, P_{th} is the sensitivity of the RFID chip, and τ is the power transmission coefficient given by:

$$\tau = 1 - \left|\Gamma\right|^2 \tag{18}$$

where: Γ is the reflection coefficient at the antenna input terminal.

The Federal Communications Commission (FCC) allows a maximum EIRP about 36 dBm and the reader emitted power was set to 4W EIRP for the American band and the transmitting antenna has a gain of 6 dB. The read range is calculated for this value and can be estimated by 5.77 m.

The simulated peak gain of this antenna at the central frequency of 915 MHz is about 1.88 dB. The 2D radiation patterns are given by Figure 28 in the plane E (left) and in the plane H (right) respectively.

The E-plane (y z) is dipole like and H-plane (x-y) is slightly omnidirectional, and these patterns are consistent with those of a typical dipole antenna.

Antenna Design 2

Nowadays, performing in a complex environment including metallic objects such as cars, containers etc, is a challenging task for most of RFID system designer. The closest metallic objects influence the RFID system and can create destructive interference. It has been observed that basic dipole antenna suffers from performance degradation when used for metallic applications. The effects of these undesired phenomena

Figure 28. 2D radiation patterns in E –plane (left) and in H –plane (right) for the designed structure at resonant frequency, for 915MHz

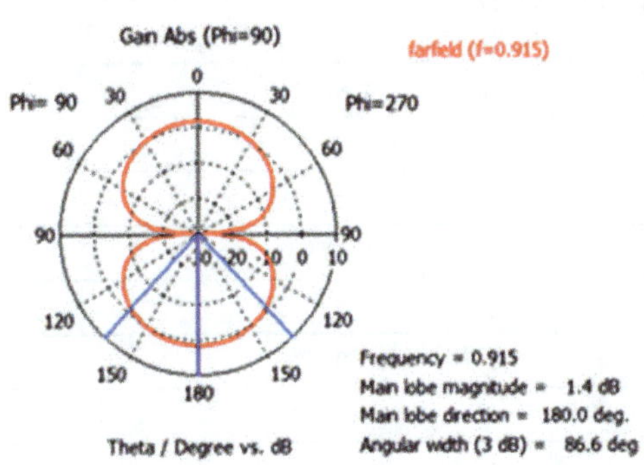

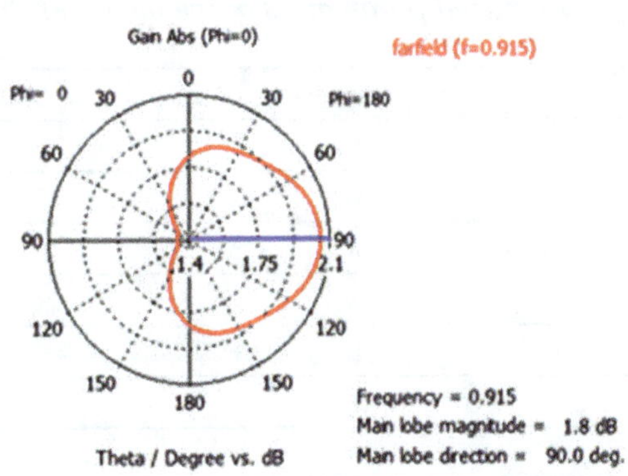

were thoroughly studied in (Mo & Zhang, 2008; Faudzi, 2014). A solution to this problem is the use of the microstip patch antenna where the metal surface act as an extension of the antenna ground plane.

In this work, a new miniaturized RFID Tag antenna for metallic objects is proposed. The initial value of the length W_P is a half wavelength and the initial length of the radius of the nested slot R is quarter wavelength ($\lambda/2 = 150$ mm at 915 MHz). By properly choosing the dimension of the inserted elliptical slot located in the middle of the radiating patch and the dimensions of the T-match circuit following the methodology described above, the conjugate matching between the antenna and the RFID chip can be easily achieved. There are no vias in the proposed antenna, which makes it easy to fabricate. The simulated bandwidth of the proposed antenna is 28 MHz (900–928 MHz). The geometry of the proposed tag antenna is depicted in Figure 29 (left).

The antenna is mounted on an FR4 substrate of 57×26 mm^2 with a dielectric constant=4.4, dielectric loss tangent $\delta = 0.025$ and thickness H=0.5 mm. For the design of this tag, a "MURATA -LXMS31ACNA – 010 " (Murata product, 2017) passive tag microchip having input impedances of 17.6-j100.9Ω at 915 MHz in North and South of America, and the minimum threshold power to activate the chip is -8 dBm. Therefore, the input impedance of the antenna has to be 17.6+j100.9Ω so that maximum power transfer can be reached. The dimensions are reported in the Table 7.

Figure 29 (right) shows the simulated reflection coefficient versus frequency of the proposed tag antenna obtained by using CST Microwave Studio and ADS softwares. We can observe a good agreement in return loss obtained with CST and ADS. The simulated bandwidth obtained by CST microwave studio and ADS simulator is 28MHz (0.900-0.928 GHz) and 26 MHz (0.900-0.926 GHz), respectively.

The Figure 30 (right) shows the configuration of the antenna in the presence of a metallic surface. The simulation of the return loss as a function of the distance h between the antenna and a metal plate (100×100 mm^2) is illustrated in the Figure 30 (left). No significant changes in the matching or bandwidth of the antenna are observed which shows that the tag is not influenced by the presence of the metallic plane.

Based on the values introduced in the first design, the calculated read range is 5.74m.

The simulated peak gain of this antenna at the central frequency of 915 MHz is about 1.6dB.

As shown in Figure 31, the E-plane (y-z) is directional due to the presence of the ground plane acting as a reflector and the antenna radiate bi-directionally in the H-plane (x-y).

To showcase the interest of the developed structures compared to those presented in bibliography, authors present in Table 8 a performance comparison based on relevant parameters characterizing the choice of antennas specifically size, gain and return loss. It can be easily deduced from the results of

Table 7. Optimized parameters of the meander dipole (unit in:mm)

Parameter	Length
L_P	26
W_P	57
W_1	3
W_2	5.6
L_1	9.8
R	8
W_C	4

Figure 29. Structure of the nested slot antenna (left) and comparison of simulated return loss using CST and ADS (right)

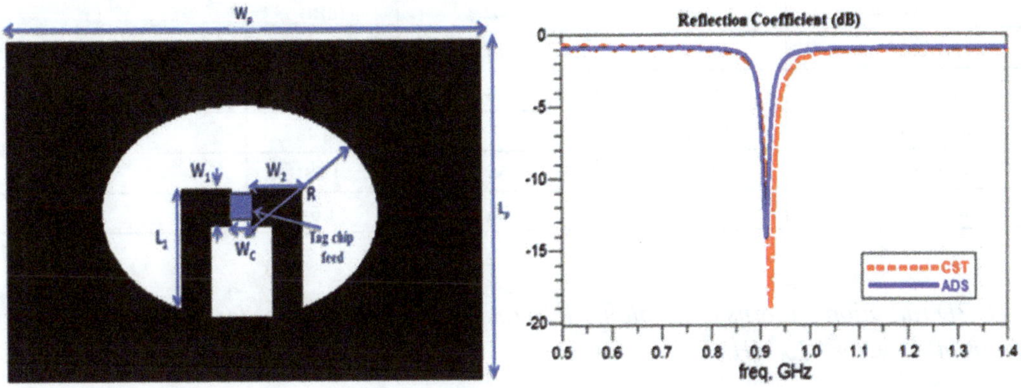

Figure 30. The simulated return loss for different separation value (left) and Structure of the antenna standing on a metallic plate (right)

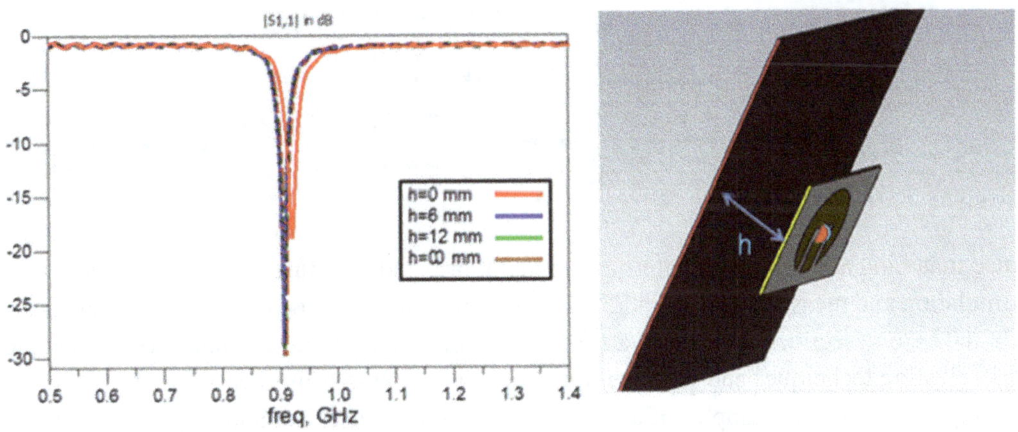

Table 8 that the antennas conceived in the second part of the chapter have serious advantages over other works mentioned in references.

CONCLUSION

In this work, four new antennas based on microstrip technology were presented with a bibliography synthesis on microstrip feeding, multiband behavior and miniaturization techniques used behind. The optimization of the designs is based on the two electromagnetic solvers CST Microwave studio and ADS software "Advanced Design System". The chapter is divided into two sections. In the first section a study of theoretical and practical aspects of multiband antennas has been done, including some common used techniques for this purpose, to illustrate this, two new dual band antennas structures are designed and validated in the ISM bands at 2.45 GHz and 5.8 GHz showing a good performance in terms of input

Table 8. Performances comparison between the two designed antennas and some existing antenna

Published Literature Versus Proposed Work	Antenna Size (mm^2)	Bandwidth (MHz)	Gain (dBi)	Read Range in Free Space (m)
(Bashri, 2015)	130×63	40	-6.05	6.18
(Lukas, 2008)	60×60	30	-	0.14
(Ghiotto, 2008)	99.6×99.5	7	-3.4	0.93
Antenna 1	72×26	55	1.88	5.78
Antenna 2	57×26	28	1.6	5.7

Figure 31. 2D radiation patterns in E –plane (left) and in H –plane (right) for the designed structure at resonant frequency, for 915 MHz

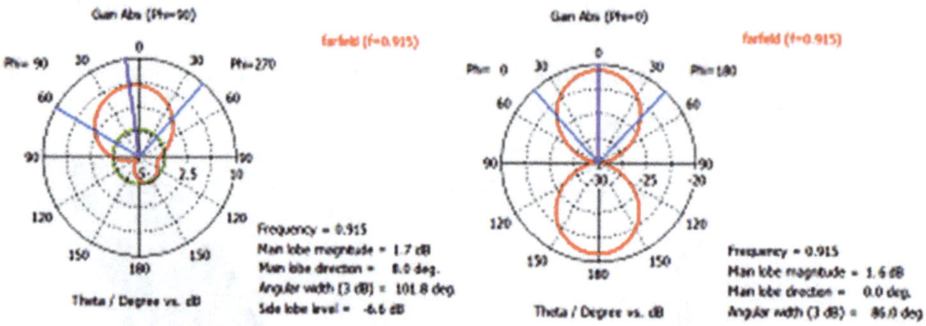

impedance matching and radiation pattern. The final fabricated structures show a good accordance between simulation and measurement results. These two antennas are good candidates for handheld RFID reader. In the second section, the state of art for the design of RFID Tag antenna is described including the miniaturization techniques and some common matching methods used between the antenna and the IC microchip and then two examples of antennas are presented. The first antenna based on flexible HP photo-paper substrate is validated into simulation in the UHF RFID in North and South of America (902-928 MHz); it presents a good input impedance matching and radiation pattern. The second structure is a microstrip slotted patch antenna using FR4 substrate, with the presence of the metallic ground plane, this antenna is capable to operate well on metallic objects.

REFERENCES

Ahson, S., & Ilyas, M. (2008). RFID Handbook Applications, Technology, Security and Privacy. Taylor & Francis Group. doi:10.1201/9781420055009 doi:10.1201/9781420055009

Bahl, J., & Bhartia, P. (1981). Microstrip Antennas. Artech House.

Balanis, C. A. (2003). Antenna theory: Analysis and Design. J. Wiley & Sons.

Bashri, M., & Ibrahimy, M. (2015). Design of a wideband inductively coupled loop feed patch antenna for UHF RFID tag. Wuxiandian Gongcheng.

Brown, D. (2007). RFID Implementation. New York: McGraw-Hill.

Chandra, K. N. (2010). Handbook of Smart Antennas for RFID Systems. John Wiley & Sons.

Chen, H. D., & Chen, H. T. (2004). A CPW-Fed Dual-Frequency Monopole Antenna. IEEE Transactions on Antennas and Propagation, 52(4), 978–982. doi:10.1109/TAP.2004.825620 doi:10.1109/TAP.2004.825620

Chu, L. J. (1948). *Physical limitations on omni-directional antennas.* Massachusetts Institute of Technology. Technical Report No. 64.

Dehbashi, R., Atlasbaf, Z., & Forooraghi, K. (2006). New Compact Size Microstrip Antennas With Harmonic Rejection. IEEE Antennas and Propagation Letters, 5, 395–398. doi:10.1109/LAWP.2006.881918 doi:10.1109/LAWP.2006.881918

Dobkin, D. M. (2008). *The RF in RFID Passive UHF RFID in Practice.* Elsevier Inc.

Ennajih, A., & Zbitou, J. (2017). Dual Band Metamaterial Printed Antenna Based on CSRR For RFID Applications. *International Journal of Microwave and Optical Technology.*

Eunni, M. B. (2004). *A Novel Planar Microstrip Antenna Design for UHF RFID* (Master's Thesis). Madras University.

Faudzi, N. M., & Ali, M. T. (2014). Metal mountable UHF-RFID tag antenna with meander feed line and double T-match. *IEEE International Symposium on Technology Management and Emerging Technologies.* doi:10.1109/ISTMET.2014.693647310.1109/ISTMET.2014.6936473

Finkenzeller, K. (2010). RFID Handbook: Fundamentals and Applications in Contactless Smart Cards and Identification. New York: John Wiley and Sons. doi:10.1002/9780470665121 doi:10.1002/9780470665121

Foster, P., & Bueberry, R. (1999). Antenna Problems in RFID Systems. IEE Coll. on RFID Technology, 1(3), 1–5.

Ghione, G., & Naldi, C. (1984). Analytical Formulas for Coplanar Lines in Hybrid and Monolithic Mics. Electronics Letters, 20(4), 179–181. doi:10.1049/el:19840120 doi:10.1049/el:19840120

Ghiotto, A., & Cantalice, S. F. (2008). Miniaturized Patch Antenna for the Radio Frequency Identification of Metallic Objects. *Proceedings of the IEEE MTT-S International Microwave Symposium (IMS2).*

Ghiotto, A. (2008). *Conception d'Antennes de Tags RFID UHF, Application à la Réalisation par Jet de Matière* (Doctoral Thesis). Institut Polytechnique de Grenoble.

Harrington, R. F. (1958). On Gain and Beamwidth of Directional Antennas. IRE Transactions on Antennas and Propagation, 6(3), 219–225. doi:10.1109/TAP.1958.1144605 doi:10.1109/TAP.1958.1144605

Hebib, S. (2008). *Nouvelle Topologie d'antennes Multi-bandes pour Applications Spatiales* (Doctoral dissertation). Université de Toulouse.

Huang, Y., & Boyle, K. (2008). Antennas: from Theory to Practice. Chichester, UK: John Wiley & Sons Ltd. doi:10.1002/9780470772911 doi:10.1002/9780470772911

Ihamji, A. (2017). Design of Compact Tri-Band Fractal Antenna for RFID Readers. *International Journal of Electrical and Computer Engineering*.

Impinj, Inc. (2006). *Monza-1a EPC global Generation 2 RFID*. Retrieved from http://www.orangetags.com/rfid-chipdatasheet/impinj

Lukas, I., & Qing, X. (2008). UHF Omnidirectional Bent Dipole Antenna for Item-level RFID Applications. *IEEE Antennas and Propagation Society International Symposium*. doi:10.1109/APS.2008.461958910.1109/APS.2008.4619589

Maicas, J.P. (2011). *Recent Developments in Mobile Communications – A Multidisciplinary Approach*. InTechOpen.

Marrocco, G. (2003). Gain-Optimized Self-Resonant Meander Line Antennas for RFID Applications. IEEE Antennas and Wireless Propagation Letters, 2, 302–305. doi:10.1109/LAWP.2003.822198 doi:10.1109/LAWP.2003.822198

Marrocco, G. (2008). The Art of UHF RFID Antenna Design: Impedance-matching and Size-reduced Techniques. IEEE Transactions on Antennas and Propagation, 50(11), 66–79. doi:10.1109/MAP.2008.4494504 doi:10.1109/MAP.2008.4494504

McLean, J. S. (1996). A Re-Examination of the Fundamental Limits on the Radiation Q of Electrically Small Antennas. IEEE Transactions on Antennas and Propagation, 44(5), 672–676. doi:10.1109/8.496253 doi:10.1109/8.496253

Mo, L., Zhang, H., & Zhou, H. (2008). Broadband UHF RFID tag antenna with a pair of U slots mountable on metallic objects. Electronics Letters, 44(20), 1173–1174. doi:10.1049/el:20089813 doi:10.1049/el:20089813

Molisch, A. F. (2011). Wireless Communications. John Wiley & Sons Ltd.

Murata Co. Ltd. (2017). *Murata -LXMS31ACNA – 010, EPC Global Gen2*. Retrieved from http://www.murata.com//en-eu/products/rfid/rfid/uhf

Necibi, O., & Gharsallah, A. (2016). A novel RFID antenna with HIS structure for Ka band. International Journal on Communications Antenna and Propagation, 6(6), 30–38.

Nieto, A. L. (2011). RFID Design Fundamentals and Applications. Taylor and Francis Group.

Paret, D. (2008). RFID en Ultra et Super Hautes Fréquences UHF-SHF, Théorie et mise en œuvre. Dunod.

Pillai, V. (2006). Impedance Matching in RFID Tags: To Which Impedance to Match? *Proceedings of The IEEE Antennas and Propagation Society International Symposium*. doi:10.1109/APS.2006.171137310.1109/APS.2006.1711373

Rao, S., Nikitin, P. V., & Lam, S. F. (2005). Antenna design for UHF RFID Tags: A review and a practical application. IEEE Transactions on Antennas and Propagation, 53(12), 3870–3876. doi:10.1109/TAP.2005.859919 doi:10.1109/TAP.2005.859919

Salama, A., & Quboa, K. M. (2008). A New Fractal Loop Antenna For Passive UHF RFID Tags Applications. *Proceedings of The 3rd International Conference on Information and Communication Technologies: From Theory to Applications*. doi:10.1109/ICTTA.2008.453014110.1109/ICTTA.2008.4530141

Salama, A. (2008). Fractal dipoles as meander line antennas for passive UHF RFID tags. In *Proceedings of The 5th International Multi-Conference on Systems, Signals and Devices*. Philadelphia University. doi:10.1109/SSD.2008.463284310.1109/SSD.2008.4632843

Simons, R. N. (2001). Coplanar Waveguide Circuits, Components and Systems. Wiley Interscience. doi:10.1002/0471224758 doi:10.1002/0471224758

Son, H.-W., & Pyo, C.-S. (2005). Design of RFID tag antennas using an inductively coupled feed. Electronics Letters, 41(18), 994–996. doi:10.1049/el:20051536 doi:10.1049/el:20051536

Su, Y., & Lai, X. Z. (2008). Research on Fractal Tree RFID TagAntenna on Papery Substrate. *Proceedings of 4th IEEE International Conference on Wireless Communications, Networking and Mobile Computing.*

Tang, F. M., & Li, L. C. (2015). Meander Dipole Antenna design for Passive UHF RFID Tags. *Proceedings of The 2nd International Conference on Electrical, Computer Engineering and Electronics (ICECEE)*. doi:10.2991/icecee-15.2015.15010.2991/icecee-15.2015.150

Turcu, C. (2009). *Development and Implementation of RFID Technology*. InTechOpen.

Yang, L., & Basat, S. (2006). Design and Development of Novel Miniaturized UHF RFID Tags on on Ultra-lowcost Paper-based Substrates. *Proceedings of Asia-Pacific Microwave Conference (APMC).*

Yang, T., & Wen, G. (2014). Compact multi-band printed antenna with multi-triangular ground plane for WLAN/ WiMAX/RFID applications. International Journal of Microwave and Wireless Technologies.

Yasar, A. (2013). *Printable Green RFID Antennas for Embedded Sensors* (Doctoral Thesis). KTH Royal Institute of Technology, Stockholm, Sweden.

Youbok, L. (2003). *Antenna Circuit Design for RFID Applications.* Microchip Technology Inc. Application Note AN710.

Zhang, J., Zhang, X. M., Liu, J.-S., Wu, Q.-F., Ying, T., & Jin, H. (2009). Dual-band Bidirectional High Gain Antenna for WLAN 2.4 and 5.8 GHz applications. Electronics Letters, 45(1), 6–7. doi:10.1049/el:20092617 doi:10.1049/el:20092617

KEY TERMS AND DEFINITIONS

Coplanar Waveguide (CPW) Fed: A feeding technique for mircostrip antenna where the signal and ground currents are etched on the same layer.

Fractal Structure: Shapes used to describe naturally occurring objects and produce similar patterns at increasingly small scales.

ISM: The industrial, scientific, and medical band is composed of multiple frequency bands dedicated for industrial, scientific, and medical purposes without a license in most countries. The two commonly used bands are (2.4-2.5 GHz) and (5.725-5.875 GHz).

Meandering Technique: Miniaturization strategy that consists on folding the arms of a dipole antenna along a meandered path.

Multi-Band Antenna: A kind of antenna capable of operating with similar features, both in terms of radiation and impedance matching, at two or more separate frequencies.

Read Range: The maximum distance from which the tag can be detected. In other words, the read range expresses the maximum distance from which the tag receives just enough power to be activated and send back data respond to the reader.

Return Loss: A measure of the loss of power in the signal reflected by a transmission line to a load, such as an antenna usually expressed in decibels (dB) as the logarithmic ratio of relative magnitudes of input power and reflected power.

RFID: Radio frequency identification is one method for automatic identification and data capture with the purposes of automatically identifying and tracking tags attached to objects.

Chapter 10
Design of Miniaturized Antenna for RFID Applications

Mohamed Ihamji
University of Hassan 1st, Morocco

Elhassane Abdelmounim
University of Hassan 1st, Morocco

Hamid Bennis
Moulay Ismail University, Morocco

Mohamed Latrach
ESEO, France

ABSTRACT

This chapter presents the design of some miniature antenna for RFID application, in the ISM (industrial, scientific, and medical) band at 915 MHz and 2.45 GHz, by using two techniques. The first technique is the use of slots inserted into the microstrip antenna, and the second technique is the use of the fractal structure. In the end, both techniques are used together in one structure to get the benefit of each technique at the same time. These antennas are designed for RFID system. They can be used in a variety of fields such as access control, transport, banks, health, and logistics. One major consideration for handheld and portable RFID system applications is the compact size. Therefore, the design of miniature RFID antennas is important, and the microstrip antenna is a good choice because they are known to be low-profile, low weight, easy to make, and mechanically robust.

INTRODUCTION

This Chapter is divided into two sections. First section is dedicated to microstrip antenna and miniaturization techniques, and second section presents four new miniature antennas designed on FR4 substrate. The first section describes the microstrip antenna with their analysis methods, and their feeding methods; at the end of this section, the miniaturization techniques and the state of the art of RFID antenna

DOI: 10.4018/978-1-5225-7539-9.ch010

are presented. In the second section, four new miniature RFID antennas with different miniaturization techniques are designed and compared, in the band 915 MHz and 2.45 GHz.

Radio Frequency Identification (RFID) is the wireless use of electromagnetic field to identify tagged objects and is used in a variety of fields such as access control, transport, banks, health, and logistic. One major consideration for handheld and portable RFID system applications is the compact size (Finkenzeller, 2010).

RFID technology developments started as early as World War II, where airplanes used to be identified as "friend or foe" using this technology. The real explosion of passive RFID technology was at the end of the 1980s and was made possible by price of semiconductor technologies, current consumption of the circuitries, and the improved size. This enabled an acceptable RFID performance (communication distance) for passive systems under acceptable investment. The first generations of RFID tags were only commercially used in animal tracking in the United States with only a fixed identification code stored into the tag's memory. There was mainly a one way communication with the tag communicating back its memory content when triggered by reader activation (Turcu, 2009).

Now RFID systems are widely used in applications as identification device, and there are also new applications with higher security and computation as payment card (Karmakar, 2010).

An RFID system is generally composed of a reader, tags, and data processing system (Kitsos & Zhang, 2008), as shown in Figure 1.

The RFID data processing system stores related information such as product information, tracking logs and reader location. Since information retrieving and storing can be performed easily and speedily from RFID tags.

The Reader consists of a control unit, the transmitter module and the receiver module. The control unit contains the firmware and the hardware that control the reader activities such as communications with a host computer and the tag, as well as data processing. The transmitter generates the RF signal (data and power level) which is connected to the antenna resonance circuit. The receiver part receives the RF signal generated by the tag, demodulates and decodes the data, and sends the binary data to the control unit for further processing.

The reader has also an external interface, that are communication ports such as USB, RS232, Wireless network interfaces, and local area network (LAN) ports.

An example of reader block diagram, MFRC522 from NXP Semiconductors, is shown in the following figure:

Figure 1. RFID system architecture

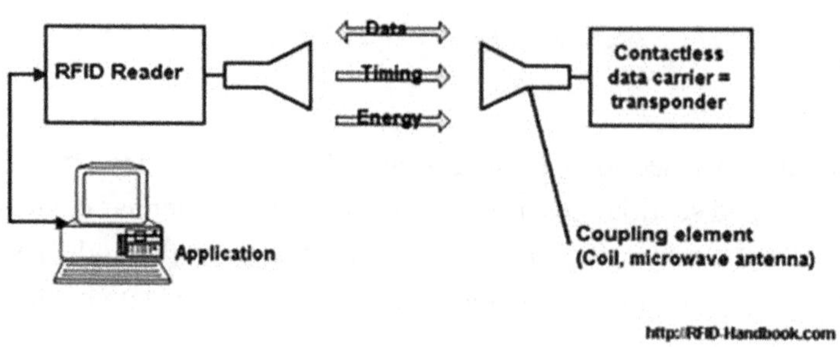

Figure 2. Block diagram of MFRC522 IC from NXP datasheet

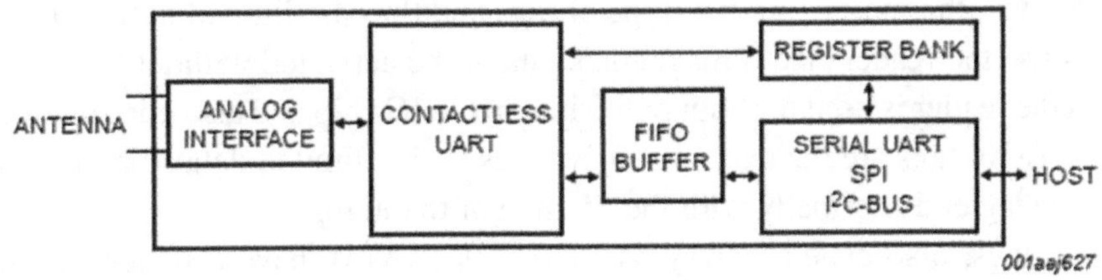

In this example, the analog interface handles the modulation and demodulation of the RF signals. The contactless UART manages the protocol communication requirements. The FIFO buffer ensures fast and convenient data transfer to and from the host and the contactless UART and vice versa. Various host interfaces are implemented.

Tags are small chips with antenna; Data is stored in the chips and transmitted through the antenna. The tag chip contains three major parts: Analog RF Interface, Digital Controller, and EEPROM. The analog part provides power supply and demodulates data received from the reader for processing by the digital part. Furthermore, the modulation block of the analog part transmits data back to the reader.

The digital section includes the state machines, processes the protocol and handles communication with the EEPROM, which contains a unique ID and user data.

An example of tag chip block diagram, SL3ICS3001 IC from NXP Semiconductors, is shown in the following figure:

The tag can be active or passive according to the supply of electronic power.

- Active tags are powered by a battery. The battery life and the tag's life can be extended when a sleep mode is used to consume the least power in the idle stage. The memory capacity in active

Figure 3. Block diagram of SL3ICS3001 IC from NXP datasheet

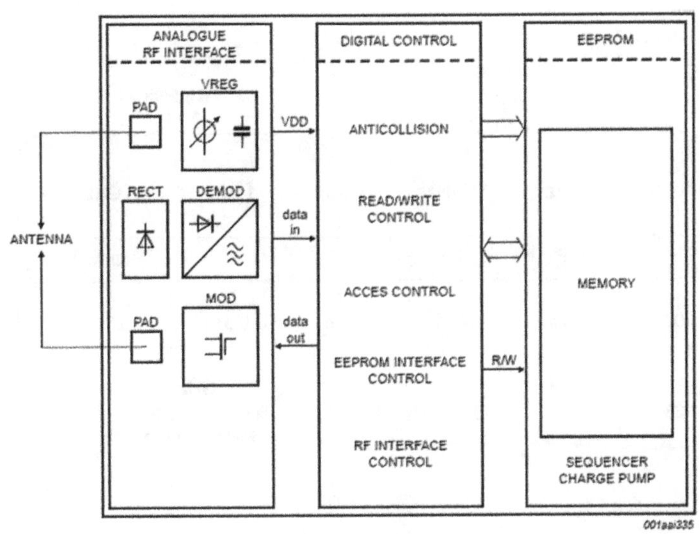

tags is the highest. Longest read range of up to kilometers can be reached by these tags. The active tags are, however, the most expensive type of tags and thus are limited in their usage.

- Passive tags use the reader electromagnetic signal to be activated without batteries. Power transmitted from the reader is used for both powering up the IC chip and also communicating back with the reader. Passive tags are the least expensive tags with a limited data storage capability and the lifetime of the tag ends normally with the lifetime of the item.
- Semi-passive tags, also called battery assisted tags (BATs), have a battery-assisted, and communicate with the reader by backscatter coupling. Longer read ranges up to 100 ft are possible compared to the passive tags. These tags have more memory capacity than the passive tags. Semi-passive tags are more expensive, bigger in size, and the life of the tag is determined by the life of the battery.

The communication between the reader and the tag is defined by a protocol which the principal characteristic is the operating frequency. The most important ISM bands for RFID systems are following frequency bands: 125/134 kHz, 13.56 MHz, 868/915 MHz, 2.45 GHz, and 5.8 GHz, as shown in Figure 4.

There is not an ideal operating frequency, which can cover all application requirements. Every frequency has its pros and cons, which are given by the properties of the frequency itself but also given from the regulation restrictions (level and bandwidth).

The choice of each operating frequency is essentially defined by the performance asked for the RFID system application. The following table summarizes the characteristics of each frequency:

The most important RFID Standards defining communications cover many applications (Finkenzeller, 2010):

Low Frequency Standards 134.2/125 kHz:

- ISO11784 – Animal ID code structure
- ISO11785 – Animal ID technical concept
- ISO14223 – Animal; Advanced transponders
- ISO18000-2 – Item management; Air interface and protocol structure for LF

Figure 4. Frequencies range

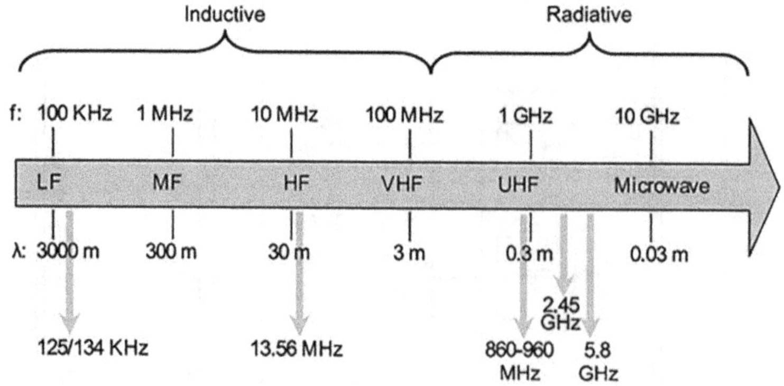

Table 1. RFID frequency characteristics

Frequency	Characteristics	Applications
Low Frequency 125 Khz	• Middle reading distance: (10 to 150 cm) • Middle reading speed	• Animal identification • No reading writing • No anti-collision management
High Frequency 13,56 Mhz	• Low reading distance • Few centimeters (with equal transmit power)	• Access control • Reading easier to write
Ultra High Frequency 900 Mhz	• Great reading distance, Up to 5 meters. • Important reading speed	• Logistics, multiple inventory management, collision-free • Pallet management
Microwave Frequency 2,4 Ghz, 5.8 Ghz	• Very fast reading speed • Very large detection distance (> 10 meters)	• Highway toll booth • Tag powered (active)

High Frequency Standards 13.56 MHz:

- ISO/IEC15693 – Vicinity cards;
- ISO/IEC14443 – Proximity cards;
- ISOIEC18000-3 – Item management; Air interface and protocol structure for HF

Ultra High Frequency Standards 868/915 MHz:

- ISO/IEC18000-6 – Air interface and protocol structure
- EPC Gen 2 – Air interface and protocol structure (ISO18000-6C)

The main RFID applications covered by these standards are:

- Animal identification
- Item management
- Logistics
- Access control
- Smart cards
- Payment
- e-Passports
- Waste management

MICROSTRIP ANTENNA

Antenna Introduction

The antenna is defined in The IEEE Standard Definitions Terms for Antennas (IEEE Std 145–1983), as "a means for radiating or receiving radio waves".

A transmission-line Thevenin equivalent of the antenna system in the transmitting mode is shown in Figure 5, where the source is represented by an ideal generator, the transmission line is represented by a line with characteristic impedance Zc, and the antenna is represented by a load ZA [ZA = (RL + Rr) + jXA] connected to the transmission line.

The energy generated by the source should be totally transferred to the radiation resistance Rr in ideal conditions, but in real condition, there are many conduction-dielectric losses due to the transmission line and the antenna, and also to reflections (mismatch) losses at the interface between the line and the antenna (Guha & Yahia, 2011). Many various types of antenna exist, as shown in Figure 6, and briefly described here below:

- Wire antennas are famous and implemented in many applications as in automobiles, buildings, ships, and aircraft. There are various shapes of wire antennas such as a straight wire (dipole), loop, and helix which are shown in Figure 6. They may take the form of many configurations as a square, rectangle, or ellipse.
- Aperture antennas are mostly used in higher frequencies, for aircraft and spacecraft applications. They have sophisticated forms as pyramidal horn, as shown in Figure 6. They can be used in hard environment conditions.

Figure 5. Transmission-line Thevenin equivalent of antenna in transmitting mode

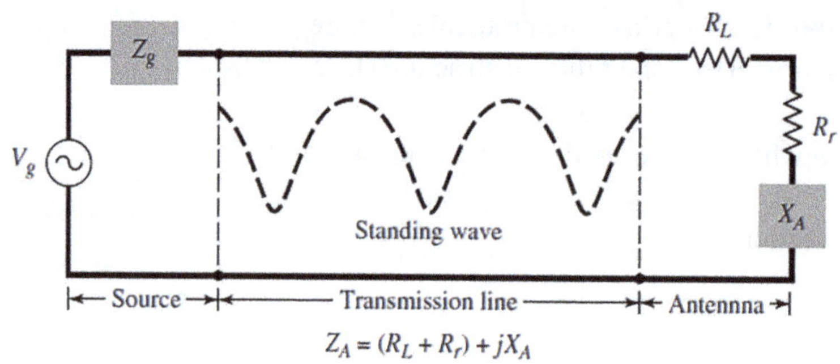

Figure 6. Antenna types

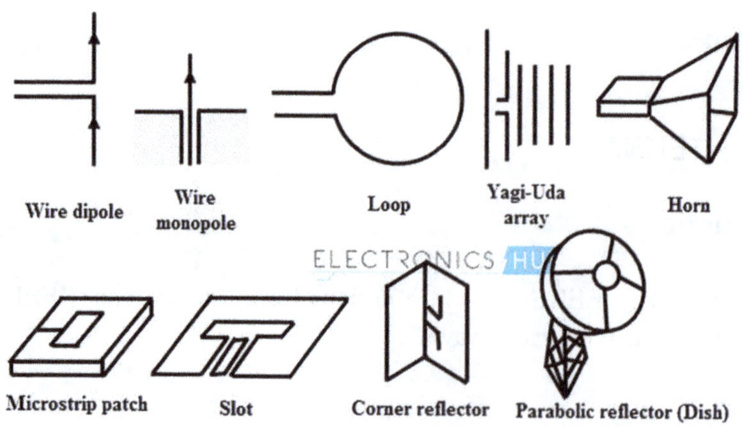

- Reflector antenna can be used to cover great distances. The parabolic reflector shown in Figures 6 is the very common antenna form. These antennas have sophisticated forms with large dimensions, used in order to transmit and receive signals that had to travel millions of miles.
- Microstrip antennas consist of a metallic patch on a grounded substrate. The microstrip antennas are known to be low-profile, low weight, ease of fabrication, conformable to planar and nonplanar surfaces and mechanically robust. Rectangular patch is the basic and the most used microstrip antenna in wireless communication.

Microstrip Antenna Description

Microstrip antennas have three parts, as shown in Figure 7, which are radiating element, substrate and ground (Fang, 2010).

They can be designed on many structure shapes such as rectangular, elliptic, circular and triangular, etc (Asrokin, Abd. Rahim, Jamaluddin, & Ahmad, 2006; Berrich, & Zenkouar, 2016; Lim, Son, Oh, Kim, & Yu, 2008; Marrocco 2003; Munson, 1974), as shown in Figure 8.

Figure 7. Microstrip antenna structure

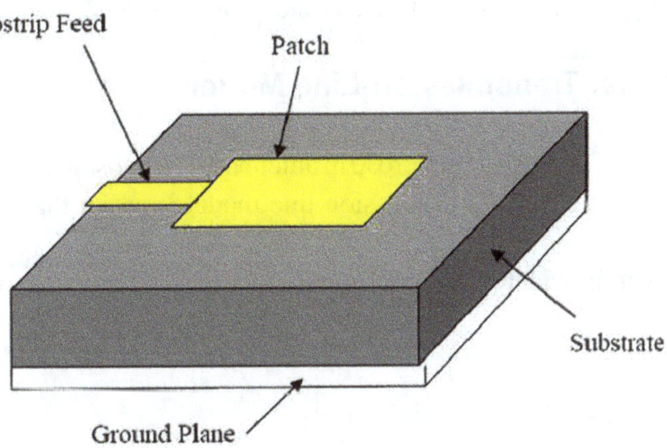

Figure 8. Microstrip antenna types

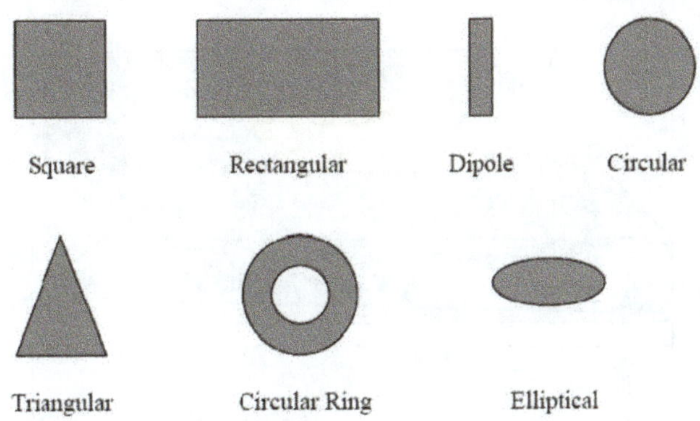

The microstrip antennas are widely used in wireless communication due to their small size, low weight, low cost, high performance, and ease installation. These antennas are simple and inexpensive to manufacture using modern printed-circuit technology, mechanically robust when mounted on rigid surfaces (Nasimuddin, 2011).

Feeding Methods

Many configurations are used to feed microstrip antennas. As shown in Figure 9 to 12, the most popular are the microstrip line, coaxial probe, aperture coupling, proximity coupling, and coplanar wave guide (Balanis, 2003; Arulaalan, & Nithyanandan, 2014; Bisht, Saini, & Prakash, Nautiyal, 2014; Soh, Rahim, Asrokin, & Abdul Aziz, 2007; Irfan, Yagoub, & Hettak, 2012).

The microstrip feed line is a conducting strip connected directly to the edge of the microstrip patch, as shown in Figure 9.

The outer conductor of the coaxial cable is connected to the ground plane, and the center conductor is attached to the patch antenna, as shown in Figure 10.

The aperture coupling consists of two substrates separated by a ground plane. The energy of a microstrip feed line, placed on the bottom side of the lower substrate, is coupled to the patch through a slot on the ground plane, as shown in Figure 11.

The proximity coupling consists of two substrates with a feed line placed between the substrates; the radiating patch is on top of the upper substrate, as shown in Figure 12.

Methods of Analysis: Transmission-Line Model

There are many methods of analysis for microstrip antennas. The most popular models are the transmission-line, cavity, and full wave. The transmission-line model is easier than others methods, and gives good physical insight.

The rectangular patch is widely used configuration, and it is very easy to analyze using both the transmission-line model.

Figure 9. Microstrip line feed

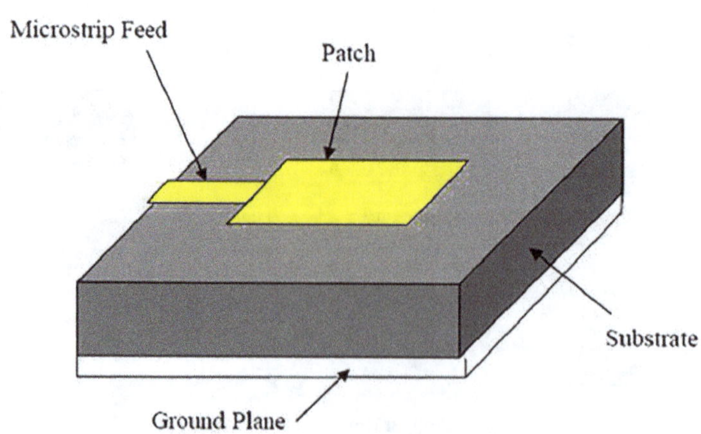

Figure 10. Coaxial probe feed

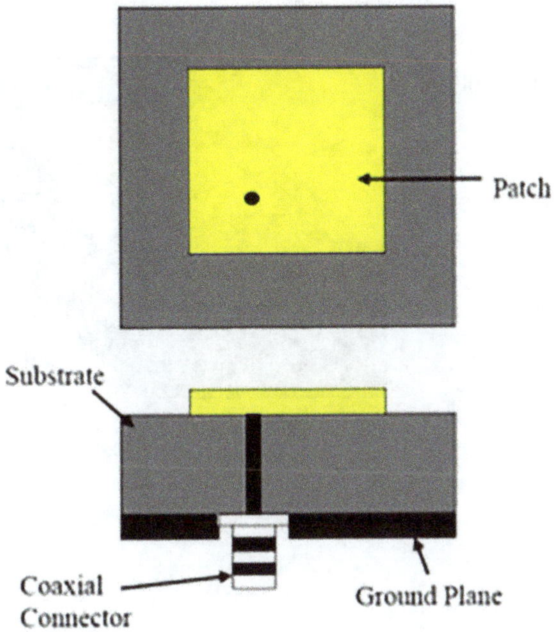

Figure 11. Aperture coupling feed

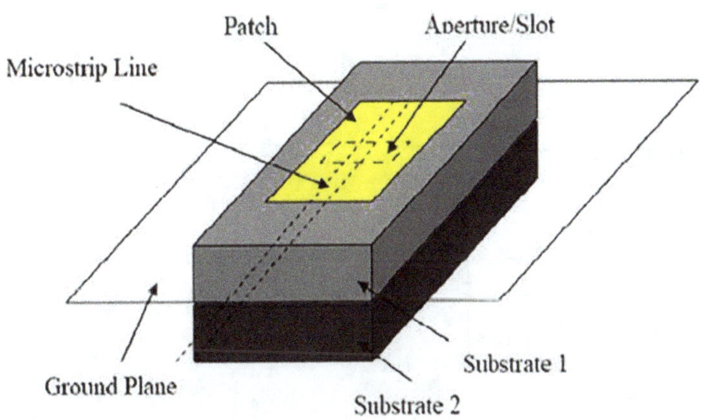

Rectangular patch antenna has two dimensions, the length Lpatch and width Wpatch, which are related to the resonant frequency, to the permittivity and to the thickness of substrate. Using the transmission line model, they can be calculated by the following conventional Equations (1) and (2) (Balanis, 2003; Derneryd, 1978):

$$W_{\text{Patch}} = \frac{c}{2f\sqrt{\frac{\varepsilon_r + 1}{2}}}. \tag{1}$$

Figure 12. Proximity coupling feed

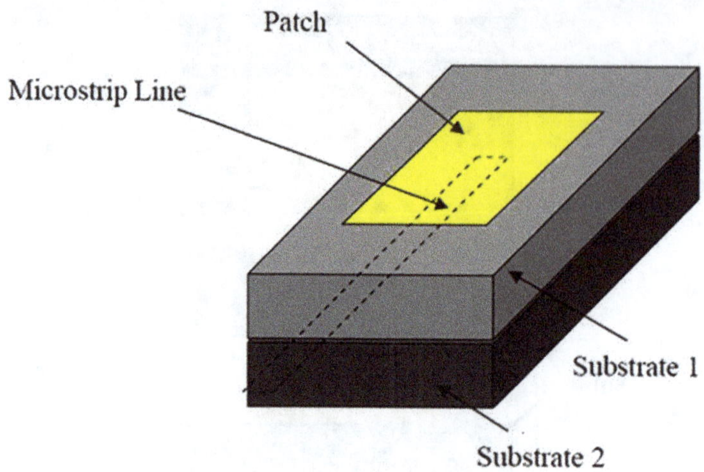

$$L_{\text{Patch}} = L_{\text{eff}} - 2 \times \Delta L. \tag{2}$$

Where c is the speed of light, f is the resonant frequency and ε_r .s the substrate's dielectric constant L_{eff} is the effective length given by:

$$L_{\text{eff}} = \frac{c}{2f\sqrt{\varepsilon_{\text{eff}}}}. \tag{3}$$

and ΔL is the length extension, given by:

$$\Delta L = 0.412h \frac{\left(\varepsilon_{\text{eff}} + 0.3\right) \times \left(\frac{W}{h} + 0.264\right)}{\left(\varepsilon_{\text{eff}} - 0.258\right) \times \left(\frac{W}{h} + 0.8\right)}. \tag{4}$$

Where, h is the height of substrate and ε_{eff} .is the effective dielectric constant which can be determined by:

$$\varepsilon_{\text{reff}} = \frac{\varepsilon_r + 1}{2} + \frac{\varepsilon_r - 1}{2}\left(1 + 12\left(\frac{h}{W}\right)\right)^{-\frac{1}{2}}. \tag{5}$$

The following table summarizes some patch dimension according to operating frequency, for FR4 substrate with a dielectric constant (4.4), loss tangent (0.025), thickness H = 1.58 mm, and a metal thickness t= 0.035 mm.

Table 2. Conventional rectangular antenna dimensions

Frequency	Lpatch (mm)	Wpatch (mm)
13.56 MHz	5271	6732
915 MHz	76	99
2.45 GHz	26.86	37

Miniaturization Techniques

Many techniques are used to miniaturize the antenna size, as the slot technique, the fractal structure, and Defected Ground Structure (DGS) (Ghiotto, Cantalice, Vuong, Pouzin, Fontgallandt, & Tedjini, 2008; Pigeon, 2011; Reynolds, 2012; Ihamji, Abdelmounim, Bennis, Hefnawi, & Latrach, 2017; Ihamji, Abdelmounim, Zbitou, Bennis, & Latrach, 2016; Ihamji, Abdelmounim, Zbitou, Bennis, & Latrach, 2016; Kapoor & Parkash, 2012; Rahmadani & Munir, 2011; Sofi, Roky, & Hadj Baraka, 2014; Yu, Huang, Zheng, & Wu, 2014; Ahmed, Fekar, Seladji-Hassaine, & Marouf, 2016; Chiu, Shum, & Chan, 2007; Li, Dejean, Tentzeris, & Laskar, 2004; Ukkonen, Sydanheimo, & Kivikoski, 2005; Deleruyelle, Pannier, Bergeret, & Bourdel, 2008; Hua, & Ma, 2007; Kim, Shin, Park, Choi, & Seong, 2006; Kumar & Ray, 2003; Kuo & Wong, 2003; Rambabu, Ramesh, & Kalghatgi, 1997).

The slot technique consists in using certain approaches to manipulate the current distribution of the structure to change its resonant frequency. When the slot is inserted into the patch antenna, as shown in Figure 13, the current is forced to flow through a long path around the slots and therefore change the resonance frequency (Anscy, 2013; Chandrappa, Ambresh, Vani, & Hunagund, 2014; Chang & Lin, 2007; Irsadi Aksun, & Chuang, 1990; Kwon & Lee, 2005; Lin, Chen, & Mittra, 2010; Nithisopa, Nakasuwan1, Songthanapitak, Anantrasirichai, & Wakabayashi, 2007; Tabakh, Jorio, El Amrani El Idrissi, & Mazri, 2016; Zhao, Wang, & Lai, 2008).

The fractal technique is based on the idea of realization of the operational characteristics of the antenna by repeating an initial radiating structure in arbitrary or regular scales. There are many popular fractal geometries, referred in the literature, as the Koch fractal, the Sierpinski fractal, the Hilbert fractal, the Minkowski and the Square Curve fractals, as shown in Figure 14 and 15.

The Koch fractal microstrip patches are commonly used in virtue of their attractive properties: they have small size, a single feeding port is enough, multi-frequency performance, satisfactory polarization properties and gain (Abed Al-Husain, Ahmed, & Osman, 2014; Benyetho, El Abdellaoui, Zbitou, Bennis, Tribak, & Latrach, 2017; Borja, & Romeu, 2003; Elavarasi, & Shanmuganantham, 2017; Gianvittorio, & Rahmat, 2000; Guesmi, Ferchichi, & Gharsallah, 2014; Kumar, Malathi, & Shinde, 2007; Mahatthanajatuphat, Saleekaw, Akkaraekthalin, & Krairiksh, 2009; Mehdipour, Rosca, Sebak, Trueman, & Hoa, 2010; Oraizi, & Hedayati, 2012; Puente & Pous, 1996; Puente, Romeu, Pous, Ramis, & Hijazo, 1998; Puente, Romeu, & Cardama, 2000; Reha, & Elamri, 2014; Reha, El Amri, Benhmammouch, Oulad Said, El Ouadih, & Bouchouirbat, 2015; Suganthi, Tharini, Sarankumar, Raghavan, & Kumar, 2011; Zahraoui, Errkik, Zbitou, Tajmouati, & Mediavilla, 2016; Zainud-Deen, Malhat, & Awadalla, 2009; Ferchichi, & Gharsallah, 2013).

The Defected Ground Structures consist of slots or defects integrated on the ground plane metal of a microstrip (Arya, Patnaik, & Kartikeyan, 2011), as shown in Figure 16.

Figure 13. Microstrip antenna with two slots

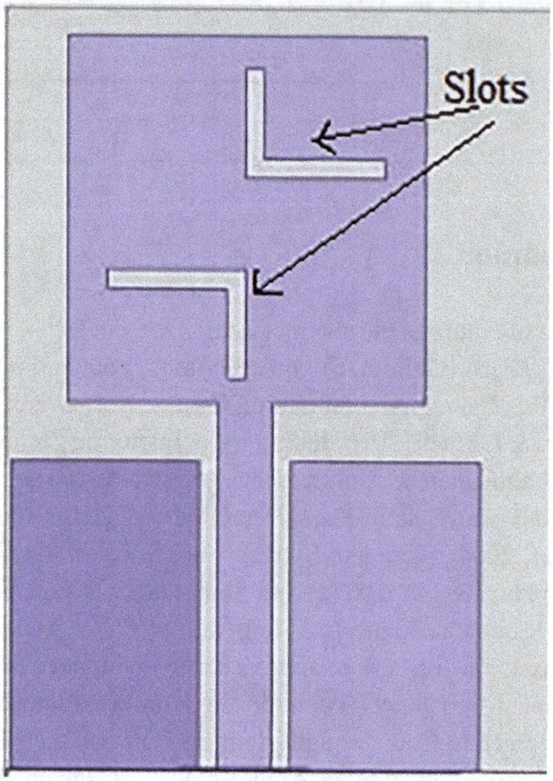

Figure 14. Koch fractal structure

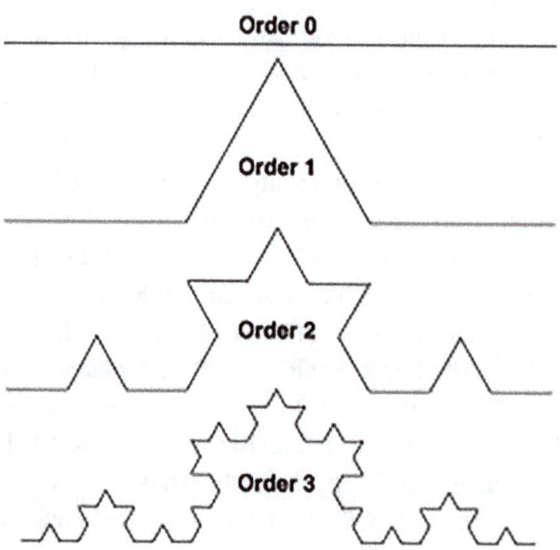

Figure 15. Sierpinski fractal structure example

Figure 16. Example of DGS structure

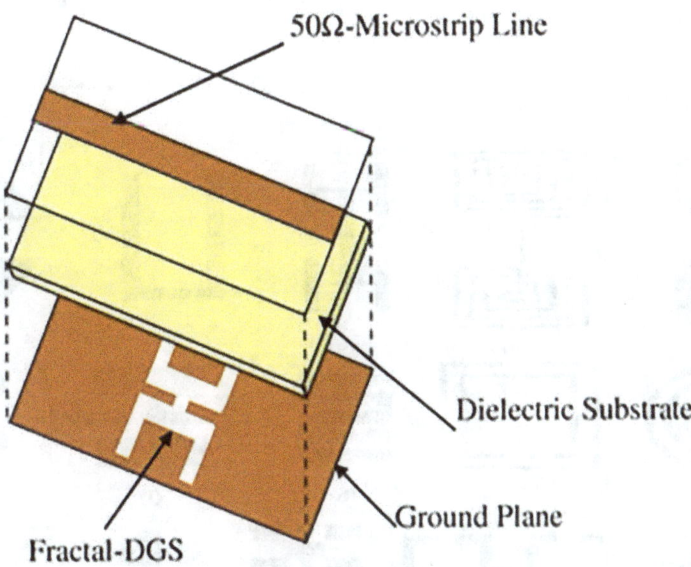

They have different geometries, as shown in Figure 17, and sizes with different frequency responses. The presence of a DGS under a printed transmission line actually perturbs the current distribution in the ground plane and thus modifies the equivalent line parameters over the defected region.

State of the Art of RFID Antennas

There are several companies that develop RFID system; some of them are presented in the following table, and classified by location (Turcu, 2009). Some companies are focalized on RFID systems as Advanced ID, Avery Dennison, Brady or Intermec.

Some RFID antennas that used in industrial applications are summarized in this Table 4 with their characteristics (Datasheets: Impinj Brickyard; BIS U-301-C1-TNCB; PEL90206; Impinj Mini-Guardrail; A901; Outdoor antenna 304020; Indoor antenna 304021; Puck antenna ADA-A2450-S; Patch antenna APAKN2504-S2448-T; Dual band antenna 2118060-1; Patch antenna APAKN1304-S5517-T). This table shows that the antenna dimension affects the gain level; the gain has a low level when the dimension is small.

Many researches are developed around the miniaturization of RFID antenna. Some of these researches are presented in Table 5 (Ghiotto, Vuong, Yagoub, & Tedjini, 2007; Monti, Catarinucci, & Tarricone, 2009; Ennajih, Zbitou, Latrach, Errkik, El Abdellaoui, & Tajmouati, 2018; Ennajih, Zbitou, Latrach, Errkik, El Abdellaoui, & Tajmouati, 2017; Rao, Lam, & Nikitin, 2010; Rao, Nikitin, & Lam, 2005; Son, Lim, Lee, Min, & Yu, 2010 ; Seladji-Hassaine, & Bendimerad, 2016; Er-rebyiy, Zbitou, Latrach, Tajmouati, Errkik, & Abdellaoui, 2017; El Hamraoui, Abdelmounim, Zbitou, Bennis, & Latrach, 2015; El Hamraoui, Abdelmounim, Zbitou, Bennis, & Latrach, 2018; El Hamraoui, Abdelmounim, Zbitou, Errkik, Bennis, & Latrach, 2018; Chen, & Tsao, 2010; Chen & Hsu, 2005; Chen & Hsu, 2004). Different miniaturization techniques are used into these antennas as fractal technique, slot technique, or DGS (Defected Ground Structures) technique. At frequency 915 MHz, the compact square quadrifilar spiral antenna has the smaller dimension and it is designed with a square quadrifilar spiral structure.

Figure 17. DGS geometries

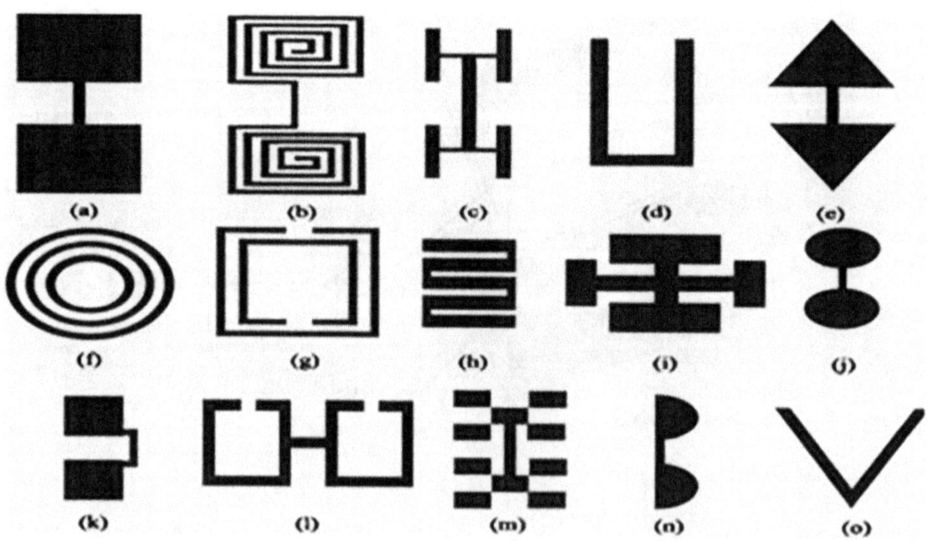

Table 3. RFID Companies, Business Descriptions, and RFID applications

Name (SYMBOL)	Location	Business	Description
3M (MMM)	USA	Creates innovative products and services in diversified areas	Tracking and management of physical files and library materials
Advanced ID(AIDO.OB)	Canada	Markets RFID components	Animal and biological sciences, bio-security and food safety
Atmel (ATML)	USA	Designs, manufactures, & sells RF semiconductors	Security and access control, manufacturing and logistics, and animal identification
Avery Dennison (AVY)	USA	Markets RFID components	Retail industry
Axcess Int (AXSI.OB)	USA	Provider of security and asset management systems	Personnel and vehicle access control and automatic asset tracking and protection
Brady (BRC)	USA	Markets RFID components	General
CCL Label (CCL-A.TO)	USA	Provider of RFID labels	Packaging, promotional and pharmaceutical industries
Digital Angel (DIGA)	USA	Develops, manufactures, & markets RFID devices	Security for people, animals, food supply, private area, and commercial assets
Infineon Tech (IFX)	Germany	Designs, develops, & sells semiconductors and systems	Telecommunication industry
Innovision Res & Tech (INN.L)	UK	Designs, develops, & licenses RFID solutions	General
Intermec (IN)	USA	Designs, manufactures & markets RFID products and systems	General
RF Micro Devices (RFMD)	USA	Designs, manufactures & markets RF semiconductors	Mobile communications
SIRIT (SI.TO)	Canada	Designs, develops, manufactures & sells RFID technology	Supply Chain, Cashless Payment, Inventory Control, Asset Tracking

The miniature hybrid loop RFID antenna has a lower size at frequency 2.45 GHz, and a good gain by using an inner loop.

DESIGN AND OPTIMIZATION OF MINIATURE MICROSTRIP ANTENNA

In this work, two methods are explored to design miniature microstrip antennas, the slots techniques and the fractal structure. The proposed antennas are based on the rectangular radiating patch and uses an FR4 substrate with dielectric constant $\varepsilon_r = 4.4$, loss tangent $\tan(\delta) = 0.025$, thickness H = 1.60 mm, and metal thickness t = 0.035 mm.

Four miniature antennas are presented in this paragraph; they are developed for reader and tag systems, in the band 915 MHz and 2.45 GHz. the following Table 6 summarizes the characteristics of these antennas:

Table 4. Industrial antennas with impedance of 50 Ohms

Name	Company	Frequency	Gain (dBi)	Dimensions (mm^3)
Impinj Brickyard (With packaging)	Cisper	902-928 MHz	6	300 x 300 x 60
BIS U-301-C1-TNCB (With packaging)	Balluff	902-928 MHz	5.5	133 x 133 x 18.4
PEL90206 (Without packaging)	Laird	902-928 MHz	6	120 x 120 x 7
Impinj Mini-Guardrail (With packaging)	Cisper	902-928 MHz	-20	70 x 133 x 19
A901- Patch antenna (Without packaging)	Idtronic	918-926 MHz	4	90 x 90 x 6
Outdoor directional antenna 304020 (With packaging)	GAO RFID	2.4-2.5 GHz	14	240 x 240 x 40
Indoor directional antenna 304021 (With packaging)	GAO RFID	2.4-2.5 GHz	6	80 x 80 x 158
Puck antenna ADA-A2450-S (With packaging)	Adatcus	2.40-2.50 GHz	-5	40.5 x 38 x 12.3
Patch antenna APAKN2504-S2448-T (With packaging)	Abracon	2.41-2.48 GHz	7	25 x 25 x 4
Dual band antenna 2118060-1 (Without packaging)	TE connectivity	2.30-5.87 GHz	2	29.6 x 41.2 x 0.30
Patch antenna APAKN1304-S5517-T (With packaging)	Abracon	5.15-5.85 GHz	6	13 x 13 x 4

Table 5. Some RFID antennas developed by researcher

Antenna	Authors	Frequency	Gain (dBi)	Technique	Dimensions (mm^2) (Reduction Size Compared to Theoretic Antenna)
Conventional Patch Antenna	Theoretic	915 MHz	-	Without	101 x 97.2 (0%)
Miniaturized Patch Antenna for the Radio Frequency Identification	A. Ghiotto	915 MHz	4.3	Slot	99.6 x 99.5 (1%)
Compact Microstrip Antenna For RFID Applications	G. Monti	866 MHz	-	Fractal	76 x 67 (48%)
New Miniature Microstrip Antenna Based On Metamaterial	A. Ennajih	915 MHz - 2.45 GHz	2.2	Metamaterial	85 x 50 (57%)
UHF RFID Tag antenna for Metal Containers	K.V.S. Rao Intermec	915 MHz	3.3	Slot	62 x 62 (61%)
Compact Square Quadrifilar Spiral Antenna with circular polarization	Wang-Ik Son	915 MHz	2.5	Quadrifilar spiral antenna	60 x 60 (63%)
Conventional Patch Antenna	Theoretic	2.45 GHz	1.76	Without	39 x 45.2 (0%)
Dual Band Metamaterial Printed Antenna	A. Ennajih	2.45 – 5.8 GHz	2.68	DGS	36 x 36 (26%)
Miniaturized Dual Band Triangular Microstrip Antenna with Sierpinski Fractal Ground	N. Seladji	2.45 – 4.48 GHz	3.4	Fractal	35 x 35 (30%)
Miniature Planar Microstrip Antenna Using DGS	R. Er-rebyiy	2.5 GHz	1.75	DGS	34 x 34 (34%)
Compact Low Cost Dual-Band Slot Antenna	A. El Hamraoui	2.45 – 5.8 GHz	2.32	Slot	35 x 33 (34%)
Miniature hybrid loop RFID antenna	S.Y. Chen	2.45 GHz	2.2	Inner loop	34 x 15 (71%)

Table 6. Characteristics of the proposed antennas

Antenna	Frequency	Miniaturization Technique	RFID System
Miniature microstrip antenna with 4 symmetric slots	2.45 GHz	Slot technique	Reader
Miniature microstrip antenna with 6 symmetric slots	915 MHz	Slot technique	Reader
Miniature fractal microstrip antenna	2.45 GHz	Fractal technique	Reader
Miniature fractal microstrip antenna with L slots	2.45 GHz	Fractal and slot techniques	Tag

The proposed antennas are designed and simulated using the ADS software based on the Method of Moment (MoM), and the CST Studio Microwave software based on Finite Integration Technique (ADS; CST Studio Microwave, 2010). The prototypes of proposed antennas are achieved by using a standard FR4 substrate with the conventional Printed Circuit Board (PCB) techniques.

Miniature Microstrip Antenna With Symmetric Slots at 2.45GHz for RFID Reader

The design of this antenna is based on a rectangular microstrip structure with the length L and the width W. These dimensions are, as it is explained into the paragraph of microstrip antenna, related to the resonant frequency, to the permittivity and to the thickness of substrate. These antennas are fed by 50 ohms microstrip line with the length L_{feed} and the width W_{feed}..

Using the slots technique to minimize the patch size, two antenna structures are proposed, as shown in Figure 18, the first with two symmetric slots, and the second with four symmetric slots. The slots are inserted symmetrically into the patch antenna to force the current to flow through a long path around these slots and thereby change the resonant frequency.

The following Table 7 indicates the dimensions of each antenna:

The result obtained of the two patch antennas, summarized in the following Table 8, shows that the miniaturization of antenna dimensions increases when the number of the slots inserted into the patch rises.

A prototype of the antenna with four slots is realized, as shown in Figure 19, by using a standard FR4 substrate, with conventional Printed Circuit Board (PCB) techniques. The return loss was measured by using the Vectorial Network Analyzer (VNA) PNA-X from Agilent Technologies. The measured results are compared to the results obtained from simulation.

This miniature antenna with four slots has a stable radiation pattern as presented in Figure 20, with directional patterns in the E-plane, and bidirectional patterns in the H-plane. The simulated values of gain and directivity have slowly decreased from 1.76 dBi to -1 dBi for gain, and from 6.37 dB to 6.19 dB for directivity; these values remain acceptable for RFID application.

Figure 21 shows the simulated and measured return loss results of the antennas which are in acceptance agreement. The measured return loss is approximately equal to -10 dB at the operating frequency 2.45 GHz.

This antenna is achieved and measured for RFID reader system; it shows an acceptable agreement between simulation and measurement results as show in the following Table 9. The validated antenna has good matching input impedance.

Figure 18. Patch antenna with two and four slots

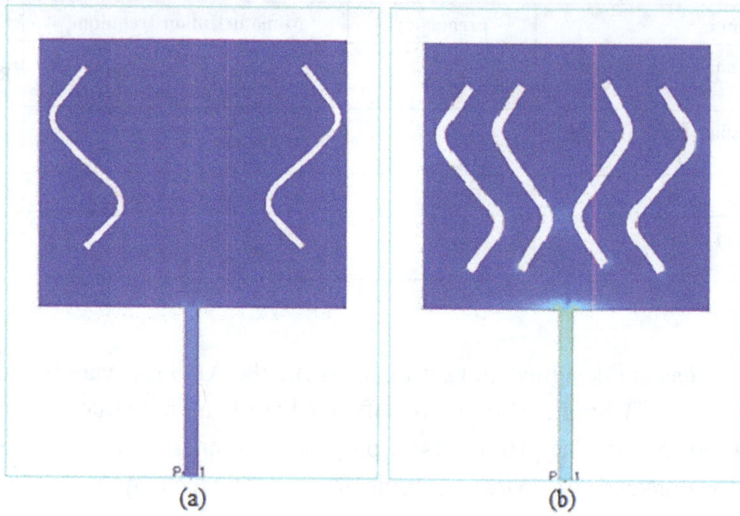

Table 7. Patch antenna dimensions

Patch Antenna 2.45 GHz	W (mm)	L (mm)	L_{feed} (mm)	W_{feed} (mm)
Patch antenna with 2 slots	30	25.23	16.01	1.4
Patch antenna with 4 slots	30	25.23	16.01	1.4

Table 8. Patch antennas results

Patch Antenna 2.45 GHz	S11 (dB)	W (mm)	L (mm)	Patch Size Reduction (WxL)
Without slot	- 43.5	37	28.3	-
With two slots	- 30.2	30	25.23	**27.5%**
With four slots	- 29.7	27.18	24.8	**35.6%**

The total area of the final circuit is Wsub x Lsub = 27.18 x 40.81 mm^2; a size reduction of 37% has been obtained with respect to a conventional rectangular patch as shown in Table 10. This miniaturization of antenna structure affects the gain and it is very small in case of ordinary ground plane (rectangular ground). The structure of ground plane will be modified in the next proposed antenna, to improve the gain.

Miniature Microstrip Antenna With Symmetric Slots at 915MHz for RFID Reader

As seen in the previous antenna example with four slots, the miniaturization of antenna dimensions increases when the number of the slots inserted into the patch rises; so in this antenna six symmetric slots are inserted, as shown in Figure 22.

Figure 19. The antenna prototype

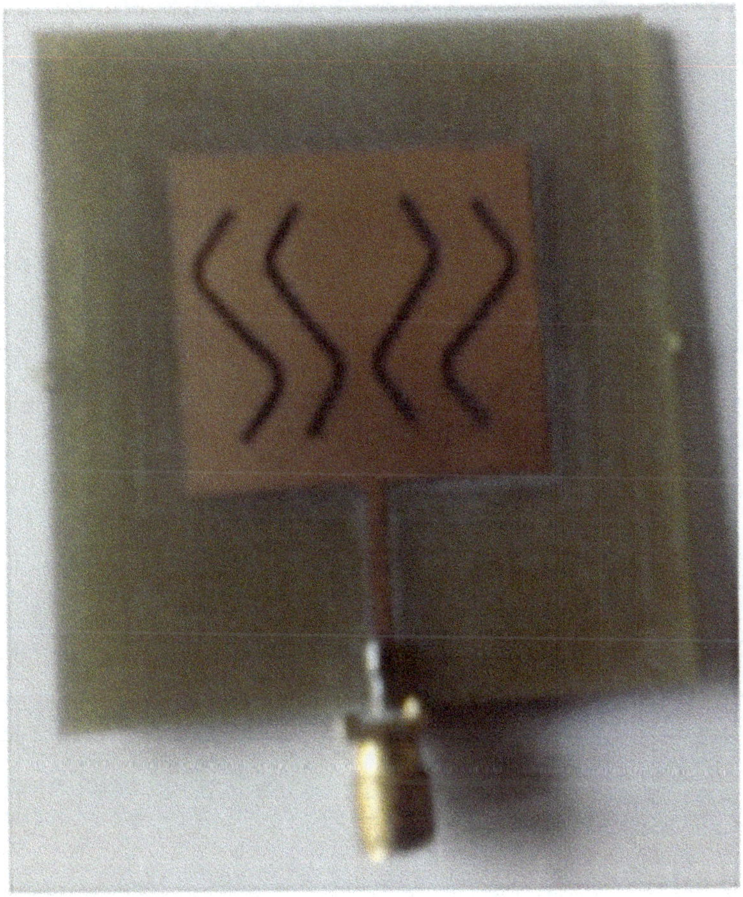

Figure 20. 2D radiation pattern at 2.45 GHz in the: (a) E-plane, (b) H-plane

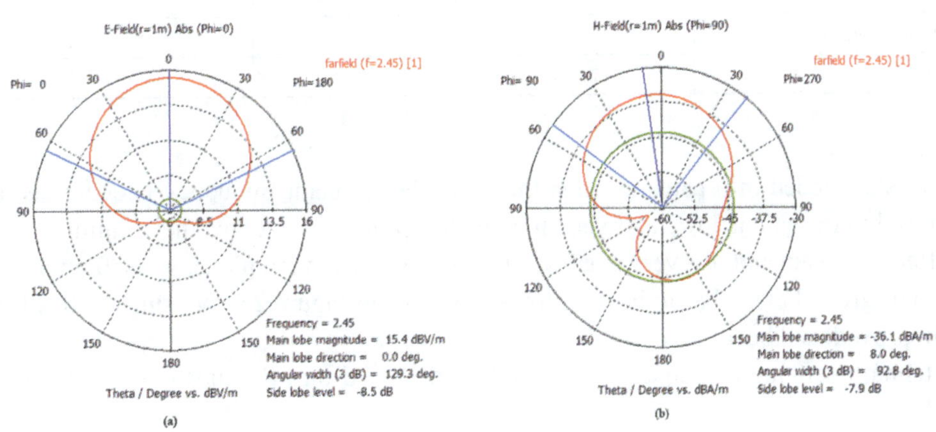

Figure 21. Measured and simulated return loss of the proposed antenna

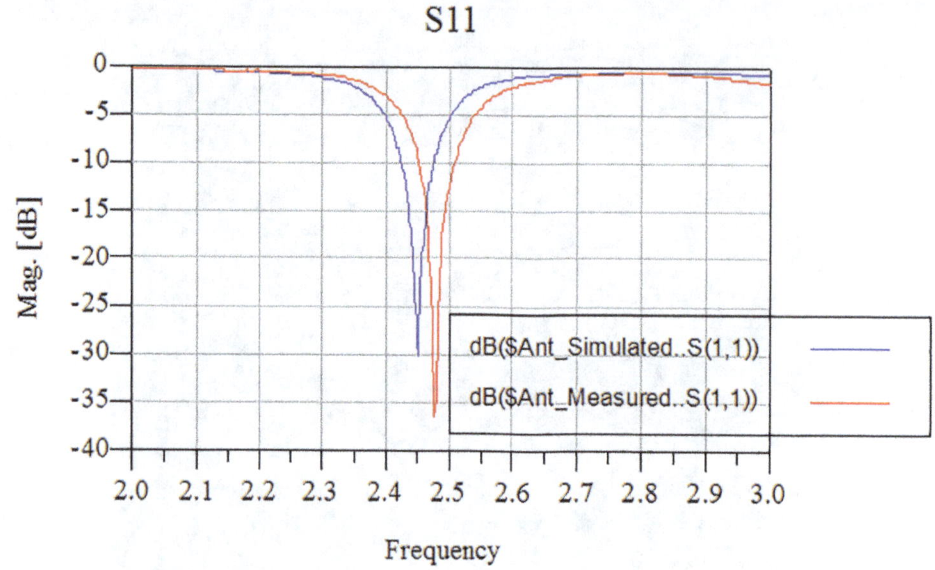

Table 9. Simulated and measured results

	Resonance Frequency (GHz)	Bandwidth (MHz)	S11 (dB)
Simulated	2.450	50	-29.7
Measured	2.478	57	-36.5

Table 10. Antenna size results

Patch Antenna 2.45 GHz	Wsub (mm)	Lsub (mm)	Antenna Size Reduction (Wsub x Lsub)
Conventional patch antenna	39	45.2	**0%**
Proposed antenna	27.18	40.81	**37%**

The more slots added into patch antenna the more the resonant frequency is decreased and patch size is reduced but the gain is lower. As shown in the last proposed antenna, this minimizing of antenna structure affects the gain and it is very small with a rectangular ground plane, so as to improve the gain, the structure of ground plane has to be optimized as shown in Figure 23, by using ADS software; gain result has improved, and becomes acceptable.

This antenna, operating at frequency 915MHz, is designed on FR4 substrate and fed by 50 ohms microstrip line.

The proposed patch antenna has the length L and width W. It is connected to a feed line with the width of W_f and the length of L_f. The conducting ground plane with length L_{gnd} placed, on the other side of the substrate, is optimized to have a good gain. The width W_f of the microstrip feed line is fixed at 2.73 mm.

The following Table 11 indicates the antenna dimensions:

Figure 22. Top face of microstrip antenna

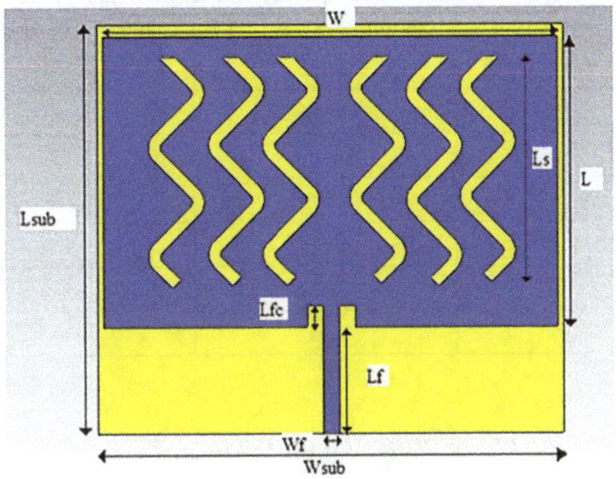

Figure 23. Back face of microstrip antenna

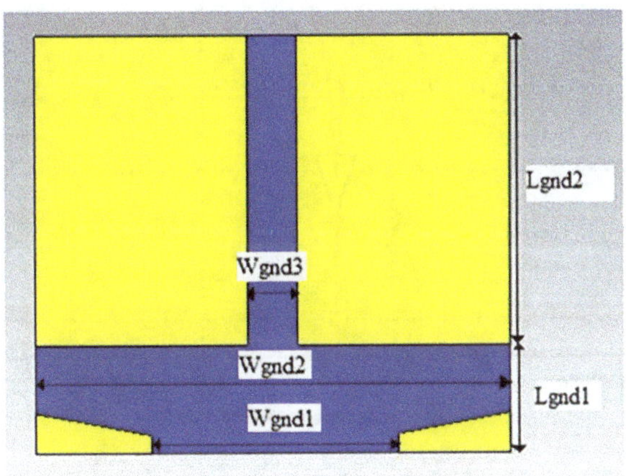

Figure 24 shows the return loss S11 result for the antenna, obtained by the CST Studio Microwave simulator. It has good matching input impedance in the ISM band.

A return loss (S11) equal to -37.02 dB is obtained with this patch antenna. The bandwidth value is approximately 11 MHz. The radiation pattern is omni-directional as presented in Figure 25. The gain and directivity are respectively 1.04 dB and 2.41 dB.

The results obtained of this patch antenna with six symmetric slots, summarized in the following Table 12, show that the patch size reduction is increased compared to the previous antenna with four slots.

This antenna is designed for RFID reader system on FR4 substrate and simulated performing CST Microwave Studio. It provides appropriate characteristics, with a return loss equal to -37.02 dB, a bandwidth of 11 MHz, a directivity of 2.41 dB, and a gain of 1.04 dB.

The total area of the final circuit is Wsub x Lsub = 67.72 x 78.75 mm^2; a size reduction of 45.7% has been obtained with respect to a conventional rectangular patch as shown in Table 13.

Table 11. Antenna dimensions

Antenna Dimensions	Optimized Value (mm)	Antenna Dimensions	Optimized Value (mm)
L	48.12	Lgnd1	17.6
W	76.74	Lgnd2	50.12
Lsub	67.72	Lf	17.6
Wsub	78.75	Lfc	3.61
Wgnd1	40.93	Wf	2.73
Wgnd2	78.74	Lslot	37.83
Wgnd3	8.11		

Figure 24. Return loss (S11) value of the proposed antenna

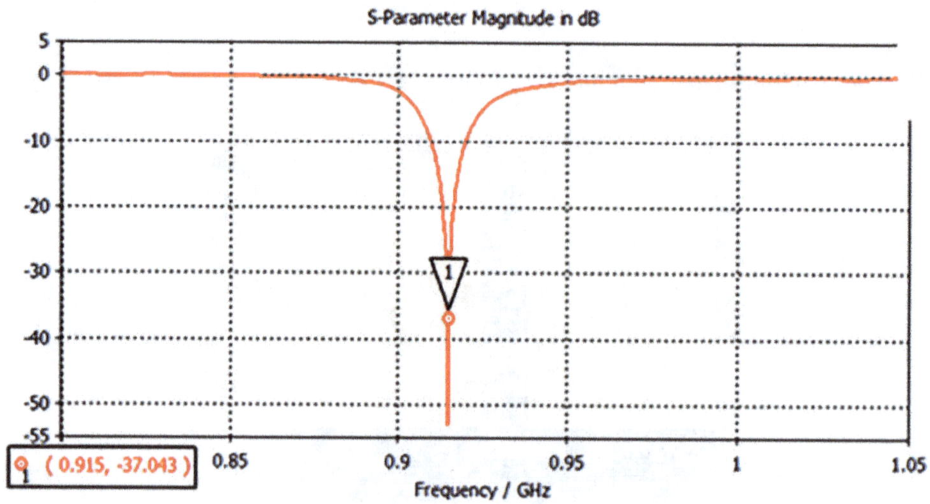

Figure 25. 3D radiation pattern of proposed antenna: (a) Gain, (b) Directivity

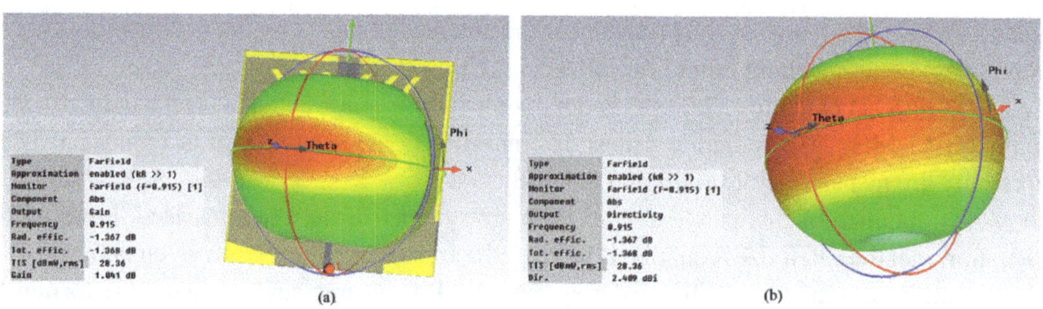

Table 12. Patch antennas results

Patch Antenna 915 MHz	S11 (dB)	W (mm)	L (mm)	Patch Size Reduction (WxL)
Without slot	- 40	99	77.6	
With six slots	- 37.02	76.74	48.12	**51.9%**

Table 13. Antenna size results

Patch Antenna 915 MHz	Wsub (mm)	Lsub (mm)	Antenna Size Reduction (Wsub x Lsub)
Conventional patch antenna	101	97.2	**0%**
Proposed antenna	78.75	67.72	**45.7%**

Miniature Fractal Microstrip Antenna at 2.45 GHz for RFID Reader

This antenna is designed using the fractal structure. The non radiating edges of patch are designed with the Koch curve which is commonly used in virtue of their attractive properties.

The size reduction of the proposed microstrip antenna is achieved by etching the patch edges according to Koch curve. Multiple iterations of the Koch fractal are presented in Figure 26. To form the first iteration, the original line segment is partitioned into n=7 equal line segments of 1/4 the original length. This corresponds to a reduction of 7/4 for the first iteration, $(7/4)^2$ for the second iteration, and so on.

As described before into microstrip antenna section, there are many methods to feed the microstrip antennas, as the microstrip line, the slot line or the CPW (Coplanar Wave Guide) line. The advantage of CPW fed antenna is their wide bandwidth compared to other feds which is approved by research (Reha, El Amri, Benhmammouch, Oulad Said, El Ouadih, & Bouchouirbat, 2015; Zahraoui, Errkik, Abdelmounim, Bennis, El Abdellaoui, & Sanchez Mediavilla, 2015). This antenna is fed by 50 ohms Coplanar Wave Guide line to have a wide bandwidth.

Figure 26. The 3 iterations of the KOCH Curve

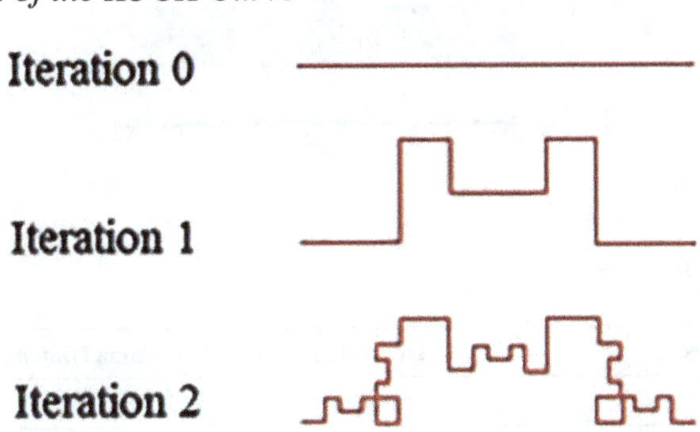

This antenna with the fractal side has the length L and width W, as shown in Figure 27. It is connected to a Coplanar Wave Guide line with the width of W_f and the length of L_f. There is no conducting ground plane placed on the other side of the substrate. This circuit is validated with CST-MW Studio simulator.

The following Table 14 indicates the antenna dimensions of the proposed antenna:

A prototype of this antenna is realized, by using a standard FR4 substrate, with conventional Printed Circuit Board (PCB) techniques. The return loss was measured by using the Vectorial Network Analyzer (VNA) PNA-X from Agilent Technologies. The measured results are compared to the results obtained from simulation.

Figure 28 presents the simulated and measured return loss of the antenna. The results show good agreement with a return loss of approximately -21 dB at frequency 2.45 GHz.

Figure 29 respectively illustrates the 2D radiation patterns at E-plane and H-plane. The proposed antenna has an omni-directional radiation pattern for both H-plane and E-plane. The angular width is 85.8 degree at 2.45GHz.

Figure 27. Top face of microstrip antenna

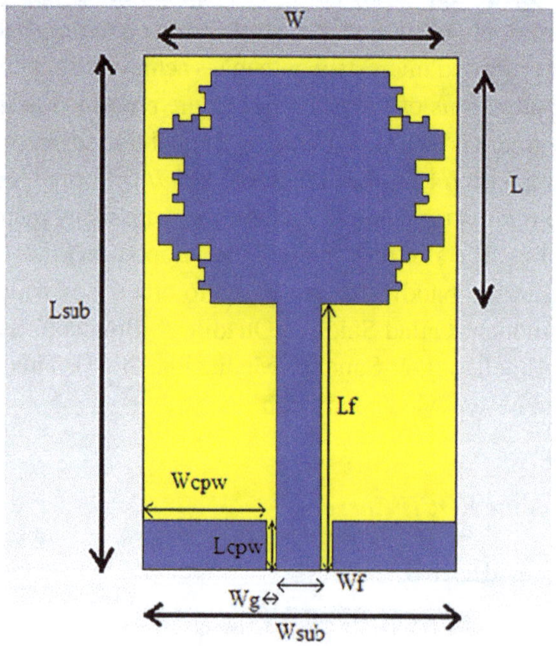

Table 14. Antenna dimensions

Antenna Dimensions	Optimized Value (mm)	Antenna Dimensions	Optimized Value (mm)
L	14.45	Lf	16.6
W	19.225	Wf	3
Lsub	32.05	Wg	0.75
Wsub	21.225		

Figure 28. Measured (Blue) and simulated (Red) return loss of the proposed antenna

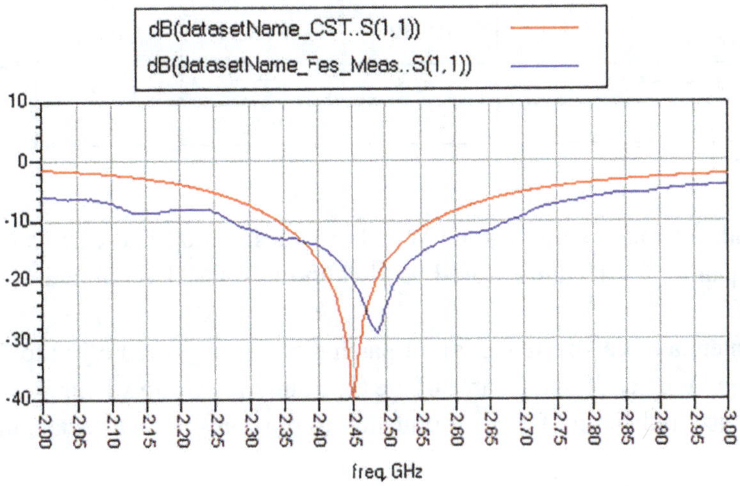

Figure 29. 2D radiation pattern at 2.45 GHz in the: (a) H-plane, (b) E-plane

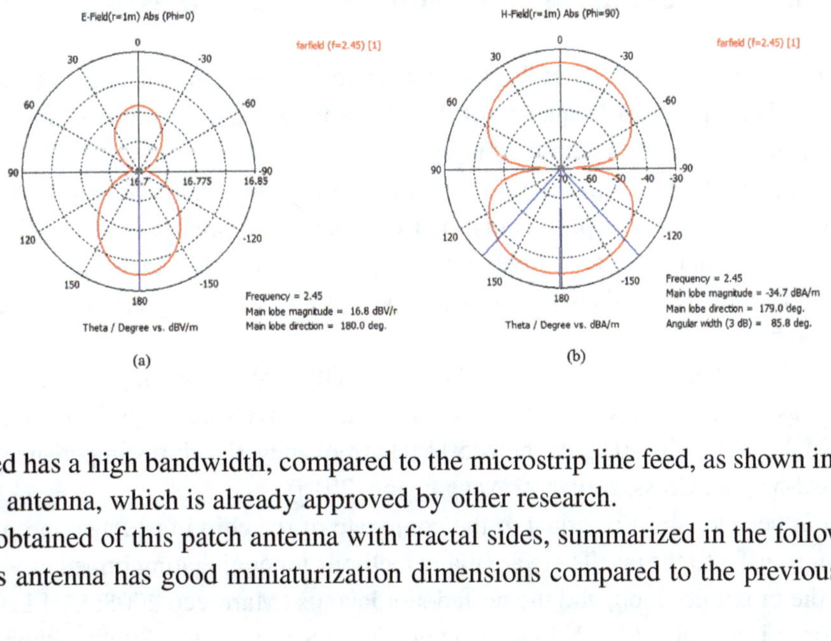

A CPW-feed has a high bandwidth, compared to the microstrip line feed, as shown in Table 15 with ordinary patch antenna, which is already approved by other research.

The result obtained of this patch antenna with fractal sides, summarized in the following Table 16, shows that this antenna has good miniaturization dimensions compared to the previous antenna with six slots.

Table 15. Antenna results

Patch Antenna 2.45 GHz	S11 (dB)	Bandwidth (MHz)	Gain (dBi)
Conventional Patch with microstrip fed	- 43.5	60	1.76
Proposed Antenna	- 35.9	327	2.2

Table 16. Patch antennas results

Patch Antenna 2.45 GHz	S11 (dB)	W (mm)	L (mm)	Patch Size Reduction (WxL)
Conventional Patch with microstrip fed	- 43.5	37	28.3	-
Proposed Antenna	- 35.9	19.22	14.45	**73.5%**

This antenna with fractal structure and CPW fed provides appropriate characteristics, with a measured return loss equal to -21 dB, a measured bandwidth of 400 MHz, a directivity value of 1.91 dB, and a gain of 2.2 dBi.

The proposed antenna has a size reduction of patch dimension (WxL) equal to 73.5% compared to conventional rectangular patch. The total area of the final circuit is Wsub x Lsub = 32.05 x 21.225 mm^2, representing a total size reduction of 61.4% compared to the conventional rectangular patch as shown in Table 17.

The antenna has been designed on a standard FR4 substrate and realized with conventional Printed Circuit Board (PCB) techniques.

Miniature Fractal Microstrip Antenna with L Slots at 2.45 GHz for RFID Tag

Another antenna structure is developed by using both techniques: fractal and slot techniques. This microstrip antenna is proposed at frequency of 2.45 GHz for RFID tag, and is designed by using the symmetric L-slots inside the patch with fractal geometry.

As shown in Figure 30. The patch antenna, with the length L and width W, has two L-slots, inserted symmetrically on each edge of the patch, such that it affects the resonant mode by changing the current distribution of the patch. The L-slot element is repeated inside the patch structure symmetrically in regular scales in order to have the fractal effect. There is no conducting ground plane placed on the other side of the substrate.

The chip named UCODE HSL IC, SL3ICS3001 (UCODE High Frequency Smart Label) from NXP Semiconductor was chosen because it can be used in small passive smart tags. The chip can operate at 867 MHz, 915 MHz, and 2450 MHz. At the 2450 MHz frequency, the chip's impedance is 11.5 – j 295 Ohms (Numan-Al-Mobin, Cross, Kellar, & Anagnostou, 2015).

The antenna impedance must be equal to the conjugate of the chip impedance in order to have the maximum power transfer to the tag. There are many methods for conjugating impedance matching such as the T-mach, the proximity-loop, and the nested-slot layouts (Marrocco 2008; C H Loo, et al., 2008; Choi et al., 2006; Evizal et al., 2013; K Dongho et al., 2012; S K Kuo et al. 2008; Ukkonen et al., 2004;

Table 17. Antenna size results

Patch Antenna 2.45 GHz	Wsub (mm)	Lsub (mm)	Antenna Size Reduction (Wsub x Lsub)
Conventional patch antenna	39	45.2	**0%**
Proposed antenna	32.05	21.225	**61.4%**

Xi et al., 2010). The nested slot method is used into this proposed antenna to conjugate the impedance matching with the tag chip.

The final circuit is validated by the simulator of CST-MW Studio.

The following Table 18 indicates the antenna dimensions of the proposed antenna:

Figure 31 shows the return loss S11 result for the antenna, it has good matching input impedance in the ISM band. The return loss (S11) is equal to -36.45 dB with this patch antenna. The bandwidth value is equal approximately to 44 MHz.

Figure 30. Geometry of the proposed antenna: Top Face

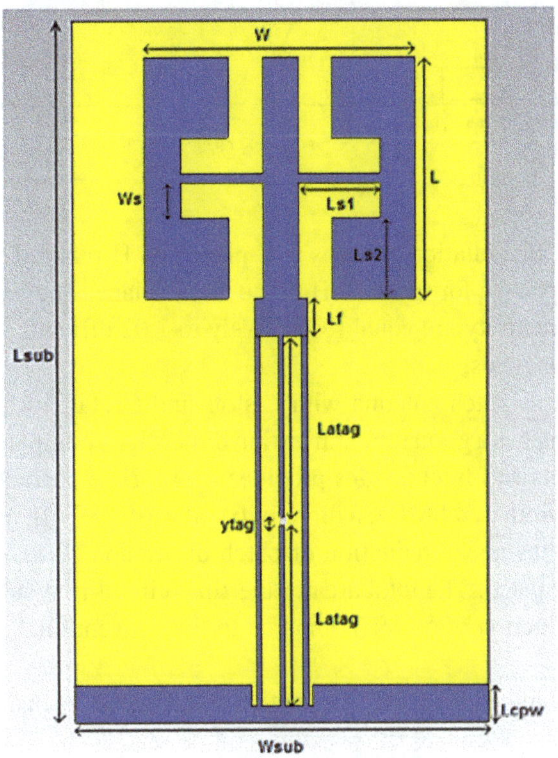

Table 18. Antenna dimensions

Antenna Dimensions	Optimized Value (mm)	Antenna Dimensions	Optimized Value (mm)
L	13	Ws	1.9
W	15	Ls1	4.33
Lsub	37.83	Ls2	4.45
Wsub	23	Latag	9.74
Lf	2	Ytag	0.5
Wf	2.87		

Figure 31. Return loss (S11) value of the proposed antenna

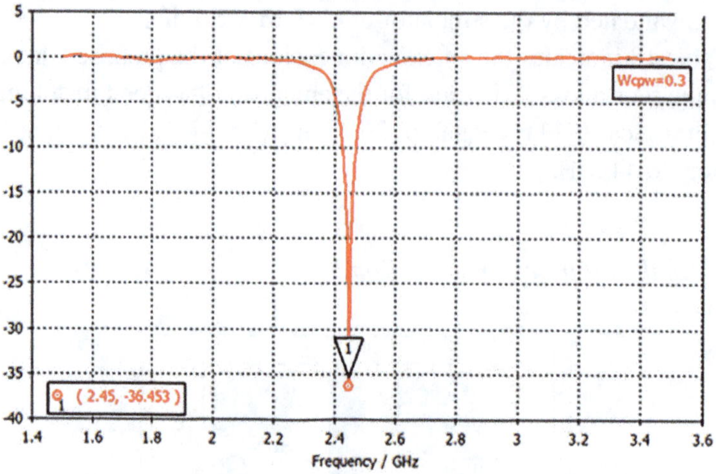

Figure 32 illustrates the 2D radiation patterns at E-plane and H-plane. The proposed antenna has an omni-directional radiation pattern for the both H-plane and E-plane. The angular width is 96 degree at 2.45 GHz. The gain and directivity are equal respectively to 1.02 dBi and 1.89 dB, this gain is correct for RFID application in near fields.

The result obtained of this patch antenna with L slots and fractal sides, summarized in following table, shows that this antenna has a good miniaturization dimensions compared to the previous antennas.

This antenna with L slots and fractal sides provides appropriate characteristics, with a return loss equal to -36.45 dB, a bandwidth of 44 MHz, a directivity value of 1.89 dB, and a gain equal to 1.02 dBi.

This proposed antenna has a size reduction of patch dimension (WxL) equal to 81.3% compared to conventional rectangular patch. The total area of the final circuit is Wsub x Lsub = 23 x 37.8 mm^2, representing a total size reduction of 50.6% compared to the conventional rectangular patch as shown in Table 20.

The antenna has been designed for RFID Tag with a standard FR4 substrate.

Figure 32. 2D radiation pattern at 2.45 GHz in the: (a) E-plane, (b) H-plane

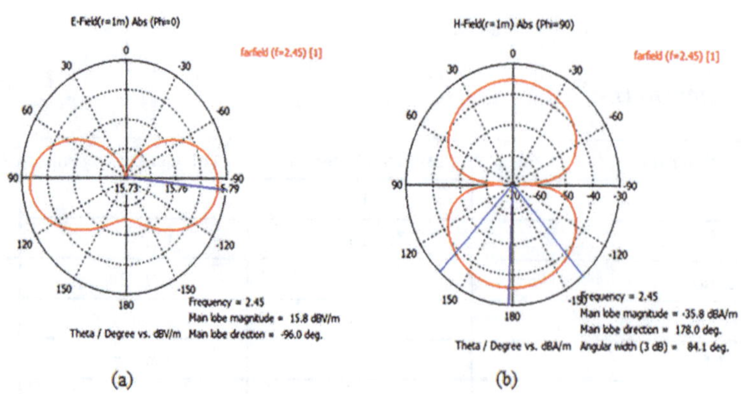

Table 19. Patch antennas results

Patch Antenna 2.45 GHz	S11 (dB)	W (mm)	L (mm)	Patch Size Reduction (WxL)
Conventional Patch antenna	- 43.5	37	28.3	-
Proposed antenna	- 36.45	15	13	**81.3%**

Table 20. Antenna size results

Patch Antenna 2.45 GHz	Wsub (mm)	Lsub (mm)	Antenna Size Reduction (Wsub x Lsub)
Conventional patch antenna	39	45.2	**0%**
Proposed antenna	23	37.8	**50.6%**

Table 21. Antenna miniaturization results

Antenna	Frequency	W (mm)	L (mm)	Patch Size Reduction (WxL) / Ordinary Patch Antenna
Miniature microstrip antenna with 2 symmetric slots	2.45 GHz	30	25.23	**27.5%**
Miniature microstrip antenna with 4 symmetric slots	2.45 GHz	27.18	24.8	**35.6%**
Miniature microstrip antenna with 6 symmetric slots	915 MHz	76.74	48.12	**51.9%**
Miniature fractal microstrip antenna	2.45 GHz	19.22	14.45	**73.5%**
Miniature fractal microstrip antenna with L slots	2.45 GHz	15	13	**81.3%**

Table 22. RFID antennas developed by researcher

Antenna	Authors	Frequency	Gain (dBi)	Technique	Dimensions (mm^2) (Reduction Size Compared to Theoretic Antenna)
Miniature microstrip antenna with 6 symmetric slots	M. Ihamji	915 MHz	1.04	Slot	**78.7 x 67.7 (46%)**
Compact Microstrip Antenna For RFID Applications	G. Monti	866 MHz	-	Fractal	76 x 67 (48%)
Compact Square Quadrifilar Spiral Antenna with circular polarization	Wang-Ik Son	915 MHz	2.5	Quadrifilar spiral antenna	60 x 60 (63%)
Miniature Planar Microstrip Antenna Using DGS	R. Er-rebyiy	2.5 GHz	1.75	DGS	34 x 34 (34%)
Compact Low Cost Dual-Band Slot Antenna	A. El Hamraoui	2.45 – 5.8 GHz	2.32	Slot	35 x 33 (34%)
Miniature fractal microstrip antenna with L slots	M. Ihamji	2.5 GHz	1.02	Fractal + Slot	**23 x 37.8 (51%)**
Miniature fractal microstrip antenna	M. Ihamji	2.5 GHz	2.2	Fractal	**32 x 21 (61%)**
Miniature hybrid loop RFID antenna	S.Y. Chen	2.45 GHz	2.2	Inner loop	34 x 15 (71%)

CONCLUSION

In this chapter, new miniature antennas based on slots techniques and fractal geometry are developed with a bibliography synthesis on microstrip antenna and miniaturization techniques. Various microstrip antennas with slots and fractal sides have also been studied and compared to the conventional microstrip antenna size.

As shown in Table 21, the miniaturization of antenna dimensions increases when the number of the slots inserted into the patch rises, as shown in the example of microstrip antennas with 2, 4 and 6 symmetric slots. The miniaturization increases also when the number of fractal structure iteration rises, as described in the example of fractal microstrip antenna.

This miniaturization is better when both techniques, fractal structure and slots technique, are used into the microstrip antenna, as is done with the case of fractal microstrip antenna with L slots.

The three last proposed antennas present a good size miniaturization compared to antennas developed by others researchers, as described in the following Table 22.

Other techniques, as DGS technique or Inner Loop, can be investigated and used with fractal and slot techniques, in order to have a very good miniaturization for the future RFID antennas.

REFERENCES

A901-Patch antenna Datasheet from IDtronic Company. (n.d.). Retrieved from https://www.en.idtronic-rfid.com/d/1278572/d/uhf_patch_antennas.pdf

Abed Al-Husain, Z. A., Ahmed, S. K., & Osman, A. A. (2014). Design of Ultra-Wideband Micrstrip Antennas with Slots and Fractal Based Ground Plane. *Journal of Al Rafidain University College, 34*, 244-258. Retrieved from http://www.home.agilent.com/agilent/home.js

Ahmed, M., Fekar, M., Seladji-Hassaine, N., & Marouf, F. (2016). Analysis of Patch and Printed Antennas integrated into UHF RFID Passive Tags. *Int. J. of Microwave and Opt. Tech, 11*(6), 460–468.

Anscy, S. (2013). Slot microstrip antenna for 2.4GHz RFID reader application. Department of Electronics and Communication Engineering Karunya University. *International Journal of Advanced Research in Electronics and Communication Engineering, 2*(5).

Arulaalan, M., & Nithyanandan, L. (2014). Analysis of Microstrip Line Feed Triangular Patch Antenna for WLAN (2.4 GHz). *Proc. of Int. Conf. on Advances in Communication, Network, and Computing, CNC,* 802-806.

Arya, A., Patnaik, A., & Kartikeyan, M. (2011). Microstrip patch antenna with skew-F shaped DGS for dual band operation. *Progress in Electromagnetics Research, 19*, 147–160. doi:10.2528/PIERM11052305

Asrokin, A., Abd. Rahim, M.K., Jamaluddin, M.H., & Ahmad, M.R. (2006). Triangular and Circular Dual Band Microstrip Antenna for WLAN Applications. *International RF and Microwave Conference Proceeding*, 53-57. 10.1109/RFM.2006.331036

Balanis, A. (2003). *Antenna Theory Analysis and Design* (2nd ed.). John Wiley & Sons, Inc.

Benyetho, T., El Abdellaoui, L., Zbitou, J., Bennis, H., Tribak, A., & Latrach, M. (2017). A New Dual Band Planar Fractal Antenna Array for ISM Band. *International Journal of Microwave and Optical Technology*, *12*(3).

Berrich, L., & Zenkouar, L. (2016). Microruban Dipole Antenna for RFID Applications at 2.45 GHz. *International Journal of Electrical and Computer Engineering*, *6*(6). Retrieved from http://publications.balluff.com/mongodb_pdb/public_pdb_material.php?id=221392&revision=0.5&pg=PG185676&pf=PF185742&con=en&ws=approval

Bisht, S., Saini, S., Prakash, V., & Nautiyal, B. (2014). Study The Various Feeding Techniques of Microstrip Antenna Using Design and Simulation Using CST Microwave Studio. *International Journal of Emerging Technology and Advanced Engineering*, *4*(9).

Borja, C., & Romeu, J. (2003). On the behavior of Koch island fractal boundary microstrip antenna. *IEEE Transactions on Antennas and Propagation*, *51*(6), 1281–1291. doi:10.1109/TAP.2003.811479

Chandrappa, D. N., Ambresh, P. A., Vani, R. M., & Hunagund, P. V. (2014). Multi-slots reconfigurable microstrip antenna with capacitive loading technique. *International Journal of Innovative Research in Computer and Communication Engineering*, *2*(1). Retrieved from www.ijircce.com

Chang, T. N., & Lin, J. M. (2007). A novel circularly polarized patch antenna with a serial multislot type of loading. *IEEE Transactions on Antennas and Propagation*, *55*(11), 3345–3348. doi:10.1109/TAP.2007.908846

Chen, H.-D., & Tsao, Y.-H. (2010). Low-Profile PIFA Array Antennas for UHF Band RFID Tags Mountable on Metallic Objects. *IEEE Transactions on Antennas and Propagation*, *58*(4), 1087–1092. doi:10.1109/TAP.2010.2041158

Chen, S., & Hsu, P. (2005). Hybrid Loop Antenna For 2.4-Ghz RFID Tags. *Proceedings of ISAP2005*.

Chen, S.-Y., & Hsu, P. (2004). CPW-Fed Folded-Slot Antenna for 5.8-GHz RFID Tags. *Electronics Letters*, *40*(24), 1516–1517. doi:10.1049/el:20046780

Chiu, C. Y., Shum, K. M., & Chan, C. H. (2007). A tunable via-patch loaded PIFA with size reduction. *IEEE Transactions on Antennas and Propagation*, *55*(1), 65–71. doi:10.1109/TAP.2006.888389

Choi, W., Son, H. W., Bae, J.-H., Choi, G. Y., Pyo, C. S., & Chae, J.-S. (2006). An RFID Tag Using a Planar Inverted-F Antenna Capable of Being Stuck to Metallic Objects. *ETRI Journal*, *28*(2), 216–218. doi:10.4218/etrij.06.0205.0082

CST Studio Suite Version. (2010). Retrieved from http://www.cst.com/Content/Products/DS/Overview.aspx

Deleruyelle, T., Pannier, P., Bergeret, E., & Bourdel, S. (2008). Dual band UHF and microwave RFID antenna. *Proceedings of the 38th European Microwave Conference*. 10.1109/EUMC.2008.4751818

Derneryd, A. G. (1978). A Theoretical Investigation of the Rectangular Microstrip Antenna Element. *IEEE Transactions on Antennas and Propagation*, *26*(4), 532–535. doi:10.1109/TAP.1978.1141890

Dongho, K., & Junho, Y. (2012). Dual-Band Long-Range Passive RFID Tag Antenna Using an AMC Ground Plane. *IEEE Transactions on Antennas and Propagation, 60*(6), 2620–2626. doi:10.1109/TAP.2012.2194638

Dual band antenna 2118060-1 Datasheet from TE Connectivity Company. (n.d.). Retrieved from http://www.farnell.com/datasheets/2357390.pdf

El Hamraoui, A., Abdelmounim, E., Zbitou, J., Bennis, H., & Latrach, M. (2015). A new design of a compact CPW-Fed dual-band printed antenna for RFID applications. Wireless Networks and Mobile Communications (WINCOM). *IEEE Conference Publications.*

El Hamraoui, A., Abdelmounim, E., Zbitou, J., Errkik, A., Bennis, H., & Latrach, M. (2018). A Dual-Band Microstrip Slot Antenna Reader for UHF and Microwave RFID Applications. *TELKOMNIKA Journal, 16*(1), 94–101. doi:10.12928/telkomnika.v16i1.7310

El Hamraoui, A., Abdelmounim, E.H., Zbitou, J., Bennis, H., & Latrach, M. (2018). A compact low cost dual-band slot antenna for RFID readers. *International Journal of Microwave and Optical Technology, 13*(1).

Elavarasi, C., & Shanmuganantham, T. (2017). SRR Loaded Iterated Pythagoras Fractal Antenna for Multiple Functional Resonance Frequencies. *International Journal of Microwave and Optical Technology, 12*(3).

Ennajih, A., Zbitou, J., Latrach, M., Errkik, A., El Abdellaoui, L., & Tajmouati, A. (2017). Dual Band Metamaterial Printed Antenna Based on CSRR For RFID Applications. *International Journal of Microwave and Optical Technology, 12*(2).

Ennajih, A., Zbitou, J., Latrach, M., Errkik, A., El Abdellaoui, L., & Tajmouati, A. (2018). Dual New Miniature Microstrip Antenna Based On Metamaterial For RFID Applications. Telkomnika Journal, 16(1).

Er-rebyiy, R., Zbitou, J., Latrach, M., Tajmouati, A., Errkik, A. L., & Abdellaoui, E. (2017). New Miniature Planar Microstrip Antenna Using DGS for ISM applications. Telkomnika Journal, 15(3).

Evizal, E., Abdul Rahman, T., & Abdul Rahim, S. (2013). UHF RFID Tag Antenna for Vehicle License Plate Number. TELKOMNIKA Journal, 11(2).

Fang, D. G. (2010). *Antenna Theory and Microstrip Antennas.* Taylor and Francis Group, LLC.

Ferchichi, A., & Gharsallah, A. (2013). A Novel Small Sierpenski Antennas. *International Journal of Electrical and Computer Engineering, 3*(4).

Finkenzeller, K. (2010). *RFID Handbook Fundamentals and Applications in Contactless Smart Cards, Radio Frequency Identification and Near-Field Communication* (3rd ed.). Giesecke & Devrient GmbH.

Ghiotto, A. (2008). *Tag UHF RFID antenna design, application by jet of material* (Ph.D. dissertation). Institut Polytechnique, Grenoble, France.

Ghiotto, A., Cantalice, S. F., Vuong, T. P., Pouzin, A., Fontgallandt, G., & Tedjini, S. (2008). Miniaturized Patch Antenna for the Radio Frequency Identification of Metallic Objects. *IEEE MTT-S International Microwave Symposium (IMS2008)*, 583-586. 10.1109/MWSYM.2008.4633233

Ghiotto, A., Vuong, T. P., Yagoub, M. C. E., & Tedjini, S. (2007). *On the Design and Miniaturization of Passive UHF RFID Transponder Antenna*. Presented at the *Int. Conf: URSI2007*, Ottawa, Canada.

Gianvittorio, J. P., & Rahmat, Y. (2000). Fractal Element Antennas: A Compilation of Configurations with Novel Characteristics. *IEEE Antennas and Propagation Society International Symposium*. 10.1109/APS.2000.874566

Guesmi, C., Ferchichi, A., & Gharsallah, A. (2014). A Modified Fractal Bow Tie Antenna for an RFID Reader. *International Journal of Electrical and Computer Engineering*, *4*(3).

Guha, D., & Yahia, M.M (2011). *Microstrip and Printed Antennas: New Trends, Techniques and Application*. John Wiley & Sons, Ltd.

Hua, R. C., & Ma, T. G. (2007). A printed dipole antenna for ultra high frequency (UHF) radio frequency identification (RFID) handheld reader. *IEEE Transactions on Antennas and Propagation*, *55*(12), 3742–3745. doi:10.1109/TAP.2007.910521

Ihamji, M., Abdelmounim, E., Bennis, H., Hefnawi, M., & Latrach, M. (2017). Design of Compact Tri-Band Fractal Antenna for RFID Readers. *International Journal of Electrical and Computer Engineering, 7*(1).

Ihamji, M., Abdelmounim, E., Zbitou, J., Bennis, H., & Latrach, M. (2016). Novel Design of a Miniature Fractal Microstrip CPW Fed Antenna for RFID Reader. *IEEE Conference Publications, The International Conference on Wireless Networks and Mobile Communications (WINCOM2016)*. 10.1109/WINCOM.2016.7777197

Ihamji, M., Abdelmounim, E., Zbitou, J., Bennis, H., & Latrach, M. (2016). Novel design of a miniature L-Slot microstrip CPW-fed Antenna for RFID reader. *IEEE Conference Publications, The 11th International Conference on Intelligent Systems: Theories and Applications (SITA2016)*.

Impinj Mini-Guardrail Datasheet from Cisper Company. (n.d.). Retrieved from https://www.cisper.nl/rfid/downloads/Impinj_Mini-Guardrail%20Data%20Sheet.pdf

Indoor directional antenna 304021 Datasheet from GAO RFID Company. (n.d.). Retrieved from http://gaorfid.com/RFID-brochures/2.4GHz_6_dBi_Indoor_Directional_Antenna_304021.pdf

Irfan, N., Yagoub, M. C. E., & Hettak, K. (2012). Design of a microstrip-line-fed inset patch antenna for RFID applications. *IACSIT International Journal of Engineering and Technology*, *4*(5), 558–561. doi:10.7763/IJET.2012.V4.432

Irsadi Aksun, M., & Chuang, S.-H. (1990). On slot-coupled microstrip antennas and their applications to CP operation-theory and experiment. IEEE Transactions on Antenna & Propagation, 38(8).

Jian-bin, X., Fengjie, H., & Dong, W. (2015). Analysis and Design of Tag Antenna Based UHF RFID for Libraries. *Indonesian Journal of Electrical Engineering and Computer Science, 13*(3).

Kapoor, S., & Parkash, D. (2012). Miniaturized triple band microstrip patch antenna with defectedground structure for wireless communication applications. *International Journal of Computers and Applications*, 57.

Karmakar, N. C. (2010). Handbook of Smart Antennas for RFID systems. John Wiley & Sons, Inc. doi:10.1002/9780470872178

Kim, J. S., Shin, K. H., Park, S. M., Choi, W. K., & Seong, N. S. (2006). Polarization and space diversity antenna using inverted-f antennas for RFID reader applications. *IEEE Antennas and Wireless Propagation Letters*, *5*, 265–268. doi:10.1109/LAWP.2006.875892

Kitsos, P., & Zhang, Y. (2008). *RFID Security: Techniques, Protocols and System-on-Chip Design.* Springer Science+Business Media.

Kumar, G., & Ray, K. P. (2003). *Broadband Microstrip Antennas*. Artech House.

Kumar, R., Malathi, P., & Shinde, J. P. (2007). Design of miniaturized fractal antenna. *37th Europ. Microw. Conf.*, 474-477.

Kuo, S. K., Chen, S.-L., & Lin, C.-T. (2008). An accurate method for impedance measurement of RFID tag antenna. *Progress in Electromagnetics Research*, *83*, 93–106. doi:10.2528/PIER08042104

Kuo, Y. L., & Wong, K. L. (2003). Printed Double-T Monopole Antenna for 2.4/5.2 GHz Dual-Band WLAN Operations. *IEEE Transactions on Antennas and Propagation*, *51*(9).

Kwon, H., & Lee, B. (2005). Compact slotted planar inverted-F RFID tag mountable on metallic objects. *Electronics Letters*, *41*(24), 1308–1310. doi:10.1049/el:20052940

Li, R. L., Dejean, G., Tentzeris, M. M., & Laskar, J. (2004). Development and analysis of a folded shorted-patch antenna with reduced size. *IEEE Transactions on Antennas and Propagation*, *52*(2), 555–562. doi:10.1109/TAP.2004.823884

Lim, W.-G., Son, W.-I., Oh, K.-S., Kim, W.-K., & Yu, J.-W. (2008). Compact integrated antenna with circulator for UHF RFID system. *IEEE Antennas and Wireless Propagation Letters*, *7*, 673–675. doi:10.1109/LAWP.2008.921348

Lin, K., Chen, S., & Mittra, R. (2010). A Capacitively Coupling Multifeed Slot Antenna for Metallic RFID Tag Design. *IEEE Antennas and Wireless Propagation Letters*, *9*, 447–450. doi:10.1109/LAWP.2010.2048991

Loo, C. H., Elmahgoub, K., Yang, F., Elsherbeni, A. Z., Kajfez, D., Kishk, A. A., ... Ruuskanen, P. (2008). Chip impedance matching for UHF RFID tag antenna design. *Progress in Electromagnetics Research*, *81*, 359–370. doi:10.2528/PIER08011804

Mahatthanajatuphat, C., Saleekaw, S., Akkaraekthalin, P., & Krairiksh, M. (2009). A rhombic patch monopole antenna with modified minkowski fractal geometry for UMTS,WLAN, and mobile WiMAX application. *Progress in Electromagnetics Research*, *89*(1), 57–74. doi:10.2528/PIER08111907

Marrocco, G. (2003). Gain-optimized self-resonant meander line antennas for RFID applications. *IEEE Antennas and Wireless Propagation Letters*, *2*(21), 302–305. doi:10.1109/LAWP.2003.822198

Marrocco, G. (2008). The Art of UHF RFID Antenna Design: Impedance-Matching and Size-Reduction Techniques. *IEEE Antennas & Propagation Magazine*, *50*(1), 66–79. doi:10.1109/MAP.2008.4494504

Mehdipour, A., Rosca, I. D., Sebak, A., Trueman, C. W., & Hoa, S. V. (2010). Full-composite Fractal Antenna using Carbon Nanotubes for Multiband Wireless Applications. *IEEE Antennas and Wireless Propagation Letters*, *9*, 891–894. doi:10.1109/LAWP.2010.2076342

Monti, G., Catarinucci, L., & Tarricone, L. (2009). Compact Microstrip Antenna For Rfid Applications. *Progress In Electromagnetics Research Letters.*, *8*, 191–199. doi:10.2528/PIERL09042803

Munson, R. E. (1974). Conformal microstrip antennas and microstrip phased arrays. *IEEE Transactions on Antennas and Propagation*, *22*(1), 74–78. doi:10.1109/TAP.1974.1140723

Nasimuddin. (2011). *Microstrip Antennas*. Rijeka, Croatia: InTech.

Nithisopa, K., Nakasuwan, J., Songthanapitak, N., Anantrasirichai, N., & Wakabayashi, T. (2007). Design CPW fed slot antenna for wideband applications. Piers Online, 3(7).

Numan-Al-Mobin, A., Cross, W. M., Kellar, J. J., & Anagnostou, D. E. (2015). *RFID Integrated QR Code Tag Antenna*. Retrieved from https://www.researchgate.net/publication/308559377

Oraizi, H., & Hedayati, S. (2012). *Miniaturization of Microstrip Antennas by the Novel Application of the Giuseppe Peano Fractal Geometries. In Antennas and Propagation* (pp. 3559–3567). IEEE Transactions.

Outdoor directional antenna 304020 Datasheet from GAO RFID Company. (n.d.). Retrieved from http://gaorfid.com/RFID-brochures/2.4GHz_14_dBi_Outdoor_Directional_Antenna_304020.pdf

Patch antenna APAKN1304-S5517-T Datasheet from Abracon Company. (n.d.). Retrieved from http://www.farnell.com/datasheets/2611067.pdf

Patch antenna APAKN2504-S2448-T Datasheet from Abracon Company. (n.d.). Retrieved from http://www.farnell.com/datasheets/2611070.pdf

PEL90206 Datasheet from Laird Company. (n.d.). Retrieved from https://www.lairdtech.com/sites/default/files/public/solutions/Industrial%20Wireless%20RFID%20Antenna%20Solutions.pdf

Pigeon, M. (2011). Design of ultra-compact antennas based on metamaterials, application to the fabrication of a miniature GNSS antenna (Ph.D. dissertation). Institut National Polytechnique, Toulouse, France.

Puck antenna ADA-A2450-S Datasheet from Adatcus Company. (n.d.). Retrieved from http://www.farnell.com/datasheets/1391861.pdf

Puente, C., & Pous, R. (1996). Fractal design of multiband and low side-lobe arrays. *IEEE Transactions on Antennas and Propagation*, *44*(5), 730–739. doi:10.1109/8.496259

Puente, C., Romeu, J., & Cardama, A. (2000). The Koch Monopole: A Small Fractal Antenna. *IEEE Transactions on Antennas and Propagation*, *48*(11), 1773–1781. doi:10.1109/8.900236

Puente, C., Romeu, J., & Cardama, R. (1998). On the Behavior Of the Sierpinski Multiband Fractal Antenna. IEEE Transactions on Antenna and Propagation, 46(4).

Puente, C., Romeu, J., Pous, R., Ramis, J., & Hijazo, A. (1998). Small but Long Koch Fractal Monopole. *Electronics Letters*, *34*(1), 9–10. doi:10.1049/el:19980114

Rahmadani & Munir. (2011). Microstrip patch antenna miniaturization using artificial magnetic conductor. *6th International Conference on Telecommunication Systems, Services, and Applications (TSSA)*, 219-223.

Rambabu, K., Ramesh, M., & Kalghatgi, A. T. (1997). Design method for patch antennas short circuited at one end. *Asia Pacific Microw. Conf.*, 233-236. 10.1109/APMC.1997.659347

Rao, K., Nikitin, P., & Lam, S. (2005). Antenna design for UHF RFID tags: A review and a practical application. *IEEE Transactions on Antennas and Propagation*, *53*(12), 3870–3876. doi:10.1109/TAP.2005.859919

Rao, K. V. S., Lam, S. F., & Nikitin, P. V. (2010). UHF RFID Tag for Metal Containers. *Proceedings of Asia-Pacific Microwave Conference.*

Reha, A., El Amri, A., Benhmammouch, O., Oulad Said, A., El Ouadih, A., & Bouchouirbat, M. (2015). CPW-fed H-tree fractal antenna for WLAN, WIMAX, RFID, C-band, HiperLAN, and UWB applications. *International Journal of Microwave and Wireless Technologies*, 1–8.

Reha, A., & Elamri, A. (2014). Fractal antennas a novel miniaturization technique for wireless networks. *Transactions on Networks and Communications, 2*(5), 166.

Reynolds, N. D. (2012). *Long range ultra-high frequency (UHF) radio frequency identification (RFID) antenna design* (Ph.D. dissertation). Purdue University, Fort Wayne, IN.

Seladji-Hassaine, N., & Bendimerad, F. T. (2016). Miniaturized Dual Band Triangular Microstrip Antenna with Sierpinski Fractal Ground. *International Journal of Microwave and Optical Technology*, *11*(5).

Sofi, A., Roky, K., Bouhorma, M., & Hadj Baraka, I. (2014). Novel Antennas for UHF RFID Tags: Design and Miniaturization. *International Journal of Electrical and Computer Engineering, 4*(1).

Sofi, A., Roky, K., & Hadj Baraka, I. (2014). Antenna Design for UHF RFID Tags Using Two Coupled Patches. *International Journal of Information and Network Security, 3*(1).

Soh, P.J., Rahim, M.K.A., Asrokin, A., & Abdul Aziz, M.Z.A. (2007). Design Modeling and performance comparison of feeding techniques for a microstrip patch antenna. *Journal Technologi*, *47*(D), 103-120.

Son, W., Lee, H., Lee, M., Min, S., & Yu, J. (2010). Compact Square Quadrifilar Spiral Antenna with circular polarization for UHF mobile RFID Reader. *Microwave Conference Proceedings (APMC).*

Son, W.-I., Lim, W.-G., Lee, M.-Q., Min, S.-B., & Yu, J.-W. (2010, May). Design of compact quadrifilar inverted-f antenna with circular polarization for GPS receiver. *IEEE Transactions on Antennas and Propagation*, *58*(5), 1503–1510. doi:10.1109/TAP.2010.2044344

Suganthi, S., Tharini, K. S., Sarankumar, P. S., Raghavan, S., & Kumar, D. (2011). *Design and simulation of planar Minkowski fractal antennas*. Paper presented at 2nd International Conference on Wireless Communication, Vehicular Technology, Information Theory and Aerospace & Electronic Systems Technology (Wireless VITAE), Chennai, India. 10.1109/WIRELESSVITAE.2011.5940935

Tabakh, I., Jorio, M., El Amrani El Idrissi, N., & Mazri, T. (2016). Design and Optimization of a New Slotted Patch Antenna for RFID Applications. International Journal on Communications Antenna and Propagation, 6(1).

Turcu, C. (2009). *Development and Implementation of RFID Technology*. In-Tech. doi:10.5772/100

Ukkonen, L., Sydanheimo, L., & Kivikoski, M. (2004). *A novel Tag Design Using Inverted-F Antenna for Radio Frequency Identification of Metallic Objects*. Princeton, NJ: IEEE Sarnoff Symp. Advances Wired and Wireless Communication. doi:10.1109/SARNOF.2004.1302848

Ukkonen, L., Sydanheimo, L., & Kivikoski, M. (2005). Effects of Metallic Plate Size on the Performance of Microstrip PatchType Tag Antennas for Passive RFID. *IEEE Antennas and Propagation Letters*, *4*, 410–413. doi:10.1109/LAWP.2005.860212

Xi, J., Zhu, H., & Ye, T. (2010). Platform-tolerant PIFA-type UHF RFID tag antenna. *IEEE RFID Conference*, 174–180.

Yu, Y., Huang, Z., Zheng, C., & Wu, Y. (2014). A Miniature Antenna for 2.45 GHz RFID Tag. *Journal of Computers*, *9*(2), 404–411. doi:10.4304/jcp.9.2.404-411

Zahraoui, I., Errkik, A., Abdelmounim, E., Bennis, H., El Abdellaoui, L., & Sanchez Mediavilla, A. (2015). A New Design of a CPW Fed Triple-Band for GPS, UMTS and WiMAX Applications. International Journal on Communications Antenna and Propagation, 5(6).

Zahraoui, I., Errkik, A., Zbitou, J., Tajmouati, A., & Mediavilla, A. S. (2016). A Novel Design of a Fractal Antenna for IMT and WiMAX Applications. *International Journal of Microwave and Optical Technology*, *11*(6).

Zainud-Deen, S. H., Malhat, H. A., & Awadalla, K. H. (2009). Fractal Antenna for Passive UHF RFID Applications. *Progress in Electromagnetics Research*, *16*, 209–228. doi:10.2528/PIERB09051506

Zhao, W., Wang, G., & Lai, X. (2008). Active E-Plate with Slot Antenna. *International Conference on Wireless Communications, Networking and Mobile Computing*, 1-3.

KEY TERMS AND DEFINITIONS

Circular Polarization: The circular polarization of an electromagnetic wave is a polarization in which the electric field of the passing wave does not change strength but only changes direction in a rotary manner.

CPW: A coplanar wave guide is a type of electrical planar transmission line which can be fabricated using printed circuit board technology, and is used to convey microwave-frequency signals.

FR4: The FR-4 is a common material for printed circuit boards (PCBs). A thin layer of copper foil is laminated to one or both sides of an FR-4 glass epoxy panel. FR4 (i.e. flame retardant 4) is a NEMA grade designation for glass-reinforced epoxy laminate material. FR-4 is a composite material composed of woven fiberglass cloth with an epoxy resin binder that is flame resistant.

ISM: The industrial, scientific, and medical radio band (ISM band), also called unlicensed bands, refers to a group of radio bands or parts of the radio spectrum that are internationally reserved for the use of radio frequency (RF) energy intended for scientific, medical and industrial requirements rather than for communications. ISM bands are generally open frequency bands, which vary according to different regions and permits. The most used bands are 2.4-2.5 GHz and 5.725-5.875 GHz.

Microstrip Antenna: Is a type of electrical transmission line that can be fabricated using printed circuit board technology. It consists of a conducting strip separated from a ground plane by a dielectric layer known as the substrate. An individual microstrip antenna consists of a patch of metal foil of various shapes (a patch antenna) on the surface of a PCB, with a metal foil ground plane on the other side of the board.

Omnidirectional Antenna: It is a class of antenna which have an axis about which radio wave power is radiated symmetrically, and, upon that axis, is zero. this is different from an isotropic antenna, which power is radiated symmetrically about any axis, having a spherical radiation pattern.

Patch Antenna: A patch antenna (also known as a rectangular microstrip antenna) is a type of radio antenna with a low profile, which can be mounted on a flat surface. It consists of a flat rectangular sheet or "patch" of metal, mounted over a larger sheet of metal called a ground plane.

PCB: A printed circuit board (PCB) is an electronic circuit used in devices to provide mechanical support and a pathway to its electronic components. It is made by combining different sheets of non-conductive material, such as fiberglass or plastic, that easily holds copper circuitry. PCBs can be single-layer for simple electronic devices.

Section 3

Microwave and RF Active Circuits

Chapter 11
Design and Analysis of a New Configuration of Microwave Power Amplifier

Amine Rachakh
University of Hassan 1st, Morocco

Larbi El Abdellaoui
University of Hassan 1st, Morocco

Mohamed Latrach
ESEO, France

ABSTRACT

The chapter has the objective of introducing and explaining the concept of a microwave power amplifier (PA). PA is one of the blocks that has a large effect on the overall performance of communication systems, especially in transmitter systems, and their design is decided by the parameters of the transistor selected. This chapter is divided into three parts, which will be as follows: Part 1 provides background theory relevant to our research. Part 2 describes the matching techniques for PAs. Part 3 utilizes the tools developed in Parts 1 and 2 to analyze and design the proposed microwave power amplifier with a microstrip technology intended for wireless applications.

INTRODUCTION

Modern radio frequency (RF) and microwave communication systems is a dynamic and exciting field, owing in wide part to the coherence between recent advances in modern electronic device technology and the development in demand for video, voice and data communication capacity that begin in the 1990s and continues through the present (Haynes, 2007). Before to this advancement in communications, microwave technology was the almost chosen domain of the defence industry; the recent and dramatic growth in demand for communication systems for like applications as broadcast video, wireless paging, mobile telephony, and tethered or untethered computer networks has driven the industry to evolve. These

DOI: 10.4018/978-1-5225-7539-9.ch011

communication systems are utilized through a large range of environments, including industrial and manufacturing facilities, corporate offices, private homes, as well as infrastructure for municipalities. The variety of applications and operational environments has led, across the accompanying tall production volumes, to enormous progress in cost-efficient manufacturing capabilities of radio frequency and microwave products. This, in turn, has reduced the implementation cost of a host of novel and cost-effective wireless as well as wired microwave and radio frequency services. Automotive collision-avoidance radar, Inexpensive handheld GPS navigational aids, and widely available broadband digital service access are amongst these. Microwave technology is generally adapted for these appear applications in sensing and communications, since the taller operational frequencies allow both wide numbers of independent channels for the large diversity of uses envisioned as well as considerable disposable bandwidth per channel for high-speed communication (Pozar, 2009; Steer, 2009).

In modern microwave communication systems, Amplifiers are a key part of all transmitter systems. The microwave power amplifier (PA) is responsible for the most of the system's power consumption. When the amplifier is operating below 50% DC-DC efficiency, it is more an electrical heater than an oscillator, necessitating high power DC supplies and heat dissipation systems. Heat damage can wreak havoc on satellite and radar communication systems, demanding sophisticated cooling structures to protect not only the PA, but also the surrounding electronics. With wireless communication systems, hand-held battery operated mobile devices suffer the most from the PAs high power consumption, resulting in a decreased battery life and shorter talk time. Reported results on power amplifiers, typically operating above 85%, reliance on precise multi-harmonic impedance terminations on both the input and output terminals at very high levels of device compression, both of which results in highly nonlinear and narrow operating bandwidths (Raab, 2001; Srirattana, Raghavan, Heo, Allen, & Laskar, 2005). PA has been concentrated on specified RF bandwidths of 10% or less. Systems like worldwide interoperability for microwave access (WiMAX) and 4G necessitate wide instantaneous bandwidths, and future systems will likely require even more, not just owing to wider spectral allocations, but the base bandwidth of the signals themselves may well exceed 100 MHz (Choi, Kang, Kim, Park, Jin, & Kim, 2009; Chowdhury, Hull, Degani, Wang, & Niknejad, 2009). In addition, satellite and radar communication systems have yet to benefit from recent improvements in PA output power and efficiency because of their much wider bandwidth requirements with high power gain.

The need for power amplifiers capable of achieving high gain and good impedance matching across a broad range of frequencies has fueled extensive previous research and development of innovative power amplifier architectures. While primarily now an industry-driven field, microwave PA design still has a place in an academic setting as communication systems continue to push the limits of data rates in narrow channels.

In this chapter, basic concepts on power amplifiers will be introduced. A special emphasis on differentiating PAs from conventional small-signal amplifiers will be necessary in order to comprehend the specific requirements for this type of circuit.

Unlike in the small-signal case, the load line is a very important step for power amplifiers. It represents the large signal regime of the transistor. However, the load line does not define exclusively the type of amplifier, because this can be influenced by the dynamic behaviour of the device.

In order to start the description and characterization of microwave power amplifiers, a general configuration of a single-stage microwave amplifier is presented in Figure 1. This conventional circuit is applicable either for small or large-signal amplifiers. It comprises an input matching network, the active device, and the output matching network.

Figure 1. General block diagram of a single stage microwave amplifier

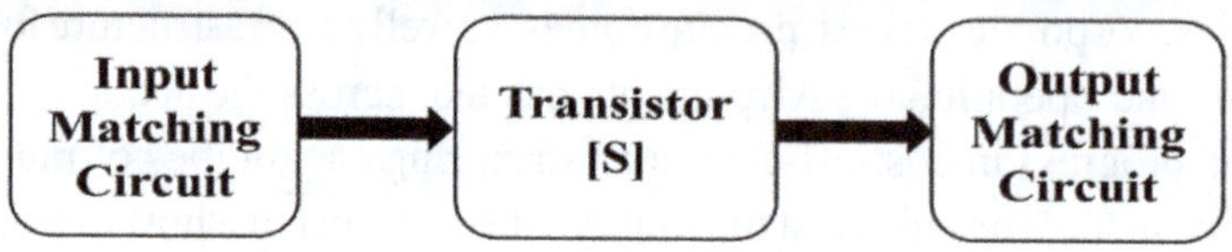

Matching networks are designed in order to make the device behave in an optimal way according to a specific goal. As an example, these matching networks can be tailored such that a maximum gain is realized, or they can be tailored such that the device is driven over its maximum rating, and thence maximum output power is obtained (power amplifier), or they can be tailored in order to make the circuit capable of amplifying extremely small signals with very low noise at the output (low noise amplifier). This matching network can be implemented physically by utilizing lumped components, distributed elements or a combination of both, depending on the application and/or range of frequencies. From the point of view of modelling, these matching networks, which comprise passive elements only, are basically characterized by intrinsic linear elements, and therefore they are considered to be linear networks per se (Student & INDIA, n.d.).

Unlike the matching networks, the active device can be characterized as either a linear or a non-linear network, depending on the electrical power range of the signals involved in the process. If the power level of the input signal is very small as illustrated in Figure 2 case (a), such that device characteristic and output approach respectively a linear system and linear output, i.e., when scaling properties and superposition are fulfilled (Couch, Kulkarni, & Acharya, 1997), consequently, therefore, the circuit can be modeled in terms of two-port S- parameters. This is the general approach utilized in PAs and LNAs. This Scattering parameter description is only valid under a specific bias point and does not conceptually rely on the level of the input signal. The linear concept can be announced from the frequency viewpoint as well. If the number of harmonic components at the output is the same as at the input, i.e., there is no harmonic generation, thus the system is linear.

Contrary to the small signal approach, under large signal levels operation, the time domain waveform at the output shows some distortion with respect to the input signal as illustrated in Figure 2 case (b). Very complex models intended to represent the non-linear characteristics of the active devices or amplifiers are necessary. From the frequency viewpoint, this means that besides the harmonic components of the exciting signal, additional harmonic components are generated due to the device or amplifier non-linearity (Vendelin, Pavio, & Rohde, 2005).

Figure 2. Linear and non-linear mode of the power amplifier

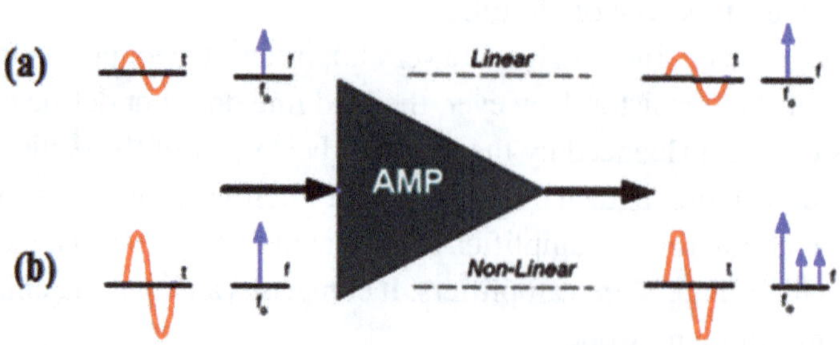

This chapter is organized into three parts will be as follows: Part 1 provides background theory relevant to our research. Part 2 describes the matching techniques for PAs. Part 3 Introduces the analysis and design of the proposed microwave power amplifier with a microstrip technology intended to wireless applications.

POWER AMPLIFIER FUNDAMENTALS

This part provides the fundamental knowledge and aspects of microwave power amplifier design. Several performance metrics, like power amplifier classification, power gain, load line, output power and stability are introduced.

Power Amplifier Classification

There is more than one way to classify PAs. A popular way is to label each type of PA with roman letters such as A, B, AB, C, D, E, F, etc. This is done depend on either passive components in the biasing network or the output network that supports the drain current or voltage into a certain form (Hella & Ismail, 2006). These PAs can be classified as either linear amplifiers or non-linear amplifiers. The class A, class B, and class AB are normally labelled as linear amplifiers as opposed to the class C, class D, class E, and class F that are categorized nonlinear amplifiers. A short introduction to each class of power amplifier is illustrated below.

A class-A illustrated in Figure 3 is a popular linear amplifier. It is biased in such a way that the transistor of the power amplifier conducts throughout the complete cycle. Thus, the major advantage of the class-A amplifier is that it has very good linearity. However, the disadvantage is its low efficiency, which is ideally only about 20% (Cripps, 2006).

A class-B amplifier illustrated in Figure 4 is also a linear type. The way it works is similar to the class-A. However, the output network of the class-B PA only works for half the sinusoidal cycle. Thus, it is popular to utilize the class-B amplifiers as a pair in a push-pull configuration. One conducts in the negative cycle and the other conduct in the positive cycle (Barber & Salter, 1978).

The class-B is more efficient than the class-A, The efficient rise up about 50%, But has problems with linearity at the crossover point where one transistor turns on and the other transistor is turning off and vice versa.

Figure 3. Class-A Amplifier

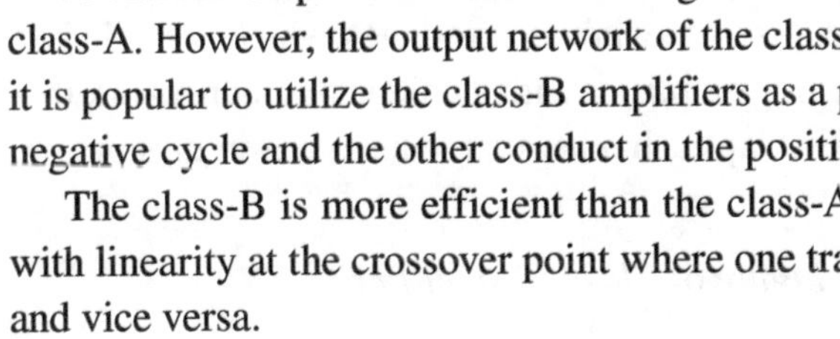

Figure 4. Class-B Amplifier

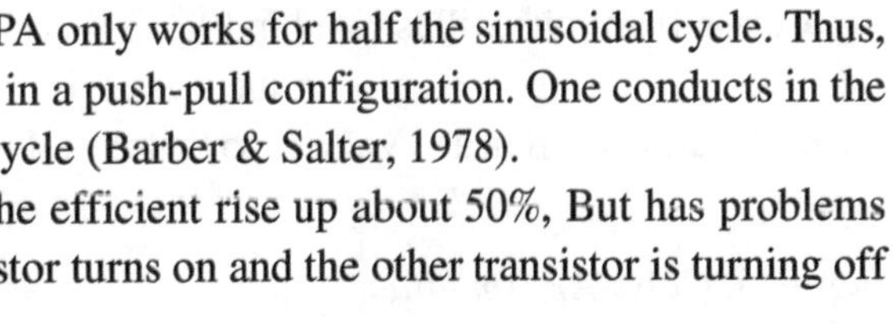

A class-AB amplifier is counted a linear amplifier illustrated in Figure 5. The class-AB is a combination of the class A and the class-B amplifiers. The main difference between the class-AB and the class-B is that the class-AB permits two transistors to be on at the same time near crossover but for a very short amount of time. Thence, each transistor conducts for more than half a cycle but less than a full cycle. Due to this, better efficiency and linearity are both achieved at a balanced point. The linearity of the class-AB is better than the class-B, and the efficiency of the class-AB is higher than the class-A (Aniktar, Sjoland, Mikkelsen, & Larsen, 2005; Bingham, 1985).

A class-C amplifier is illustrated in Figure 6. The transistor is biased in a way that the device only works less than half cycle of the input signal. According to (Raab, 2001), the class-C amplifier can attain an efficiency of 85% theoretically. However, the class-C amplifiers do not have good linearity. The class-C amplifier is not favorable for audio amplifiers because of its high distortion. Thus, they are normally utilized for high power output applications at RF/microwave frequencies and distortion owing to higher harmonics can be filtered out.

A class-D amplifier, illustrated in Figure 7, is considered to be a PWM, or switching amplifier. It is generally utilized as an audio amplifier. Compared to linear amplifiers, the power loss in the class-D amplifiers can be significantly less due to the transistor ML or MH is either totally off or on. The active devices, ML and MH, cannot be turned on at the same time. When ML or MH is fully on, there is no power dissipation ideally in the switching transistors because of no path from the power supply to the ground when one of the active devices is fully off. Thence, efficiency can possibly be as high as 90% to 95% theoretically (Guy, 2009; McCorkle, 1992). Normally a PWM carrier signal that is driving output devices is modulated by an audio signal. A low-pass filter can be positioned in the output stage network to take off the high-frequency components of the PWM signal (Guy, 2009).

A class-E amplifier illustrated in Figure 8. It is considered a nonlinear or switching amplifier. The distinctive feature of the class-E is that the drain voltage and current of the switching device are created in such a way that they do not happen simultaneously. This results in less power consumption. Therefore, the theoretical efficiency is very high (Raab, 1977; Raab, 2001).

A class-F amplifier illustrated in Figure 9 is another type of switching mode PA which is nonlinear. Just such as the class-E amplifier, the class-F is a switching amplifier with a unique output network that forms the drain current and voltage so that they do not overlap each other that reduces power dissipation.

Figure 5. Class-AB Amplifier

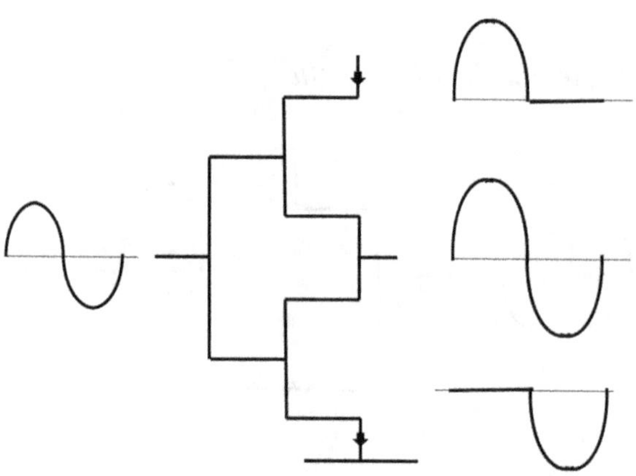

Figure 6. Class-C Amplifier

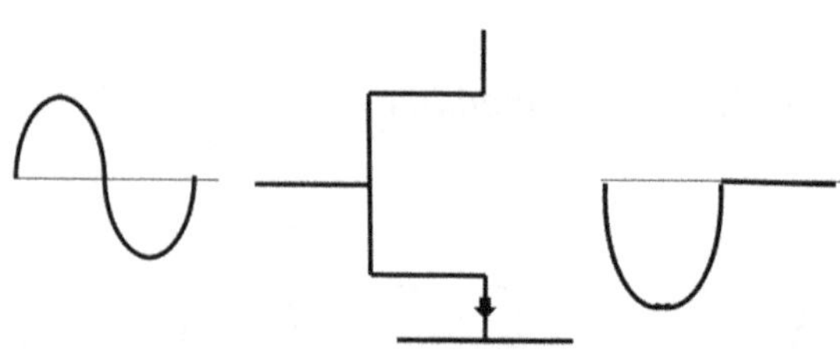

Figure 7. Class-D Amplifier

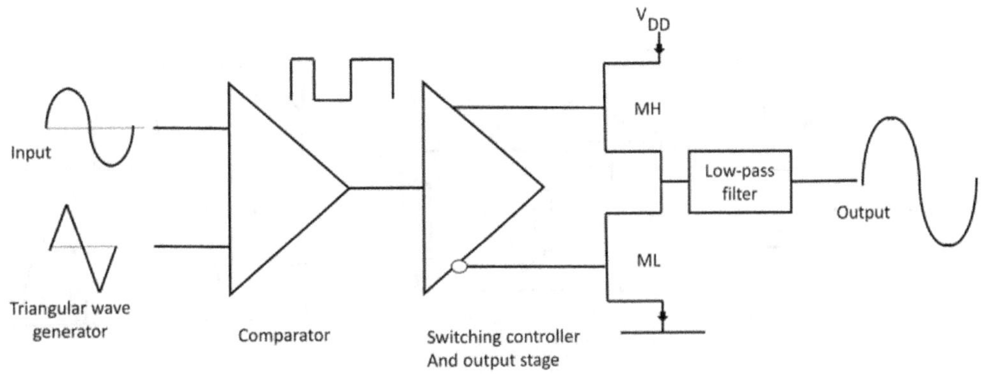

Figure 8. Class-E Amplifier

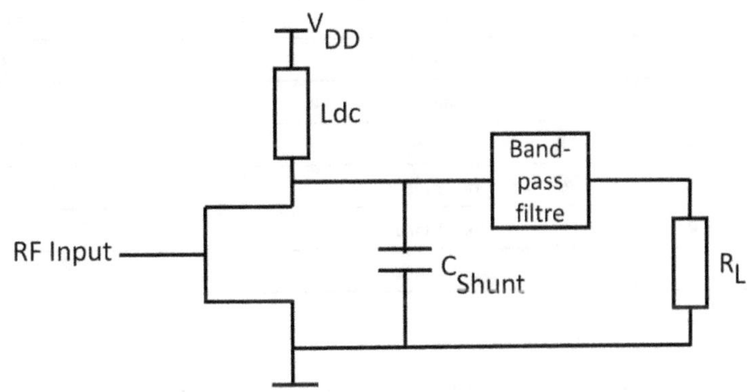

Compared to the class-E power amplifier, the main difference is in the output network. The class-F power amplifier contains two parallel LC resonant tanks. One tank functions as a matching network tuned at the fundamental frequency. The other tank serves as a harmonic tuning network tuned at the third order harmonic. Therefore, a short circuit can be seen at the even order harmonics, and an open circuit at the 3rd order harmonic. By doing so, ideally, a square wave drain voltage and a half bridge rectified sine wave drain current are created but do not overlap each other for less power dissipation (Raab, 2001).

Both non-linear and linear PAs have positives and negatives. On one hand, the non-linear PAs generally have good efficiency but have poor linearity. On the other hand, the linear PAs generally have good linearity but poor efficiency. Therefore, a trade-off between efficiency and linearity becomes a major factor for selecting an appropriate power amplifier. In Table 1, a comparison of the linearity and theoretical efficiency among the discussed PA types is presented (Colantonio, Giannini, & Limiti, 2009).

Single-Stage Power Amplifier Design

A simple configuration of microwave power amplifier consists of a single active device, a field-effect transistor (FET) in common source configuration, for example, connected to a source and load both having the system characteristic impedance, Z_0 as represented in Figure 10.

Figure 9. Class-F Amplifier

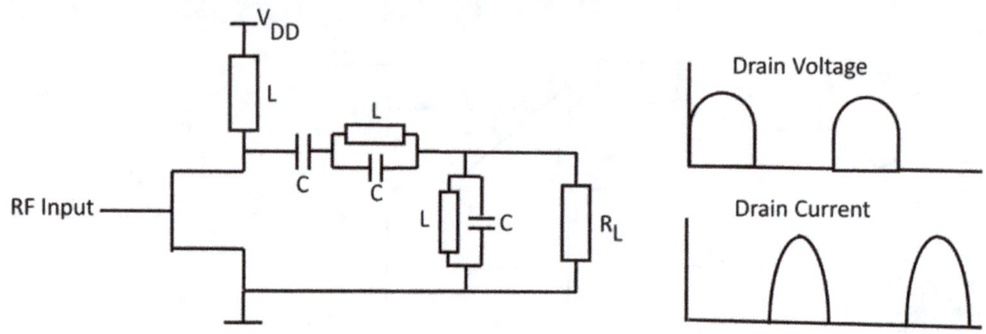

Table 1. Comparison in efficiency among different amplifier classes

Class	Operation	Theoretical Efficiency
A	Linear	50%
B	Linear	78%
AB	Linear	50% -78%
C	Non-Linear	85%
D	Non- Linear	90% or 100%
E	Non- Linear	100%
E	Non- Linear	88%

If we consider the definition of the Scattering-parameters for any two-port, we can write the relationship between the incident and the reflected power waves at the input and output ports of the simple microwave PA (see Figure 10) as follows (Pozar, 2009; Vendelin, Pavio, & Rohde, 2005):

$$b_1 = a_1 S_{11} + a_2 S_{12} \tag{1}$$

Figure 10. A simple configuration of microwave power amplifier

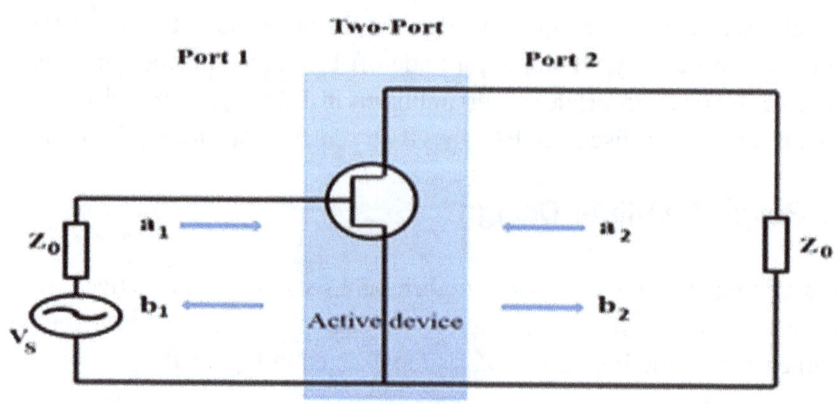

$$b_2 = a_1 S_{21} + a_2 S_{22} \tag{2}$$

The power gain of this circuit (PA) is a ratio of the power entering at port 1 to the power emerging from port 2, with no power being incident at port 2. Therefore, if we set $a_2 = 0$ in Equation (2) we can express the power gain of Figure 10 as:

$$G = \frac{\left|b_2\right|^2}{\left|a_1\right|^2} = \left|S_{21}\right|^2 \tag{3}$$

In effect, this gain parameter is called to as the Transducer Power Gain and is symbolized by G_T.

We will now present the simple case above to account for the more factual situation of an arbitrary source and load ports, as exposed in Figure 11.

In this case, the transducer gain Equation (3) needs to be changed to account for the effects of impedance mismatch at both the input/output ports. We now can express the equation of the gain for the PA of Figure 11 as follows:

$$G_T = M_S * \left|S_{21}\right|^2 * \mathrm{M}_L \tag{4}$$

Where M_S is the source mismatch factor and M_L is the load mismatch factor (Gonzalez, 1997; Mason, 1957). These "mismatch factors" describe the amount of power actually transferred to the respective loads as a proportion of power available from the respective sources, in other words:

$$M_S = \frac{P_{in}}{P_{AVS}} \tag{5}$$

$$M_L = \frac{P_L}{P_{AVN}} \tag{6}$$

Figure 11. A simple configuration of microwave PA with arbitrary terminations

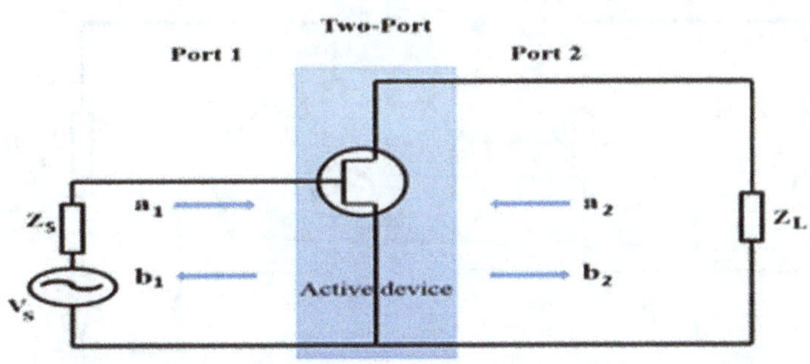

Where P_{AVS} is the power available from the source and P_{AVN} is the power available from the two-port network, respectively.

To state the mismatch factors in terms of termination impedances, we necessity to consider the power flow at the input/output of the circuit (amplifier) separately, as illustrated in Figure 12.

We will first treat the case at the input port, as represented in Figure 12-(a). The power available from the source is known as the power transferred into a conjugate matched load. If we count the source in Figure 12-(b), as being conjugate matched to the input of the active device, namely, $Z_S = Z_{in}^*$. Hence the average power available from the source is (Drogi & Tomasz, 2012; Pozar, 2005):

$$P_{AVS} = \frac{1}{2}\frac{\left|V_S\right|^2}{R_S} \tag{7}$$

The power transferred to the input of the transistor in Figure 12-(a) is determined by:

$$P_{in} = \frac{R_{in}}{2}\left|\frac{V_S}{Z_{in} + Z_S}\right|^2 \tag{8}$$

By replacing the Equation (7) into Equation (8), with some manipulation we get:

$$P_{in} = \frac{\left|V_S\right|^2}{8R_S}\left(\frac{4R_{in}R_S}{\left|Z_{in} + Z_S\right|^2}\right) = P_{AVS}\left(\frac{4R_{in}R_S}{\left|Z_{in} + Z_S\right|^2}\right) \tag{9}$$

Take into consideration the equation indicated above (5), therefore, the source mismatch factor is given by:

$$M_S = \frac{4R_{in}R_S}{\left|Z_{in} + Z_S\right|^2} \tag{10}$$

Figure 12. Mismatch factors: (a) Input port matching, and; (b) Output port matching

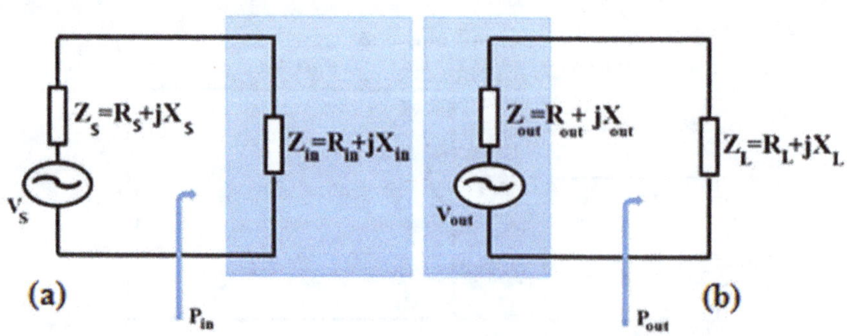

We will now convert Equation (10) to return loss. First, we write R_S in terms of as Γ_S follows:

$$R_S = \frac{\left(Z_S + Z_S^*\right)}{2} = \frac{Z_0}{2}\left[\frac{1+\Gamma_S}{1-\Gamma_S} + \frac{1+\Gamma_S^*}{1-\Gamma_S^*}\right] \tag{11}$$

$$= Z_0 \frac{1-\left|\Gamma_S\right|^2}{\left|1-\Gamma_S\right|^2}$$

Therefore, we can express the real part of Z_{in} in terms of Γ_{in} :

$$R_{in} = Z_0 \frac{1-\left|\Gamma_{in}\right|^2}{\left|1-\Gamma_{in}\right|^2} \tag{12}$$

By replacing Equations (11) and (12) into Equation (10) and substituting Z_S and Z_{in} in Equation (10) by their return loss equivalents, we get the following expression for M_S in return loss terms:

$$M_S = \frac{\left(1-\left|\Gamma_S\right|^2\right)\left(1-\left|\Gamma_{in}\right|^2\right)}{\left|1-\Gamma_S\Gamma_{in}\right|^2} \tag{13}$$

With the same reasoning applied to the output port of Figure 12-(b). We can express the load mismatch factor as follows:

$$M_L = \frac{\left(1-\left|\Gamma_L\right|^2\right)\left(1-\left|\Gamma_{out}\right|^2\right)}{\left|1-\Gamma_L\Gamma_{out}\right|^2} \tag{14}$$

According to Equation (4), we can notice that for $M_L = 1$ when $\Gamma_L = \Gamma_{out}^*$. And for the Equation (13) will similarly show that $M_S = 1$ when $\Gamma_S = \Gamma_{in}^*$. This supports the axiomatic understanding that the maximum available gain will be gotten from the device when we conjugate match the input and output ports.

For two-port, linear and non-unilateral ($S_{12} \neq 0$), Γ_{in} and Γ_{out} will be dependent on the load and source terminations, respectively. This signifies that for a no unilateral device the mismatch factors M_S and M_L are interdependent. The ports cannot be conjugate matched independently as altering the value of Γ_L will change the value of Γ_{in} and vice versa (Mason, 1957; Vendelin, Pavio, & Rohde, 2005).

The mutual relations between the input and output ports of a non-unilateral two-port signifies that if we merge Equations (13) and (14) together in a single gain term, as in Equation (4), we can ignore one of either Γ_{in} or Γ_{out}. Thus, there are two possible expressions for the transducer power gain that we get by replacing Equations (13) and (14) into Equation (4) as follows:

$$G_T = \frac{\left(1-\left|\Gamma_S\right|^2\right)}{\left|1-S_{11}\Gamma_S\right|^2}\left|S_{21}\right|^2\frac{\left(1-\left|\Gamma_L\right|^2\right)}{\left|1-\Gamma_L\Gamma_{out}\right|^2} \tag{15}$$

Or

$$G_T = \frac{\left(1-\left|\Gamma_S\right|^2\right)}{\left|1-\Gamma_S\Gamma_{in}\right|^2}\left|S_{21}\right|^2\frac{\left(1-\left|\Gamma_L\right|^2\right)}{\left|1-S_{22}\Gamma_L\right|^2} \tag{16}$$

In the case of a unilateral circuit $S_{12} = 0$, we remind that the two alternative expressions for transducer power gain, given by equations (15) and (16), consist of either the input reflection coefficient or the output reflection coefficient. In this case, we setting $S_{12} = 0$ in same equations, means that $\Gamma_{in} = S_{11}$ and $\Gamma_{out} = S_{22}$

Thence we get a single term for Unilateral Transducer Power Gain:

$$G_{TU} = \frac{\left(1-\left|\Gamma_S\right|^2\right)}{\left|1-\Gamma_S S_{11}\right|^2}\left|S_{21}\right|^2\frac{\left(1-\left|\Gamma_L\right|^2\right)}{\left|1-\Gamma_L S_{22}\right|^2} \tag{17}$$

As we observed, Unilateral Transducer Power Gain can be determined as the product of three gain.

$$G_{TU} = G_S * G_0 * G_L \tag{18}$$

Where:

$$G_S = \frac{\left(1-\left|\Gamma_S\right|^2\right)}{\left|1-\Gamma_S S_{11}\right|^2} \tag{19}$$

$$G_0 = \left|S_{21}\right|^2 \tag{20}$$

$$G_L = \frac{\left(1 - \left|\Gamma_L\right|^2\right)}{\left|1 - \Gamma_L S_{22}\right|^2} \tag{21}$$

In this case, G_S and G_L are independent of each other, which signifies that the input and output ports of the transistor can be matched independently. It is a simple to show that the maximum unilateral gain is got when we set $\Gamma_S = S_{11}^*$ and $\Gamma_L = S_{22}^*$ in Equation (17). Then we get the following expression:

$$G_{TU_{\max}} = \frac{1}{\left|1 - \left|S_{11}\right|^2\right|} \left|S_{21}\right|^2 \frac{1}{\left|1 - \left|S_{22}\right|^2\right|} \tag{22}$$

Load Line Analysis

The concept of a load line is tightly attached to power amplifiers since it yields a clear depiction of the capabilities of a device for maximum output power application. It illustrates the trajectory of all the instantaneous values of voltage (V_{DS}) and current (I_{DS}) of a device when this one is worked under a specific load and at a suitable bias point. Moreover, plotting the load line facilitates to visualize and optimize the power amplifier for maximum performance (Jain, Kang, Soin, & Xi, 2002).

In Figure 13: a basic single-stage amplifier model is presented, which will help to state the basic equation of a load line. This power amplifier is feeded through a bias voltage V_{DD} and V_{GG} with an RF choke inductor. The DC blocking capacitors at the input and the output stages will help a steady-state voltage V_{DD}, taking into account that the output voltage v_0 swings feeble and the capacitances value is huge enough to protect the voltage supply over the whole RF cycle.

Based on the figure above. It can deduce the following equations for this basic power amplifier model (In steady state).

$$V_{DS} = V_{DD} + v_0 \tag{23}$$

Figure 13. A basic single-stage amplifier model for load line definition

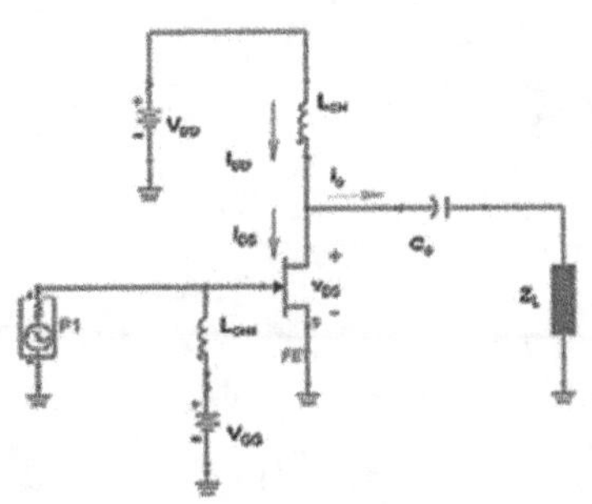

$$I_{DD} = i_{DS} + i_0 \tag{24}$$

Wherever the values of the variables correspond to the DC plus AC components, for example v_{DS} corresponds to the total drain-source DC voltage plus AC voltage.

Additionally, at the output of the device, the load impedance can be utilized to link v_0 and i_0.

$$Z_L = \frac{v_0}{i_0} \tag{25}$$

According to the three latter Equations (23), (24) and (25), a relationship among v_{DS} and i_{DS} can be determined as follows:

$$i_{DS} = I_{DD} - \frac{v_{DS} - V_{DD}}{Z_L} \tag{26}$$

This last formula defines exactly the load line of an amplifier, which allows determining a suitable bias point (V_{DD}, I_{DD}) and Z_L (load impedance), it defines the trajectory of all the combination of values (v_{DS}, i_{DS}) of the amplifier under specific functioning condition. When the value of the impedance Z_L is a real value R_L,

The Equation (26) becomes the equation of a straight line with a slope $-1/R_L$. If v_{DS} is supposed to be the independent variable. Then v_{DS} could have any value and thence could produce any infinite value for i_{DS}. Despite this is not possible owing to the real physical limitation of the device, which can be visualized in the DC (voltage & current) characteristics of the device (Jain, Kang, Soin, & Xi, 2002). That is the reason why the load line is always depicted on a DC (voltage and current) plot, as a way to confine the load line excursion inside the physical maximum rating of the device. Generally, there are

Figure 14. The load line of amplifier

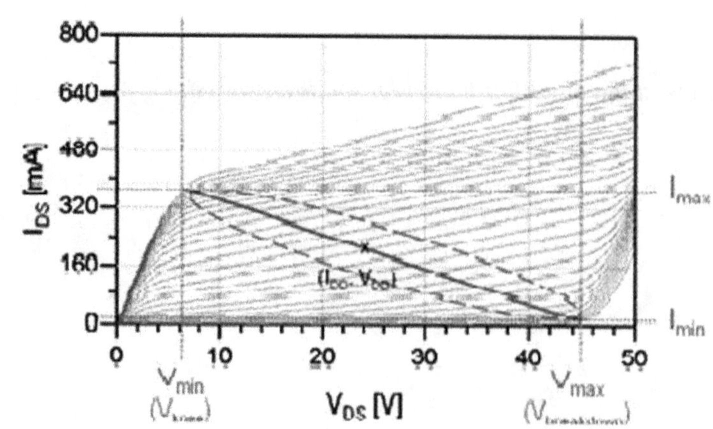

four figures that limit the excursion of the load line: zero value current (minimum i_{DS}) and the device maximum current (maximum i_{DS}), the knee voltage (minimum v_{DS}), the breakdown voltage (maximum v_{DS}) (Snider, 1967). Figure 14 represents an example of a load line with the real load impedance.

When the load is complex, equation (26) becomes the term of a rotated and shifted ellipse (Salib, Dawson, & Hahn, 1975), as illustrated in Figure 14. From this, it can be established that by noticing the load line we can deduce if the device is driven by a real load or a complex load. Nevertheless, if a microwave transistor is connected only with a 50Ω load, without matching network, the intrinsic load line noticed will be an ellipse instead of a straight line, which is owing to the effect of the intrinsic output capacitance of the device that combines with the real 50Ω impedance into a complex load. Moreover, these load line traces utilize as a technique to know when the device output capacitance is annulled by the matching circuit and the intrinsic device only sees a fully real load at its terminals.

Output Power

To define output power, we consider the basic amplifier with a matching network as depicted in Figure 15, which shows a power amplifier connected to a load, RL. The output power is defined as the active power transferred to the load, i.e. the antenna, at the fundamental frequency. Assuming that the load is purely resistive at the frequencies of concern we can represent the load as a resistor. Therefore, Pout is calculated by the fundamental components of the load current i_{out} and load voltage V_{out} that can be expressed by Equation (27) (Bahl, 2009; Vendelin, Pavio, & Rohde, 2005):

$$P_{out} = \frac{1}{2} real(\mathrm{V}_{out} * \mathrm{i}^{*}_{out})_{|f=f_0} \quad (27)$$

Stability Analysis

Generally, stability analysis in microwave power amplifiers is performed based on two-port Scattering-parameters. The detailed derivation of the method can be found in (Gonzalez, 1997; Kurokawa, 1965). With the two-port small-signal Scattering-parameters at a specific frequency and a particular bias condition, the input/output stability circles can be derived. The stability circles can be graphically represented

Figure 15. The basic amplifier with matching network

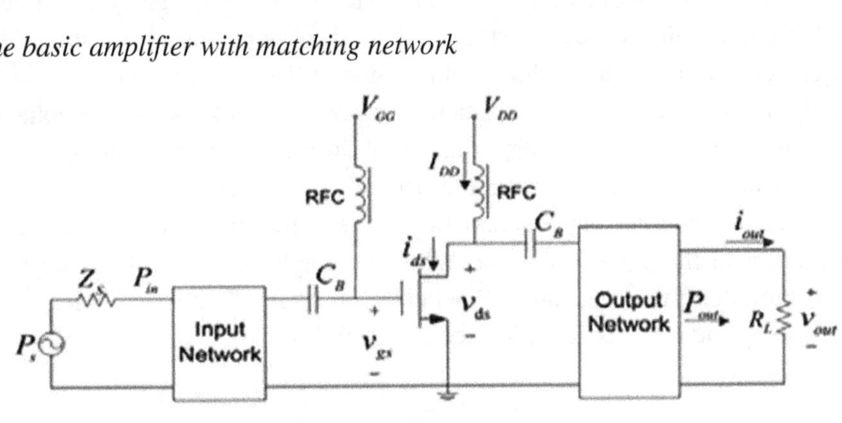

on the Smith chart. These circles afford a set of source and load impedance such that the magnitudes of the input and the output reflection coefficients are less than 1. The centers and the radii of the stability circles are given in (28) – (31).

Input stability circle:

$$r_S = \left| \frac{S_{12}S_{21}}{|S_{11}|^2 - |\Delta|^2} \right| \tag{28}$$

$$C_S = \left| \frac{\left(S_{11} - \Delta S_{22}^*\right)}{|S_{11}|^2 - |\Delta|^2} \right| \tag{29}$$

Output stability circle:

$$r_L = \left| \frac{S_{12}S_{21}}{|S_{22}|^2 - |\Delta|^2} \right| \tag{30}$$

$$C_L = \left| \frac{\left(S_{22} - \Delta S_{11}^*\right)}{|S_{22}|^2 - |\Delta|^2} \right| \tag{31}$$

Where

$$|\Delta| = |S_{11}S_{22} - S_{12}S_{21}| \tag{32}$$

The stability circles and the stability factor are helpful tools for stability analysis of linear amplifiers; they have many limitations when utilized in stability analysis of nonlinear power amplifiers. The linear analysis technique does not account for the instability due to the nonlinear elements in the transistor. This is particularly true for the variable capacitances in power amplifiers. As the voltage at the drain and the gate of the transistors change, the capacitances of the transistors show strong nonlinearity. As a result, the oscillation may occur due to the nonlinearity of the capacitors.

The stability circles concentrate on the interaction of the source and load impedances with the input and output impedances of the power amplifier. When the amplifier is potentially unstable, the result of the interaction shows negative resistance. Practically, an array of stability circles can be drawn by sweeping the frequency and the bias conditions of the amplifier.

The stability circles analysis does not provide information about the nature of the oscillation. There is no prediction of the amplitude and the phase of the oscillation. The amplitude and the phase of the oscillation provide useful information about the nature of the oscillation.

In order to achieve the unconditionally stable condition, it preferred to use the stability factor (K) and for a clearer result using too a stability measure (B1).

Stability factor (K) and stability measure (B1) analysis have several practical advantages. It is very simple to calculate, and it takes very little time on today's computers. The analysis is available in many commercial microwave circuit simulation packages. The analysis does not require accurate nonlinear models of the active devices, which are often difficult to find. Instead, it only requires one set of S-parameter measurement data. If a quick and simple analysis is needed, Stability factor (K) and stability measure (B1) analysis may still be the best choice.

$$k = \frac{1 - |S_{11}|^2 - |S_{22}|^2 + |\Delta|^2}{2|S_{12}S_{21}|} > 1 \tag{33}$$

$$B_1 = 1 + |S_{11}|^2 - |S_{22}|^2 - |\Delta|^2 > 0 \tag{34}$$

IMPEDANCE MATCHING NETWORK

In the previous part, we proved that the gain performance of a transistor (active device) is a function of the source and load terminations, Z_S and Z_L, and that these impedances can be optimized to extract maximum power gain from the transistor. In the fact, we do not always have control over terminations with which we are presented. We need a way to translate a given source and load to the terminating values we need at the transistor ports in order to attain our specific performance requirements.

The maximum transfer of signal power is one of the fundamental objectives in RF/microwave circuit design. Power pass through a network from a source to a load ordinarily by a cascade, or at least a sequence, of two-port networks. In order to get maximum power transfer, the device must be matched. More specifically, the load ought to be the complex conjugate of the impedance that the load sees looking back to the source. If the impedances are resistive, a magnetic transformer can be used to match the impedances. This will not work at RF/microwave frequencies, as the design problem is generally to match complex impedances (Boesch & Conroy, 1999).

Around a few megahertz, lumped elements (capacitors and inductors) can be utilized in matching networks (MNs), where can be used in cellular phone handsets. While for high frequencies, distributed parasitic can make lumped-element networks impractical. Therefore, segments of transmission lines are applied. However, they are also utilized at lower frequencies when loss must be kept to a minimum or power levels are high. They are also utilized at lower frequencies when power levels are high or loss must be kept to a minimum. This is due to the fact of the loss of the transmission line component is generally less than the loss of a lumped element. There are many factors affect the choice of the components of a matching network. The most important of these are bandwidth, adjustability, size, and complexity.

The function of a transmission line element is directly related to its length compared to a wavelength. the quarter wavelength is a basic functional length of a distributed element (transmission line), so that a characteristic length at 1GHz is around 7 cm modified by the square root of the permittivity of the material being used. A circuit board, for example, with a relative permittivity of 4 has a one-quarter wavelength of 3 cm. This is far too large to fit in consumer wireless products. Thus, at a few gigahertz, lumped-element matching networks are favored in terminal units (Beurrier, 1973; Boesch & Conroy, 1999).

In the last years, the performance of lumped elements (inductors and capacitors) has dramatically increased hugely in response to the pressure to manage dimension. Performance generally mentions to both the self-resonant frequency of an element and loss. Inductors are a particular case in point. A current inductor must be modeled employing capacitive elements, capturing interwinding capacitance as well as the primary inductive element. At some frequency, the capacitive and inductive elements will establish and resonate the self-resonant frequency of the element (Drogi & Tomasz, 2012; Student & INDIA, n.d.). The self-resonant frequency is the most common metric utilized to specify the maximum operating frequency of the element. With regards to complexity, the simplest designs are most often favored. A simpler MN is generally more dependable, cheaper and less loss than a complex design. The pursuit of wider bandwidth, however, may require that the simplest circuit not be chosen. Any type of MN can ideally give, overlooking resistive loses, a perfect match at a single frequency. Away from this center frequency, the match will not be perfect.

For achieving a reasonable match nearly a 5% bandwidth based on a single-frequency design is not that difficult. However, it is often preferable to match a load over a larger band of frequencies. So MN design has a lot in common with filter design, and many designers often choose to utilize a filter design approach. Design becomes increasingly more complex as the desired bandwidth increases. Many of the evolving communication systems demand multiple functionalities, which in turn needs adjustability. In some applications, the MN may demand adjustment to match a variable load impedance. Some types of MNs are more adjustable than others (Boesch & Conroy, 1999; Lin, Zhang, Wang, & Yu, 2016). Such designs demand variable components, thus matching network circuit can be challenging and a source of competitive advantage.

Depending on the frequency and application, there are several ways to design a matching network., it may consist of:

- Lumped elements (inductors and capacitors) only. These are the easiest networks but have the most stringent limit on the frequency of working. The relatively high resistive losses of inductors may reduce their performance.
- Transmission lines (distributed elements) only. These have a good performance, but their size restricts their employ to above a few gigahertz.
- A combination of transmission lines and lumped elements, mostly small sections of transmission lines with capacitors. The line lengths are smaller than with the transmission lines only, but higher performance than lumped elements (inductors and capacitors) on their own.

This section deals with impedance matching networks, especially with a lumped element (inductors and capacitors) type, but all the cases attach to matching will be explored. The focus here is the development of design equations and synthesis of required results. An alternative process, applied by some designers, is to select a circuit topology and then utilize a circuit optimizer to arrive at circuit values that yield the required characteristics. This is sometimes satisfactory design technique but is not a perfect solution in modern designs as it does not provide sufficient information and does not facilitate the choice of simple topologies.

The Quality Factor

Capacitors and inductors, which are the main components in lumped element matching networks, also have parasitic elements and loss. With inductors, there is both series resistance and shunt capacitance mainly from interwinding capacitance, while with capacitors there will be shunt resistance and series inductance generally. A practical lumped element will be limited to operation what is called the self-resonant frequency. The impact of the loss is quantified by the quality factor (or Q factor). The quality factor is loosely related to bandwidth in general and is based on the response of a parallel or series connection of a capacitor (C), an inductor (L), and a resistor (R). The response of an RLC network is expressed by a second order differential equation with the conclusion that the 3-dB fractional bandwidth of the response is 1/Q; when the power response is at its half-power level below its peak response. The fractional bandwidth is ${\Delta f}/{f_0}$ where f_r is the resonant frequency at the center of the band and Δf is the 3dB bandwidth. This is not true for any configuration other than a second-order circuit, but as a guiding principle, whilst while networks with higher quality factor will have a narrower bandwidth (Bahl, 2003; Danesh, Long, Hadaway, & Harame, 1998).

Definition

The quality factor of a component or circuit is known as the ratio of 2π times the maximum energy stored in the energy lost per cycle. Figure 16-(a) represents a series resonant circuit with elements R, L, and C, the quality factor of a series resonant circuit is expressed by:

$$Q = \frac{\omega_r L}{R} = \frac{1}{\left(\omega_r CR\right)} \tag{35}$$

And for a parallel resonant circuit, as depicted in Figure 16-(b), with G, L and C elements, the expression of the quality factor is:

$$Q = \frac{\omega_r C}{G} = \frac{1}{\left(\omega_r LG\right)} \tag{36}$$

These resonant circuits have a band-pass transfer characteristic as shown in Figure 17. The quality factor is approximately the inverse of the fractional bandwidth of the resonator. The fractional bandwidth, Δf, is measured at the half-power points (see Figure 17) then:

$$Q = \frac{f_r}{\Delta f} \tag{37}$$

Figure 16. (a) A parallel resonant circuits; (b) A series resonant circuits

Where:

- $f_r = \frac{\omega_r}{2\pi}$ Is the resonant frequency and is the frequency at which the maximum amount of energy is stored.
- Δf Is also referred to as the two-sided −3 dB bandwidth.

Therefore, the quality factor is a measure of the sharpness of the band-pass frequency response. The pairs of definitions—Equations (35) and (37), and Equations (36) and (37)—are identical for series RLC and parallel RLC circuits, respectively.

The determination of the quality factor employing the measurement of bandwidth together with Equation (37) is often not very accurate, so another definition that utilizes the much more sensitive phase change at resonance is preferred. With ϕ being the phase (in radians) of the transfer characteristic, the definition of quality factor is now:

$$Q = \frac{\omega_r}{2}\left|\frac{d\phi}{d\omega}\right| \tag{38}$$

This is another equivalent definition of quality factor for series or parallel resonant circuits. However, the quality factor is utilized for more complicated circuits and then its meaning is always a ratio of the energy stored to the energy dissipated. Therefore, it is useful to talk about the quality factor of circuits other than three element RLC circuits (Lee, Jung, Suh, & Hyun, 1997). The Q factor of these structures can no longer be determined precisely by bandwidth or by the rate of phase change considerations.

The Quality Factor of Lumped Elements

The quality factor is used to characterize the loss of lumped element (capacitors and inductors). Inductors have a series resistance and the main loss mechanism of a capacitor is a shunt conductance G (see Figure 18) (Bahl, 2003).

The quality factor of an inductor at frequency $f = \omega/2\pi$ with a series inductance L and resistance R is:

Figure 17. Transfer characteristic of a resonant circuit

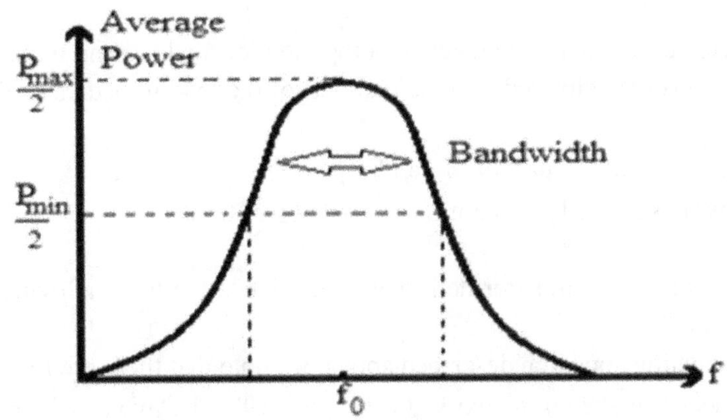

Figure 18. Loss elements of practical capacitors and inductors: (a) An inductor has a series resistance R, and; (b) For a capacitor has a shunt conductance G

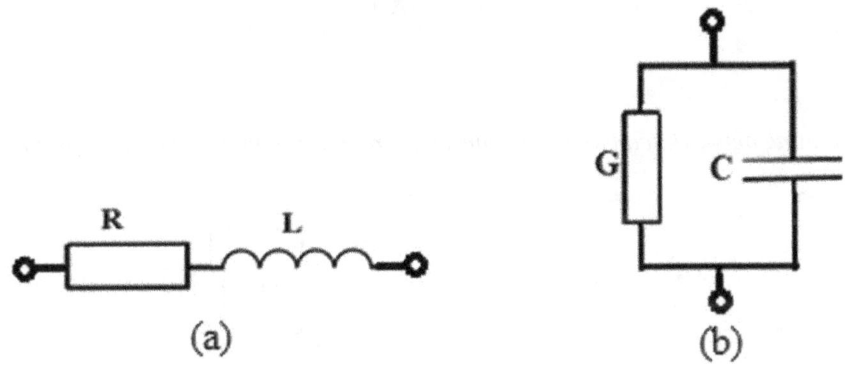

$$Q_L = \frac{\omega L}{R} \tag{39}$$

Since R is approximately constant for an inductor, the quality factor will vary with frequency. The quality factor of a capacitor with a shunt capacitance C and conductance G is:

$$Q_C = \frac{\omega C}{G} \tag{40}$$

G is owing mainly to relaxation loss mechanisms of the dielectric of the capacitor and so varies linearly with frequency. Thus, the quality factor of a capacitor is almost constant with respect to frequency.

The L Matching Network

These networks are used to eliminate the reactive component of the load and transform the real part so that the full available power is delivered into the real part of the load impedance (Kikkert, 2013).

- Absorb or resonate imaginary part of Z_S and Z_L.
- Transform real part as needed to obtain maximum power transfer.

Therefore, a resistance can be transformed to any resistive value by employing a LC transforming circuit.

Formalizing the matching approach described above, we note that there are four possible configurations with a two-element L matching networks (see Figure 19). The two possible cases, $R_S \prec R_L$ and $R_S \succ R_L$, will be considered in the following subsections.

1. (*Low-pass* $R_S \prec R_L$ $Q = \sqrt{\frac{R_L}{Rs - 1}} = \frac{|X_L|}{R_S} = \frac{R_L}{|X_c|}$

Figure 19. L matching networks consisting of one series reactive element and one shunt reactive element

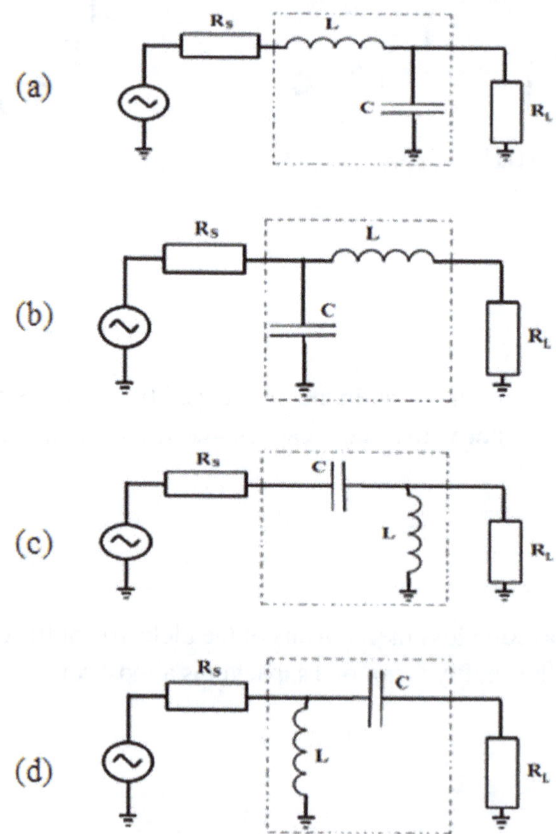

2. *Low-pass* $R_S \succ R_L$ $Q = \sqrt{\dfrac{R_S}{R_L - 1}} = \dfrac{|X_L|}{R_L} = \dfrac{R_L}{|X_c|}$

3. *High-pass* $R_S \prec R_L$ $Q = \sqrt{\dfrac{R_L}{R_S - 1}} = \dfrac{|X_C|}{R_S} = \dfrac{R_L}{|X_L|}$

3. *High-pass* $R_S \succ R_L$ $Q = \sqrt{\dfrac{R_S}{R_L - 1}} = \dfrac{|X_C|}{R_L} = \dfrac{R_S}{|X_L|}$

- **Design Equations for** $R_S \prec R_L$

Based the matching network topology of Figure 20. The input impedance is expressed as follows:

$$Z_{in} = \frac{R_L(\mathrm{j}X_L)}{R_L + jX_L} = \frac{X_L^2 R_L}{R_L^2 + X_L^2} + j\frac{R_L^2 X_L}{R_L^2 + X_L^2} \tag{41}$$

So we can write the input impedance as $Z_{in} = R_S - jX_S$ Where:

$$R_S = \frac{R_L\,\mathrm{X}_L^2}{R_L^2 + X_L^2};\ X_S = -\frac{X_L R_L^2}{R_L^2 + X_L^2} \tag{42}$$

From these,

$$\frac{R_S}{R_L} = \frac{1}{(\frac{R_L}{X_L})^2 + 1};\ \frac{-X_S}{R_S} = \frac{R_L}{X_L} \tag{43}$$

Introducing the quantities:

- Q_S = The quality factor of the series leg = $\dfrac{X_S}{R_S}$
- Q_P = The quality factor of the shunt leg = $\dfrac{R_L}{X_L}$

Leads to the final design equations for $R_S \prec R_L$:

$$|Q_S| = |Q_P| = \sqrt{\frac{R_L}{R_S} - 1} \tag{44}$$

$$-Q_S = Q_P;\ Q_S = \frac{X_S}{R_S} \text{ and } Q_P = \frac{R_L}{X_L} \tag{45}$$

The L matching network principle is that X_S and X_L will be either inductive or capacitive and they will have the opposite sign (i.e., the L matching network comprises one capacitor and one inductor). Also, once R_S and R_L are given, the quality factor of the network is defined; with the L network, the designer does not have a choice of circuit Q.

- **L Network Design for** $R_S \succ R_L$

For $R_S \succ R_L$, the topology illustrated in Figure 21 is used. The design equations for the L network design are similarly derived and are;

$$|Q_S| = |Q_P| = \sqrt{\frac{R_S}{R_L} - 1} \tag{46}$$

$$-Q_S = Q_P;\ Q_S = \frac{X_S}{R_L},\ \text{and } Q_P = \frac{R_S}{X_P} \tag{47}$$

Figure 20. Two-element matching network topology for $R_S \prec R_L$

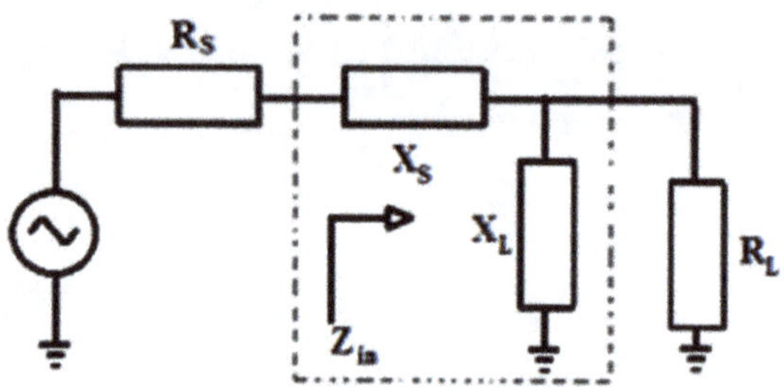

Figure 21. Two-element matching network topology for $R_S \succ R_L$

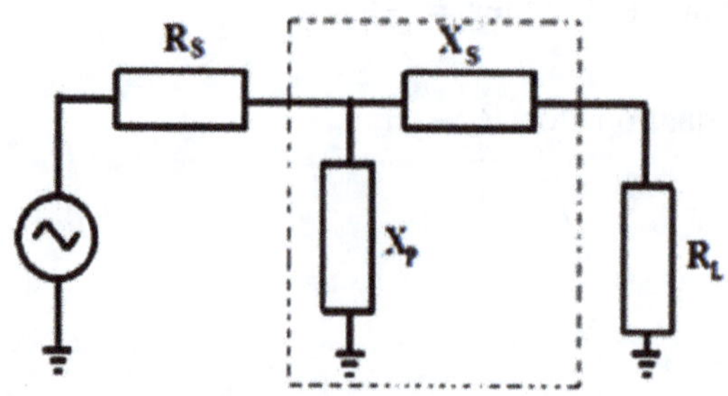

Multi-Element Matching

With the two-element matching (i.e. L network), the quality factor is fixed once the source and load resistances:

$$Q = \sqrt{\frac{R_L}{R_S} - 1} \left(R_L \succ R_S\right) \tag{48}$$

Consequently, the designer does not have a choice of circuit Q. Introducing a third element in the matching network provides the extra degree of freedom in the design for adjusting Q, and hence help to get a bandwidth (Bahl, 2003; Kikkert, 2013).

Generally, there are two configurations of three-element matching networks, the Pi network and the T network, are illustrated in Figure 22. Whether a T network or a Pi network is utilized depends on:

- The realization constraints associated with the specific design.
- The type of the reactive parts of the source impedance and load impedance (the nature of network that facilitates the handling of these reactive parts utilizing resonance or absorption should be chosen).

Distributed Matching

Matching utilizing lumped elements leads to parallel and series lumped elements (capacitors and inductors). The parallel elements can be implemented employing shunt transmission lines, as a short length (less than one-quarter wavelength long) of a short section of open-circuited transmission line looks as a capacitor and a short-circuited transmission line looks as an inductor. However, in most distributed matching it is not possible to realize the series elements as lengths of transmission lines. While it has been shown that a short length of distributed matching is inductive, replacing series inductors by a length of transmission line of high characteristic impedance is not the preferred approach to realizing networks. The solution is to utilize the lengths of distributed matching (transmission line) together with shunt elements. If space is not at a premium, this is an optimum solution, as distributed matching have a much feeble loss than a lumped inductor. The series transmission lines rotate the return loss on the

Figure 22. (a) T network; and (b) Pi network

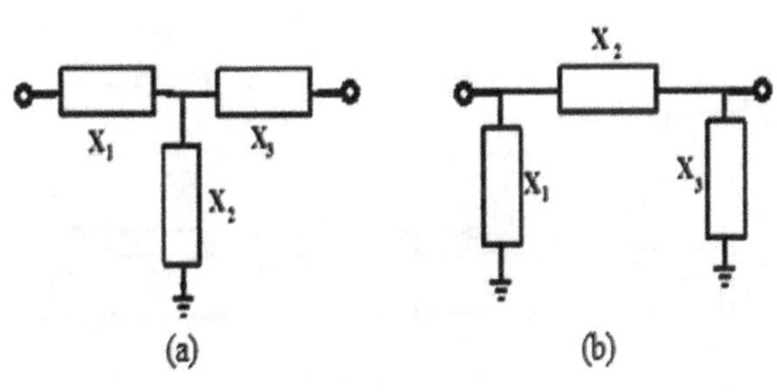

Smith chart. It is also possible to utilize lumped capacitors as lumped elements together with lengths of distributed matching (transmission line) (Beurrier, 1973; Gonzalez, 1996; Pozar, 2009). As with all design matching, using distributed matching begins with a topology in mind (Boesch & Conroy, 1999; Vendelin, Pavio, & Rohde, 2005). Various topologies are shown in Figure 23. Figure 23-(a) is the top view of a microstrip matching network with a series transmission line and stub realized as an open-circuited transmission line. Figure 23-(b) shows a schematic for this circuit. Matching network design then becomes a problem of choosing the characteristic impedances and lengths of the lines. The stub here is utilized to realize a capacitive shunt element. This circuit corresponds to 2-element matching with a shunt capacitor. The value of the shunt capacitance can be turn up utilizing a dual stub, as shown in Figure 23-(c), where the capacitive input impedance of each stub are in parallel. The dual circuit is illustrated in Figure 23-(d) together with its schematic in Figure 23-(e). This network has a short-circuited stub, which realizes a shunt inductance. Mixing a distributed matching (transmission line) with lumped capacitors, as illustrated in Figure 23-(f), achieves a much more space-efficient network design. There are several variations to stub-based matching network design, including the two-stub design as shown in Figure 23-(g). The double-stub tuner is illustrated schematically in Figure 23-(h). The triple-stub tuner illustrated in Figure 23-(i). All of these designs necessitate the transmission line lengths and impedances to be designed.

MICROWAVE POWER AMPLIFIER DESIGN

The design process of the microwave power amplifier is presented in the flow chart in Figure 24. From the beginning, it is important to be clear in the design specification and aim of the design, which is helpful in achieving a successful design.

Here is a list of design specifications that were known before designing the microwave PA.

- **Operating Frequency:** 2.6 GHz (LTE band)
- **Power Supply:** $V_{ds} = 4$ V
- **Input/Output Impedance:** 50 Ω
- Unconditional stable
- High power gain and low return loss

Figure 23. Matching networks with transmission line elements

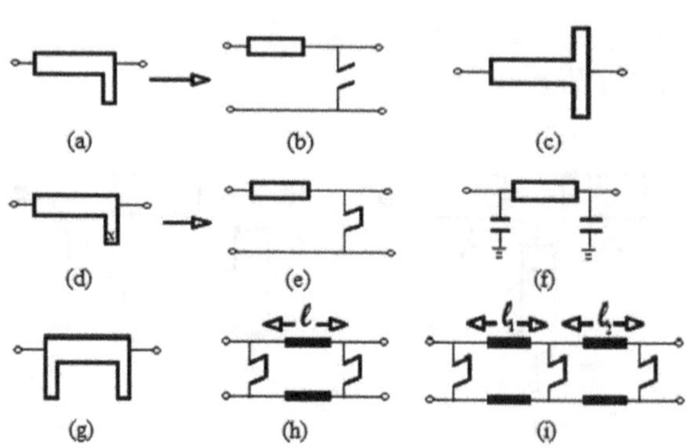

Figure 24. Power amplifier design flow

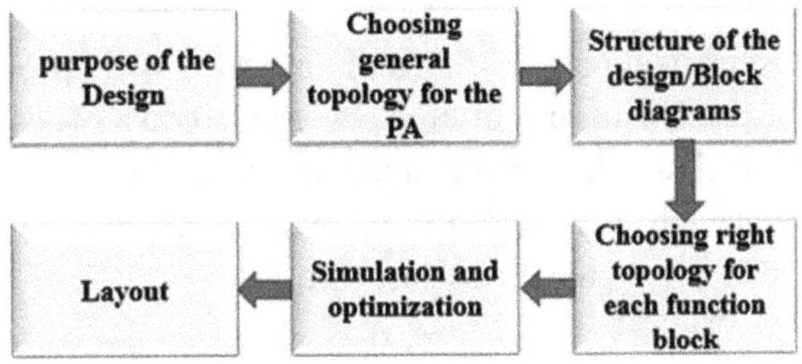

Choosing a Suitable Topology

As mentioned in the previous part, nonlinear and linear PAs each have their own negatives and positives advantage when it comes to a specific application. As far as this low power design is concerned, output power is the most important performance parameter in selecting an appropriate topology while other parameters, such as power supply consumption, stability, S-parameters, etc.... Another factor to be considered is the linearity. Linearity of the PA is critical. Therefore, linear power amplifiers (class A, B, AB, and C amplifiers) are selected and investigated to see how effective the linear PAs can be.

When selecting a suitable topology among linear power amplifiers, there are numerous considerations to evaluate. The Class-A is preferred more than others due to several factors (Cripps, 2006). Class-A PAs are a popular choice for transceiver designs in the research field for wireless applications where linearity is important. Although the Class-A PAs have the lower theoretical efficiency compared to Class-B and AB. Moreover, class-AB designs are generally utilized for audio PAs, which is not the purpose of this design project. Class C has a little more complex output network that utilizes more inductors, which is clearly not desirable for this design in integrated form and given that, the process being utilized has limited inductor capability.

A Conceptual Model for Microwave Power Amplifier

The basic topology of a Class-A PA is normally utilized in output stage of power amplifier design. The microwave PA topology consists of an active component and a passive component; the active device used in this design is a field effect transistor (FET), and the passive component includes a DC feed inductor (RF-choke) and capacitors (dc block capacitor) in both sides of the transistor.

The block diagram of microwave power amplifier is shown in figure 15. It consists of an input bias network, output bias network, input matching network (IMN), an output matching network (OMN), and the main power device (transistor). The IMN and OMN are used for transforming input and output impedances of the transistor to the source and load impedances, respectively in order to achieve maximum power transfer and minimize the reflection (Raab, Asbeck, Cripps, Kenington, Popovic, Pothecary, ... Sokal, 2002). The gate and drain bias networks provide DC voltages to operate the transistor while preventing the RF signals to go into the DC sources. By doing this, the transistor can act as a voltage controlled current source.

Circuit Design Process and *Functional Blocks*

The circuit design is the most critical part of the whole process. To improve circuit performance, the microstrip technology will be used. This amplifier is constructed on a FR4 substrate having the following specification: a substrate thickness h=1.6mm, a metallization thickness t=0.035mm, a tangent loss tan (δ) =0.025 and a relative permittivity =4.4. The microwave PA circuit was designed and simulated using Agilent's Advanced Design System (ADS) software.

Device Selection and DC Analysis

One of the first steps in the design process is to choose a suitable active device. The active device utilized for this example is the ATF-21170, a Field Effect Transistor (FET) packaged in a standard SOT-70 plastic package. The chosen device has acceptable performance for wireless applications at a higher frequency, and offers various other advantages: it is accurate models are readily available, and extremely low cost. As a rule of thumb in high-volume, low-cost applications is to utilize the least expensive device that will accomplish the job, the ATF-21170 is a good choice for this amplifier.

The next analysis that needs to be performed is the DC analysis to find out the right bias points for the PA. Figure 25-(a) illustrates the DC analysis setup and figure 25-(b) illustrates the DC analysis results for the same, According to the class-A operation, the drain-source and gate-source bias voltages are set to 4V and 0.5V, respectively.

Bias Network Design

Bias network design for power amplifier is dependent on the frequency range in which PA needs to be designed, that is to say, if PA needs to be designed for low-frequency application then an inductor (RF-choke) is utilized but getting discrete inductors at microwave frequencies is difficult so a high-impedance quarter wavelength line (λ/4) at desired frequency is the best possible choice which designers can select to design bias network. For improved and more elaborate biasing, a radial stub directly after a quarter-wavelength line (λ/4) was selected; a high impedance bias line which will facilitate to obtain appropriate

Figure 25. (a) DC Analysis setup for ATF-21170 ;(b) DC Analysis results

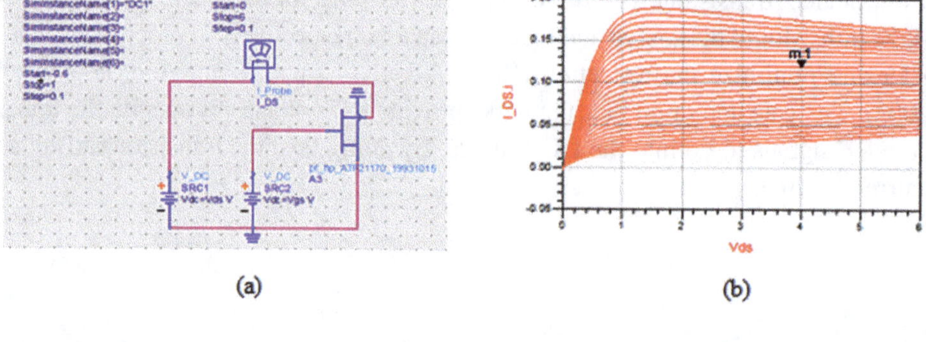

(a) (b)

isolation at desired RF frequency band and to make the role of an RF choke (Rachakh, El Abdellaoui, Mandry, Errkik, Tajmouati, & Latrach, 2017).

Figure 26-(a) illustrates the proposed bias circuit. For further clarification layout of the bias circuit is illustrated in Figure 26-(b).

Figures 27-(a) and 27-(b) illustrates the results for the proposed bias circuit and it can be observed that this design is playing a role as a near perfect bias circuit at 2.6GHz.

Impedance Matching Circuit Design

After the circuit is feeded in desired frequency, now we can begin the design of the matching networks (input and output impedance matching) so that we could achieve the desired specification of the power amplifier. The selection of the topology of the matching network mainly depends on the operating frequency of the PA so that designer can select between the lumped element and distributed elements matching networks. Figure 28 below presents input and output impedances on the Smith chart that needs to be matched with the 50-ohm impedance.

Figure 26. (a). Bias network; (b). Layout of bias network

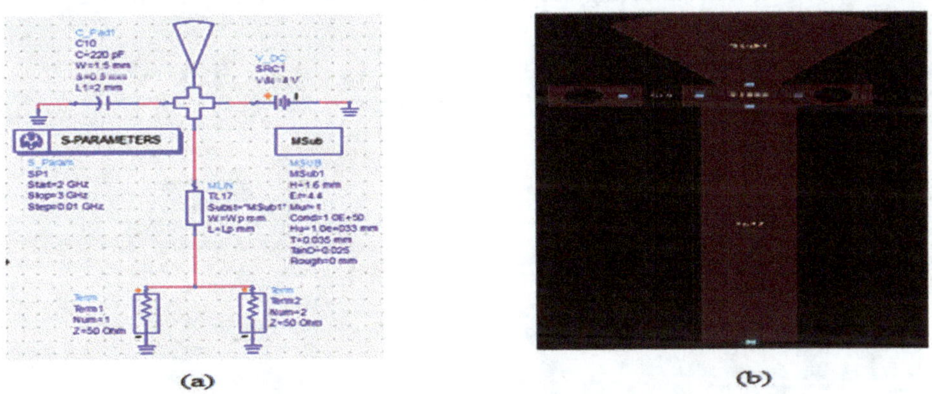

(a) (b)

Figure 27. (a.). Reflection coefficient ; (b). Transmission coefficient

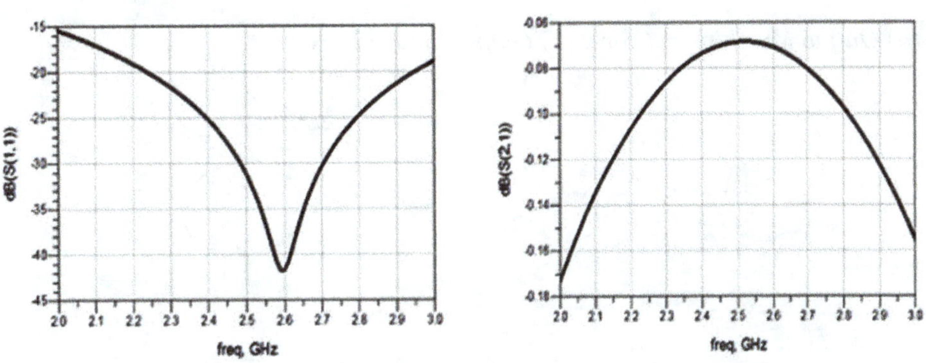

(a) (b)

Figure 28. Input and Output impedances on Smith Chart

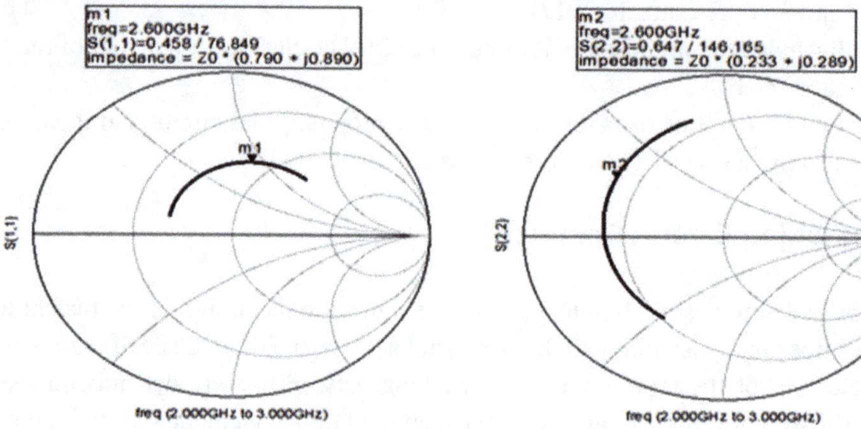

For the present microwave power amplifier design a distributed elements approach was utilized to design the matching networks to attain the best possible input and output return losses. Figure 29-(a) presents the input matching networks and Figure 29-(b) presents the results after connecting input matching networks to the PA circuit. While Figure 30-a below presents the output matching networks and Figure 30-b presents the results after connecting output matching networks to the PA circuit.

Figure 29. (a) Input Matching network; (b) Input Return Loss

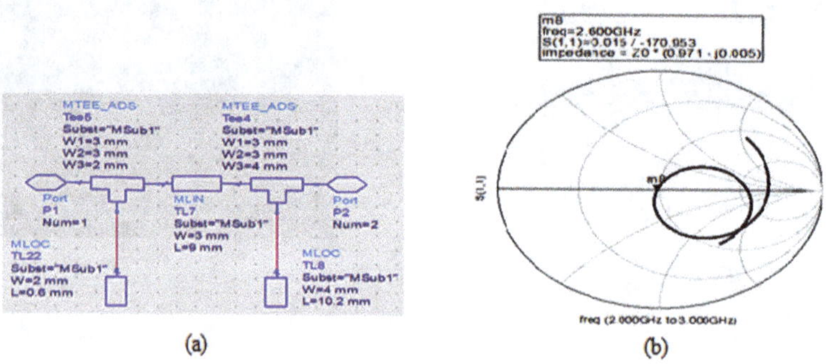

Figure 30. (a) Output Matching network; (b) Output Return Loss

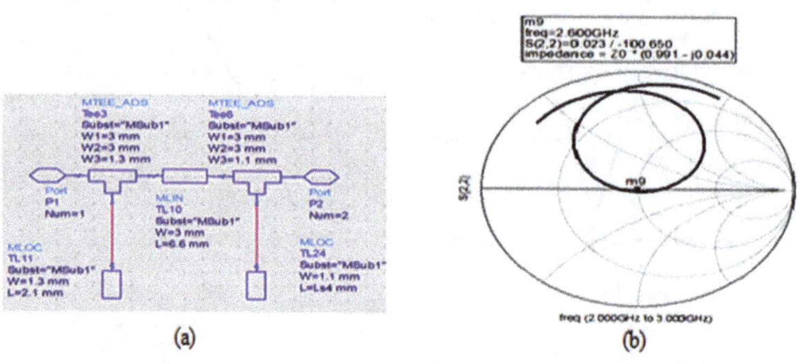

Stability Analysis

Stability analysis is a quite important aspect of any active circuit design and it is equally important in power amplifier design too. There are several stability configurations, which could be utilized to stabilize the circuit, the most preferable being using a taper line. At the source path, a taper line was added in series to improve the stability and then the value of taper line was tuned to achieve the proper stability. Figure 31 illustrates one of the configurations to stabilize the circuit with matching networks. Figure 32 below presents the results after stabilization.

The stability was tested from 2GHz up to 3GHz. Figures 32 shows Stability Factor (K) is 1.165 and Stability Measure (B1) is 0.679 Therefore, it can be said that circuit is very much stable around 2.6GHz frequency range.

Figure 31. Stabilized Circuit with matching networks

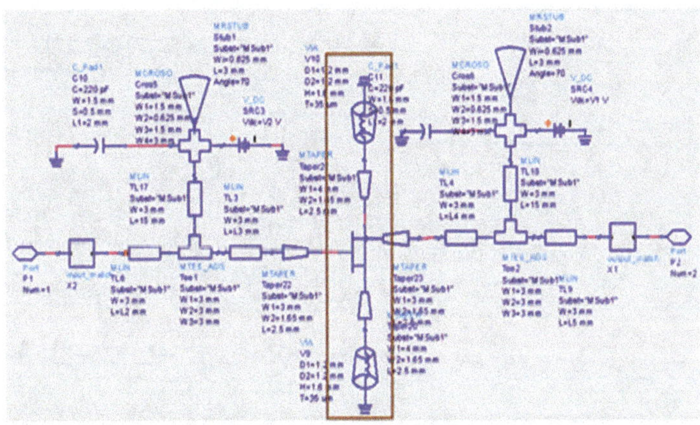

Figure 32. (a) Stability Factor versus frequency characteristics; (b) Stability Measure versus frequency characteristics

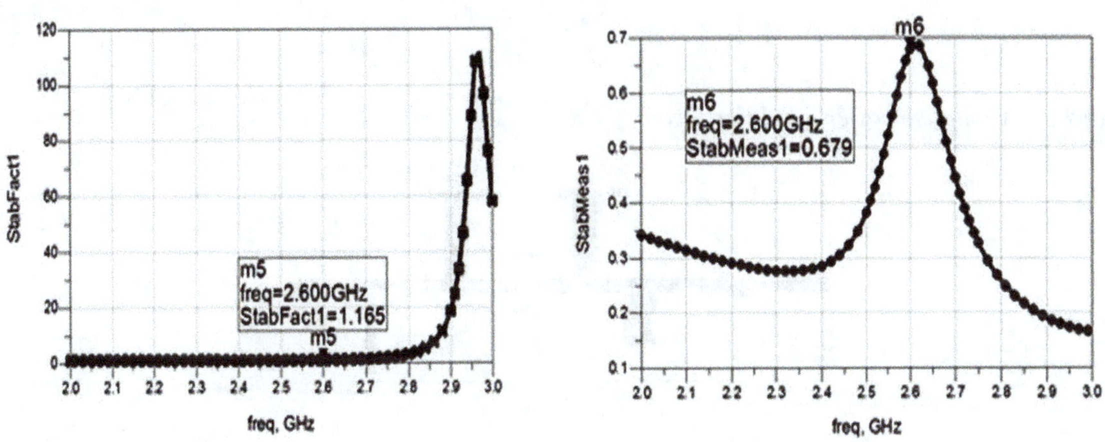

(a) (b)

Overall Power Amplifier Performance

The only thing remaining now in PA design is to relate all the sub-networks together and see the overall amplifier performance and to optimize the overall circuit if necessary. The complete designed power amplifier is shown in Figure 33, Figure 34 depicts the complete layout of the designed power amplifier, Figure 35 illustrates the power amplifier results, and these were obtained after minimal manual tuning of the matching transmission lines and stub lengths to attain the desired results after each of the blocks together.

From the simulation S-parameter. As can be observed, the power amplifier has -36.6dB input return loss (S11) and – 32.9dB output return loss (S22) at 2.6GHz. While the power gain (S21) has a maximum value of 18.779dB with a unilateral transmission coefficient (S12) less than -23.712dB in the operating frequency band.

Figure 33. Complete power amplifier schematic

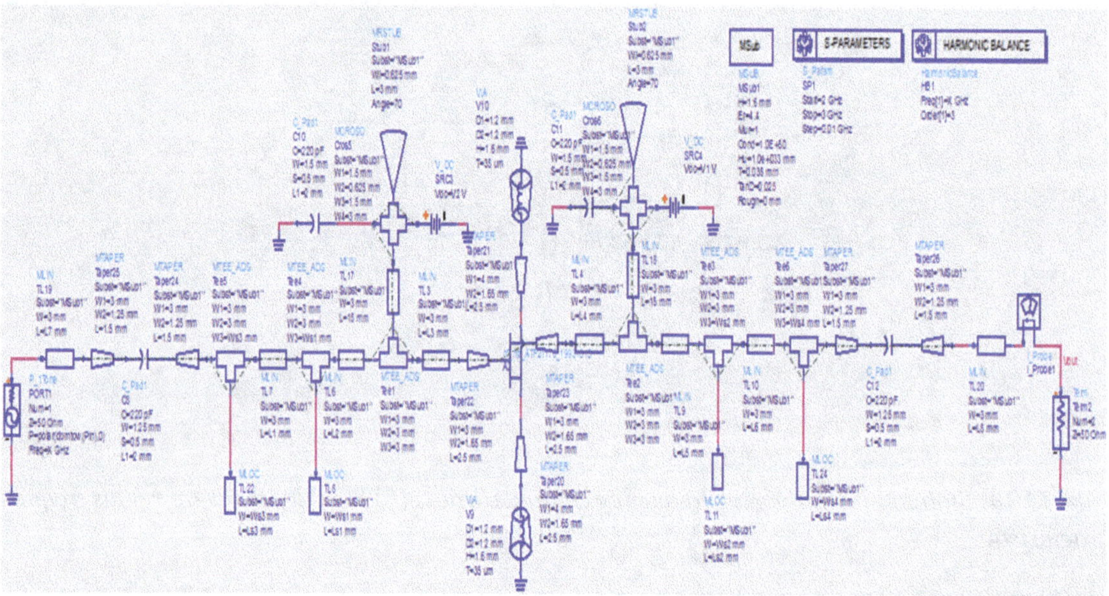

Figure 34. Complete power amplifier layout

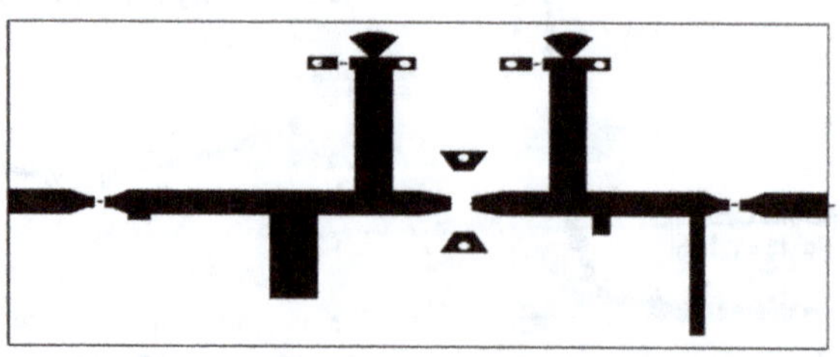

Figure 35. Power amplifier results

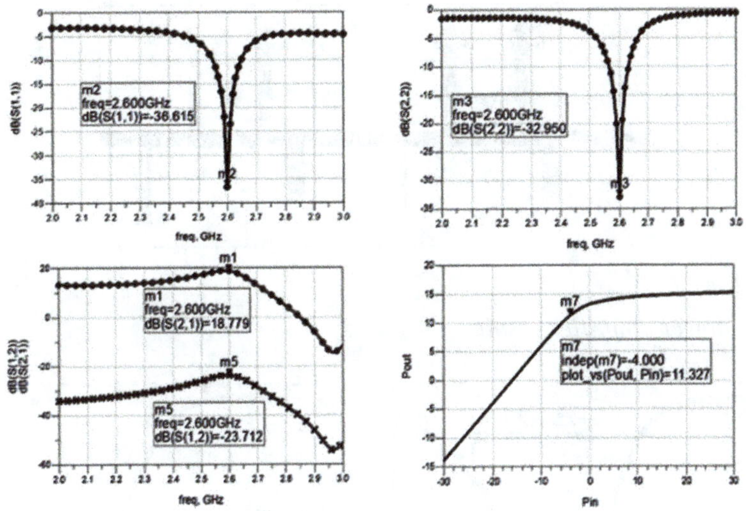

Cascade Power *Amplifier*

For most systems, a single-stage amplifier does not provide sufficient bandwidth or gain or will not have the correct input or output impedance matching. The solution is to combine two stages of amplification. to improve the performance of the designed power amplifier; a second stage based on the same transistor is added in cascade. The complete designed power amplifier (cascade) is shown in Figure 36. Figure 37 depicts the complete layout of the designed cascade power amplifier and Figure 38 illustrates the cascade power amplifier results.

The simulation S-parameter result of cascade power amplifier illustrated in figures above. As depicted Figure 38 presents the variation of the power gain S21 and the unilateral transmission (S12) between 2.4GHz and 2.7GHz. It can be seen that this power gain varies between 37dB and 20dB. Thus, these values indicate a good amplification of the microwave amplifier over a wideband (300MHz) and for the

Figure 36. Complete cascade power amplifier schematic

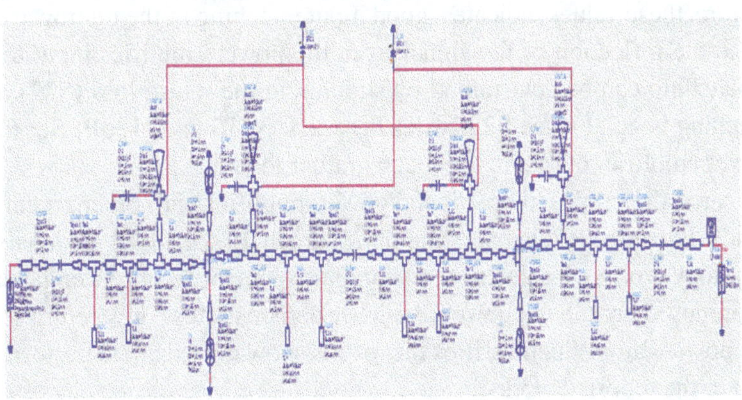

Figure 37. Complete cascade power amplifier layout

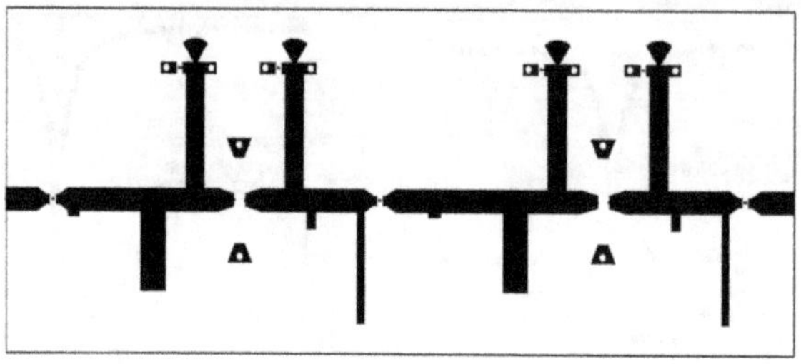

Figure 38. Cascade power amplifier results

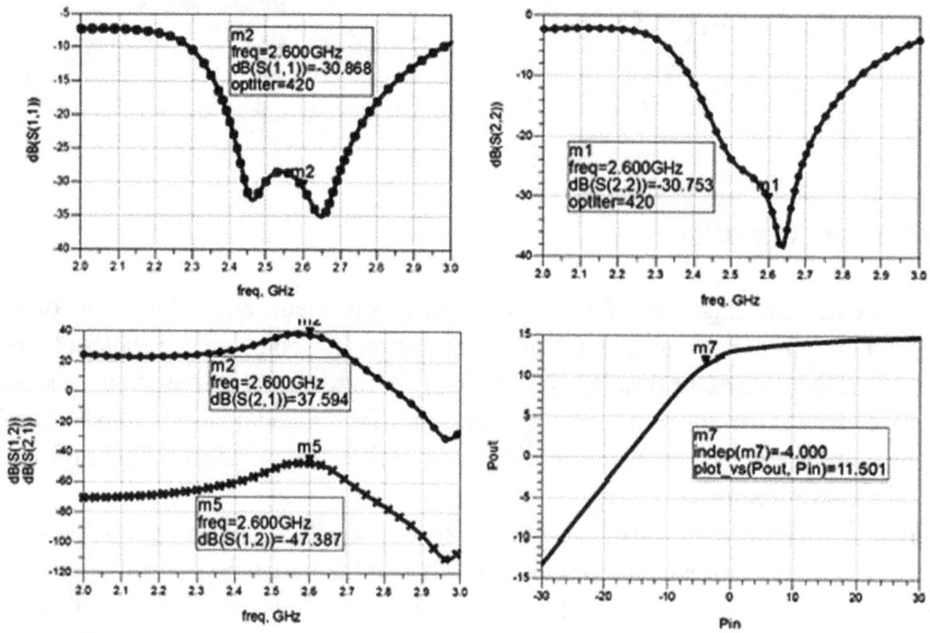

unilateral transmission; we notice that this parameter varies between -47dB and -56dB. These values are feeble. Hence, the reflection of the power from load to source is without effect.

For return loss. We observed that the parameter S11 varied between -35dB and -23dB between 2.4GHz and 2.7GHz. Therefore, these values indicate a good input matching of the microwave cascade amplifier over a wideband and the reflection of the signal from the power amplifier input to the source is low. While the output matching can be determined by simulating the return loss (S22). It can be observed that the output matching is good since S22 varies between -39dB and -13dB. So; the reflection of the signal from the power amplifier output to the load is rather feeble.

Finally, we compared the proposed broadband power amplifier with other recent research (Aniktar, Sjoland, Mikkelsen, & Larsen, 2017; Ribate, Mandry, Latrach, Errkik, & El Abdellaoui, 2017) designed with different technology process. Table 2 gives a comparison in terms of power gain and return loss. As it can be seen, the proposed broadband power amplifier demonstrates a low return loss with broadband frequency and high power gain. Which verifies the results show the excellent performance of our design approach compared to the reported works.

Table 2. Comparision The proposed amplifier with other recently amplifiers

Parameter	Rachakh, El Abdellaoui, Zbitou, et al., (2017)	Ribate et al., (2017)	Aniktar et al., (2005)	The Proposed BPA
Process	BJTs	GaAs FET	CMOS	GaAs FET
Frequency (GHz)	1.75 to 2.15	1 to 4	1.65 to 2	2.4 to 2.7
Power Gain (dB)	10.5-11.5	6-12	5.1-10	20-37
Input Return Loss(dB)	-22	-24	-21	-29
Output Return Loss(dB)	-18	-18	-8	-26

CONCLUSION

The focus of this author's work has been on the study and design of microwave power amplifier. To begin the study, we conjure up the fundamental concepts of microwave power amplifiers. A theoretical revision of different PA classes from class A to class F amplifiers as well as the load line concept has been presented. The section of this chapter also explains several figures of merit used for quantifying PA performance; such as power gain, output power and stability analysis. In the second section, impedance matching network based on a lumped element and transmission lines has been discussed in detail. The last section of this chapter presents a detailed process for designing a new structure of microwave single-stage power amplifier, intended for LTE applications at 2.6GHz. This is performed by employing a GaAs FET. The use of Pi matching technique at the input and the output stages has contributed the best performance for the amplifier. In order to get a better performance of designed power amplifier, a second stage based on the same active device is added in cascade. The design has been adjusted utilizing the optimization tools applied on matching networks, such that the final design is improved in both power gain and return loss. The complete designed (two stages amplifier) shows a peak power gain of 37dB, a low return loss and unconditionally stable over a desired range of frequencies and also shows high output power.

REFERENCES

Aniktar, H., Sjoland, H., Mikkelsen, J. H., & Larsen, T. (2005). A Class-AB 1.65 GHz-2GHz broadband CMOS medium power amplifier. In *NORCHIP Conference, 2005. 23rd* (pp. 269–272). Academic Press. 10.1109/NORCHP.2005.1597041

Bahl, I. (2009). *Fundamentals of RF and microwave transistor amplifiers*. John Wiley & Sons. doi:10.1002/9780470462348

Bahl, I. J. (2003). *Lumped elements for RF and microwave circuits*. Artech House.

Beurrier, H. (1973). *Impedance-matched amplifiers*. Google Patents.

Boesch, R. D., & Conroy, K. P. (1999). *Impedance matching circuit for power amplifier*. Google Patents.

Choi, J.-S., Kang, D.-H., Kim, D.-S., Park, J.-M., Jin, B.-S., & Kim, B.-M. (2009). Power amplifiers and transmitters for next generation mobile handsets. *JSTS: Journal of Semiconductor Technology and Science*, *9*(4), 249–256. doi:10.5573/JSTS.2009.9.4.249

Chowdhury, D., Hull, C. D., Degani, O. B., Wang, Y., & Niknejad, A. M. (2009). A fully integrated dual-mode highly linear 2.4 GHz CMOS power amplifier for 4G WiMax applications. *IEEE Journal of Solid-State Circuits*, *44*(12), 3393–3402. doi:10.1109/JSSC.2009.2032277

Cripps, C. (2006). *RF Power Amplifiers For Wireless Communications*. Academic Press.

Danesh, M., Long, J. R., Hadaway, R. A., & Harame, D. L. (1998). A Q-factor enhancement technique for MMIC inductors. In *Microwave Symposium Digest, 1998 IEEE MTT-S International* (*Vol. 1*, pp. 183–186). Academic Press. 10.1109/MWSYM.1998.689352

Drogi, S. F., & Tomasz, M. (2012). *RF power amplifier circuit with mismatch tolerance*. Google Patents.

Gonzalez, G. (1996). *Microwave transistor amplifiers: analysis and design*. Prentice-Hall, Inc.

Gonzalez, G. (1997). *Microwave Transistor Amplifiers Analysis and Design (2nd ed.)*. Prentice Hall.

Iqbal, M., & Piacibello, A. (2016). A 5W class-AB power amplifier based on a GaN HEMT for LTE communication band. In *Microwave Symposium (MMS), 2016 16th Mediterranean* (pp. 1–4). Academic Press. 10.1109/MMS.2016.7803827

Jain, P. K., Kang, W., Soin, H., & Xi, Y. (2002). Analysis and design considerations of a load and line independent zero voltage switching full bridge DC/DC converter topology. *IEEE Transactions on Power Electronics*, *17*(5), 649–657. doi:10.1109/TPEL.2002.802181

Kikkert, C. J. (2013). *RF Electronics: design and simulation*. James Cook University.

Kurokawa, K. (1965). Power waves and the scattering matrix. *IEEE Transactions on Microwave Theory and Techniques*, *13*(2), 194–202. doi:10.1109/TMTT.1965.1125964

Lee, B.-K., Jung, J.-W., Suh, B.-S., & Hyun, D.-S. (1997). A new half-bridge inverter topology with active auxiliary resonant circuit using insulated gate bipolar transistors for induction heating applications. In *Power Electronics Specialists Conference, 1997. PESC'97 Record., 28th Annual IEEE* (Vol. 2, pp. 1232–1237). IEEE.

Lin, L., Zhang, L., Wang, Y., & Yu, Z. (2016). A 150GHz amplifier based on coupled transmission line neutralization with 22dB gain in 65nm CMOS. In *Solid-State and Integrated Circuit Technology (ICSICT), 2016 13th IEEE International Conference on* (pp. 1576–1578). IEEE.

Mason, S. (1957). Some properties of three-terminal devices. *IRE Transactions on Circuit Theory*, *4*(4), 330–332. doi:10.1109/TCT.1957.1086413

Pozar, D. M. (2005). Microwave Engineering 3e. *Tramsmission Lines and Waveguides*, 143–149.

Pozar, D. M. (2009). *Microwave engineering*. John Wiley & Sons.

Raab, F. (1977). Idealized operation of the class E tuned power amplifier. *IEEE Transactions on Circuits and Systems*, *24*(12), 725–735. doi:10.1109/TCS.1977.1084296

Raab, F. H. (2001). Class-E, class-C, and class-F power amplifiers based upon a finite number of harmonics. *IEEE Transactions on Microwave Theory and Techniques, 49*(8), 1462–1468. doi:10.1109/22.939927

Raab, F. H., Asbeck, P., Cripps, S., Kenington, P. B., Popovic, Z. B., Pothecary, N., ... Sokal, N. O. (2002). Power amplifiers and transmitters for RF and microwave. *IEEE Transactions on Microwave Theory and Techniques, 50*(3), 814–826. doi:10.1109/22.989965

Rachakh, A., El Abdellaoui, L., Mandry, R., Errkik, A., Tajmouati, A., & Latrach, M. (2017). A New Design of a Microstrip Microwave Power Amplifier for PCS BAND. In *Proceedings of the 2nd International Conference on Computing and Wireless Communication Systems* (p. 3). Academic Press. 10.1145/3167486.3167489

Rachakh, A., El Abdellaoui, L., Zbitou, J., Errkik, A., Tajmouati, A., & Latrach, M. (2017). A novel design of a microstrip microwave broadband power amplifier for DCS, PCS and UMTS bands. In *Wireless Technologies, Embedded and Intelligent Systems (WITS), 2017 International Conference on* (pp. 1–5). Academic Press.

Ribate, M., Mandry, R., Latrach, M., Errkik, A., & El Abdellaoui, L. (2017). GaAs FET Broadband Power Amplifier for L and S Bands Applications. In *Proceedings of the 2nd International Conference on Computing and Wireless Communication Systems* (p. 42). Academic Press. 10.1145/3167486.3167530

Salib, M. L., Dawson, D. E., & Hahn, H. K. (1975). Load-line analysis in the frequency domain with distributed amplifier design examples. In *Microwave Symposium Digest, 1987 IEEE MTT-S International* (Vol. 2, pp. 575–578). IEEE.

Snider, D. M. (1967). A theoretical analysis and experimental confirmation of the optimally loaded and overdriven RF power amplifier. *IEEE Transactions on Electron Devices, 14*(12), 851–857. doi:10.1109/T-ED.1967.16120

Srirattana, N., Raghavan, A., Heo, D., Allen, P. E., & Laskar, J. (2005). Analysis and design of a high-efficiency multistage Doherty power amplifier for wireless communications. *IEEE Transactions on Microwave Theory and Techniques, 53*(3), 852–860. doi:10.1109/TMTT.2004.842505

Steer, M. (2009). Microwave and RF design: A systems approach. *Sciences et Techniques (Paris).*

Vendelin, G. D., Pavio, A. M., & Rohde, U. L. (2005). *Microwave circuit design using linear and nonlinear techniques*. John Wiley & Sons. doi:10.1002/0471715832

Compilation of References

A901-Patch antenna Datasheet from IDtronic Company. (n.d.). Retrieved from https://www.en.idtronic-rfid.com/d/1278572/d/uhf_patch_antennas.pdf

Abed Al-Husain, Z. A., Ahmed, S. K., & Osman, A. A. (2014). Design of Ultra-Wideband Micrstrip Antennas with Slots and Fractal Based Ground Plane. *Journal of Al Rafidain University College*, *34*, 244-258. Retrieved from http://www.home.agilent.com/agilent/home.js

Abidin, Z. Z. (2012). *Design, modelling and implementation of antennas using electromagnetic bandgap material and defected ground planes. Surface Meshing Analysis and Genetic Algorithm Optimisation on EBG and Defected Ground Structures for Reducing the Mutual Coupling between*. University of Bradford.

Abla, B. (2010). *Contribution à l'étude des antennes microstrip Supraconducteur de forme triangulaire*. Mémoire de Magistère en Micro-ondes. Université Mentouri.

Ahmed, M., Fekar, M., Seladji-Hassaine, N., & Marouf, F. (2016). Analysis of Patch and Printed Antennas integrated into UHF RFID Passive Tags. *Int. J. of Microwave and Opt. Tech*, *11*(6), 460–468.

Ahn, D., Park, J. S., Kim, C. S., Kim, J., Qian, Y., & Itoh, T. (2001). A design of the low-pass filter using the novel microstrip defected ground structure. *IEEE Transactions on Microwave Theory and Techniques*, *49*(1), 86–93. doi:10.1109/22.899965

Ahson, S., & Ilyas, M. (2008). RFID Handbook Applications, Technology, Security and Privacy. Taylor & Francis Group. doi:10.1201/9781420055009 doi:10.1201/9781420055009

Ajioka, J. S., Joe, D. M., & McFarland, J. L. (1984). Slot radiators in septated waveguide. *IEEE Transactions on Antennas and Propagation*, *32*(3), 247–251. doi:10.1109/TAP.1984.1143295

Al Husseini, Ramadan, El Hajj, & Kabalan. (2009). Design of a Compact and Low-Cost Fractal-Based UWB PCB Antenna. *Sciences of Electronic, Technologies Information and Telecommunications*.

Alam. (n.d.). Design and performance analysis of microstrip array antenna. *Session 5AP*.

Alboni, E., & Cerretelli, M. (2001). *Microstrip Patch Antenna for GPS application*. Retrieved from http://www.elettromagnetismo.it/atti_rinem/2002S09A01.pdf

Alexandrea, P. (2017). *3Dnatives*. Retrieved from https://www.3dnatives.com/en/arizton-additive-manufacturing-280920174/

Alqadami, A. S. M., Jamlos, M. F., Soh, P. J., Rahim, S. K. A., Vandenbosch, G. A. E., & Narbudowicz, A. (2016). Miniaturized dual-band antenna array with double-negative (DNG) metamaterial for wireless applications. *Applied Physics. A, Materials Science & Processing*, *123*(1), 22. doi:10.100700339-016-0678-3

Alsath, M. G. N., Kanagasabai, M., & Balasubramanian, B. (2013). Implementation of Slotted Meander-Line Resonators for Isolation Enhancement in Microstrip Patch Antenna Arrays. *IEEE Antennas and Wireless Propagation Letters, 12*, 15–18. doi:10.1109/LAWP.2012.2237156

Amir, S., Saghanezhad, H., & Atlasbaf, Z. (2015). Miniaturized Dual-Band CPW-Fed Antennas Loaded With U-Shaped Metamaterials. *IEEE Antennas and Wireless Propagation Letters, 14*, 658–661. doi:10.1109/LAWP.2014.2376554

Aniktar, H., Sjoland, H., Mikkelsen, J. H., & Larsen, T. (2005). A Class-AB 1.65 GHz-2GHz broadband CMOS medium power amplifier. In *NORCHIP Conference, 2005. 23rd* (pp. 269–272). Academic Press. 10.1109/NORCHP.2005.1597041

Anscy, S. (2013). Slot microstrip antenna for 2.4GHz RFID reader application. Department of Electronics and Communication Engineering Karunya University. *International Journal of Advanced Research in Electronics and Communication Engineering, 2*(5).

Ansys Inc. (2011). *HFSS. v13.0*. Retrieved from http://www.ansys.com/products/electronics/ansys-hfss

Arand, B. A., & Bazrkar, A. (2015). Gain enhancement of a tetra-band square-loop patch antenna using an AMC–PEC substrate and a superstrate. *Wireless Personal Communications, 84*(1), 87–97. doi:10.100711277-015-2595-8

Aris, M. A., Ali, M. T., Rahman, N. H. A., & Pasya, I. (2016). Development of antenna array using Defected Ground Structure. In *2016 10th European Conference on Antennas and Propagation (EuCAP)* (pp. 1–5). Academic Press. 10.1109/EuCAP.2016.7481577

Aris, M., & Ali, M. T. (2015). *Frequency Reconfigurable Aperture-Coupled Microstrip Patch Antenna Using Defected Ground Structure*. 2015 IEEE International RF and Microwave Conference (RFM 2015), Kuching, Malaysia.

Arulaalan, M., & Nithyanandan, L. (2014). Analysis of Microstrip Line Feed Triangular Patch Antenna for WLAN (2.4 GHz). *Proc. of Int. Conf. on Advances in Communication, Network, and Computing, CNC*, 802-806.

Arya, A., Patnaik, A., & Kartikeyan, M. (2011). Microstrip patch antenna with skew-F shaped DGS for dual band operation. *Progress in Electromagnetics Research, 19*, 147–160. doi:10.2528/PIERM11052305

Asrokin, A., Abd. Rahim, M.K., Jamaluddin, M.H., & Ahmad, M.R. (2006). Triangular and Circular Dual Band Microstrip Antenna for WLAN Applications. *International RF and Microwave Conference Proceeding*, 53-57. 10.1109/RFM.2006.331036

Azimi Fashi, A., Kamyab, M., & Barati, M. (2010). A microstrip small-sized array antenna based on the meta-material zeroth-order resonator. *Progress in Electromagnetics Research, 14*, 89–101. doi:10.2528/PIERC10032302

Baccarelli, P., Paulotto, S., & Di Nallo, C. (2007). Full-wave analysis of bound and leaky modes propagating along 2D periodic printed structures with arbitrary metallisation in the unit cell. *IET Microwaves, Antennas & Propagation, 1*(1), 217–225. doi:10.1049/iet-map:20050321

Baena, J. D., Bonache, J., Martin, F. R., Sillero, M., Falcone, F., Lopetegi, T., ... Sorolla, M. (2005). Equivalent-circuit models for split-ring resonators and complementary split-ring resonators coupled to planar transmission lines. *IEEE Transactions on Microwave Theory and Techniques, 53*(4), 1451–1461. doi:10.1109/TMTT.2005.845211

Baena, J. D., Marqués, R., Medina, F., & Martel, J. (2004). Artificial magnetic metamaterial design by using spiral resonators. *Physical Review B: Condensed Matter and Materials Physics, 69*(1). doi:10.1103/PhysRevB.69.014402

Bahl, J., & Bhartia, P. (1981). Microstrip Antennas. Artech House.

Bahl, I. (2009). *Fundamentals of RF and microwave transistor amplifiers*. John Wiley & Sons. doi:10.1002/9780470462348

Bahl, I. J. (2003). *Lumped elements for RF and microwave circuits*. Artech House.

Bahl, I. J., & Bhartia, P. (1980). *Microstrip antennas*. Dedham, MA: Artech House.

Bahl, I. J., & Bhartia, P. E. (1980). *Microstrip antenna*. Dedham, MA: Artech House.

Baik, J. W., Han, S. M., Jeong, C., Jeong, J., & Kim, Y. S. (2008). Compact Ultra-Wideband Bandpass Filter With EBG Structure. *IEEE Microwave and Wireless Components Letters*, *18*(10), 671–673. doi:10.1109/LMWC.2008.2003456

Balanis, C. A. (2003). Antenna theory: Analysis and Design. J. Wiley & Sons.

Balanis, C. A. (2005). Antenna theory: analysis and design (3rd ed.). Academic Press.

Balanis, C. A. (2005). Antenna Theory: Analysis and Design. Hoboken, NJ: John Wiley & Sons Inc.

Balanis, A. (2003). *Antenna Theory Analysis and Design* (2nd ed.). John Wiley & Sons, Inc.

Balanis, C. A. (2005). *Antenna Theory: Analysis and Design*. Hoboken, NJ: John Wiley & Sons Inc.

Baliarda, C. P., Romeu, J., & Cardama, A. (2000, November). The Koch monopole: A small fractal antenna. *Antennas and Propagation, IEEE Transactions*, *48*(11), 1773–1781. doi:10.1109/8.900236

Bancroft, R. (2009). *Microstrip and printed antenna design*. SciTech Publishing.

Barlevy, A. S., & Rahmat-Samii, Y. (2001, June). Characterization of electromagnetic band-gaps composed of multiple periodic tripods with interconnecting vias concept, analysis, and design. *IEEE Transactions on Antennas and Propagation*, *49*(3), 343–353. doi:10.1109/8.918607

Bashri, M., & Ibrahimy, M. (2015). Design of a wideband inductively coupled loop feed patch antenna for UHF RFID tag. Wuxiandian Gongcheng.

Basil, Jeemon, Shambavi, & Alex. (2013). A Multi-fractal Antenna for WLAN and WiMAX application. *Proceedings of IEEE Conference on Information and Communication Technologies (ICT 2013).*

Basilio, L. I., Khayat, M. A., Williams, J. T., & Long, S. A. (2001). The dependence of the input impedance on feed position of probe and microstrip line-fed patch antennas. *IEEE Transactions on Antennas and Propagation*, *49*(1), 45–47. doi:10.1109/8.910528

Baudrand, H., Tao, J., & Atechian, J. (1988). Study of radiation properties of open-ended rectangular waveguides. *IEEE Transactions on Antennas and Propagation*, *36*(8), 1071–1077. doi:10.1109/8.7219

Baumeister, P. (1958). Design of multilayer filters by successive approximations. *JOSA*, *48*(12), 955–958. doi:10.1364/JOSA.48.000955

Bayatmaku, N., Lotfi, P., Azarmanesh, M., & Soltani, S. (2011). Design of simple multi-band patch antenna for mobile communication applications using new E-shape fractal. *IEEE Antennas and Wireless Propagation Letters*, *99*, 1.

Beiranvand, E., Afsahy, M., & Sharbati, V. (2017). Reduction of the mutual coupling in patch antenna arrays based on EBG by using a planar frequency-selective surface structure. *International Journal of Microwave and Wireless Technologies*, *9*(2), 349–355. doi:10.1017/S1759078715001440

Benyetho, T., El Abdellaoui, L., Zbitou, J., Bennis, H., Tribak, A., & Latrach, M. (2017). A New Dual Band Planar Fractal Antenna Array for ISM Band. *International Journal of Microwave and Optical Technology*, *12*(3).

Bernhard, J. T. (2003). Reconfigurable antennas and apertures: State-of-the-art and future outlook. *Proceedings of SPIE Conference on Smart Electronics, MEMS, BioMEMS, and Nanotechnology*, *5055*, 1–9.

Bernhard. (2003). Reconfigurable antennas and apertures: State-of-the-art and future outlook. *Proceedings of SPIE Conference on Smart Electronics, MEMS, BioMEMS, and Nanotechnology, 5055*, pp. 1–9.

Berrich, L., & Zenkouar, L. (2016). Microruban Dipole Antenna for RFID Applications at 2.45 GHz. *International Journal of Electrical and Computer Engineering, 6*(6). Retrieved from http://publications.balluff.com/mongodb_pdb/public_pdb_material.php?id=221392&revision=0.5&pg=PG185676&pf=PF185742&con=en&ws=approval

Best, S. R. (2002). A Comparison of the Performance Properties of the Hilbert Curve Fractal and Meander Line Monopole Antennas. *Microwave and Optical Technology Letters, 35*(4), 258–262. doi:10.1002/mop.10576

Beurrier, H. (1973). *Impedance-matched amplifiers.* Google Patents.

Bisht, S., Saini, S., Prakash, V., & Nautiyal, B. (2014). Study The Various Feeding Techniques of Microstrip Antenna Using Design and Simulation Using CST Microwave Studio. *International Journal of Emerging Technology and Advanced Engineering, 4*(9).

Biswas, S., Guha, D., & Kumar, C. (2013). Control of Higher Harmonics and Their Radiations in Microstrip Antennas Using Compact Defected Ground Structures. *IEEE Transactions on Antennas and Propagation, 61*(6), 3349–3353. doi:10.1109/TAP.2013.2250240

Björck, Å. (2016). *Numerical methods in matrix computations.* Springer.

Bodnar, D. G., & Paris, D. T. (1970). New variational principle in electromagnetic. *IEEE Transactions on Antennas and Propagation, 18*(2), 216–223. doi:10.1109/TAP.1970.1139642

Boesch, R. D., & Conroy, K. P. (1999). *Impedance matching circuit for power amplifier.* Google Patents.

Booker, H. G. (1946). Slot aerials and their relation to complementary wire aerials. J. IEE (London), 93, 620–626.

Borja, C., & Romeu, J. (2003). On the behavior of Koch island fractal boundary microstrip antenna. *IEEE Transactions on Antennas and Propagation, 51*(6), 1281–1291. doi:10.1109/TAP.2003.811479

Boybay, M. S., & Ramahi, O. M. (2012). Material Characterization Using Complementary Split-Ring Resonators. *IEEE Transactions on Instrumentation and Measurement, 61*(11), 3039–3046. doi:10.1109/TIM.2012.2203450

Brown, D. (2007). RFID Implementation. New York: McGraw-Hill.

Brown, E. R. (2001). on the gain of a reconfigurable-aperture antenna. *IEEE Transactions on Antennas and Propagation, 49*(10), 1357–1362. doi:10.1109/8.954923

Brown, E. R., Parker, C. D., & Yablonovitch, E. (1993). Radiation properties of a planar antenna on a photonic-crystal substrate. *JOSA B, 10*(2), 404–407. doi:10.1364/JOSAB.10.000404

Busch, K., Von Freymann, G., Linden, S., Mingaleev, S. F., Tkeshelashvili, L., & Wegener, M. (2007). Periodic nanostructures for photonics. *Physics Reports, 444*(3-6), 101–202. doi:10.1016/j.physrep.2007.02.011

Cadman, D., Zhang, S., & Vardaxoglou, Y. (2016). Fused deposition modeling for microwave circuits and antenna. *International Symposium on Antenna and Propagation (ISAP)*, Okinawa, Japan.

Caloz and Itoh. (2005). *Electromagnetic Metamaterials: 'Transmission Line Theory and Microwave Applications.* Piscataway, NJ: Wiley- IEEE.

Carver, K. R., & Mink, J. W. (1981). Microstrip Antenna Technology. *IEEE Transactions on Antennas and Propagation, AP-29*(1), 2–24. doi:10.1109/TAP.1981.1142523

Chandra, K. N. (2010). Handbook of Smart Antennas for RFID Systems. John Wiley & Sons.

Chandrappa, D. N., Ambresh, P. A., Vani, R. M., & Hunagund, P. V. (2014). Multi-slots reconfigurable microstrip antenna with capacitive loading technique. *International Journal of Innovative Research in Computer and Communication Engineering, 2*(1). Retrieved from www.ijircce.com

Chang, C. W., Chen, K. M., & Qian, J. (1997). Nondestructive determination of electromagnetic parameters of dielectric materials at X-band frequencies using a waveguide probe system. *IEEE Transactions on Instrumentation and Measurement, 46*(5), 1084–1092. doi:10.1109/19.676717

Chang, I. C. I., & Lee, B. L. B. (2002). Design of defected ground structures for harmonic control of active microstrip antenna. *IEEE Antennas and Propagation Society International Symposium (IEEE Cat. No.02CH37313), 2*, 852–855. 10.1109/APS.2002.1016779

Chang, T. N., & Lin, J. M. (2007). A novel circularly polarized patch antenna with a serial multislot type of loading. *IEEE Transactions on Antennas and Propagation, 55*(11), 3345–3348. doi:10.1109/TAP.2007.908846

Chaurasia, Pimpalgaonkar, Raval, & Upadhyaya. (2016). Electrically small antennas: The art of miniaturization. *International Journal of Advanced Research in Computer Engineering & Technology, 5*(2).

Chen, H. D., & Chen, H. T. (2004). A CPW-Fed Dual-Frequency Monopole Antenna. IEEE Transactions on Antennas and Propagation, 52(4), 978–982. doi:10.1109/TAP.2004.825620 doi:10.1109/TAP.2004.825620

Cheney, E., & Kincaid, D. (2012). *Numerical mathematics and computing*. Nelson Education.

Cheng, Y. F., Ding, X., Shao, W., & Wang, B. Z. (2017). Reduction of Mutual Coupling Between Patch Antennas Using a Polarization-Conversion Isolator. *IEEE Antennas and Wireless Propagation Letters, 16*, 1257–1260. doi:10.1109/LAWP.2016.2631621

Chen, H.-D., & Tsao, Y.-H. (2010). Low-Profile PIFA Array Antennas for UHF Band RFID Tags Mountable on Metallic Objects. *IEEE Transactions on Antennas and Propagation, 58*(4), 1087–1092. doi:10.1109/TAP.2010.2041158

Chen, S., & Hsu, P. (2005). Hybrid Loop Antenna For 2.4-Ghz RFID Tags. *Proceedings of ISAP2005.*

Chen, S.-Y., & Hsu, P. (2004). CPW-Fed Folded-Slot Antenna for 5.8-GHz RFID Tags. *Electronics Letters, 40*(24), 1516–1517. doi:10.1049/el:20046780

Cheype, C., Serier, C., Thèvenot, M., Monédière, T., Reineix, A., & Jecko, B. (2002). An electromagnetic bandgap resonator antenna. *IEEE Transactions on Antennas and Propagation, 50*(9), 1285–1290. doi:10.1109/TAP.2002.800699

Chitra, J., & Nagarajan, V. (2014). Frequency Reconfigurable Antenna Using PIN Diodes. *IEEE Transactions on Antennas and Propagation, 53*(2).

Chiu, C. Y., Shum, K. M., & Chan, C. H. (2007). A tunable via-patch loaded PIFA with size reduction. *IEEE Transactions on Antennas and Propagation, 55*(1), 65–71. doi:10.1109/TAP.2006.888389

Choi, J.-S., Kang, D.-H., Kim, D.-S., Park, J.-M., Jin, B.-S., & Kim, B.-M. (2009). Power amplifiers and transmitters for next generation mobile handsets. *JSTS: Journal of Semiconductor Technology and Science, 9*(4), 249–256. doi:10.5573/JSTS.2009.9.4.249

Choi, W., Son, H. W., Bae, J.-H., Choi, G. Y., Pyo, C. S., & Chae, J.-S. (2006). An RFID Tag Using a Planar Inverted-F Antenna Capable of Being Stuck to Metallic Objects. *ETRI Journal, 28*(2), 216–218. doi:10.4218/etrij.06.0205.0082

Choubani, M., Choubani, F., Gharsallah, A., David, J., & Mastorakis, N. E. (2010). Analysis and design of electromagnetic band gap structures with stratified and inhomogeneous media. *Optics Communications, 283*(22), 4499–4504. doi:10.1016/j.optcom.2010.04.078

Chowdhury, D., Hull, C. D., Degani, O. B., Wang, Y., & Niknejad, A. M. (2009). A fully integrated dual-mode highly linear 2.4 GHz CMOS power amplifier for 4G WiMax applications. *IEEE Journal of Solid-State Circuits, 44*(12), 3393–3402. doi:10.1109/JSSC.2009.2032277

Chu, L. J. (1948). *Physical limitations on omni-directional antennas.* Massachusetts Institute of Technology. Technical Report No. 64.

Chu, L. (1948). Physical Limitations of Omni-Directional Antennas. *Journal of Applied Physics, 19*(12), 1163–1175. doi:10.1063/1.1715038

Collin & Rothchild. (1964). Evaluation of Antenna Q. *IEEE Trans. AP, AP-12*, 23–27.

Collin, R. E. (1960). *Field theory of guided waves.* Academic Press.

Collin, R. E. (1985). *Antennas and Radiowave Propagation.* McGraw-Hill.

Combes, P. F. (1997). *Micro-ondes: Circuits passifs, propagation, antennes: cours et exercices.* Dunod.

Compton, R. T. Jr. (1963). *The aperture admittance of a rectangular waveguide radiating into a lossy half-space. Technical Report, 1691-1.* Columbus, OH: Ohio State University.

Constantine, A. B., & ... (2005). *Antenna theory: analysis and design. In Microstrip Antennas* (3rd ed.). John wiley & sons.

Constantine, A. B. (2005). Antenna theory: analysis and design. In *Microstrip Antennas* (3rd ed.). John Wiley & Sons.

Costantine, J. (2009). *Design, optimization and analysis of reconfigurable antennas* (Ph.D. dissertation). Electr Comput, Eng. Dept., Univ. New Mexico (UNM), Albuquerque, NM.

Craeye, C., & González-Ovejero, D. (2011). A review on array mutual coupling analysis. *Radio Science, 46*(2), 1–25. doi:10.1029/2010RS004518

Cripps, C. (2006). *RF Power Amplifiers For Wireless Communications.* Academic Press.

Croswell, W. F. (2007). Slot Antennas. In J. L. Volakis (Ed.), *Antenna Engineering Handbook* (pp. 8-1–8-16). McGraw-Hill.

CST microwave studio. (2011). Retrieved from https://www.cst.com /products/cstmws

CST Studio Suite Version. (2010). Retrieved from http://www.cst.com/Content/Products/DS/Overview.aspx

Cui, T. J., Smith, D. R., & Liu, R. (2010). *Metamaterials Theory Design and Applications.* Springer.

Da Silva, E. F., & McPhun, M. K. (1978). Calibration techniques for one-port measurement. *Microwave Journal, 21*(6), 97–100.

Dalia, Elsheakh, Elsadek, & Abdallah. (2012). Antenna Designs with Electromagnetic Band Gap Structures Metamaterial. InTech.

Danesh, M., Long, J. R., Hadaway, R. A., & Harame, D. L. (1998). A Q-factor enhancement technique for MMIC inductors. In *Microwave Symposium Digest, 1998 IEEE MTT-S International* (*Vol. 1*, pp. 183–186). Academic Press. 10.1109/MWSYM.1998.689352

Darwin, C. (1859). On the origin of species. John Murray.

Das, B. N., Ramakrishna, J., & Sarap, B. (1984). Resonant conductance of inclined slots in the narrow wall of a rectangular waveguide. *IEEE Transactions on Antennas and Propagation, 32*(7), 759–761. doi:10.1109/TAP.1984.1143401

Dehbashi, R., Atlasbaf, Z., & Forooraghi, K. (2006). New Compact Size Microstrip Antennas With Harmonic Rejection. IEEE Antennas and Propagation Letters, 5, 395–398. doi:10.1109/LAWP.2006.881918 doi:10.1109/LAWP.2006.881918

Deleruyelle, T., Pannier, P., Bergeret, E., & Bourdel, S. (2008). Dual band UHF and microwave RFID antenna. *Proceedings of the 38th European Microwave Conference*. 10.1109/EUMC.2008.4751818

Delogne, P. P., & Deryck, L. (1980). Underground use of a coaxial cable with leaky sections. *IEEE Transactions on Antennas and Propagation*, *28*(6), 875–883. doi:10.1109/TAP.1980.1142426

Denlinger, E. J. (1969, April). Radiation from microstrip radiators. *IEEE Transactions on Microwave Theory and Techniques*, *17*(4), 235–236. doi:10.1109/TMTT.1969.1126937

Derneryd, A. G. (1978). A Theoretical Investigation of the Rectangular Microstrip Antenna Element. *IEEE Transactions on Antennas and Propagation*, *26*(4), 532–535. doi:10.1109/TAP.1978.1141890

Dey, S., & Mittra, R. (1997). Compact microstrip patch antenna. *Microwave and Optical Technology Letters*, *13*(1), 12–14. doi:10.1002/(SICI)1098-2760(199609)13:1<12::AID-MOP4>3.0.CO;2-Q

Dhevi, B. L., Vishvaksenan, K. S., & Rajakani, K. (2018). Isolation Enhancement in Dual-Band Microstrip Antenna Array Using Asymmetric Loop Resonator. *IEEE Antennas and Wireless Propagation Letters*, *17*(2), 238–241. doi:10.1109/LAWP.2017.2781907

Ding, K., Fang, X., Chen, A., & Wang, Y. (2013). A Novel Parallel-Series Feeding Network Based on Three-Way Power Divider for Microstrip Antenna Array. *IEEE Antennas and Wireless Propagation Letters*, *12*, 757–760. doi:10.1109/LAWP.2013.2270938

Dixit, L., & Pourush, P. K. (2009), Radiation characteristics of switchable ferrite microstrip array antenna. *Inst. Electr Eng Proc V Microw Antennas Propag.*, *147*(2), 151–155.

Dobkin, D. M. (2008). *The RF in RFID Passive UHF RFID in Practice*. Elsevier Inc.

Dongho, K., & Junho, Y. (2012). Dual-Band Long-Range Passive RFID Tag Antenna Using an AMC Ground Plane. *IEEE Transactions on Antennas and Propagation*, *60*(6), 2620–2626. doi:10.1109/TAP.2012.2194638

Drogi, S. F., & Tomasz, M. (2012). *RF power amplifier circuit with mismatch tolerance*. Google Patents.

Dual band antenna 2118060-1 Datasheet from TE Connectivity Company. (n.d.). Retrieved from http://www.farnell.com/datasheets/2357390.pdf

Eaton, J. E., Eyces, L. J., & Macfarlane, G. G. (1983). Linear-Array Antennas and Feeds. In S. Silver (Ed.), Microwave Antenna Theory and Design (pp. 257 - 333). The Institution of Engineering and Technology (IET).

Ebadi, S., & Semnani, A. (2014). Mutual coupling reduction in waveguide-slot-array antennas using electromagnetic bandgap (EBG) structures. *IEEE Antennas & Propagation Magazine*, *56*(3), 68–79. doi:10.1109/MAP.2014.6867683

El Hamraoui, A., Abdelmounim, E.H., Zbitou, J., Bennis, H., & Latrach, M. (2018). A compact low cost dual-band slot antenna for RFID readers. *International Journal of Microwave and Optical Technology*, *13*(1).

El Hamraoui, A., Abdelmounim, E., Zbitou, J., Bennis, H., & Latrach, M. (2015). A new design of a compact CPW-Fed dual-band printed antenna for RFID applications. Wireless Networks and Mobile Communications (WINCOM). *IEEE Conference Publications*.

El Hamraoui, A., Abdelmounim, E., Zbitou, J., Errkik, A., Bennis, H., & Latrach, M. (2018). A Dual-Band Microstrip Slot Antenna Reader for UHF and Microwave RFID Applications. *TELKOMNIKA Journal*, *16*(1), 94–101. doi:10.12928/telkomnika.v16i1.7310

El Maleky, O., Ben Abdelouahab, F., Essaaidi, M., & Ajana, L. (2017). *Reconfigurable T-Shaped Antenna for S-Band Applications.* IEEE Xplore.

Elavarasi, C., & Shanmuganantham, T. (2017). SRR Loaded Iterated Pythagoras Fractal Antenna for Multiple Functional Resonance Frequencies. *International Journal of Microwave and Optical Technology, 12*(3).

Elftouh, H., Touhami, N., Aghoutane, M., El Amrani, S., Tazon, A., & Boussouis, M. (2014). Miniaturized Microstrip Patch Antenna with Defected Ground Structure. *Progress In Electromagnetics Research C, 55*, 25–33.

Enjiu, R. K., & Perotoni, M. B. (2013). Slotted waveguide antenna design using 3D EM simulation. *Microwave Journal, 56*(7), 72–84.

Ennajih, A., & Zbitou, J. (2017). Dual Band Metamaterial Printed Antenna Based on CSRR For RFID Applications. *International Journal of Microwave and Optical Technology.*

Ennajih, A., Zbitou, J., Latrach, M., Errkik, A., El Abdellaoui, L., & Tajmouati, A. (2017). Dual Band Metamaterial Printed Antenna Based on CSRR For RFID Applications. *International Journal of Microwave and Optical Technology, 12*(2).

Ennajih, A., Zbitou, J., Latrach, M., Errkik, A., El Abdellaoui, L., & Tajmouati, A. (2018). Dual New Miniature Microstrip Antenna Based On Metamaterial For RFID Applications. Telkomnika Journal, 16(1).

Er-rebyiy, R., Zbitou, J., Latrach, M., Errkik, A., Tajmouati, A., & Abdellaoui, L. E. L. (2017b). A Novel Structure of a Reconfigurable Printed Antenna for Wireless Applications. International Review of Electrical Engineering, 12(5), 387-392.

Er-rebyiy, R., Zbitou, J., Latrach, M., Tajmouati, A., Errkik, A. L., & Abdellaoui, E. (2017). New Miniature Planar Microstrip Antenna Using DGS for ISM applications. Telkomnika Journal, 15(3).

Er-rebyiy, R., Zbitou, J., Latrach, M., Errkik, A., Tajmouati, A., & Abdellaoui, L. E. L. (2017a). New Design of a Miniature Microstrip Patch Antenna Integrating Defected Ground Structure. *International Journal of Microwave And Optical Technology, 12*(5), 354–359.

Er-rebyiy, R., Zbitou, J., Latrach, M., Tajmouati, A., Errkik, A., & Abdellaoui, L. E. L. (2017, September). New Miniature Planar Microstrip Antenna Using DGS for ISM applications. *TELKOMNIKA, 15*(3), 1149–1154. doi:10.12928/telkomnika.v15i3.6864

Eunni, M. B. (2004). *A Novel Planar Microstrip Antenna Design for UHF RFID* (Master's Thesis). Madras University.

Evizal, E., Abdul Rahman, T., & Abdul Rahim, S. (2013). UHF RFID Tag Antenna for Vehicle License Plate Number. TELKOMNIKA Journal, 11(2).

Falcone F., Lopetegi T., Laso M. A. G., Baena J. D., Bonache J., Beruete M., Marqués R., Martín F., & Sorolla M. (2004). Babinet principle applied to the design of metasurfaces and metamaterials. *Physical Review Letters, 93*(19), 197 401-1–197 401-4.

Falcone, F., Lopetegi, T., Baena, J. D., Marques, R., Martin, F., & Sorolla, M. (2004). Effective Negative – ε Stopband Microstrip Lines Based on Complementary Split Ring Resonators. *IEEE Microwave and Wireless Components Letters, 14*(6), 280–282. doi:10.1109/LMWC.2004.828029

Fang, D. G. (2010). *Antenna Theory and Microstrip Antennas.* Taylor and Francis Group, LLC.

Farahani, H. S., Veysi, M., Kamyab, M., & Tadjalli, A. (2010). Mutual coupling reduction in patch antenna arrays using a UC-EBG superstrate. *IEEE Antennas and Wireless Propagation Letters, 9*, 57–59. doi:10.1109/LAWP.2010.2042565

Farzami, F., Forooraghi, K., & Norooziarab, M. (2012). Miniaturization of a microstrip antenna using a compact and thin magneto-dielectric substrate'. *IEEE Antennas and Wireless Propagation Letters*, *10*, 1540–1542. doi:10.1109/LAWP.2011.2181968

Fattal, D., Li, J., Peng, Z., Fiorentino, M., & Beausoleil, R. G. (2010). Flat dielectric grating reflectors with focusing abilities. *Nature Photonics*, *4*(7), 466–470. doi:10.1038/nphoton.2010.116

Faudzi, N. M., & Ali, M. T. (2014). Metal mountable UHF-RFID tag antenna with meander feed line and double T-match. *IEEE International Symposium on Technology Management and Emerging Technologies*. doi:10.1109/ISTMET.2014.693647310.1109/ISTMET.2014.6936473

Ferchichi, A., & Gharsallah, A. (2013). A Novel Small Sierpenski Antennas. *International Journal of Electrical and Computer Engineering*, *3*(4).

Fezai, F., Menudier, C., Thevenot, M., Arnaud, E., & Mondiere, T. (2015). Reconfigurable Parasitic Element Antenna Using Reflection Phase Shifters. *IEEE Antennas and Wireless Propagation Letters*, *14*, 775–778. doi:10.1109/LAWP.2014.2377592

Finkenzeller, K. (2010). RFID Handbook: Fundamentals and Applications in Contactless Smart Cards and Identification. New York: John Wiley and Sons. doi:10.1002/9780470665121 doi:10.1002/9780470665121

Finkenzeller, K. (2010). *RFID Handbook Fundamentals and Applications in Contactless Smart Cards, Radio Frequency Identification and Near-Field Communication* (3rd ed.). Giesecke & Devrient GmbH.

Fleuriault, J.-M. (1996). *Synthèse du diagramme de rayonnement d'un réseau de sources*. These l'Universitede Rennes I.

Foged, L. J., Giacomini, A., Saccardi, F., Tancioni, L. M., Schirosi, V., Di Cintio, A., ... Baracco, J.-M. (2015). Miniaturized Array Antenna Using Artificial Magnetic Materials for Satellite-Based AIS System. *IEEE Transactions on Antennas and Propagation*, *63*(4), 1276–1287. doi:10.1109/TAP.2015.2398462

Foltz, H. D., McLean, J. S., & Crook, G. (1998, December). Disk-Loaded Monopoles with Parallel Strip Elements. *IEEE Transactions on Antennas and Propagation*, *46*(12), 1894–1896. doi:10.1109/8.743844

Foroozesh, A., & Shafai, L. (2009). Effects of Artificial Magnetic Conductors in the Design of Low-Profile High-Gain Planar Antennas With High-Permittivity Dielectric Superstrate. *IEEE Antennas and Wireless Propagation Letters*, *8*, 10–13. doi:10.1109/LAWP.2008.916684

Foster, P., & Bueberry, R. (1999). Antenna Problems in RFID Systems. IEE Coll. on RFID Technology, 1(3), 1–5.

Ganchev, S. I., Bakhtiari, S., & Zoughi, R. (1992). A novel numerical technique for dielectric measurement of generally lossy dielectrics. *IEEE Transactions on Instrumentation and Measurement*, *41*(3), 361–365. doi:10.1109/19.153329

Gao, S. C., Li, L. W., Leong, M. S., & Yeo, T.-S. (2001). Analysis of an h-shaped patch antenna by using the FDTD method. *Progress in Electromagnetics Research*, *34*, 165–187. doi:10.2528/PIER01060501

Garg, R. (2001). *Microstrip antenna design handbook*. Boston: Artech House.

Ghaloua, A., Zbitou, J., El Abdellaoui, L., Latrach, M., Tajmouati, A., & Errkik, A. (2018). Reduction of Mutual Coupling between Closely Spaced Microstrip Antennas Arrays using Electromagnetic Band-gap (2D-EBG) Structures. *Telkomnika (Telecommunication Computing Electronics and Control), 16*(1).

Ghaloua, A., Zbitou, J., Latrach, M., El Abdellaoui, L., Errkik, A., & Mandry, R. (2017). Mutual coupling reduction and miniaturization arrays antennas using new structure of EBG. In *2017 International Conference on Wireless Technologies, Embedded and Intelligent Systems (WITS)* (pp. 1–5). Academic Press. 10.1109/WITS.2017.7934593

Ghaloua, A., Zbitou, J., Latrach, M., El Abdellaoui, L., Tajmouati, A., & Abounaima, M. C. (2017). Miniaturization and reduction of mutual coupling for four arrays antennas using new structure of EBG. In *2017 International Conference on Wireless Technologies, Embedded and Intelligent Systems (WITS)* (pp. 1–5). Academic Press. 10.1109/WITS.2017.7934595

Ghen, Z., Ney, M. M., & Hoefer, W. J. (1993). Absorbing and connecting boundary conditions for the TLM method. *IEEE Transactions on Microwave Theory and Techniques*, *41*(11), 2016–2024. doi:10.1109/22.273429

Ghione, G., & Naldi, C. (1984). Analytical Formulas for Coplanar Lines in Hybrid and Monolithic Mics. Electronics Letters, 20(4), 179–181. doi:10.1049/el:19840120 doi:10.1049/el:19840120

Ghiotto, A. (2008). *Conception d'Antennes de Tags RFID UHF, Application à la Réalisation par Jet de Matière* (Doctoral Thesis). Institut Polytechnique de Grenoble.

Ghiotto, A. (2008). *Tag UHF RFID antenna design, application by jet of material* (Ph.D. dissertation). Institut Polytechnique, Grenoble, France.

Ghiotto, A., & Cantalice, S. F. (2008). Miniaturized Patch Antenna for the Radio Frequency Identification of Metallic Objects. *Proceedings of the IEEE MTT-S International Microwave Symposium (IMS2).*

Ghiotto, A., Cantalice, S. F., Vuong, T. P., Pouzin, A., Fontgallandt, G., & Tedjini, S. (2008). Miniaturized Patch Antenna for the Radio Frequency Identification of Metallic Objects. *IEEE MTT-S International Microwave Symposium (IMS2008)*, 583-586. 10.1109/MWSYM.2008.4633233

Ghiotto, A., Vuong, T. P., Yagoub, M. C. E., & Tedjini, S. (2007). *On the Design and Miniaturization of Passive UHF RFID Transponder Antenna.* Presented at the *Int. Conf: URSI2007*, Ottawa, Canada.

Ghosh, A., Chakraborty, S., Chattopadhyay, S., Nandi, A., & Basu, B. (2016). Rectangular microstrip antenna with dumbbell shaped defected ground structure for improved cross polarised radiation in wide elevation angle and its theoretical analysis. *IET Microwaves, Antennas & Propagation*, *10*(1), 68–78. doi:10.1049/iet-map.2015.0179

Ghosh, I. S., Hilgers, A., Schlenker, T., & Porath, R. (2001). T.shlenker, R.Porah, "Ceramic microwave antennas for mobile applications. *Journal of the European Ceramic Society*, *21*(15), 2621–2628. doi:10.1016/S0955-2219(01)00327-2

Gianvittorio, J. P., & Rahmat, Y. (2000). Fractal Element Antennas: A Compilation of Configurations with Novel Characteristics. *IEEE Antennas and Propagation Society International Symposium.* 10.1109/APS.2000.874566

Gilbert, R. A. (2007). Waveguide Slot Antenna Arrays. In J. L. Volakis (Ed.), *Antenna Engineering Handbook* (pp. 9-1–9-36). McGraw-Hill.

Godara. (1997). Performance improvement feasibility and System considerations. *IEEE, 85*(7).

Goldberg, D. E. (1989). *Genetic Algorithms in Search, Optimization & Machine Learning.* Addison Wesley.

Goldstone, L. O., & Oliner, A. A. (1959). Leaky-wave antennas I: Rectangular waveguides. *IRE Transactions on Antennas and Propagation*, *7*(4), 307–319. doi:10.1109/TAP.1959.1144702

Gonzalez, G. (1996). *Microwave transistor amplifiers: analysis and design.* Prentice-Hall, Inc.

Gonzalez, G. (1997). *Microwave Transistor Amplifiers Analysis and Design (2nd ed.).* Prentice Hall.

Goswami, S., Sarmah, K., Sarma, A., Sarma, K. K., & Baruah, S. (2016, January). *Slot Loaded Square Patch Antenna with CSRR at Ground Plane.* Paper presented at the IEEE International Conference on Microelectronics, Computing and Communication (MicroCom), India. 10.1109/MicroCom.2016.7522558

Goswami, S., Sarmah, K., Sarma, A., Sarma, K. K., & Baruah, S. (2017). Design of a CSRR based Compact Microstrip Antenna for Image Rejection in RF Down-converter based WLAN Receivers. *AEÜ. International Journal of Electronics and Communications*, *74*, 128–134. doi:10.1016/j.aeue.2017.02.004

Green, H. E. (1989). Determination of the cutoff of the first higher order mode in a coaxial line by the transverse resonance technique. *IEEE Transactions on Microwave Theory and Techniques*, *1*(10), 1652–1653. doi:10.1109/22.41018

Grenier, D. (2016). *Antennes et Propagation radio (PhD thesis).* Université Laval Québec Canada.

Gronau, G., & Wolff, I. (1986, May). Aperture-coupling of a rectangular microstrip resonator. *Electronics Letters*, *22*(10), 554–556. doi:10.1049/el:19860378

Guesmi, C., Ferchichi, A., & Gharsallah, A. (2014). A Modified Fractal Bow Tie Antenna for an RFID Reader. *International Journal of Electrical and Computer Engineering*, *4*(3).

Guha, Biswas, & Kumar. (1999). *Printed Antenna Designs Using Defected Ground Structures: A Review of Fundamentals and State-of-the-Art Developments.* Forum for Electromagnetic Research Methods and Application Technologies (FERMAT).

Guha, D., & Yahia, M.M (2011). *Microstrip and Printed Antennas: New Trends, Techniques and Application.* John Wiley & Sons, Ltd.

Guha, D., & Yahia, M.M. (2011). *Microstrip and Printed Antennas: New Trends, Techniques and Application.* John Wiley & Sons, Ltd.

Guha, D., & Antar, Y. M. M. (2011). *Microstrip and printed antennas: new trends, techniques and applications.* John Wiley & Sons.

Habashi, A., Nourinia, J., & Ghobadi, C. (2011). Mutual Coupling Reduction Between Very Closely Spaced Patch Antennas Using Low-Profile Folded Split-Ring Resonators (FSRRs). *IEEE Antennas and Wireless Propagation Letters*, *10*, 862–865. doi:10.1109/LAWP.2011.2165931

Haider, Sabbagh, Mahmood, Edwards, & Brister. (2012). A UWB Fractal Antenna for Body Area Network Applications. Loughborough Antennas and Propagation Conference, Loughborough, UK.

Hajj, M., Rodes, E., Serhal, D., Chantalat, R., Thevenot, M., Monediere, T., & Jecko, B. (2007, September). Metallic EBG sectoral antenna for base stations. In *Applied Electromagnetics and Communications, 2007. ICECom 2007. 19th International Conference on* (pp. 1-4). IEEE. 10.1109/ICECOM.2007.4544413

Hall, P. S., Wood, C., & James, J. R. (1981). *Microstrip antenna theory and design.* London: IEE Press.

Hansen, R. C. (1992). Array pattern control and synthesis. *Proceedings of the IEEE*, *80*(1), 141–151. doi:10.1109/5.119573

Hansen, R. C., & Burke, M. (2000). Antennas with magneto-dielectrics. *Microwave and Optical Technology Letters*, *26*(2), 75–78. doi:10.1002/1098-2760(20000720)26:2<75::AID-MOP3>3.0.CO;2-W

Hao, Z. C., & Li, B. W. (2017). Developing Wideband Planar Millimeter-Wave Array Antenna Using Compact Magneto-Electric Dipoles. *IEEE Antennas and Wireless Propagation Letters*, *16*, 2102–2105. doi:10.1109/LAWP.2017.2697903

Harrington, R. F. (1958). On Gain and Beamwidth of Directional Antennas. IRE Transactions on Antennas and Propagation, 6(3), 219–225. doi:10.1109/TAP.1958.1144605 doi:10.1109/TAP.1958.1144605

Harrington, R. (1960). Effects of Antenna Size on Gain, Bandwidth, and Efficiency. *J Nat Bur. Stand.*, *64-D*, 1–12.

Hashmi, R. M., Zeb, B. A., & Esselle, K. P. (2014). Wideband high-gain EBG resonator antennas with small footprints and all-dielectric superstructures. *IEEE Transactions on Antennas and Propagation, 62*(6), 2970–2977. doi:10.1109/TAP.2014.2314534

Hebib, S. (2008). *Nouvelle Topologie d'antennes Multi-bandes pour Applications Spatiales* (Doctoral dissertation). Université de Toulouse.

Heddebaut, M. (2009). Leaky waveguide for train-to-wayside communication-based train control. *IEEE Transactions on Vehicular Technology, 58*(3), 1068–1076. doi:10.1109/TVT.2008.928635

Herscovici, N., Osorio, M., & Peixeiro, C. (2001). Minimization of a rectangular patche using genetic algorithms. *Antennas and Propagation Society International Symposium IEEE, 4*, 34 – 37.

He, Y., Chahal, P., & Papapolymerou, J. (2017). A lego-like reconfigurable cavity using 3-D Polyjet Technology. *International Microwave Symposium*, Honolulu, HI. 10.1109/MWSYM.2017.8058616

Hildebrand, L. T., & McNamara, D. A. (1995). A guide to implementational aspects of the spatial-domain integral equation analysis of microstrip antennas. *Applied Computational Electromagnetics Journal, 10*(1), 40–51.

Hirohide, S. (2005). Radiation for a flanged rectangular waveguide. *IEEE Transactions on Antennas and Propagation, 53*(12), 3953–3962. doi:10.1109/TAP.2005.859748

Holland, J. H. (1992). Genetic algorithms. *Scientific American, 267*(1), 66–72. doi:10.1038cientificamerican0792-66

Homayoon, O., & Shahram, H. (2012). Miniaturization of Microstrip Antennas by the Novel Application of the Giuseppe Peano Fractal Geometrices. *IEEE Transactions on Antennas and Propagation, 60*(8), 3559–3567. doi:10.1109/TAP.2012.2201070

Howell, J. Q. (1972). Microstrip antennas. *IEEE International Symposium on Antennas and Propagation*, 177–180.

Huang, Y., & Boyle, K. (2008). Antennas: from Theory to Practice. Chichester, UK: John Wiley & Sons Ltd. doi:10.1002/9780470772911 doi:10.1002/9780470772911

Hua, R. C., & Ma, T. G. (2007). A printed dipole antenna for ultra high frequency (UHF) radio frequency identification (RFID) handheld reader. *IEEE Transactions on Antennas and Propagation, 55*(12), 3742–3745. doi:10.1109/TAP.2007.910521

Huff, G. H., & Bernhard, J. T. (2006). Integration of packaged RF-MEMS switches with radiation pattern reconfigurable square spiral microstrip antennas. *IEEE Transactions on Antennas and Propagation, 54*(2), 464–469. doi:10.1109/TAP.2005.863409

Hum, S. V., & Xiong, H. Y. (2010). *Analysis and design of a differentially-fed frequency agile microstrip patch antenna. IEEE Transactions on Antennas and Propagation.*

Huque, Hosain, Islam, & Chowdhury. (2011). Design and Performance Analysis of Microstrip Array Antennas with Optimum Parameters for X-band Applications. *International Journal of Advanced Computer Science and Applications, 2*(4).

Hu, W., Ismail, M. Y., Cahill, R., Encinar, J. A., Fusco, V., Gamble, H. S., ... Rea, S. P. (2007). Liquid-crystal-based reflect array antenna with electronically switchable monopulse patterns. *Electronics Letters, 43*(14), 744. doi:10.1049/el:20071098

Huynh, M.-C., & Stutzman, W. (2003). Ground plane effects on planar inverted-f antenna (pifa) performance. *Microwaves, Antennas and Propagation, IEE Proceedings, 150*(4), 209–213. 10.1049/ip-map:20030551

Hwangbo, S., Yang, H. Y., & Yoon, Y. K. (2017). Mutual Coupling Reduction Using Micromachined Complementary Meander-Line Slots for a Patch Array Antenna. *IEEE Antennas and Wireless Propagation Letters, 16*, 1667–1670. doi:10.1109/LAWP.2017.2663114

Ihamji, A. (2017). Design of Compact Tri-Band Fractal Antenna for RFID Readers. *International Journal of Electrical and Computer Engineering*.

Ihamji, M., Abdelmounim, E., Bennis, H., Hefnawi, M., & Latrach, M. (2017). Design of Compact Tri-Band Fractal Antenna for RFID Readers. *International Journal of Electrical and Computer Engineering, 7*(1).

Ihamji, M., Abdelmounim, E., Zbitou, J., Bennis, H., & Latrach, M. (2016). Novel Design of a Miniature Fractal Microstrip CPW Fed Antenna for RFID Reader. *IEEE Conference Publications, The International Conference on Wireless Networks and Mobile Communications (WINCOM2016)*. 10.1109/WINCOM.2016.7777197

Ihamji, M., Abdelmounim, E., Zbitou, J., Bennis, H., & Latrach, M. (2016). Novel design of a miniature L-Slot microstrip CPW-fed Antenna for RFID reader. *IEEE Conference Publications, The 11th International Conference on Intelligent Systems: Theories and Applications (SITA2016)*.

Ihamji, M., Abdelmounim, E., Latrach, M., & Bennis, B. (2017, August). Design of Compact Tri-Band Fractal Antenna for RFID Readers. *Iranian Journal of Electrical and Computer Engineering, 7*(4), 2036–2044.

Ikonen, M., & Tretyakov. (2006). Antennas with Artificial Magneto-Dielectric substrates: The Dominant Role of Frequency Dispersion. In *Antennas and Propagation Society International Symposium*. IEEE.

Ikonen, P. M. T., Rozanov, K. N., Osipov, A. V., Alitalo, P., & Tretyakov, S. A. (2006). Magnetodielectric Substrates in Antenna Miniaturization: Potential and Limitations. *Antennas and Propagation, IEEE Transactions, 54*(11), 3391–3399. doi:10.1109/TAP.2006.884303

Impinj Mini-Guardrail Datasheet from Cisper Company. (n.d.). Retrieved from https://www.cisper.nl/rfid/downloads/Impinj_Mini-Guardrail%20Data%20Sheet.pdf

Impinj, Inc. (2006). *Monza-1a EPC global Generation 2 RFID*. Retrieved from http://www.orangetags.com/rfid-chipdatasheet/impinj

Indoor directional antenna 304021 Datasheet from GAO RFID Company. (n.d.). Retrieved from http://gaorfid.com/RFID-brochures/2.4GHz_6_dBi_Indoor_Directional_Antenna_304021.pdf

Iqbal, M., & Piacibello, A. (2016). A 5W class-AB power amplifier based on a GaN HEMT for LTE communication band. In *Microwave Symposium (MMS), 2016 16th Mediterranean* (pp. 1–4). Academic Press. 10.1109/MMS.2016.7803827

Irfan, N., Yagoub, M. C. E., & Hettak, K. (2012). Design of a microstrip-line-fed inset patch antenna for RFID applications. *IACSIT International Journal of Engineering and Technology, 4*(5), 558–561. doi:10.7763/IJET.2012.V4.432

Irsadi Aksun, M., & Chuang, S.-H. (1990). On slot-coupled microstrip antennas and their applications to CP operation-theory and experiment. IEEE Transactions on Antenna & Propagation, 38(8).

Islam, M. T., & Samsuzzaman, M. (2014). *Miniaturized Dual Band Multislotted Patch Antenna on Polytetrafluoroethylene Glass Microfiber Reinforced for C/X Band Applications. The Scientific World Journal*.

Jain, P. K., Kang, W., Soin, H., & Xi, Y. (2002). Analysis and design considerations of a load and line independent zero voltage switching full bridge DC/DC converter topology. *IEEE Transactions on Power Electronics, 17*(5), 649–657. doi:10.1109/TPEL.2002.802181

James, J. R., & Hall, P. S. (1989). *Handbook of Microstrip Antennas* (Vols. 1–2). Academic Press.

Jena, M. R., Mangaraj, B. B., & Mishra, D. (2013). Bandwidth and Gain Enhancement of Multiband Fractal Antenna Based on The Sierpinski CARPET GEOMETRY. *ICTACT Journal on Communication Technology*, *4*(1), 669–674. doi:10.21917/ijct.2013.0095

Jian-bin, X., Fengjie, H., & Dong, W. (2015). Analysis and Design of Tag Antenna Based UHF RFID for Libraries. *Indonesian Journal of Electrical Engineering and Computer Science, 13*(3).

Jin, Z., & Mortazawi, A. (2003). An L-band tunable microstrip antenna using multiple varactors. In *IEEE Antennas and Propagation Society International Symposium. Digest. Held in conjunction with: USNC/CNC/URSI North American Radio Sci. Meeting (Cat. No.03CH37450)* (Vol. 4, pp. 524–527). IEEE. 10.1109/APS.2003.1220325

Joannopoulos, J. D., Johnson, S. G., Winn, J. N., & Meade, R. D. (2011). *Photonic crystals: molding the flow of light.* Princeton University Press.

Johan, S., William, F., Aurélien, P., Olivier, T., Nicolas, D., Bila, S., ... Gramond, R. P. (2016). Plastic and Metal Additive Manufacturing Technologies for Hyper Frequency Passive Components up to Ka-band. *46th European Microwave Conference*, London, UK.

John, S. (1987). Strong localization of photons in certain disordered dielectric superlattices. *Physical Review Letters*, *58*(23), 2486–2489. doi:10.1103/PhysRevLett.58.2486 PMID:10034761

Johnson, S. G., & Joannopoulos, J. D. (2000). Three-dimensionally periodic dielectric layered structure with omnidirectional photonic band gap. *Applied Physics Letters*, *77*(22), 3490–3492. doi:10.1063/1.1328369

Josefsson, L. G. (1991). *A waveguide antenna which includes a slotted hollow waveguide*. U.S. Patent No. EP0545873A2.

Joseph, S., Paul, B., Mridula, S., & Mohanan, P. (2013). A Novel Planar Fractal Antenna with CPW Feed for Multiband Applications. *Radioengeneering*, *22*(4), 1262–1266.

Joshi & Das. (1982). Impedance of a radiating slot in the ground plane of a microstripline. IEEE Trans. Antenna and Propagation, AP-30.

Kaabal, A., Ahyoud, S., Asselman, A., & Faize, A. (2013). Design Of High Gain Ultra Wide-Band Antenna For Wireless Communication Using EBG Structures. *European Scientific Journal, 9*(30).

Kabiri, A., & Ramahi, O. (2009, June). Limitations of artificial magnetic materials with negative permeability. Paper Presented at the *2009 IEEE International Symposium on Antennas and Propagation and USNC-URSI Radio Science Meeting*. 10.1109/APS.2009.5171634

Kapoor, S., & Parkash, D. (2012). Miniaturized triple band microstrip patch antenna with defectedground structure for wireless communication applications. *International Journal of Computers and Applications*, 57.

Karkkainen, M., & Ikonen, P. (2005). Patch Antenna with Stacked Split-Ring Resonators as Artificial Magneto Dielectric Substrate. *Microwave and Optical Technology Letters*, *46*(6), 554–556. doi:10.1002/mop.21048

Karmakar, N. C. (2010). Handbook of Smart Antennas for RFID systems. John Wiley & Sons, Inc. doi:10.1002/9780470872178

Katehi, P. B., & Alexopoulos, N. G. (1984, November). On the modeling of electromagnetically coupled microstrip antennas-the printed strip dipole. *IEEE Transactions on Antennas and Propagation*, *AP-32*(11), 1179–1186. doi:10.1109/TAP.1984.1143235

Katsibas, K. D., Balanis, C. A., Tirkas, P. A., & Birtcher, C. R. (1998). Folded Loop Antenna for Mobile Handheld Units. *IEEE Transactions on Antennas and Propagation*, *46*(2), 260–266. doi:10.1109/8.660971

Khandelwal, M. K., Kanaujia, B. K., & Kumar, S. (2017). Defected ground structure: Fundamentals, analysis, and applications in modern wireless trends. *International Journal of Antennas and Propagation.*

Khodaei, G. F., Nourinia, J., & Ghobadi, C. (2008). A practical miniaturized u-slot patch antenna with enhanced bandwidth. *Progress In Electromagnetics Research B, 3*, 47–62. doi:10.2528/PIERB07112201

Kikkert, C. J. (2013). *RF Electronics: design and simulation.* James Cook University.

Kim, D. O., Jo, N. I., Jang, H.-A., & Kim, C. Y. (2011). Design of the ultrawideband antenna with a quadruple-band rejection characteristics using a combination of the complementary split ring resonators. *Progress in Electromagnetics Research, 112*, 93–107. doi:10.2528/PIER10111607

Kim, J. H., Enkhbayar, B., Bang, J. H., & Ahn, B. C. (2010). New formulas for the reflection coefficient of an open-ended rectangular waveguide radiating into air including the effect of wall thickness or flange. *Progress In Electromagnetics Research M., 12*, 143–153. doi:10.2528/PIERM10033104

Kim, J. S., Shin, K. H., Park, S. M., Choi, W. K., & Seong, N. S. (2006). Polarization and space diversity antenna using inverted-f antennas for RFID reader applications. *IEEE Antennas and Wireless Propagation Letters, 5*, 265–268. doi:10.1109/LAWP.2006.875892

Kim, S. T., Yun, G. H., & Park, H. K. (1998). Numerical analysis of the propagation characteristics of multiangle multislot coaxial cable using moment method. *IEEE Transactions on Microwave Theory and Techniques, 46*(3), 269–279. doi:10.1109/22.661714

King, K. (2017). *Data Management Solutions Review.* Retrieved from https://solutionsreview.com/data-management/idc-data-creation-to-reach-163-zettabytes-by-2025/

Kitsos, P., & Zhang, Y. (2008). *RFID Security: Techniques, Protocols and System-on-Chip Design.* Springer Science+Business Media.

Koch, H. V. (1906). Une méthode géométrique élémentaire pour l'étude de certaines questions de la théorie des courbes planes. *Acta Mathematica, 30*(1), 145–174. doi:10.1007/BF02418570

Kshetrimayum R. S., & Karthikeyan S. S. (n.d.). A parametric study on the stopband characteristics of CSRRs. *International Journal of Recent Trends in Engineering, 1*(3), 274-276.

Kula, J., Psychoudakis, D., Liao, W., Chen, C., Volakis, J., & Halloran, J. (2006). Patch antenna miniaturization using recently available ceramic substrates. *IEEE Antennas & Propagation Magazine, 48*(6), 13–20. doi:10.1109/MAP.2006.323335

Kula, Psychoudakis, & Liao, Chen, Volakis, & Halloran. (2006). Patch Antenna Miniaturization Using Thick Truncated Textured Ceramic Substrates. *IEEE Antennas & Propagation Magazine, 48*(6).

Kumar & Ray. (2003). *Broadband Microstrip Antennas.* London: Artech House.

Kumar, R., Cousin, J. C., & Huyart, B. (2014). 3D Indoor Localization System using Single Reference/Anchor Position based on Multistatic FMCW Radar System. European Microwave Week (EuMW2014), Rome, Italy.

Kumar, R., Louzir, A., & Naour, J. Y. L. (2017). Flexible RFID-Helix Antenna for Sensing the Home Electrical Device Activity. *European Conference on Antenna & Propagation (EuCAP2017)*, Paris, France.

Kumar, R., Malathi, P., & Shinde, J. P. (2007). Design of miniaturized fractal antenna. *37th Europ. Microw. Conf.*, 474-477.

Kuo, S. K., Chen, S.-L., & Lin, C.-T. (2008). An accurate method for impedance measurement of RFID tag antenna. *Progress in Electromagnetics Research, 83*, 93–106. doi:10.2528/PIER08042104

Kuo, Y. L., & Wong, K. L. (2003). Printed Double-T Monopole Antenna for 2.4/5.2 GHz Dual-Band WLAN Operations. *IEEE Transactions on Antennas and Propagation*, *51*(9).

Kurokawa, K. (1965). Power waves and the scattering matrix. *IEEE Transactions on Microwave Theory and Techniques*, *13*(2), 194–202. doi:10.1109/TMTT.1965.1125964

Kushwaha, N., & Kumar, R. (2014). Study of different shape electromagnetic band gap (EBG) structures for single and dual band applications. *Journal of Microwaves, Optoelectronics and Electromagnetic Applications*, *13*(1), 16–30. doi:10.1590/S2179-10742014000100002

Kwon, H., & Lee, B. (2005). Compact slotted planar inverted-F RFID tag mountable on metallic objects. *Electronics Letters*, *41*(24), 1308–1310. doi:10.1049/el:20052940

Lacher, R. (2009). *Development of Status of Next Generation Automotive Radar in EU.* ITS Forum, Tokyo, Japan.

Lee, Yook, & Park. (2002). Broadband planar antenna having round corner rectangular wide slot. *IEEE Antennas and Propagation Society International Symposium.*

Lee, B., Harackiewicz, F., Jung, B., & Park, M. J. (2005). Cavity-backed slot antenna array for the repeater system of a satellite digital multimedia broadcasting service. *IEEE Antennas and Wireless Propagation Letters*, *4*, 389–392. doi:10.1109/LAWP.2005.857967

Lee, B.-K., Jung, J.-W., Suh, B.-S., & Hyun, D.-S. (1997). A new half-bridge inverter topology with active auxiliary resonant circuit using insulated gate bipolar transistors for induction heating applications. In *Power Electronics Specialists Conference, 1997. PESC'97 Record., 28th Annual IEEE* (Vol. 2, pp. 1232–1237). IEEE.

Lee, Y., Lu, X., Hao, Y., Yang, S., Ubic, R., Evans, J. R. G., & Parini, C. G. (2007). Directive millimetre-wave antenna based on freeformed woodpile EBG structure. *Electronics Letters*, *43*(4), 195–196. doi:10.1049/el:20073798

Leger, L. (2004). *Nouveaux développements autour des potentialités de l'antenne BIE planaire* (Doctoral dissertation). Limoges.

Leger, L., Monediere, T., & Jecko, B. (2005). Enhancement of gain and radiation bandwidth for a planar 1-D EBG antenna. *IEEE Microwave and Wireless Components Letters*, *15*(9), 573–575. doi:10.1109/LMWC.2005.855373

Leger, L., Serier, C., Chantalat, R., Thevenot, M., Monedière, T., & Jecko, B. (2004, March). 1D dielectric electromagnetic band gap (EBG) resonator antenna design. *Annales des Télécommunications*, *59*(3-4), 242–260.

Levine, H., & Papas, C. H. (1951). Theory of the circular diffraction antenna. *Journal of Applied Physics*, *22*(1), 29–43. doi:10.1063/1.1699816

Levy, Kumar, & Dinh. (2013). A novel fractal UWB antenna for earthquake and tsunami prediction application (LETPA). *IEEE Canadian Conference of Electrical And Computer Engineering (CCECE).*

Li, J., Jiang, T., Cheng, C., & Wang, C. (2013). Hilbert Fractal Antenna for UHF Detection of Partial Discharges in Transformers. *IEEE Transactions on Dielectrics and Electrical Insulation*, *20*(6), 2017–2025. doi:10.1109/TDEI.2013.6678849

Li, M., Li, Q. L., Wang, B., Zhou, C. F., & Cheung, S. W. (2018). Miniaturized Dual-Band Base Station Array-Antenna Using Band Notch Dipole Antenna Elements and AMC Reflectors. *IEEE Transactions on Antennas and Propagation*, *1*. doi:10.1109/TAP.2018.2819890

Lim, J.-S., Kim, C.-S., Ahn, D., Jeong, Y.-C., & Nam, S. (2005). Design of low-pass filters using defected ground structure. *IEEE Transactions on Microwave Theory and Techniques*, *53*(8), 2539–2545. doi:10.1109/TMTT.2005.852765

Lim, W.-G., Son, W.-I., Oh, K.-S., Kim, W.-K., & Yu, J.-W. (2008). Compact integrated antenna with circulator for UHF RFID system. *IEEE Antennas and Wireless Propagation Letters*, *7*, 673–675. doi:10.1109/LAWP.2008.921348

Lin, L., Zhang, L., Wang, Y., & Yu, Z. (2016). A 150GHz amplifier based on coupled transmission line neutralization with 22dB gain in 65nm CMOS. In *Solid-State and Integrated Circuit Technology (ICSICT), 2016 13th IEEE International Conference on* (pp. 1576–1578). IEEE.

Li-na, W., Shu, L., Run-nan, C., Guan-long, H., & Wen-bin, Z. (2011). *Multiband Printed Monopole Antenna with Square-Nested Fractal.* Paper presented at 6th International ICST Conference on Communications and Networking (CHINACOM), China.

Lin, K., Chen, S., & Mittra, R. (2010). A Capacitively Coupling Multifeed Slot Antenna for Metallic RFID Tag Design. *IEEE Antennas and Wireless Propagation Letters*, *9*, 447–450. doi:10.1109/LAWP.2010.2048991

Li, R. L., Dejean, G., Tentzeris, M. M., & Laskar, J. (2004). Development and analysis of a folded shorted-patch antenna with reduced size. *IEEE Transactions on Antennas and Propagation*, *52*(2), 555–562. doi:10.1109/TAP.2004.823884

Liu, B., Gong, X., & Chappell, W. J. (2004). Layer-by-layer polymer stereolithography fabrication for three-dimensional RF components. In *International Microwave Symposium Digest IEEE MTT-S* (vol. 2, pp. 481-484). IEEE.

Liu, H., Li, Z., Sun, X., & Mao, J. (2005). Harmonic suppression with photonic bandgap and defected ground structure for a microstrip patch antenna. *IEEE Microwave and Wireless Components Letters*, *15*(2), 55–56. doi:10.1109/LMWC.2004.842809

Liu, L., & Langley, J. R. (2008). Liquid crystal tunable microstrip patch antenna. *Electronics Letters*, *44*(20), 1179–1180. doi:10.1049/el:20081995

Li, Y. (2009, December). Investigation of minimum cavity height of small EBG-resonator antennas for maximum directivity. In *Microwave Conference, 2009. APMC 2009. Asia Pacific* (pp. 2687-2690). IEEE. 10.1109/APMC.2009.5384121

Li, Y., Iskander, M. F., Zhang, Z., & Feng, Z. (2013). A new low-cost leaky wave coplanar waveguide continuous transverse stub antenna array using metamaterial-based phase shifters for beam steering. *IEEE Transactions on Antennas and Propagation*, *61*(7), 3511–3518. doi:10.1109/TAP.2013.2257649

Li, Y., Liu, Y., & Gong, S. (2008). Microstrip antenna using ground-cut slots for low RCS with size miniaturization techniques. *Progress In Electromagnetics Research Letters*, *1*, 211–220.

Li, Y., & Wang, J. H. (2010). Polarization property of leaky coaxial cable with overlapped triangle slots. *IEEE Antennas and Wireless Propagation Letters*, *9*, 1049–1052. doi:10.1109/LAWP.2010.2090332

Li, Y., Wang, J. H., Zhang, Z., & Chen, M. (2012). Empirical formula for the electric field calculation in rectangular slots on leaky coaxial cable. *IEEE Transactions on Antennas and Propagation*, *60*(12), 5825–5833. doi:10.1109/TAP.2012.2208929

Li, Z. Y., & Lin, L. L. (2003). Photonic band structures solved by a plane-wave-based transfer-matrix method. *Physical Review. E*, *67*(4), 046607. doi:10.1103/PhysRevE.67.046607 PMID:12786509

Lizzi, L., & Oliveri, G. (2011). *Hybrid Design of a Fractal-Shaped GSM/UMTS Antenna*. Technical Report, DISI-11-089.

Lo, J. (2012). *Étude de la reconfigurabilité d'une structure à bande interdite électromagnétique (BIE) métallique par plasmas de décharge* (Doctoral dissertation). Université Paul Sabatier-Toulouse III.

Loo, C. H., Elmahgoub, K., Yang, F., Elsherbeni, A. Z., Kajfez, D., Kishk, A. A., ... Ruuskanen, P. (2008). Chip impedance matching for UHF RFID tag antenna design. *Progress in Electromagnetics Research, 81*, 359–370. doi:10.2528/PIER08011804

Low, Z. N., Cheong, J. H., & Law, C. L. (2005). Low-Cost PCB Antenna for UWB Applications. IEEE Antennas and Wireless Propagation Letters, 4.

Lo, Y. T., Solomon, D., & Richards, W. F. (1979). Theory and experiment on microstrip antennas. *IEEE Transactions on Antennas and Propagation, AP-27*(2), 137–149. doi:10.1109/TAP.1979.1142057

Lukas, I., & Qing, X. (2008). UHF Omnidirectional Bent Dipole Antenna for Item-level RFID Applications. *IEEE Antennas and Propagation Society International Symposium*. doi:10.1109/APS.2008.461958910.1109/APS.2008.4619589

Luk, K., Chair, R., & Lee, K. (1998). Small rectangular patch antenna. *Electronics Letters, 34*(25), 2366–2367. doi:10.1049/el:19981643

Lu, L., & Coetzee, J. C. (2005). Reduced-size microstrip patch antenna for Bluetooth applications. *IEEE Electronics Letters, 41*(17), 944–945. doi:10.1049/el:20052312

Mahatthanajatuphat, C., Saleekaw, S., Akkaraekthalin, P., & Krairiksh, M. (2009). A rhombic patch monopole antenna with modified minkowski fractal geometry for UMTS,WLAN, and mobile WiMAX application. *Progress in Electromagnetics Research, 89*, 57–74. doi:10.2528/PIER08111907

Mahmoud, S. F. (2004). A New Miniaturized Annular Ring Patch Resonator Partially Loaded by A Metamaterial Ring with Negative Permeability and Permittivity. *IEEE Antennas and Wireless Propagation Letters, 3*, 19–22. doi:10.1109/LAWP.2004.825092

Maicas, J.P. (2011). *Recent Developments in Mobile Communications – A Multidisciplinary Approach*. InTechOpen.

Mailloux, R. J. (2007). Electronically Scanned Arrays. *Synthesis Lectures on Antennas, 2*(1), 1–82. doi:10.2200/S00081ED1V01Y200612ANT006

Mandelbrot, B.B. (1975). *Les objets fractals: forme hasard et dimension*. Nouvelle bibliothèque scientifiques.

Manimegalai, B., Raju, S., & Abhaikumar, V. (2009). A multi fractal cantor antenna for multiband wireless applications. *IEEE Antennas and Wireless Propagation Letters, 8*, 359–362. doi:10.1109/LAWP.2008.2000828

Manke, I., Kardjilov, N., Strobl, M., Hilger, A., & Banhart, J. (2008). Investigation of the skin effect in the bulk of electrical conductors with spin-polarized neutron radiography. *Journal of Applied Physics, 104*(7), 076109. doi:10.1063/1.2992516

Marcuvitz, N. (1986). Waveguide Handbook. The Institution of Engineering and Technology (IET). doi:10.1049/PBEW021E

Marques, R. (2007). Metamaterials with Negative Parameters Theory, Design, and Microwave Applications. Academic Press.

Marques, R., Mesa, F., Martel, J., & Medina, F. (2003). Comparative Analysis of Edge- and Broadside- Coupled Split Ring Resonators for Metamaterial Design—Theory and Experiments. *IEEE Transactions on Antennas and Propagation, 51*(10), 2572–2581. doi:10.1109/TAP.2003.817562

Marrocco, G. (2003). Gain-Optimized Self-Resonant Meander Line Antennas for RFID Applications. IEEE Antennas and Wireless Propagation Letters, 2, 302–305. doi:10.1109/LAWP.2003.822198 doi:10.1109/LAWP.2003.822198

Marrocco, G. (2008). The Art of UHF RFID Antenna Design: Impedance-matching and Size-reduced Techniques. IEEE Transactions on Antennas and Propagation, 50(11), 66–79. doi:10.1109/MAP.2008.4494504 doi:10.1109/MAP.2008.4494504

Mason, S. (1957). Some properties of three-terminal devices. *IRE Transactions on Circuit Theory*, *4*(4), 330–332. doi:10.1109/TCT.1957.1086413

McLean, J. S. (1994). The radiative properties of electrically-small antennas. IEEE Int Sympos on Electromagnetic Compatibility, 320–324.

McLean, J. S. (1996). A re-examination of the fundamental limits on the radiation Q of electrically small antennas. *IEEE Trans. AP*, *44*(5), 672–676. doi:10.1109/8.496253

McSpadden, J. O., Yoo, T., & Chang, K. (1992). Theoretical and experimental investigation of a rectenna element for microwave power transmission. *IEEE Transactions on Microwave Theory and Techniques*, *40*(12), 2359–2366. doi:10.1109/22.179902

McVay, J., & Hoorfar, A. (2007). Miniaturization of top-loaded monopole antennas using Peano-curves. *IEEE Radio and Wireless Symp*, 253-256. 10.1109/RWS.2007.351816

Mehdipour, A., Rosca, I. D., Sebak, A., Trueman, C. W., & Hoa, S. V. (2010). Full-composite Fractal Antenna using Carbon Nanotubes for Multiband Wireless Applications. *IEEE Antennas and Wireless Propagation Letters*, *9*, 891–894. doi:10.1109/LAWP.2010.2076342

Mendez, T., Jardon, J., Aguilar, H., & Iturbide-Sanchez, F. (2006). Application of the defected microstrip structure as a tuning technique for rectangular printed antennas. *Microwave and Optical Technology Letters*, *48*, 16–19.

Milligan, T. A. (2005). *Modern antenna design*. John Wiley & Sons. doi:10.1002/0471720615

Mishra, S., & Satpathy, S. (2003). One-dimensional photonic crystal: The Kronig-Penney model. Physical Review B, 68(045121).

Misilmani, H. M. E., Husseini, M. A., & Kabalan, K. Y. (2015). Design of slotted waveguide antennas with low sidelobes for high power microwave applications. *Progress In Electromagnetics Research C.*, *56*, 15–28. doi:10.2528/PIERC14121903

Miwave Techno Ventures. (2009). *Slotted Broad Wall Microwave Antenna*. Retrieved from https://www.indiamart.com/proddetail/slotted-broad-wall-microwave-antenna-3736329988.html

Mizeikis, V., Seet, K. K., Juodkazis, S., & Misawa, H. (2004). Three-dimensional woodpile photonic crystal templates for the infrared spectral range. *Optics Letters*, *29*(17), 2061–2063. doi:10.1364/OL.29.002061 PMID:15455780

Mo, L., Zhang, H., & Zhou, H. (2008). Broadband UHF RFID tag antenna with a pair of U slots mountable on metallic objects. Electronics Letters, 44(20), 1173–1174. doi:10.1049/el:20089813 doi:10.1049/el:20089813

Moghaddam, B. T., Entezar, S. R., & Adl, H. P. (2013). Nonlinear properties of a graded-index photonic heterostructure. *Pramana*, *80*(5), 887–894. doi:10.100712043-013-0523-6

Molisch, A. F. (2011). Wireless Communications. John Wiley & Sons Ltd.

Monti, G., Catarinucci, L., & Tarricone, L. (2009). Compact Microstrip Antenna For Rfid Applications. *Progress In Electromagnetics Research Letters.*, *8*, 191–199. doi:10.2528/PIERL09042803

Morente, J. A., Porti, J. A., & Khalladi, M. (1992). Absorbing boundary conditions for the TLM method. *IEEE Transactions on Microwave Theory and Techniques*, *40*(11), 2095–2099. doi:10.1109/22.168768

Mosallaei, M., & Sarabandi, K. (2002). Periodic meta-material structures in electromagnetics: Concept, analysis, and applications. *IEEE AP-S International Symposium*. 10.1109/APS.2002.1016103

Mosig, J. R., & Gardiol, F. E. (1983). Analytical and numerical techniques in the Green's function treatment of microstrip antennas and scatterers. *IEE Proceedings, 130*(2), 175–182. 10.1049/ip-h-1.1983.0029

Mosig, J. R., & Gardiol, F. E. (1985). General integral equation formulation for microstrip antennas and scatterers. *IEE Proceedings, 132*(7), 424–432. 10.1049/ip-h-2.1985.0076

Mosig, J. R. (1988, February). Arbitrarily shaped microstrip structures and their analysis with a mixed potential integral equation. *IEEE Transactions on Microwave Theory and Techniques*, *36*(2), 314–323. doi:10.1109/22.3520

Mu, J., Wang, H., Jiang, X., Xu, D., & Huang, Y. (2016, October). *Design of Low-RCS and Gain Enhancement Microstrip Antenna Based on Miniaturized Polarisation-Dependent AMC.* Paper presented at the International Symposium on Antennas and Propagation (ISAP), Japan.

Munson, R. E. (1974). Conformal microstrip antennas and microstrip phased arrays. *IEEE Transactions on Antennas and Propagation*, *22*(1), 74–78. doi:10.1109/TAP.1974.1140723

Murad, N., Esa, Y. M., & Ali, S. (2006). Hilbert Curve Fractal Antenna for RFID Application. *Proceedings of IEEE International RF and Microwave Conference,* 182-186. 10.1109/RFM.2006.331065

Murata Co. Ltd. (2017). *Murata -LXMS31ACNA – 010, EPC Global Gen2.* Retrieved from http://www.murata.com// en-eu/products/rfid/rfid/uhf

Naji, Aziz, & Fyath. (n.d.). Design and Simulation of RFID Aperture Coupled Fractal Antennas. *International Journal of Radio Frequency Identification & Wireless Sensor Networks.*

Nasimuddin. (2011). *Microstrip Antennas.* Rijeka, Croatia: InTech.

Necibi, Ferchichi, Vuong, & Gharsallah. (2014). Miniaturized CSRR TAG Antennas for 60GHz Applications. *International Journal of Electrical and Computer Engineering, 4*(1), 64-74.

Necibi, O., & Gharsallah, A. (2016). A novel RFID antenna with HIS structure for Ka band. International Journal on Communications Antenna and Propagation, 6(6), 30–38.

Nguyen, H. T., Noghanian, S., & Shafai, L. (2005). Microstrip patch antenna miniaturization by slots loading. *IEEE Antennas and Propagation Society International Symposium*, 215-218. 10.1109/APS.2005.1551525

Nieto, A. L. (2011). RFID Design Fundamentals and Applications. Taylor and Francis Group.

Nithisopa, K., Nakasuwan, J., Songthanapitak, N., Anantrasirichai, N., & Wakabayashi, T. (2007). Design CPW fed slot antenna for wideband applications. Piers Online, 3(7).

Numan-Al-Mobin, A., Cross, W. M., Kellar, J. J., & Anagnostou, D. E. (2015). *RFID Integrated QR Code Tag Antenna.* Retrieved from https://www.researchgate.net/publication/308559377

Oline, A. A., & Jackson, D. R. (2007). Leaky-Wave Antennas. In J. L. Volakis (Ed.), *Antenna Engineering Handbook* (pp. 11-1–11-56). McGraw-Hill.

Oraizi, H., & Hamidkhani, M. (2001). Design of miniaturized multi band microstrip antennas with helical slot patterns on patches. *Microwave Conference (EuMC), 41st European.*

Oraizi, H., & Hedayati, S. (2012). *Miniaturization of Microstrip Antennas by the Novel Application of the Giuseppe Peano Fractal Geometries. In Antennas and Propagation* (pp. 3559–3567). IEEE Transactions.

Oscar, A., Peverini, A., Lumia, M., Caliganano, F., Addamo, G., Lorusso, M., … Virone, G. (2017). Selective Laser Melting Manufacturing of Microwave Waveguide Devices. In *Proceeding of the IEEE* (*vol. 105*, pp. 620-631). IEEE. 10.1109/JPROC.2016.2620148

Oulhaj, Touhami, Aghoutane, & Tazon. (2016). A Miniature Microstrip Patch Antenna Array with Defected Ground Structure. *IJMOT, 11*(1).

Oulhaj, O., Touhami, N. A., Aghoutane, M., & Kchairi, A. B. (2017, August). A New Design of a Wideband Miniature Antenna Array. *Iranian Journal of Electrical and Computer Engineering*, *7*(3), 1850–1857.

Outdoor directional antenna 304020 Datasheet from GAO RFID Company. (n.d.). Retrieved from http://gaorfid.com/RFID-brochures/2.4GHz_14_dBi_Outdoor_Directional_Antenna_304020.pdf

Ozbay, E., Aydin, K., Cubukcu, E., & Bayindir, M. (2003). Transmission and reflection properties of composite double negative metamaterials in free space. *IEEE Transactions on Antennas and Propagation*, *51*(10), 2592–2595. doi:10.1109/TAP.2003.817570

Pandeeswari, R., & Raghavan, S. (2014). Microstrip antenna with complementary split ring resonator loaded ground plane for gain enhancement. *Microwave and Optical Technology Letters*, *52*(2), 292–296.

Paret, D. (2008). RFID en Ultra et Super Hautes Fréquences UHF-SHF, Théorie et mise en œuvre. Dunod.

Park, C. H., & Son, H. W. (2016). Mutual coupling reduction between closely spaced microstrip antennas by means of H-shaped conducting wall. *Electronics Letters*, *52*(13), 1093–1094. doi:10.1049/el.2016.1339

Patch antenna APAKN1304-S5517-T Datasheet from Abracon Company. (n.d.). Retrieved from http://www.farnell.com/datasheets/2611067.pdf

Patch antenna APAKN2504-S2448-T Datasheet from Abracon Company. (n.d.). Retrieved from http://www.farnell.com/datasheets/2611070.pdf

Pavithra, R., Mohanageetha, D., Mary Anita, E. A., & Subramaniam, M. (2014). A New Compact Microstrip Integrated E-Array Patch Antenna with High Gain and High Aperture Efficiency. *Wireless Personal Communications*, *78*(2), 1011–1020. doi:10.100711277-014-1798-8

PEL90206 Datasheet from Laird Company. (n.d.). Retrieved from https://www.lairdtech.com/sites/default/files/public/solutions/Industrial%20Wireless%20RFID%20Antenna%20Solutions.pdf

Pendry, J. B., Holden, A. J., Robbins, D. J., & Stewart, W. J. (1999). Magnetism from conductors and enhanced nonlinear phenomena. *IEEE Transactions on Microwave Theory and Techniques*, *47*(11), 2075–2084. doi:10.1109/22.798002

Pendry, J. B., Holden, A. J., Stewart, W. J., & Youngs, I. (1996). Extremely Low Frequency Plasmons in Metallic Mesostructures. *Physical Review Letters*, *76*(25), 4773–4776. doi:10.1103/PhysRevLett.76.4773 PMID:10061377

Peroulis, D., Sarabandi, K., & Katehi, L. P. B. (2005). Design of reconfigurable slot antennas. *IEEE Transactions on Antennas and Propagation*, *53*(2), 645–654. doi:10.1109/TAP.2004.841339

Peterson, B., & Schnaufer, D. (2018). *5G Fixed Wireless Access Array and RF Front-End Trade-Offs. Microwave Journal.*

Piazza, D., Kirsch, N. J., Forenza, A., Heath, R. W., & Dandekar, K. R. (2008). Design and evaluation of a reconfigurable antenna array for MIMO systems. *IEEE Transactions on Antennas and Propagation*, *56*(3), 869–881. doi:10.1109/TAP.2008.916908

Pigeon, M. (2011). Design of ultra-compact antennas based on metamaterials, application to the fabrication of a miniature GNSS antenna (Ph.D. dissertation). Institut National Polytechnique, Toulouse, France.

Pillai, V. (2006). Impedance Matching in RFID Tags: To Which Impedance to Match? *Proceedings of The IEEE Antennas and Propagation Society International Symposium*. doi:10.1109/APS.2006.171137310.1109/APS.2006.1711373

Pozar & Daniel. (1992). *The Analysis and Design of Microstrip Antennas and Arrays*. Academic Press.

Pozar & Sanchez. (1998). Magnetic tuning of a microstrip antenna on a ferrite substrate. *Electron Lett, 24*(12), 729–731.

Pozar, D. M. (1992). Microstrip antennas. *IEEE, 80*(1), 79-91.

Pozar, D. M. (2005). Microwave Engineering 3e. *Tramsmission Lines and Waveguides*, 143–149.

Pozar, Schaubert, & Adrian. (1989, June). Effect of microstrip antenna substrate thickness and permittivity: Comparison of theories and experiment. *IEEE Transactions on Antennas and Propagation, AP-37*(6), 677–682.

Pozar. (1992, January). Microstrip antennas. *IEEE Trans. Antennas Propagation.*

Pozar, D. M. (1992, January). Microstrip antennas. *Proceedings of the IEEE, 80*(1), 79–81. doi:10.1109/5.119568

Pozar, D. M. (2011). *Microwave engineering*. John Wiley & Sons Inc.

Puck antenna ADA-A2450-S Datasheet from Adatcus Company. (n.d.). Retrieved from http://www.farnell.com/datasheets/1391861.pdf

Puente, C., Romeu, J., & Cardama, R. (1998). On the Behavior Of the Sierpinski Multiband Fractal Antenna. IEEE Transactions on Antenna and Propagation, 46(4).

Puente, C., & Pous, R. (1996). Fractal design of multiband and low side-lobe arrays. *IEEE Transactions on Antennas and Propagation, 44*(5), 730–739. doi:10.1109/8.496259

Puente, C., Romeu, J., Pous, R., Ramis, J., & Hijazo, A. (1998). Small but Long Koch Fractal Monopole. *Electronics Letters, 34*(1), 9–10. doi:10.1049/el:19980114

Qi, H., Liu, L., Yin, X., Zhao, H., & Kulesza, W. J. (2016). Mutual Coupling Suppression Between Two Closely Spaced Microstrip Antennas With an Asymmetrical Coplanar Strip Wall. *IEEE Antennas and Wireless Propagation Letters, 15*, 191–194. doi:10.1109/LAWP.2015.2437995

Qorvo. (2018). *QPF4005 37-40.5 GaN Dual Channel FEM Datasheet*. Retrieved from www.qorvo.com/products/d/da006271

Raab, F. (1977). Idealized operation of the class E tuned power amplifier. *IEEE Transactions on Circuits and Systems, 24*(12), 725–735. doi:10.1109/TCS.1977.1084296

Raab, F. H. (2001). Class-E, class-C, and class-F power amplifiers based upon a finite number of harmonics. *IEEE Transactions on Microwave Theory and Techniques, 49*(8), 1462–1468. doi:10.1109/22.939927

Raab, F. H., Asbeck, P., Cripps, S., Kenington, P. B., Popovic, Z. B., Pothecary, N., ... Sokal, N. O. (2002). Power amplifiers and transmitters for RF and microwave. *IEEE Transactions on Microwave Theory and Techniques, 50*(3), 814–826. doi:10.1109/22.989965

Rachakh, A., El Abdellaoui, L., Mandry, R., Errkik, A., Tajmouati, A., & Latrach, M. (2017). A New Design of a Microstrip Microwave Power Amplifier for PCS BAND. In *Proceedings of the 2nd International Conference on Computing and Wireless Communication Systems* (p. 3). Academic Press. 10.1145/3167486.3167489

Rachakh, A., El Abdellaoui, L., Zbitou, J., Errkik, A., Tajmouati, A., & Latrach, M. (2017). A novel design of a microstrip microwave broadband power amplifier for DCS, PCS and UMTS bands. In *Wireless Technologies, Embedded and Intelligent Systems (WITS), 2017 International Conference on* (pp. 1–5). Academic Press.

RADIAFLEX. (2013). *The Optimal Confined-Coverage Solution. Radio Frequency Systems*. RFS.

Radisic, V., Qian, Y., Coccioli, R., & Itoh, T. (1998). Novel 2-D photonic bandgap structure for microstrip lines. *IEEE Microwave and Guided Wave Letters*, *8*(2), 69–71. doi:10.1109/75.658644

Rahim, M. K. A., Abdul Aziz, M. Z. A., & Abdullah, N. (2005). Wideband Sierpinski Carpet Monopole Antenna. *Proceedings of Asia-Pacific Conference on Applied Electromagnetics*, 62-65.

Rahmadani & Munir. (2011). Microstrip patch antenna miniaturization using artificial magnetic conductor. *6th International Conference on Telecommunication Systems, Services, and Applications (TSSA)*, 219-223.

Rama Rao, Alekhya, & Sridevi. (2013). Design of Circular with triangle Microstrip Fractal Antenna for Dual band and UWB applications. *International Journal of Scientific & Engineering Research, 4*(7).

Ramasubramanian, K., & Ginsburg, B. (2017). *AWR1243 sensor: Highly integrated 76-81 GHz radar front-end for emerging ADAS applications*. Retrieved from http://www.ti.com/lit/wp/spyy003/spyy003.pdf

Rambabu, K., Ramesh, M., & Kalghatgi, A. T. (1997). Design method for patch antennas short circuited at one end. *Asia Pacific Microw. Conf.*, 233-236. 10.1109/APMC.1997.659347

Rani, R. B., & Pandey, S. K. (2017). A CPW-fed circular patch antenna inspired by reduced ground plane and CSRR slot for UWB applications with notch band. *Microwave and Optical Technology Letters*, *59*(4), 745–749. doi:10.1002/mop.30386

Rao, S., Nikitin, P. V., & Lam, S. F. (2005). Antenna design for UHF RFID Tags: A review and a practical application. IEEE Transactions on Antennas and Propagation, 53(12), 3870–3876. doi:10.1109/TAP.2005.859919 doi:10.1109/TAP.2005.859919

Rao, K. V. S., Lam, S. F., & Nikitin, P. V. (2010). UHF RFID Tag for Metal Containers. *Proceedings of Asia-Pacific Microwave Conference*.

Rayleigh, L. (1888). XXVI. On the remarkable phenomenon of crystalline reflexion described by Prof. Stokes. *Philosophical Magazine and Journal of Science*, *26*(160), 256–265. doi:10.1080/14786448808628259

Reha, A., & Elamri, A. (2014). Fractal antennas a novel miniaturization technique for wireless networks. *Transactions on Networks and Communications, 2*(5), 166.

Reha, A., El Amri, A., Benhmammouch, O., & Oulad Said, A. (2014). Fractal Antennas: A Novel Miniaturization Technique for wireless Networks. *Transactions on Networks and Communications*, *2*(5), 165–193. doi:10.14738/tnc.25.566

Reha, A., El Amri, A., Benhmammouch, O., Oulad Said, A., El Ouadih, A., & Bouchouirbat, M. (2015). CPW-fed H-tree fractal antenna for WLAN, WIMAX, RFID, C-band, HiperLAN, and UWB applications. *International Journal of Microwave and Wireless Technologies*, 1–8.

Reynolds, N. D. (2012). *Long range ultra-high frequency (UHF) radio frequency identification (RFID) antenna design* (Ph.D. dissertation). Purdue University, Fort Wayne, IN.

Ribate, M., Mandry, R., Latrach, M., Errkik, A., & El Abdellaoui, L. (2017). GaAs FET Broadband Power Amplifier for L and S Bands Applications. In *Proceedings of the 2nd International Conference on Computing and Wireless Communication Systems* (p. 42). Academic Press. 10.1145/3167486.3167530

Roy, Choudhary, Anand, Sarkar, & Bhunia. (2014). A novel approach on miniaturization of microstrip patch antenna with loaded strips, electronics. Communication and Instrumentation (ICECI), International Conference, 1-4, 16-17.

Russell, P. S. J., Birks, T. A., & Lloyd-Lucas, F. D. (1995). Photonic Bloch waves and photonic band gaps. In *Confined electrons and photons* (pp. 585–633). Boston, MA: Springer. doi:10.1007/978-1-4615-1963-8_19

Rylander, T., Ingelström, P., & Bondeson, A. (2012). *Computational electromagnetics*. Springer Science & Business Media.

Sadhu, B., Tousi, Y., Hallin, J., Sahl, S., Reynolds, S., Renstrom, O., … Garcia, A. V. (2017). *A 28GHz 32-Element Phased-Array Transceiver IC with Concurrent Dual Polarized Beams and 1.4 Degree Beamsteering Resolution for 5G Communication.* IEEE International Solid-State Circuits Conference (ISSCC), San Francisco, CA.

Sage, G. P. L. (2016). 3D printed waveguide slot array antennas. *IEEE Access: Practical Innovations, Open Solutions, 4*, 1258–1265. doi:10.1109/ACCESS.2016.2544278

Sakoda, K. (2004). *Optical properties of photonic crystals* (Vol. 80). Springer Science & Business Media.

Salah Toubeh, M. (2011). *Etude d'antennes BIE planaires de hauteur très inférieure à la longueur d'onde dite: The ULP EBG Antennas* (Doctoral dissertation). Limoges.

Salama, A. (2008). Fractal dipoles as meander line antennas for passive UHF RFID tags. In *Proceedings of The 5th International Multi-Conference on Systems, Signals and Devices*. Philadelphia University. doi:10.1109/SSD.2008.463284310.1109/SSD.2008.4632843

Salama, A., & Quboa, K. M. (2008). A New Fractal Loop Antenna For Passive UHF RFID Tags Applications. *Proceedings of The 3rd International Conference on Information and Communication Technologies: From Theory to Applications.* doi:10.1109/ICTTA.2008.453014110.1109/ICTTA.2008.4530141

Salehi, M., & Tavakoli, A. (2006). A novel low mutual coupling microstrip antenna array design using defected ground structure. *AEÜ. International Journal of Electronics and Communications, 60*(10), 718–723. doi:10.1016/j.aeue.2005.12.009

Salib, M. L., Dawson, D. E., & Hahn, H. K. (1975). Load-line analysis in the frequency domain with distributed amplifier design examples. In *Microwave Symposium Digest, 1987 IEEE MTT-S International* (Vol. 2, pp. 575–578). IEEE.

Sarkar, D., Saurav, K., & Srivastava, K. V. (2013, March). *Design of a Novel Dual-band Microstrip Patch Antenna for WLAN/WiMAX Applications Using Complementary Split Ring Resonators and Partially Defected Ground Structure.* Paper Presented at the *Progress in Electromagnetics Research Symposium*, Taipei, Taiwan.

Sarma, A., Sarmah, K., Sarma, K. K., Goswami, S., & Baruah, S. (2016). Soft-computing based Optimization of Low Return Loss Multiband Microstrip Patch Antenna. In J. K. Mandal, S. Mukhopadhyay, & T. Pal (Eds.), *Handbook of Research on Natural Computing for Optimization Problems.* IGI Global. doi:10.4018/978-1-5225-0058-2.ch024

Sarmah, K., Goswami, S., Sarma, A., Sarma, K. K., & Baruah, S. (2016, February). *An Experimental Study on Designing of a Dual Band Antenna using CSRR Structure with a Single Band Antenna*. Paper Presented at the *2nd IEEE International Conference on Computational Electromagnetics (ICCEM)*, China. 10.1109/COMPEM.2016.7588563

Sarmah, K., Goswami, S., Sarma, K. K., & Baruah, S. (2015 December). *Design of a CSRR based Dual Band Rectangular Patch Antenna with Predictable Far Field Radiation Pattern.* Paper Presented at the *6th International Conference on Computers and Devices for Communication (CODEC)*, India.

Scardelletti, M. C., Ponchak, G. E., Merritt, S., Minor, J. S., & Zorman, C. A. (2008). Electrically small folded slot antenna utilizing capacitive loaded slot lines. *Radio and Wireless Symposium, 2008 IEEE*, 731-734. 10.1109/RWS.2008.4463596

Schaubert. (1983). *Frequency-agile polarization diversity microstrip antennas and frequency scanned arrays.* U.S. Patent 4 367 474.

Schaubert, D. H., Pozar, D. M., & Adrian, A. (1989, June). Effect of Microstrip Antenna Substrate Thickniss and Permittivity: Comparision of Theories and Experiment. *IEEE Transactions on Antennas and Propagation*, *AP-37*(6), 677–682. doi:10.1109/8.29353

Seladji-Hassaine, N., & Bendimerad, F. T. (2016). Miniaturized Dual Band Triangular Microstrip Antenna with Sierpinski Fractal Ground. *International Journal of Microwave and Optical Technology*, *11*(5).

Selvan, K. T. (1998). Simple formulas for the gain and far-field of open-ended rectangular waveguides. *IEE Proceedings. Microwaves, Antennas and Propagation*, *145*(1), 80–84. doi:10.1049/ip-map:19981552

Shahparnia, S. (2005). *Electromagnetic bandgap structures for broadband switching noise mitigation in high-speed packages*. Academic Press.

Sickinger, F., Weissbrodt, E., & Vossiek, M. (2017). 76-81 GHz LTC antenna for an automotive miniature radar frontend. *European Radar Conference EURAD2017*, Nuremberg, Germany.

Sievenpiper, D. F. (2008). Artificial impedance surfaces for antennas. *Modern Antenna Handbook*, 737-777.

Sievenpiper, D., Zhang, L., Broas, R. F. J., Alexopolous, N. G., & Yablonovitch, E. (1999). High-impedance electromagnetic surfaces with a forbidden frequency band. *IEEE Transactions on Microwave Theory and Techniques*, *47*(11), 2059–2074. doi:10.1109/22.798001

Sigalas, M. M., Chan, C. T., Ho, K. M., & Soukoulis, C. M. (1995). Metallic photonic band gap materials. *Physical Review B: Condensed Matter and Materials Physics*, *52*(11744). PMID:9980306

Simons, R. N. (2001). Coplanar Waveguide Circuits, Components and Systems. Wiley Interscience. doi:10.1002/0471224758 doi:10.1002/0471224758

Simovski, Sochava, & Tretyakov. (2004). New compact and wide-band high impedance surface. In *Antennas and propagation Society International Symposium*. IEEE.

Smith, J. K. (1999). Reconfigurable aperture antenna (RECAP). *DARPA*. Retrieved from www.darpa.mil

Smith, D. R., Padilla, W. J., Vier, D. C., Nemat-Nasser, N. C., & Schultz, S. (2000). Composite Medium with Simultaneously Negative Permeability and Permittivity. *Physical Review Letters*, *84*(18), 4184–4187. doi:10.1103/PhysRevLett.84.4184 PMID:10990641

Snider, D. M. (1967). A theoretical analysis and experimental confirmation of the optimally loaded and overdriven RF power amplifier. *IEEE Transactions on Electron Devices*, *14*(12), 851–857. doi:10.1109/T-ED.1967.16120

Sofi, A., Roky, K., & Hadj Baraka, I. (2014). Antenna Design for UHF RFID Tags Using Two Coupled Patches. *International Journal of Information and Network Security*, *3*(1).

Sofi, A., Roky, K., Bouhorma, M., & Hadj Baraka, I. (2014). Novel Antennas for UHF RFID Tags: Design and Miniaturization. *International Journal of Electrical and Computer Engineering*, *4*(1).

Soh, P.J., Rahim, M.K.A., Asrokin, A., & Abdul Aziz, M.Z.A. (2007). Design Modeling and performance comparison of feeding techniques for a microstrip patch antenna. *Journal Technologi*, *47*(D), 103-120.

Son, H.-W., & Pyo, C.-S. (2005). Design of RFID tag antennas using an inductively coupled feed. Electronics Letters, 41(18), 994–996. doi:10.1049/el:20051536 doi:10.1049/el:20051536

Son, W.-I., Lim, W.-G., Lee, M.-Q., Min, S.-B., & Yu, J.-W. (2010, May). Design of compact quadrifilar inverted-f antenna with circular polarization for GPS receiver. *IEEE Transactions on Antennas and Propagation*, *58*(5), 1503–1510. doi:10.1109/TAP.2010.2044344

Son, W., Lee, H., Lee, M., Min, S., & Yu, J. (2010). Compact Square Quadrifilar Spiral Antenna with circular polarization for UHF mobile RFID Reader. *Microwave Conference Proceedings (APMC).*

Srirattana, N., Raghavan, A., Heo, D., Allen, P. E., & Laskar, J. (2005). Analysis and design of a high-efficiency multistage Doherty power amplifier for wireless communications. *IEEE Transactions on Microwave Theory and Techniques, 53*(3), 852–860. doi:10.1109/TMTT.2004.842505

Steer, M. (2009). Microwave and RF design: A systems approach. *Sciences et Techniques (Paris).*

Stevenson, A. F. (1948). Theory of slots in rectangular wave-guides. *Journal of Applied Physics, 19*(24), 24–38. doi:10.1063/1.1697868

Stratton, J. A. (1941). *Electromagnetic Theory.* New York: McGraw-Hill.

Stutzman, W. L. (1998). Estimating directivity and gain of antennas. *Antennas and Propagation Magazine, IEEE, 40*(4), 7–11. doi:10.1109/74.730532

Su, Y., & Lai, X. Z. (2008). Research on Fractal Tree RFID TagAntenna on Papery Substrate. *Proceedings of 4th IEEE International Conference on Wireless Communications, Networking and Mobile Computing.*

Suganthi, S., Tharini, K. S., Sarankumar, P. S., Raghavan, S., & Kumar, D. (2011). *Design and simulation of planar Minkowski fractal antennas.* Paper presented at 2nd International Conference on Wireless Communication, Vehicular Technology, Information Theory and Aerospace & Electronic Systems Technology (Wireless VITAE), Chennai, India. 10.1109/WIRELESSVITAE.2011.5940935

Tabakh, I., Jorio, M., El Amrani El Idrissi, N., & Mazri, T. (2016). Design and Optimization of a New Slotted Patch Antenna for RFID Applications. International Journal on Communications Antenna and Propagation, 6(1).

Tahar, F., Barakat, A., Saad, R., Yoshitomi, K., & Pokharel, R. K. (2017). Dual-Band Defected Ground Structures Wireless Power Transfer System With Independent External and Inter-Resonator Coupling. *IEEE Transactions on Circuits and Wystems. II, Express Briefs, 64*(12), 1372–1376. doi:10.1109/TCSII.2017.2740401

Tam, W. Y. (2007). Woodpile and diamond structures by optical interference holography. *Journal of Optics. A, Pure and Applied Optics, 9*(11), 1076–1081. doi:10.1088/1464-4258/9/11/018

Tang, F. M., & Li, L. C. (2015). Meander Dipole Antenna design for Passive UHF RFID Tags. *Proceedings of The 2nd International Conference on Electrical, Computer Engineering and Electronics (ICECEE).* doi:10.2991/icecee-15.2015.15010.2991/icecee-15.2015.150

Tawk, Y., Albrecht, A. R., Hemmady, S., Balakrishnan, G., & Christodoulou, C. G. (2010). Optically Pumped Frequency Reconfigurable Antenna Design. *IEEE Antennas and Wireless Propagation Letters, 9*, 280–283. doi:10.1109/LAWP.2010.2047373

Taybi, A., Ghaloua, A., Tajmouati, A., Mandry, R., El Abdellaoui, L., & Latrach, M. (2017). A New Design of Rectenna Array with High Performances. In *Proceedings of the 2Nd International Conference on Computing and Wireless Communication Systems* (p. 1:1--1:7). New York, NY: ACM. 10.1145/3167486.3167487

Technologies, A. (2010). *Specifying calibration standards and kits for Agilent vector network analyzers: application note 1287-11.* Agilent Technologies Inc.

Thevenot, M., Cheype, C., Reineix, A., & Jecko, B. (1999). Directive photonic-bandgap antennas. *IEEE Transactions on Microwave Theory and Techniques, 47*(11), 2115–2122. doi:10.1109/22.798007

Time-Harmonic Electromagnetic Fields. (1961). *McGraw-Hill Book Co.*

Tretyakov, S. (2003). *Analytical modeling in applied electromagnetics*. Artech House.

Tsuchiya, A., & Onodera, H. (2015). Impact of anomalous skin effect on the metal wire for terahertz integrated circuit. *IEEE International Symposium on Radio-Frequency Integration Technology*, Sendai, Japan. 10.1109/RFIT.2015.7377939

Turcu, C. (2009). *Development and Implementation of RFID Technology*. InTechOpen.

Turcu, C. (2009). *Development and Implementation of RFID Technology*. In-Tech. doi:10.5772/100

Tyagi, Y., Mevada, P., Chakrabarty, S., & Jyoti, R. (2017). High-efficiency broadband slotted waveguide array antenna. *IET Microwaves, Antennas & Propagation, 11*(10), 1401–1408. doi:10.1049/iet-map.2016.0682

Ukkonen, L., Sydanheimo, L., & Kivikoski, M. (2004). *A novel Tag Design Using Inverted-F Antenna for Radio Frequency Identification of Metallic Objects*. Princeton, NJ: IEEE Sarnoff Symp. Advances Wired and Wireless Communication. doi:10.1109/SARNOF.2004.1302848

Ukkonen, L., Sydanheimo, L., & Kivikoski, M. (2005). Effects of Metallic Plate Size on the Performance of Microstrip PatchType Tag Antennas for Passive RFID. *IEEE Antennas and Propagation Letters, 4*, 410–413. doi:10.1109/LAWP.2005.860212

Vandenbosch & Vasylchenko. (2011). *A Practical Guide to 3D Electromagnetic Software Tools*. InTech.

Vendelin, G. D., Pavio, A. M., & Rohde, U. L. (2005). *Microwave circuit design using linear and nonlinear techniques*. John Wiley & Sons. doi:10.1002/0471715832

Veselago, V. G. (1968). The electrodynamics of substances with simultaneously negative values of ε and μ. *Physics Uspekhi, 10*(4), 509–514. doi:10.1070/PU1968v010n04ABEH003699

Visser, H. J. (n.d.). Array and Phased Array Antenna Basics. John Wiley & Sons. doi:10.1002/0470871199

Wakrim, L., Ibnyaich, S., & Hassani, M. M. (2014). Optimization by genetic algorithm of a PIFA antenna parameters for Wifi application. *IEEE Multimedia Computing and Systems International Conference*, 1501 – 1505. 10.1109/ICMCS.2014.6911223

Wang, J. H., Li, Y. J, Zhang, Z., Chen, M. (2011). Radiation mechanism and polarization properties of leaky coaxial cables. *Radio Science, 46*(RS2009), 1-12.

Wang, B. Z., Xiao, S., & Wang, J. (2007). Reconfigurable patch-antenna design for wide band wireless communication systems. *IET Microwaves, Antennas & Propagation, 1*(2), 414–419. doi:10.1049/iet-map:20050349

Wang, J. H., & Mei, K. K. (2001). Theory and analysis of leaky coaxial cables with periodic slots. *IEEE Transactions on Antennas and Propagation, 49*(12), 1723–1732. doi:10.1109/8.982452

Wang, J. H., & Mei, K. K. (2002). Design of leaky coaxial cables with periodic slots. *Radio Science, 37*(5), 1–9. doi:10.1029/2000RS002534

Wan, S., & Huang, K. (2018). Methods for Improving the Transmission-Conversion Efficiency From Transmitting Antenna to Rectenna Array in Microwave Power Transmission. *IEEE Antennas and Wireless Propagation Letters, 17*(4), 538–542. doi:10.1109/LAWP.2018.2801320

Wei, K., Li, J. Y., Wang, L., Xing, Z. J., & Xu, R. (2016). Mutual Coupling Reduction by Novel Fractal Defected Ground Structure Bandgap Filter. *IEEE Transactions on Antennas and Propagation, 64*(10), 4328–4335. doi:10.1109/TAP.2016.2591058

Wei, K., Li, J., Wang, L., Xing, Z., & Xu, R. (2016). S-shaped periodic defected ground structures to reduce microstrip antenna array mutual coupling. *Electronics Letters*, *52*(15), 1288–1290. doi:10.1049/el.2016.0667

Werner, D. H., & Ganguly, S. (2003, February). An Overview of Fractal Antenna Engineering Research. *IEEE Antennas & Propagation Magazine*, *45*(1), 38–57. doi:10.1109/MAP.2003.1189650

Wheeler, H. A. (1959). The radian sphere around a small antenna. Proceedings IRE, 1325-1331.

Wheeler, H. (1947). Fundamental Limitations of Small Antennas. *Proc. IRE*, 1479–1484.

White, C. R., & Rebeiz, G. R. (2009). Single and dual-polarized tunable slot-ring antennas. *IEEE Transactions on Antennas and Propagation*, *57*(1), 19–26. doi:10.1109/TAP.2008.2009664

Winn, J. N., Fink, Y., Fan, S., & Joannopoulos, J. D. (1998). Omnidirectional reflection from a one-dimensional photonic crystal. *Optics Letters*, *23*(20), 1573–1575. doi:10.1364/OL.23.001573 PMID:18091848

Won Jung, C., Lee, M., Li, G. P., & Flaviis, F. De. (2006). Reconfigurable scan-beam single-arm spiral antenna integrated with RF-MEMS switches. *IEEE Transactions on Antennas and Propagation*, *54*(2), 455–463. doi:10.1109/TAP.2005.863407

Xi, J., Zhu, H., & Ye, T. (2010). Platform-tolerant PIFA-type UHF RFID tag antenna. *IEEE RFID Conference*, 174–180.

Xing, K., Liu, B., Guo, Z., Wei, X., Zhao, R., & Ma, Y. (2017). Backlobe and Sidelobe Suppression of a Q-Band Patch Antenna Array by Using Substrate Integrated Coaxial Line Feeding Technique. *IEEE Antennas and Wireless Propagation Letters*, *16*, 3043–3046. doi:10.1109/LAWP.2017.2759909

Xue, H. G., Yang, X. X., & Ma, Z. (2015). A Novel Microstrip-CPW Fed Planar Slot Antenna With Broadband and Circular Polarization. *IEEE Antennas and Wireless Propagation Letters*, *14*, 1392–1395. doi:10.1109/LAWP.2015.2407904

Xu, H., Zhao, Z., Lv, Y., Du, C., & Luo, X. (2008). Metamaterial superstrate and electromagnetic band-gap substrate for high directive antenna. *International Journal of Infrared and Millimeter Waves*, *29*(5), 493–498. doi:10.100710762-008-9344-y

Yablonovitch, E. (1993). Photonic band-gap structures. *JOSA B*, *10*(2), 283–295. doi:10.1364/JOSAB.10.000283

Yadav, A., Agrawal, S., & Yadav, R. P. (2017). SRR and S-shape slot loaded triple band notched UWB antenna. *AEÜ. International Journal of Electronics and Communications*, *79*, 192–198. doi:10.1016/j.aeue.2017.06.003

Yaghjian, A. D. (1984). Approximate formulas for the far fields and gain of open ended rectangular waveguide. *IEEE Transactions on Antennas and Propagation*, *32*(4), 378–384. doi:10.1109/TAP.1984.1143332

Yang, L., & Basat, S. (2006). Design and Development of Novel Miniaturized UHF RFID Tags on on Ultra-lowcost Paper-based Substrates. *Proceedings of Asia-Pacific Microwave Conference (APMC)*.

Yang, T., & Wen, G. (2014). Compact multi-band printed antenna with multi-triangular ground plane for WLAN/WiMAX/RFID applications. International Journal of Microwave and Wireless Technologies.

Yang, X. D., Feng, S., Zhang, X., Ren, X. Y., & Li, W. X. (2006). Research of single mode radiation leaky coaxial cable used for ultra wide-band communication. *Proceedings, 2006 International Conference on Communications, Circuits and Systems* (vol. 2, pp. 1261-1264). Academic Press. 10.1109/ICCCAS.2006.284874

Yang, F.-R., Ma, K.-P., Qian, Y., & Itoh, T. (1999). A uniplanar compact photonic-bandgap (UC-PBG) structure and its applications for microwave circuit. *IEEE Transactions on Microwave Theory and Techniques*, *47*(8), 1509–1514. doi:10.1109/22.780402

Yang, H. Y., Alexopoulos, N. G., & Yablonovitch, E. (1997). Photonic band-gap materials for high-gain printed circuit antennas. *IEEE Transactions on Antennas and Propagation*, *45*(1), 185–187. doi:10.1109/8.554261

Yang, X. M., Liu, X. G., Zhou, X. Y., & Cui, T. J.. (2012). Reduction of Mutual Coupling Between Closely Packed Patch Antennas Using Waveguided Metamaterials. *IEEE Antennas and Wireless Propagation Letters*, *11*, 389–391. doi:10.1109/LAWP.2012.2193111

Yasar, A. (2013). *Printable Green RFID Antennas for Embedded Sensors* (Doctoral Thesis). KTH Royal Institute of Technology, Stockholm, Sweden.

Yeo, J., & Kim, D. (2009). Novel design of a high-gain and wideband Fabry-Perot cavity antenna using a tapered AMC substrate. *Journal of Infrared, Millimeter and Terahertz Waves*, *30*(3), 217–224. doi:10.100710762-008-9451-9

Yoo, T., & Chang, K. (1992). Theoretical and experimental development of 10 and 35 GHz rectennas. *IEEE Transactions on Microwave Theory and Techniques*, *40*(6), 1259–1266. doi:10.1109/22.141359

Yoshitomi, K., & Sharobim, H. R. (1994). Radiation from a rectangular waveguide with a lossy flange. *IEEE Transactions on Antennas and Propagation*, *42*(10), 1398–1403. doi:10.1109/8.320746

Youbok, L. (2003). *Antenna Circuit Design for RFID Applications*. Microchip Technology Inc. Application Note AN710.

You, K. Y. (2015). *RF Coaxial Slot Radiators: Modeling, Measurements, and Applications*. Artech House.

You, K. Y. (2017). Materials Characterization Using Microwave Waveguide System. In S. K. Goudos (Ed.), *Microwave Systems and Applications* (pp. 341–358). InTech. doi:10.5772/66230

You, K. Y., Abbas, Z., Malek, M. F. A., & Cheng, E. M. (2014). Non-destructive dielectric measurements and calibration for thin materials using waveguide-coaxial adaptors. *Measurement Science Review*, *14*(1), 16–24. doi:10.2478/msr-2014-0003

Yousefbeigi, M., Enayati, A., Shahabadi, M., & Busuioc, D. (2008). Parallel-series feed network with improved G/T performance for high-gain microstrip antenna arrays. *Electronics Letters*, *44*(3), 180–182. doi:10.1049/el:20083604

Yousefi, L., & Ramahi, O. M. (2007, March). *New artificial magnetic materials based on fractal Hilbert curves*. Paper Presented at the International Workshop on Antenna Technology (iWAT), UK. 10.1109/IWAT.2007.370119

Yu, Y., Huang, Z., Zheng, C., & Wu, Y. (2014). A Miniature Antenna for 2.45 GHz RFID Tag. *Journal of Computers*, *9*(2), 404–411. doi:10.4304/jcp.9.2.404-411

Zahraoui, I., Errkik, A., Abdelmounim, E., Bennis, H., El Abdellaoui, L., & Sanchez Mediavilla, A. (2015). A New Design of a CPW Fed Triple-Band for GPS, UMTS and WiMAX Applications. International Journal on Communications Antenna and Propagation, 5(6).

Zahraoui, I., Errkik, A., Zbitou, J., Tajmouati, A., & Mediavilla, A. S. (2016). A Novel Design of a Fractal Antenna for IMT and WiMAX Applications. *International Journal of Microwave and Optical Technology*, *11*(6).

Zainud-Deen, S. H., Malhat, H. A., & Awadalla, K. H. (2009). Fractal Antenna for Passive UHF RFID Applications. *Progress in Electromagnetics Research*, *16*, 209–228. doi:10.2528/PIERB09051506

Zarifi, D., Farahbakhsh, A., & Zaman, A. U. (2017). A Gap Waveguide-Fed Wideband Patch Antenna Array for 60-GHz Applications. *IEEE Transactions on Antennas and Propagation*, *65*(9), 4875–4879. doi:10.1109/TAP.2017.2722866

Zbitou, J., Latrach, M., & Toutain, S. (2006). Hybrid Rectenna and Monolithic Integrated Zero-Bias Microwave Rectifier. *IEEE Transactions on Microwave Theory and Techniques*, *54*(1), 147–152. doi:10.1109/TMTT.2005.860509

Zeb, B. A., Ge, Y., & Esselle, K. P. (2012, December). A single-layer thin partially reflecting surface for tri-band directivity enhancement. In *Microwave Conference Proceedings (APMC), 2012 Asia-Pacific* (pp. 559-561). IEEE. 10.1109/APMC.2012.6421663

Zeidler, E. (1995). Hilbert Spaces and Generalized Fourier Series. In *Applied Functional Analysis* (pp. 195–227). New York, NY: Springer. doi:10.1007/978-1-4612-0815-0_3

Zhang, J., Zhang, X. M., Liu, J.-S., Wu, Q.-F., Ying, T., & Jin, H. (2009). Dual-band Bidirectional High Gain Antenna for WLAN 2.4 and 5.8 GHz applications. Electronics Letters, 45(1), 6–7. doi:10.1049/el:20092617 doi:10.1049/el:20092617

Zhang, X.-X., & Yang, F. (1998). Study of a slit cut on a microstrip antenna and its applications'. *Microwave and Optical Technology Letters, 18*(4), 297–300. doi:10.1002/(SICI)1098-2760(199807)18:4<297::AID-MOP14>3.0.CO;2-1

Zhang, X., & Zhu, L. (2016). High-Gain Circularly Polarized Microstrip Patch Antenna With Loading of Shorting Pins. *IEEE Transactions on Antennas and Propagation, 64*(6), 2172–2178. doi:10.1109/TAP.2016.2552544

Zhang, Y., Hong, W., Yu, C., Kuai, Z.-Q., Don, Y.-D., & Zhou, J.-Y. (2008). Planar Ultrawideband Antennas With Multiple Notched Bands Based on Etched Slots on the Patch and/or Split Ring Resonators on the Feed Line. *IEEE Transactions on Antennas and Propagation, 56*(9), 3063–3068. doi:10.1109/TAP.2008.928815

Zhao, W., Wang, G., & Lai, X. (2008). Active E-Plate with Slot Antenna. *International Conference on Wireless Communications, Networking and Mobile Computing*, 1-3.

Zhu, F., Xu, J., & Xu, Q. (2009). Reduction of mutual coupling between closely-packed antenna elements using defected ground structure. In *2009 3rd IEEE International Symposium on Microwave, Antenna, Propagation and EMC Technologies for Wireless Communications* (pp. 1–4). IEEE. 10.1109/MAPE.2009.5355659

About the Contributors

Jamal Zbitou was born in Fes, Morocco, in June 1976. He received the Ph.D. degree in electronics from Polytech of Nantes, Nantes, France, in 2005. He is currently an associate Professor of Electronics in FST, University of Hassan 1st, Settat, Morocco and the head of Computing Networks and telecommunication in LMEET Laboratory in FSTS. He is involved in the design of hybrid, monolithic active and passive microwave electronic circuits.

Ahmed Errkik was born in July 1960 in Morocco. He received the Ph.D. degree in physics from the University of Technology Compiègne (UTC), Compiègne, France. He is currently an associate Professor of physics in FST University Hassan 1st, Settat, Morocco, and he is the head of the laboratory LMEET. He is involved in the design of hybrid, monolithic active and passive microwave electronic circuits.

* * *

El Hassane Abdelmounim received his PhD in applied Spectral analysis from Limoges University at sciences and techniques faculty, France in 1994. In 1996, he joined, as Professor, applied physics department of sciences and techniques faculty, Hassan 1st University, Settat, Morocco. His current research interests include digital signal processing and machine learning. He is currently coordinator of a Bachelor of Science in electrical engineering and researcher at System Analysis and Information Technology Laboratory.

Mohamed Aghoutane was born in Taounate, Morooco, in October 1956. He received the Ph.D degree in physics sciences from liege University, liege, Belguim, in 1994. He is currently an associate Professor of Electronics in Abdelmalek Essaadi University, Tetouan, Morocco, He is involved in Electronic and Telecomunications.

Saida Ahyoud is Professor at Abdelmalek Essaadi University, Morocco, since 2004. She received PhD degree in Electrical Engineering in 2003. Since 2015 she is Professor Ability in Electronics and Telecommunications at the Faculty of Sciences of Tetouan. She is a member of Research Unit Information Technology and Modeling Systems. Research Axis: Modeling and design of passive circuits such as antennas and filters, and signal processing.

El Hamdouni Akram was born in Safi, Morocco, in 5 Febraury 1982. He received the Ingeneer degree in telecommunications engineering from the Institut of postes and Telecomunications (INPT), Rabat, Morocco, in 2006, the Master degree from the Faculty of Sciences and Techniques of Marrakech of the University Cadi Ayyad, in 2004. AcctuallyPh.D. Studientthe LMEET Laboratoryof Mecanic, Énergétic, Electronic et Telecommunications of Faculty of Sciences and Techniques of Settat, of University Hassan 1st.

Adel Asselman was born in Tétouan, morocco, in 1957. He received the fundamental License in physics in Rabat He obtained his Dotorat in ionized gases from the Paul Sabatier University, of Toulouse, France, in 1983 and a Doctorat of State in Physical Sciences (1995). Actually Teacher Researcher since 1983 at Abdelmalek Essaadi University, Tetouan and head of the Optics and Photonics Team. Research Axis: Photonics and Solar Energy Responsible and coordinator of several research projects.

Sunandan Baruah is currently the Head of the Department of Electronics & Communication Engineering, Assam Don Bosco University, Guwahati. Prior to that, he held research positions at Uppsala University, Sweden and Asian Institute of Technology, Thailand. His research interest is in Environmental nanotechnology. His research concept is to make use of inexpensive wet-chemical methods to fabricate innovative materials and device components. He has over 80 research publications to his credit. His work has about 2594 Citations with an H-Index of 22.

Hamid Bennis was born in Meknes, Morocco, in September 1977. He received the Ph.D. degree in Computer Science and Telecommunications in 2011 from the University of Mohammed V Agdal, Rabat, Morocco. He is currently a Professor of Computer Science and Telecommunications in EST of Meknes, Moulay Ismail University, Morocco. He is involved in the design of hybrid, monolithic active and passive microwave electronic circuits.

Mustapha El Halaoui was born in Taza, morocco, in June 1987. He received the fundamental License in physics with specialization electronics, the Specialized Masters Telecommunications System Engineering, and PhD degree electronic and telecommunications in the Optics and Photonics Laboratory from the Faculty of Sciences, Abdelmalek Essadi University, Morocco, in 2009 and 2011, respectively. His major research area is antenna design for mobile applications. He has published two article and presented several conference papers. His recent work focuses on array antenna solutions for fifth generation.

Ahmed El Hamraoui was born in Berrechid, Morocco, In 30 January 1990. He received the Master degree in Electronic Systems of Aviation Safety from the Mohammed VI International Academy of Civil Aviation Morocco. He is currently working toward the Ph.D. degree in Telecommunication engineering at Faculty of Sciences and Techniques, University Hassan 1st in Settat, Morocco. He is involved in the design of microwave hybrid circuits and antennas.

Ridouane Er-Rebyiy was born in Zagora, Morocco, In June 1986.He received his Master's degree in automatic and signal processing from FST of Settat, University Hassan 1st Morocco. He is currently a Ph.D. student. His main research interests are antennas and their applications in wireless Communications.

Ahmed Ghaloua was born in Kasbat Tadla, Morocco, In July 1990. He received the master degree in electronics and telecommunications from Faculty of Sciences, University of Abdelmalek Essaidi Tetouan, Morocco. He is currently working toward the Ph.D. degree in physics and engineering sciences at Faculty of Sciences and Technologies, Settat, Morocco. His research interests include the analysis and design of antenna array using Metamaterials structures.

Sivaranjan Goswami has completed his B. Tech. in Eectronics and Communication Engineering from Assam Don Bosco University, Guwahati, Assam, India in 2013 and M. Tech. in Signal Processing and Communication from Gauhati University, Guwahati, Assam, India in 2015. He is currently working as an Assistant Professor at the Department of Electronics and Communication Technology, Gauhati University. His areas of research interest include Digital Signal Processing, Microstrip Antenna Design, and Computational Electromagnetics.

Mohamed Ihamji was born in Casablanca in 1981, Morocco. He received the diploma in electronics engineering from ENSI Caen, France, in 2006. From 2006 to 2011 he worked as development engineer for RFID application. Currently he is working towards his Ph.D. His current research activities are focused on RFID antennas.

Abdelmoumen Kaabal was born in Chefchaouen, Morocco, in 1982. He received the fundamental License in physics specialty electronics in Tetouan. He obtained his DESA of Microelectronics and Radio Frequency Techniques and a Doctorat of electronics and telecommunications from the Faculty of Sciences, Abdelmalek Essaâdi University, Tetouan, Morocco, in 2006 and 2017 respectively. Actually Teacher of qualifying secondary education since 2007. His major research area is antenna design for mobile applications. His recent work focuses on the antenna solutions for fifth generation.

Rupesh Kumar was born in Bokaro Steel City, India, in 1983. He received B.Tech degree in Electronics & Communication engineering from SMIT, India, in 2008, and MS and PhD degrees in Electronics & Communication from Institut Mines-Telecom, Paris, France, in 2011 and 2014. Currently, he is working as a Researcher at SYSTEMS RF, in Limoges, France. From 20015 to 2016, he was working at the Research and Innovation Centre of Technicolor in rennes, France. He has worked as Scientist-B with Sikkim Manipal University in a DRDO's sponsored project between 2008-2009. His researches are focused on indoor localization system, FMCW-CW-WiFi Radars, and UWB CP patch antenna, Reconfigurable Parasitic Array Antena, RFID Technology, MEMS based sensors, Innovative sensor designs.

El Abdellaoui Larbi was born in Tiflet, Morocco, in 1961. He received the Ph.D. degree in electronics from the University of Metz, in 1994, France. He is currently an associate Professor of electronics in Faculty of sciences and techniques, University Hassan 1st Settat, Morocco. Dr Larbi El Abdellaoui is involved in the design of hybrid, monolithic active and passive microwave electronic circuits.

Adnane Latif was born in Morocco in 1973, has a PHD in Telecommunications and Networks, professor at Cadi Ayyad University, National School of Applied Sciences, Marrakesh, Morocco. Researcher in the areas of Antennas, Microwaves and Wireless communications. Author of a fifties publications in international conferences and in international journals. Founder and coordinator of the Spring School 2010 "Wireless and Emerging Technologies." Founder and President of the Moroccan Association of Telecommunications and Electronics Technologies. Member organizer of several International Conferences and Spring Schools. Member of International Program Committee of several international conferences. Reviewer and Board member of several International Journals Author of two books.

Mohamed Latrach (IEEE Member and URSI-France Member) is Professor of microwave engineering at the École Supérieure d'Électronique de l'Ouest (ESEO), Angers, France, where he is the head of the RF-EMC research group. He is also an associate researcher at the IETR of the University of Rennes 1. His research interests are in design of hybrid & MMIC circuits, metamaterials, antennas, rectennas and their applications in wireless communications, RFID, sensors, connected objects, energy harvesting, etc.

Cyrille Menudier received the PhD in Telecommunications from the XLIM Research Laboratory, University of Limoges, in 2007. He then got a post-doctoral position in CNES (French Space Agency) until 2009, where he worked on reconfigurable reflect arrays. He is currently an Associate Professor in the Waves and Associated Systems Department of XLIM. Its research interests include reflector antennas, reconfigurable antennas, reflect arrays, parasitic element antennas and mutual coupling effects.

Otman Oulhaj was born in Tetuan, Morocco, In 1981. He received the B.C of Sciene in Electronic from Abdelmalek Essaadi University, Tetuan, Morocco, in 2004. He received the a Master's degree in Electronic and Telecommunication from Abdelmalek Essaadi University, Tetuan, Morocco, in 2012. Actually, he is a Ph.D student in Electronic and Telecommunication in Abdelmalek Essaadi University, Tetuan, Morocco. He is involved in the study, design and realization of miniature patch antennas array.

Amine Rachakh was born in Youssoufia, Morocco, in June 1990. He received the Master's degree in Microelectronics and Embedded Systems from the Faculty of Science Dhar El Mahrez, University Sidi Mohamed Ben Abdellah, Fes, Morocco, in 2014. He is currently prepared a Ph.D. degree in RF/Microwave and Electronics at Faculty of Sciences and Techniques, University Hassan 1st, Settat, Morocco, since 2015. He is involved in the design of hybrid, monolithic active and passive microwave electronic circuits.

Rashmi S. B. is doing her research in the field of RF CMOS design persuing her phd in analog circuit design at 60GHz published more than 35 technical paper in the field of her research interest.

Angana Sarma is currently working as Guest Lecturer in the Department of Electronics and Communication Technology, Gauhati University, Assam, India. Prior to that, she held research position at Gauhati University, Assam. She completed her BE in Electronics and Telecommunication Engineering in the year 2013 from Girijananda Chowdhury Institute of Management and Technology, Guwahati, Assam, India and her M. Tech. in Digital Signal Processing in the year 2017 from Assam Don Bosco University, Guwahati, Assam, India. Her area of research interest is Microstrip Antenna Design.

Kandarpa Kumar Sarma, currently Professor and Head, Department of Electronics and Communication Engineering, Gauhati University, Guwahati-781014, Assam, India, completed MTech in Signal Processing (2005) and Ph.D. in the area of mobile communication (2011) from IIT Guwahati, Assam India. His areas of interest include mobile communication, speech processing, machine and deep learning, antenna design and computer vision.

Kumaresh Sarmah received M.Sc degree in Electronics from Gauhati University in 2006, M.Phil degree in Electronics from Gauhati University in 2009 and M.Tech degree in Electronic Design Technology from Tezpur University in 2011. He delivered his duty as Assistant Professor in The Dept. of ECE, Assam Don Bosco University from June,2011 till March, 2013. April, 2013 onwards, he is working as Assistant Professor in The Department of Electronics and Communication Technology, Gauhati University. His research interest includes Design of Metamaterial Antenna and Nanofabricated metamaterial for wireless communication.

Abdelali Tajmouati was born in, Morocco, in 1962. He received the Ph.D. degree in science engineering from the University of Perpignan, France, in 1992. He is currently an associate Professor of Electronics, thermal transfer and thermodynamic in Faculty of sciences and techniques University Hassan 1st, Settat, Morocco. He is involved in the design of hybrid, monolithic active and passive microwave electronic circuits.

Naima Amar Touhami received DESA in Instrumentation and electronics and PhD degrees in electronics and Telecommunication from University Abdelmalek Essaadi, in 2002 and 2009 respectively. She received the AECID scholarship from the Spanish Ministry of Foreign Affairs (2005-2008) and participated in several research projects. She is Associate professor of electronics and telecommunications at University Abdelmalek Essaadi and member of EIRT. She has supervised Master and Bachelor"s Students. She has over than twenty journal papers and forty conference papers. She has participated in organization of some conferences and events for PhD students. Her research interests include synthesis of advanced high performance active and passive circuits such as antennas, filters, diplexer, amplifier, mixer.

Kok Yeow You was born in 1977. He obtained his B.Sc. Physics (Honours) degree in Universiti Kebangsaan Malaysia (UKM) in 2001. He pursued his M.Sc. in Microwave at the Faculty of Science in 2003 and his Ph.D. in Wave Propagation at the Institute for Mathematical Research in 2006 in Universiti Putra Malaysia (UPM), Serdang, Selangor, Malaysia. Recently, he is a senior lecturer at Communication Engineering Department, Faculty of Electrical Engineering, Universiti Teknologi Malaysia (UTM), Skudai, Johor, Malaysia. His main personnel research interest is in the theory, simulation, and instrumentation of electromagnetic wave propagation at microwave frequencies focusing on the development of microwave passive devices and sensors for medical and agricultural applications. For the past 16 years, Dr. You have been developed numerous projects concerning microwave waveguide measurement systems for agri/food, biomedical processing and microwave devices for biomedical treatments, as well as numerical electromagnetic modeling. Furthermore, Dr. You have been published more than 60 journals and 2 research books which related to the above-mentioned fields.

Index

www.ingramcontent.com/pod-product-compliance
Lightning Source LLC
LaVergne TN
LVHW060258011026
849161LV00011B/2736